The Physics of Waves and Oscillations

The Physics of Waves and Oscillations

N K Bajaj
St Stephen's College
University of Delhi

McGraw Hill Education (India) Private Limited

51st reprint 2024

ISBN-13: 978-0-07-451610-2
ISBN-10: 0-07-451610-8

52 53 54 55 56 APO 28 27 26 25 24

Published by McGraw Hill Education (India) Private Limited,
Anjana Complex No: 5/90 A, Butt Road, St. Thomas Mount, Chennai – 600016

Printed and bound in India at A.P. Offset Pvt. Ltd., 25/487, Zulfe Bengal, Dilshad Garden, New Delhi 110095

Visit us at: www.mheducation.co.in
Toll free in India: 1800 103 5875
Write to us at: support.india@mheducation.com
CIN: U80302TN2010PTC111532

Preface

This book is intended primarily for students pursuing degree courses in physics at the college or university undergraduate level. This textbook has originated from the lecture course which I have taught undergraduate students for a number of years. There are, of course, many other books on the physics of oscillations and waves. My reason for adding another book to the list is that I have been persuaded by my undergraduate students who successfully convinced me of the need for a book in which every formula was derived from the fundamental laws of physics (there are a few exceptions to this procedure) incorporating the details of the steps in these derivations and their logical necessity. The result is a concise textbook having sufficient mathematical detail and rigour amplified by many problems and worked-out examples covering a wide range of content and sophistication. At some places, generality and mathematical rigour have been sacrificed to make the chain of logic more distinct, clearly bringing out the physical reasoning. It has been my experience that once the student is able to grasp the physical reasoning and logic in a mathematical derivation, he can himself add what additional generality and rigour he may need. I believe that the treatment reads as physics rather than mathematics; the mathematical arguments appearing as tools to achieve physical understanding.

It is assumed that the student has a thorough knowledge of calculus and trigonometry and some acquaintance with the fundamental laws of mechanics. In the beginning only school-leaving mathematics is required. The more advanced techniques involving the use of exponential series, complex numbers and differential equations are clearly explained as and when needed. The solutions of the various differential equations encountered in the text are worked out in detail. Cartesian coordinates and SI units are used throughout.

The book has nine chapters. Each chapter is divided into sections dealing with a particular topic. The basic concepts are delineated in considerable detail. Although the stress is on an analytical approach, an attempt has been made to link theory to real examples of phenomena, illustrated wherever possible by examples from daily life. Each chapter contains a large number of solved examples. A variety of problems is appended at the end of each chapter. The student is strongly advised to work out these problems. Answers to the problems have been provided at the end.

The first four chapters are devoted to the study of free, damped and forced oscillations with one degree of freedom. The discussion is confined not only to mechanical systems but includes electrical systems as well. The presentation is quantitative and analytical rather than descriptive. The dynamical approach, rather than the conventional (and hypothetical) reference-circle approach, is used.

Chapter 5 deals with the analysis of coupled systems with two or more finite number of degrees of freedom. The presentation lays considerable emphasis on the normal modes of such discrete systems. I very strongly believe that a sound knowledge of the technique of normal mode analysis can greatly enrich the student's understanding of wave motion dealt with separately in Chapter 7. Waves naturally emerge from our analysis of a continuum of coupled oscillators. The main thesis of the book is that a medium through which energy is transmitted by wave propagation behaves essentially as a continuum of coupled oscillators. A simple oscillator is characterized by three parameters, namely inertia, elasticity and friction. The first parameter stores energy, the second is responsible for exchanging energy and the third dissipates energy. The product of the first two parameters determines the velocity of wave propagation in the medium while their ratio (in the absence of the third parameter) determines the impedance which the medium presents to waves. The behaviour of transverse waves on a string, longitudinal waves in a solid and a gas, voltage and current waves on a transmission line and electromagnetic waves in a dielectric is described.

Chapter 6 is devoted to one of the most powerful techniques of mathematical physics—that of Fourier analysis. Fourier methods occupy a central position in many of the arguments used. Specific examples of its application are described in considerable detail explaining how spatially periodic functions are Fourier analysed. The Fourier method is used again in Chapter 9 where it is applied to analyse an arbitrary pulse with a temporal periodicity. Chaper 8 deals with reflection and standing waves. The problems of impedance-matching and antireflecting layers is discussed in detail. The standing waves are identified as nothing but the normal modes of a continuous system.

The study of harmonic disturbances of different kinds finds wideranging applications in various branches of physics and engineering. This book is not intended to be a textbook on sound, electromagnetism, optics or electronic circuits, but certain topics of these subjects are treated in some detail to inject life into the general discussion. The emphasis here is on the line of common thought and underlying unity of concepts in these subjects.

I wish to express my gratitude to the students and colleagues whose thoughts, criticisms and suggestions have contributed to my understanding of the subject of this book. I also wish to thank my wife Kamlesh Bajaj for her enthusiastic efforts in typing the final manuscript.

Suggestions for the improvement of the book would be most gratefully acknowledged.

NK BAJAJ

Contents

1

Free Oscillations of Systems with One Degree of Freedom: Simple Harmonic Motion

1.1 INTRODUCTION

In everyday life we come across various things that move. The motion of physical systems can be classified into two broad categories : *translational* and *vibrational* motion. If the position of a body varies linearly with time its motion is translatory, e.g. a train moving on a straight track or a ball rolling on the ground. A motion that repeats itself in equal intervals of time is called *periodic* motion, e.g. the motion of the hands of a clock. If a body in periodic motion moves back and forth over the same path, its motion is called *vibratory* or *oscillatory*. Some examples of oscillatory motion are the oscillations of the arms of a walking person, the balance wheel of a watch, the bob of a pendulum clock, the prongs of a tuning fork, the piston of an automobile engine, beating of the heart, wings of a flying mosquito, etc. Oscillations may be very complex such as those of a piano string or those of the earth during an earthquake. It may be remarked that mechanical systems are not the only ones that can oscillate. The atoms in a solid vibrate. The electrons in a radiating or receiving antenna are in oscillation. A tuned circuit in a radio can oscillate electro-magnetically. Radio waves, microwaves and visible light are just the oscillating electric and magnetic fields. Thus the study of oscillations is essential for the understanding of various systems—mechanical, acoustical, electrical and atomic.

In this chapter we shall study the simplest and smoothest type of oscillatory motion, namely, *simple harmonic motion* (usually designated as SHM) of systems having *one degree of freedom*. A system is said to have one degree of freedom if it is completely specified by a single physical quantity. Some simple examples are a pendulum oscillating in a plane, a mass attached to a spring and an electrical circuit involving a capacitance and an inductance (Fig. 1.1). A simple pendulum oscillating in a plane can be described by angle α that the string makes with the vertical, an oscillating mass attached to a spring by the displacement x from an average (or mean) position and an LC circuit by the charge Q on the capacitor or current I in the inductance Let the symbol ψ denote the physcial quantity that characterises a system with one degree of freedom.

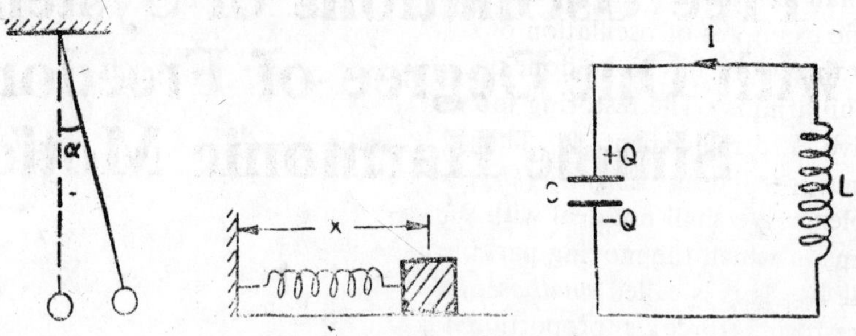

Fig. 1.1 Systems with one degree of freedom

Among the various varieties of oscillatory motions, SHM is of central importance for three basic reasons : (i) The oscillation of all physical systems is simple harmonic or a close approximation to it, if the oscillation is not too violent, (ii) We shall learn later that complex oscillations, e.g. those of a piano string, can be expressed as a superposition of harmonic motions and (iii) The study of SHM is essential for the understanding of wave motion.

1.2 WHAT CAUSES A SYSTEM TO OSCILLATE

The oscillation of a physical system results from *two* basic properties of the system, namely, *elasticity* and *inertia*. Consider a body in equilibrium so that forces on it balance. Let us displace it from its position of equilibrium (by doing work on it, i.e. applying a force) by a distance ψ. When it is released, a restoring force comes into play whose tendency is to 'restore' ψ to its original value, which is zero, by imparting to it an appropriate negative velocity $d\psi/dt$. *The magnitude of the restoring force is determined by the elastic properties of the system.* Inertia, on the other hand, tries to oppose any change in velocity. When the body reaches its equilibrium position ($\psi = 0$), the negative velocity is maximum which produces a negative displacement. The body then overshoots its position of equilibrium. The restoring force now becomes positive (i.e. it helps increase ψ) and it must now overcome the inertia of the negative velocity. Consequently the velocity keeps on decreasing until it is zero but by that time the displacement has become large and negative and the process is reversed. This process of the restoring force trying to bring ψ to zero by imparting a velocity and inertia preserving the velocity and making ψ to overshoot, repeats itself and the body oscillates.

1.3 DESCRIPTION OF SIMPLE HARMONIC MOTION

We shall first discuss the general dynamics of SHM and later analyse a few specific examples of oscillation of systems with one degree of freedom. We know that work has to be done on a system to displace it from its position of equilibrium. The restoring force F obviously depends on the work done to give a displacement ψ. Thus, F is some general function of ψ. For systems oscillating *violently* (*large* ψ) the dependence of F on ψ is very complex. We shall not deal with such systems but focus our attention on systems in which the moving part always stays close to its mean position (small ψ). This is called *small oscillation approximation*. For such systems, the restoring force is proportional to the displacement and opposes its increase. In other words,

$$F = -K\psi \tag{1.1}$$

The negative sign indicates that F opposes increase in ψ. The constant of proportionality K is called the *force constant*. In the *SI* system K is measured in newton per meter abbreviated as N m^{-1}. The magnitude of K depends on the *elastic properties of the system under study*. In the specific examples of SHM we shall compute the value of K in each case.

Equation (1.1) is a statement of *Hooke's law* for elastic forces. The general definition of *SHM is the motion in which the restoring force is proportional to the displacement from the mean position and opposes its increase*. We shall see that in such a motion the displacement varies harmonically with time.

Equation of Motion

Under the influence of a restoring force $F (= -K\psi)$ a body acquires a velocity $d\psi/dt$ and an acceleration $d^2\psi/dt^2$. If m is the mass of the body, then from Newton's law (force = mass × acceleration), the acceleration of the body is given by

$$\frac{d^2\psi}{dt^2} = -\frac{K}{m}\psi \tag{1.2}$$

Equation (1.2) is the equation of simple harmonic motion. The physical statement corresponding to this equation is that the acceleration of the body in SHM is proportional (and opposite in sign) to the displacement. In order to determine what type of motion is represented by Eq. (1.2) we need to solve this differential equation, i.e. obtain an expression of displacement ψ as a function of time t.

General Solution

The equation of motion is a homogeneous second-order ordinary differential equation with constant coefficients. Its solution is well known, but since we shall come across more difficult ones in the course of our study of oscillations, it is instructive to examine the methods of obtaining the solution. The usual procedure to solve any differential equation is to guess a solution and see if it works.

First Guess (Trignometric solution)

Equation (1.2) relates a function $\psi(t)$ (not yet known) to its second derivative $d^2\psi/dt^2$. To satisfy Eq. (1.2) we must look for a function $\psi(t)$ whose second derivative $d^2\psi/dt^2$, except for a negative constant factor $(-K/m)$, is the same as the function $\psi(t)$ itself Our knowledge of calculus tells us that sine and cosine functions have just this property, since,

$$\frac{d}{d\theta}(\sin\theta) = \cos\theta$$

$$\frac{d^2}{d\theta^2}(\sin\theta) = -\sin\theta$$

$$\frac{d}{d\theta}(\cos\theta) = -\sin\theta$$

$$\frac{d^2}{d\theta^2}(\cos\theta) = -\cos\theta$$

One can immediately verify that functions $a\sin\theta$ and $b\cos\theta$, where a and b are constants, also obey this property. Since angle θ, measured in radians, must depend on time t, we set $\theta = \omega t$, where ω is a constant to be measured in radians per second (rad s^{-1}). Thus, let us try, as a solution of Eq. (1.2)

$$\psi(t) = a\sin\omega t$$

Differentiating twice with respect to t we get,

$$\frac{d^2\psi}{dt^2} = -a\,\omega^2\sin\omega t$$

Substitution in Eq. (1.2) gives

$$-a\,\omega^2\sin\omega t = -\frac{K}{m}a\sin\omega t.$$

Therefore, if we choose constant ω such that

$$\omega = \sqrt{\frac{K}{m}} \tag{1.3}$$

then $\psi(t) = a \sin \omega t$ is, indeed, a solution of Eq. (1.2). We can similarly verify that $\psi(t) = b \cos \omega t$ is also a solution of Eq. (1.2). We can go a step further and say that

$$\psi(t) = a \sin \omega t + b \cos \omega t \tag{1.4}$$

is the general solution of Eq. (1.2) because

$$\frac{d\psi}{dt} = a\omega \cos \omega t - b\omega \sin \omega t$$

and $$\frac{d^2\psi}{dt^2} = -\omega^2 (a \sin \omega t + b \cos \omega t) = -\omega^2 \psi(t)$$

which is indeed Eq. (1.2) with the constant ω given by Eq. (1.3).

Second Guess (Power series solution)

We shall now make a much more general guess. We assume that $\psi(t)$ is a power series in t and see whether some choice of the coefficients in the series will satisfy Eq. (1.2).

Let $$\psi(t) = a_0 + a_1 t + a_2 t^2 + a_3 t^3 + a_4 t^4 + \ldots\ldots$$

Differentiating twice w.r.t. time t we get

$$\frac{d^2\psi}{dt^2} = 2a_2 + 6a_3 t + 12a_4 t^2 + 20a_5 t^3 + \ldots\ldots$$

which on substitution in Eq. (1.2) gives

$$(2a_2 + a_0\omega^2) + (6a_3 + a_1\omega^2)t + (12a_4 + a_2\omega^2)t^2 + \ldots\ldots = 0$$

For this equation to hold for every value of t, the coefficient of each power of t must be zero, giving

$$a_2 = -\frac{a_0}{2}\omega^2$$

$$a_3 = -\frac{a_1}{6}\omega^2$$

$$a_4 = -\frac{a_2\omega^2}{12} = \frac{a_0}{24}\omega^4$$

Therefore, the series that satisfies Eq. (1.2) is

$$\psi(t) = a_0\left(1 - \frac{1}{2}\omega^2 t^2 + \frac{1}{24}\omega^4 t^4 - \ldots\ldots\right)$$

$$+ a_1\left(t - \frac{1}{6}\omega^2 t^3 + \frac{1}{120}\omega^4 t^5 - \ldots\ldots\right)$$

$$= a_0\left(1 - \frac{\omega^2 t^2}{2!} + \frac{\omega^4 t^4}{4!} - \ldots\ldots\right)$$

$$+ \frac{a_1}{\omega}\left(\omega t - \frac{\omega^3 t^3}{3!} + \frac{\omega^5 t^5}{5!} - \ldots\ldots\right)$$

or
$$\psi(t) = a_0 \cos \omega t + \frac{a_1}{\omega} \sin \omega t$$

since

$$\cos x = 1 - \frac{x^2}{2!} + \frac{x^4}{4!} - \ldots\ldots$$

$$\sin x = x - \frac{x^3}{3!} + \frac{x^5}{5!} - \ldots\ldots$$

Here $x = \omega t$. Setting $a = a_1/\omega$ (ω being constant) and $b = a_0$ we recover our earlier solution (1.4).

Third Guess (Exponential solution)

Power series is not always the best guess for the solution of differential equations. The guess that is most often made to solve any differential equation is to assume an exponential solution, i.e.

$$\psi(t) = C\, e^{\alpha t}$$

Where C and α are constants. Differentiating twice with respect to t we get

$$\frac{d^2\psi}{dt^2} = \alpha^2 C\, e^{\alpha t} = \alpha^2 \psi(t)$$

Substitution in Eq. (1.2) yields

$$\alpha^2 = \omega^2$$

or
$$\alpha = \pm\, i\, \omega$$

where $i = \sqrt{-1}$. Thus $\psi(t) = C_1 e^{i\omega t}$ and $\psi(t) = C_2 e^{-i\omega t}$ are the two possible solutions of Eq. (1.2). The general solution is

$$\psi(t) = C_1 e^{i\omega t} + C_2 e^{-i\omega t} \tag{1.5}$$

One can immediately check that Eq. (1.5) satisfies Eq (1 2).

In writing the general solution as a superposition of the two possible solutions, we have used a very important principle called the superposition principle which holds only for linear differential equations (see Sec. 2 1).

Equations (1.4) and (1.5) are completely equivalent but alternative ways of writing the same solution since we can always express the exponential function in terms of trignometric functions by using

$$e^{i\theta} = \cos\theta + i \sin\theta$$

$$e^{-i\theta} = \cos\theta - i \sin\theta$$

Using these in Eq. (1.5) we get

$$\psi(t) = (C_1 + C_2) \cos \omega t + i\,(C_1 - C_2) \sin \omega t$$

If we set $C_1 + C_2 = b$ and $i\,(C_1 - C_2) = a$, we get Eq. (1.4).

In physics literature, Eq. (1.4) has been written in various alternative forms. We shall now write it in the more conventional form. Instead of writing $\psi(t)$ in terms of constants a and b we will write it in terms if other two constants A and δ which are related to a and b as

$$a = A\cos\delta$$

$$b = A\sin\delta$$

so that

$$\psi(t) = A\sin\omega t\cos\delta + A\cos\omega t\sin\delta$$

or $$\psi(t) = A\sin(\omega t+\delta) \tag{1.6}$$

where A and δ are related to a and b as

$$A = (a^2+b^2)^{1/2}$$

and $$\tan\delta = \frac{b}{a}$$

Some authors prefer to write the cosine solution by setting

$$a = -A\sin\phi$$

and $$b = A\cos\phi$$

so that

$$\psi(t) = A\cos\omega t\cos\phi - A\sin\omega t\sin\phi$$

or $$\psi(t) = A\cos(\omega t+\phi) \tag{1.7}$$

where the constants A and ϕ are related to a and b as

$$A = (a^2+b^2)^{1/2}$$

and $$\tan\phi = -\frac{a}{b}$$

In conclusion the solution of equation $d^2\psi/dt^2 = -\omega^2\psi$ may be expressed in any of the *four* forms

$$\begin{aligned}\psi(t) &= a\sin\omega t + b\cos\omega t\\ &= C_1 e^{i\omega t} + C_2 e^{-i\omega t}\\ &= A\sin(\omega t+\delta)\\ &= A\cos(\omega t+\phi)\end{aligned}$$

Each solution contains two arbitrary constants which are related as

$$a = i(C_1-C_2) = A\cos\delta = -A\sin\phi$$

$$b = C_1+C_2 = A\sin\delta = A\cos\phi$$

The equivalence of the solutions allows us to describe any SHM in any of the four forms. In much of our future analysis, however, we will find it extremely useful to decide in favour of the *cosine form.*

1.4 CHARACTERISTICS OF SHM

We shall now proceed to learn the physical meaning of the three constants A, ω and δ (or ϕ) that characterize a SHM. The three characteristics of SHM are as follows.

Amplitude

The amplitude of an SHM is the maximum (positive or negative) value of the displacement from the mean position. Since the maximum and minimum values of any cosine function are respectively $+1$ and -1, the maximum and minimum values of ψ in Eq. (1.7) are respectively $+A$ and $-A$. A is called the *amplitude* of SHM.

Time Period

The smallest time interval during which the oscillation repeats itself is called the *time period* (or simply, period) T of the oscillation. If time t in Eq (1.7) is advanced by $2\pi/\omega$, to $t' = t+2\pi/\omega$ the function becomes

$$\begin{aligned}\psi(t') &= A\cos(\omega t'+\phi)\\ &= A\cos\left\{\omega\left(t+\frac{2\pi}{\omega}\right)+\phi\right\}\\ &= A\cos(\omega t+\phi+2\pi)\\ &= A\cos(\omega t+\phi)\\ &= \psi(t)\end{aligned}$$

In other words, the displacement repeats itself in a time interval of $2\pi/\omega$. Therefore, the period T is given by

$$T = \frac{2\pi}{\omega} = 2\pi\sqrt{\frac{m}{K}} \tag{1.8}$$

Frequency of SHM, is the number of oscillations completed in a unit time interval. Therefore, by definition, frequency is the reciprocal of the time period, i e.

$$\nu = \frac{1}{T} = \frac{1}{2\pi}\sqrt{\frac{K}{m}} \tag{1.9}$$

Thus

$$\omega = \frac{2\pi}{T} = 2\pi\nu \tag{1.10}$$

Quantity ω is called the *angular frequency* of the SHM. In the SI system T is measured in s (second), ν in Hz (Hertz) and ω in rad s^{-1} (radian per second). The period (or frequency) is determined from the elastic and inertial properties namely, the force constant K and the mass m of the system under study. In Sec. 1.10 we shall compute time period of oscillations of some systems having one degree of freedom.

Phase

The argument $(\omega t+\phi)$ of the cosine function is called the *phase of the motion.* The constant ϕ of is called the *initial phase* (i. e. phase at $t = 0$) or the phase constant. The phase of an oscillating system at any instant is its state as regards its position and direction of motion at that instant. The knowledge of the phase constant enables us to find out how far from the mean position the system was at time $t = 0$. For example, if $\phi = 0$

$$\psi(t) = A \text{ cost } \omega t$$

which means that the displacement was maximum $= A$ at time $t = 0$, i.e. when the motion was started. On the other hand, if $\phi = \pi/2$

$$\psi(t) = A \cos\left(\omega t+ \frac{\pi}{2}\right) = A \sin \omega t$$

i e. the displacement was zero at time $t = 0$. In other words, the counting of time was started the moment the oscillator passed the mean position. Thus phase constant is a measure of how much time had elapsed before the oscillator last passed the mean position. Amplitude A and phase constant ϕ are determined from the *initial conditions*, i.e. the way the system is started at time $t = 0$ (see Sec. 1.11).

1.5 VELOCITY AND ACCELERATION IN SHM

It is instructive to learn how velocity and acceleration in a SHM vary with time. We know that displacement $\psi(t)$ is given by

$$\psi(t) = A \cos (\omega t+\phi)$$

Velocity V and acceleration a are given by

$$V = \frac{d\psi}{dt} = \dot{\psi} = -A \omega \sin (\omega t+\phi)$$

$$= \mp A\omega \left(1-\frac{\psi^2}{A^2}\right)^{1/2} \tag{1.11}$$

and
$$a = \frac{d^2\psi}{dt^2} \equiv \ddot{\psi} = -\omega^2 A \cos (\omega t+\phi) = -\omega^2 \psi \tag{1.12}$$

We notice that when the displacement is maximum ($+A$ or $-A$) the velocity $V = 0$, because now the oscillator has to return and velocity must change its direction. But when ψ is maximum ($+A$ or $-A$), the acceleration is also maximum ($-\omega^2 A$ and $+\omega^2 A$ respectively) and is directed opposite to the displacement. When $\psi = 0$, i.e. when $\cos(\omega t + \phi) = 0$, the velocity is maximum (ωA or $-\omega A$) and the acceleration is zero. In Fig. 1.2 ψ versus t, $\dot{\psi}$ versus t and $\ddot{\psi}$ versus t curves are plotted for a SHM taking, for simplicity, $\phi = 0$.

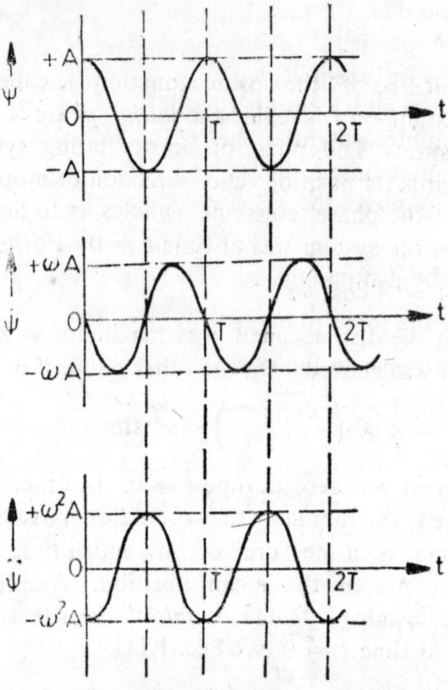

Fig. 1.2 Variation with time of displacement, velocity and acceleration in SHM, phase constant is taken as zero, for simplicity

1.6 TOTAL ENERGY IN SHM

Consider a system at rest at its position of equilibrium. When it is displaced from this position (by doing work on it) it acquires potential energy. When the system is released, it begins to move with a velocity, thus acquiring kinetic energy. At any instant of time, the kinetic energy o a system of mass m executing SHM is given by [using Eq. (1.11)]

$$\text{Kinetic energy (KE)} = \tfrac{1}{2} mV^2 = \tfrac{1}{2} m\omega^2 A^2 \sin^2(\omega t + \phi) \qquad (1.13)$$

The kinetic energy of the oscillator varies periodically. It is maximum ($= \frac{1}{2} m\omega^2 A^2$) when the velocity is maximum ($= \pm\, \omega A$) and displacement is zero. When the displacement is maximum ($= \pm A$), velocity $V = 0$ and KE $= 0$. At these extreme positions, the energy is all potential At intermediate positions (ψ lying between O and $\pm A$), the energy is partly kinetic and partly potential.

A closer look at Eq. (1 7) reveals that the total energy of the oscillator must remain constant because the maximum displacement is regained after every half cycle If no energy is dissipated (we have neglected dissipative or non-conservative forces like friction), then all the potential energy becomes kinetic and vice versa.

The energy of the oscillator may decrease not only due to friction in the system but also due to radiation. The oscillating body imparts periodic motion to the particles of the medium in which it oscillates thus producing waves For example, a tuning fork or a string produces sound waves in the medium which results in a decrease in energy.

Let us now compute potential energy at any instant of time t. Let ψ be the displacement at time t. The potential energy is given by the amount of work required to move the system from $\psi = 0$ to ψ, by applying a force

The force must be just enough to oppose the restoring force $F = -K\psi$

In other words, the force to be applied must be $K\psi$.

Work required to give an infinitesimal displacement $d\psi = K\psi\, d\psi$

Therefore, the total work done to displace the system from

$$0 \text{ to } \psi = \int_0^{\psi} K\psi d\psi = \tfrac{1}{2} K\psi^2.$$

Thus

$$\begin{aligned} \text{Potential energy (PE)} &= \tfrac{1}{2} K\psi^2 \\ &= \tfrac{1}{2} m\, \omega^2 A^2 \cos^2(\omega t + \phi) \end{aligned} \tag{1.14}$$

where we have used Eqs. (1.3) and (1.7).

Equations (1.13) and (1.14) give the instantaneous values of kinetic and potential energy.

The total energy E in SHM is, therefore, given by

$$\begin{aligned} E &= \text{KE} + \text{PE} \\ &= \tfrac{1}{2} m\omega^2 A^2 \{\sin^2(\omega t + \phi) + \cos^2(\omega t + \phi)\} \end{aligned}$$

or $$E = \tfrac{1}{2} m\omega^2 A^2$$

which is constant as we would expect. It is obvious that the maximum values of kinetic and potential energy are equal (both equal to $\frac{1}{2} m\omega^2 A^2$) indicating that the energy exchange is complete. Figure 1.3 shows how the kinetic and potential energy of the harmonic oscillator vary with time where, for simplicity, we have set $\phi = 0$.

It is instructive (and sometimes more convenient) to obtain the Eq. (1.2) of SHM using energy considerations. The total energy E of the oscillator is given by

$$E = \frac{1}{2} m \left(\frac{d\psi}{dt} \right)^2 + \frac{1}{2} K\psi^2 \tag{1.16}$$

Since E is constant, $dE/dt = 0$. Differentiating Eq. (1.16) w.r.t. time t and setting $\frac{dE}{dt} = 0$, we get.

$$\frac{d^2\psi}{dt^2} + \frac{K}{m} \psi = 0$$

which is the Eq. (1.2) we have obtained earlier.

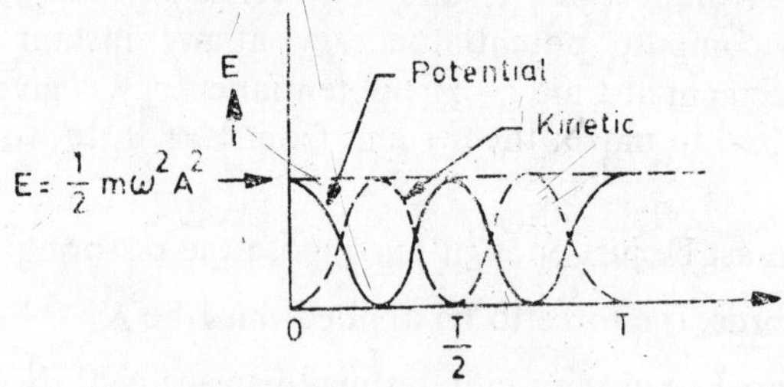

Fig. 1.3 Energy exchange in SHM

1.7 RELATION BETWEEN LINEAR SHM AND UNIFORM CIRCULAR MOTION (THE REFERENCE CIRCLE)

The expression (1.7) for the displacement in SHM has a very simple interpretation in terms of a relation between a SHM along a line and a uniform motion in a circle. This relation also gives a very simple geometric meaning to the quantities ω and ϕ.

Consider a particle P moving on a circle of radius A with a uniform speed (Fig. 1.4). Let T be the time period of this circular motion and let Q be the position of the particle at an instant of time t [Fig. 1.4 (a)]. At time $t = 0$ the particle was at P. R is the foot of the perpendicular drawn from Q on the diameter XX' of the circle. Let us study the motion of R

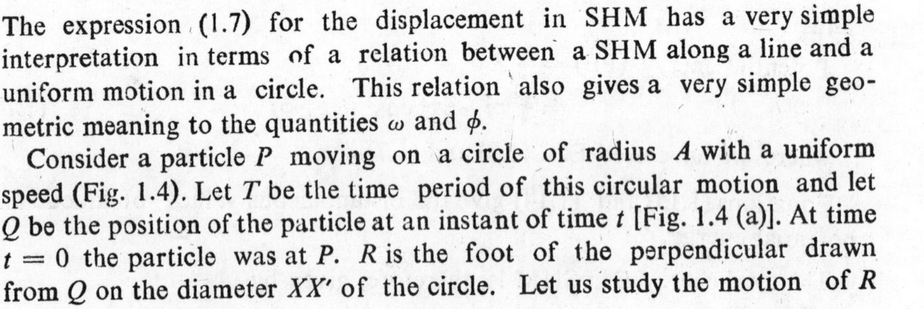

Fig. 1.4 The reference circle

as particle P moves around the circle. The point R moves from X to X' and back to X as the particle P completes one revolution around the circle. During time t when the particle P moves to Q along the circle, the radius OP sweeps an angle θ. The angular velocity of the circular motion is

$$\omega = \frac{\text{angle swept}}{\text{time taken}} = \frac{\theta}{t} = \frac{2\pi}{T}$$

since when $t = T$, $\theta = 2\pi$, or $\theta = \omega t$. Now, in $\triangle OQR$, $OR = A \cos \theta$

or $\psi = A \cos \omega t$

which agrees with Eq. (1.7) with $\phi = 0$. Therefore, the motion of R is simple harmonic. Thus *SHM can be described as the projection of a uniform circular motion on the diameter of the circle.* When this description is used, the body P is called the *reference body* and the circle along which it move is called the *reference circle.* The quantity ω of a SHM is the same as the angular velocity of the reference body.

Using this description of SHM we shall now try to understand the meaning of ϕ. In Fig. 1.4 (b) the particle P is not at X at time $t = 0$. Let the angle POX be ϕ. Now $\theta = \angle QOP$, therefore $\angle ROQ = \theta - \phi$ $= \omega t - \phi$, so that

$$\psi = A \cos (\omega t - \phi)$$

which also satisfies Eq. (1.2) and hence is simple harmonic.

If we do not start counting time when the particle is at P but a little later, as in Fig. 1.4(c) we will get,

$$\psi = A \cos (\omega t + \phi)$$

Hence phase constant ϕ is a measure of how far P is from X (or R from O) at time $t = 0$. Thus, the knowledge of ϕ helps us to know how much time had elapsed before the oscillator last passed the mean position. We can define SHM in two apparently unrelated but completely equivalent ways as follows

(i) *SHM is a motion in which the acceleration of the moving part of the system is proportional (and opposite in direction) to its displacement from the mean position.*

(ii) *SHM is the motion of the projection of a uniform circular motion on the diameter of the reference circle.*

1.8 ROTATING VECTOR REPRESENTATION OF SHM

In Section 1.7 we have used circular motion as a purely geometrical basis to describe SHM. We have seen that the projection, on the diameter of the circle (*or any straight line in the plane of the circle*) of a vector OQ rotating, in the anti-clockwise sense, at an angular frequency ω, defines SHM. We can now visualize the relationship between the alternative forms of the solution [Eqs. (1.4), (1.6) and (1.7)] more easily using the rotating vector representation. Let us represent the two terms $a \sin \omega t$ and $b \cos \omega t$ in Eq. (1.4) by the projections on the X-axis of two vectors $\vec{OP}$ and $\vec{OQ}$, rotating at an angular frequency ω in the counter-clockwise sense (Fig 1.5). The general solution ψ must be represented by the sum of the projections of $\vec{OP}$ and $\vec{OQ}$. It is quite obvious from Fig. 1.5 that the sum of projections of vectors $\vec{OP}$ and $\vec{OQ}$ is equal to the projection of a single vector $\vec{OR}$

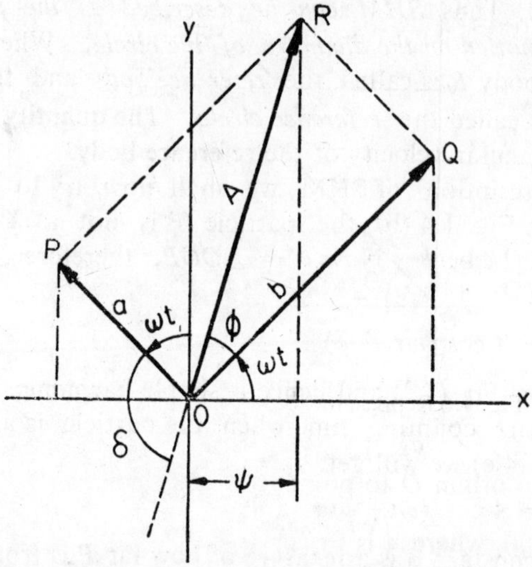

Fig 1.5 Rotating vector representation of SHM

which is the resultant of vectors $\vec{OP}$ and $\vec{OQ}$. Thus the general solution is represented equally well by the projection of $\vec{OR}$ on the x-axis. The projection of $\vec{OR}$ on x-axis given by

$$\psi = A \cos (\omega t + \phi)$$

which is solution (1.7). It is clear from the diagram that $\phi = \delta - \frac{\pi}{2}$ so that

$$\psi = A \cos \left(\omega t + \delta - \frac{\pi}{2} \right)$$

$$= A \sin (\omega t + \delta)$$

which is Eq. (1.6). The equivalence of Eqs. (1.6) and (1.7) subject to the condition that $\phi = \delta - \pi/2$ allows us to describe any SHM equally well in terms if a sine or a cosine function. As stated earlier we would prefer the cosine solution for the reason stated in the next section.

1.9 REPRESENTATION OF SHM BY A COMPLEX EXPONENTIAL

In the preceding section we have shown that the displacement in SHM may be represented by a rotating vector. A particularly useful, systematic and elegent representation of SHM results from the use of complex quantities. Consider a complex number $z = x+iy$ represented by the point R whose rectangular (cartesian) co-ordinates are (x, y) in the complex plane (Fig. 1.6). This number can also be represented by the vector

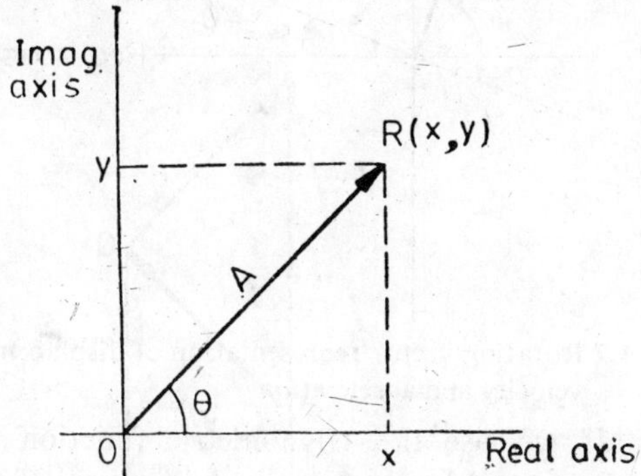

Fig. 1.6 Representation of a complex number

$\vec{OR}$ directed from origin O to point R. Let the plane polar coordinates of vector $\vec{OR}$ be (A, θ), where A is the magnitude of the vector $\vec{OR}$ and θ is the angle it makes with the real axis. Now since,

$$x = A \cos \theta$$

$$y = A \sin \theta$$

We have

$$z = x+iy = A (\cos \theta + i \sin \theta)$$

$$= Ae^{i\theta}$$

If A is constant and $\theta = \omega t+\phi$, then $A\, e^{i\theta}$ signifies a vector of constant magnitude A rotating at an angular frequency ω. Either the real or imaginary part represents a quantity varying harmonically with time.

We have seen that $Ae^{i(\omega t+\phi)}$ is a solution of Eq. (1.2) and it can be expressed as a combination of cos ωt and sin ωt by

$$z = Ae^{i(\omega t+\phi)} = A \cos (\omega t+\phi)+iA \sin (\omega t+\phi) \tag{1.17}$$

This new solution is a complex quantity. Since the results of measurements are, in general, real numbers, this new solution, at first sight, seems to have very little value. The prime merit of this solution is the special property of the exponential function, namely that the function itself reappears after every operation of differentiation or integration. In subsequent chapters we shall see that in the study of oscillations and waves we often come across equations involving periodic displacements and their

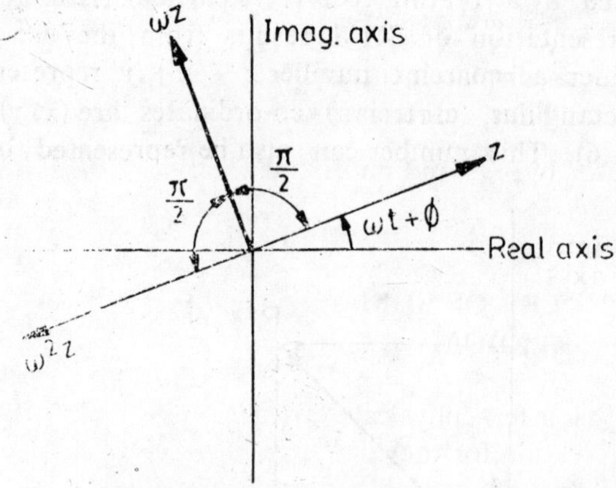

Fig. 1.7 Rotating vector representation of displacement, velocity and acceleration

time derivatives. If we use the trignometric function to describe the motion, we will have to tackle an awkward mixture of sine and cosine terms. The confusion and unnecessary work can be avoided if we use the exponential function z and adopt the convention that we use only the *real part* of the function when we are required to check physical measurements*. For example, if

$$\psi = A \cos (\omega t+\phi).$$

then $$\dot{\psi} = -A\, \omega \sin (\omega t+\phi)$$

and $$\ddot{\psi} = -\omega^2 A \cos (\omega t+\phi)$$

On the other hand, if we work with the exponential function z given by

$$z = A \cos (\omega t+\phi)+i A \sin (\omega t+\phi)$$

i.e. $$z = A\, e^{i(\omega t+\phi)}$$

with ψ = *real part of* z [written as $Re\,(z)$]

then $$\dot{z} = iA\omega e^{i(\omega t+\phi)} = i\,\omega\, z$$

and $$\ddot{z} = (i\omega)^2 A\, e^{i(\omega t+\phi)} = -\,\omega^2 z$$

where our convention requires that we take only the real part of z, $\dot{z}$ and $\ddot{z}$ to obtain respectively the displacement ψ, velocity $\dot{\psi}$, and acceleration $\ddot{\psi}$ of the motion.

---

* In Chapter 2 we will show that this is valid only for solutions of linear differential equations (i.e. equations that involve only the first power of the unknown function and its derivatives). We will prove that if a complex function is a solution of a linear differential equation then its real and imaginary parts are separately also solutions of the same equation. We could equally well adopt the convention to use only the imaginary part of the function but the convention is to use the real part.

Figure 1.7 shows the vector representations for z, $\dot{z}$ and $\ddot{z}$. Since

$$z = i\,\omega\,z = \omega\,z\,e^{i\pi/2}$$

and
$$\ddot{z} = -\omega^2 z = \omega^2 z\,e^{i\pi}$$

the velocity leads the displacement by $\pi/2$ and acceleration leads the displacement by π. Thus, differentiation involves counter-clockwise rotation of the vector by $\frac{\pi}{2}$ and multiplication by ω.

1.10 SOME EXAMPLES OF SHM OF SYSTEMS WITH ONE DEGREE OF FREEDOM

We shall now analyse a few physical systems that oscillate with SHM. We shall obtain the expression for the time period of oscillation in each case. It may be remarked that the variable ψ in SHM need not always be linear displacement. It could represent angular displacement as in the case of a simple pendulum or charge (or current) as in the case of an electrical system involving a capacitance and an inductance. In torsional oscillations ψ is the twist and in thermal oscillations it is the heat.

The Basic Mass-Spring System

When a force is applied to a spring to compress or stretch it, the resulting compression or elongation does not bear a simple relationship with the force applied. Figure 1.8 illustrates the force-displacement relationship for a spring. The relationship is, in general, not linear. Only for small

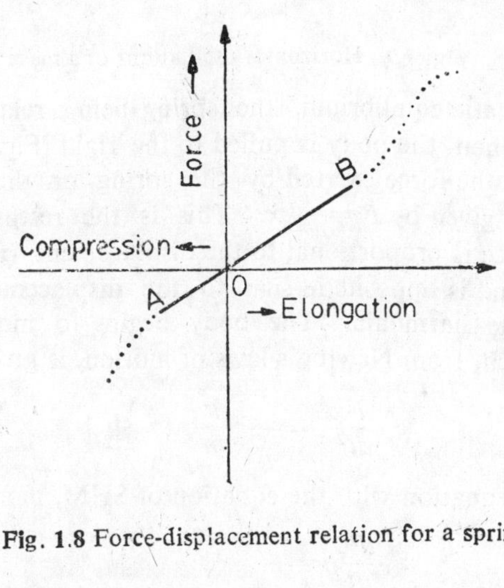

Fig. 1.8 Force-displacement relation for a spring

displacements is the relationship linear (portion AB of the curve). The elastic force produced in the linear spring is given by

$$F = -kx$$

where x is the change in the length of the spring when a force F is applied on it. The constant k is called the *spring constant* or *stiffness constant* and it is defined as the force required to produce a unit extension or compression in the spring. The unit of k in the SI system is Nm^{-1}.

Let us now analyse two simple examples of a mass oscillating on a spring.

Horizontal Oscillations

Consider a massless spring of constant k, one end of which is fixed rigidly to a wall and the other end is attached to a body of mass m which is free to move on a frictionless horizontal surface (Fig. 1.9). Figure 1.9(a) is

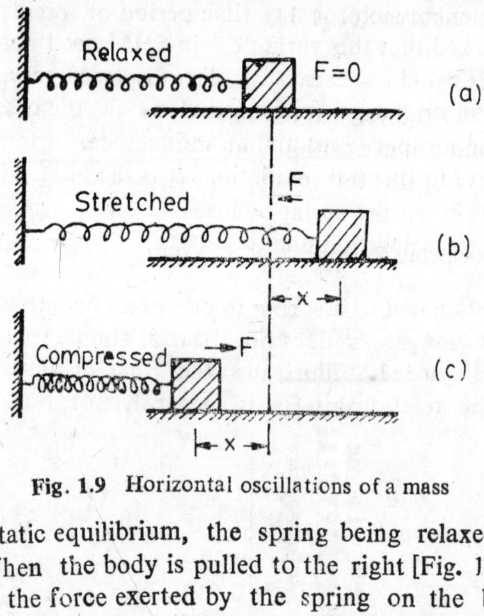

Fig. 1.9 Horizontal oscillations of a mass

the position of static equilibrium, the spring being relaxed and no force acting on it. When the body is pulled to the right [Fig. 1.9(b)] through a small distance x, the force exerted by the spring on the body is directed to the left and is given by $F = -kx$. This is the restoring force. Since the restoring force is proportional to the displacement (true only for small displacements) and is opposite in sign to the displacement, the resulting motion is simple harmonic. The body begins to move with a linear acceleration which, from Newton's laws of motion, is given by

$$\frac{d^2x}{dt^2} = -\frac{k}{m}x$$

Comparing this equation with the equation of SHM, namely,

$$\frac{d^2\psi}{dt^2} = -\omega^2\psi$$

We have

angular frequency $\omega = \sqrt{\frac{k}{m}}$

time period $T = 2\pi \sqrt{\frac{m}{k}}$

and frequency $\nu = \frac{1}{2\pi} \sqrt{\frac{k}{m}}$

The displacement x as a function of time is of the form

$$x = A \cos(\omega t + \phi)$$

Notice that angular frequency ω (or period T) is determined for all circumstances by (a) the inertia factor, mass m and (b) the elasticity factor, the spring constant k. The other two constants A and ϕ, which are required for a complete specification of the state of motion, are determined from the initial conditions (see Sec. 1.11).

Vertical Oscillations

Let us now consider the vertical oscillations of a loaded spring. Figure 1.10 illustrates the equilibrium position and force on the spring for two extreme positions of mass m. In this case, the equilibrium state of the loaded spring is the state when the spring is stretched by a distance d by the force mg where g is the acceleration due to gravity. No net force acts on the body, since

$$mg = kd$$

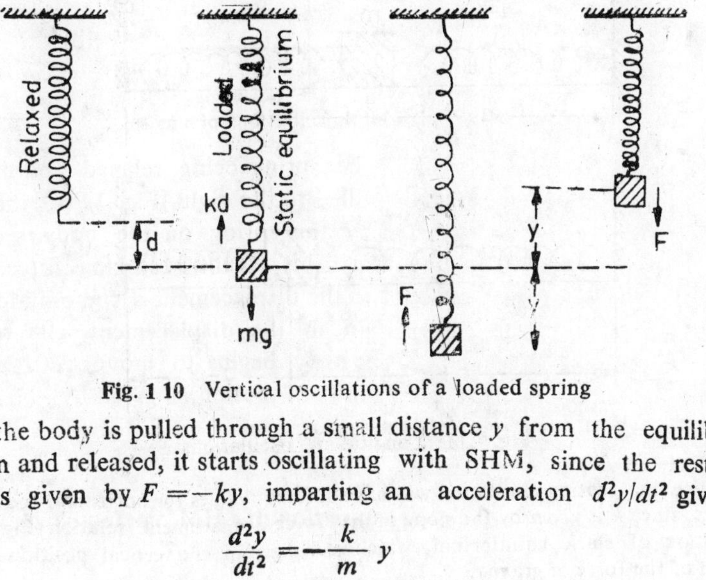

Fig. 1.10 Vertical oscillations of a loaded spring

When the body is pulled through a small distance y from the equilibrium position and released, it starts oscillating with SHM, since the restoring force is given by $F = -ky$, imparting an acceleration d^2y/dt^2 given by

$$\frac{d^2y}{dt^2} = -\frac{k}{m} y$$

which is the equation of SHM whose period is given by

$$T = 2\pi \sqrt{\frac{m}{k}}$$

which is the same as that for horizontal oscillations. The equation $mg = kd$ determines spring constant k if m, d and g are known.*

Mass and Two Springs System

Longitudinal Oscillations

Consider a body of mass m placed on a horizontal frictionless surface as shown in Fig. 1.11. It is connected to rigid walls by means of two identical massless springs, each of spring constant k and relaxed length a_o In the equilibrium position (Fig. 1.11b), each spring is stretched to a length a. The tension on each spring is givon by $T_o = k(a - a_o)$. Let the mass be displaced slightly, say to the right and released. It will execute longitudinal oscillations. Let x be the displacement of the mass at any instant of time (Fig. 1.11c). The left-hand side spring exerts a force equal to $k(a+x-a_o)$ trying to pull the mass to the left. The right-hand spring pushes the mass to the right with a force equal to $k(a-x-a_o)$. Thus, the net force in the

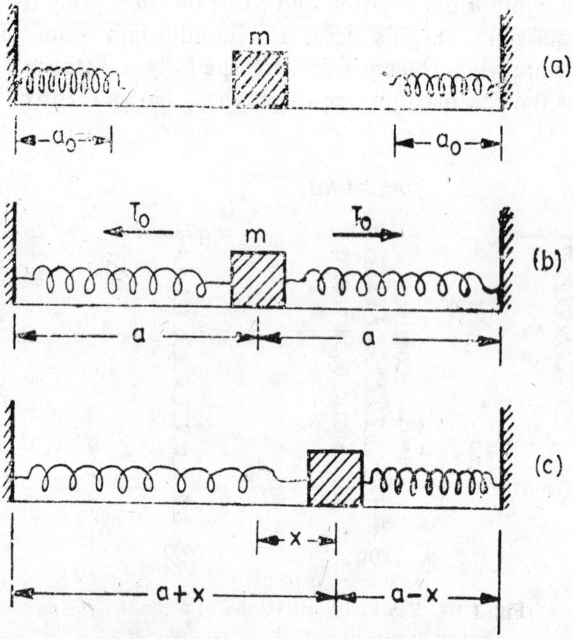

Fig. 1.11 Longitudinal oscillations

*Notice that spring constant k will have different values for horizontal and vertical positions, since k is given by the slope of the force-displacement relation (Fig. 1.8) at the position of static equilibrium, which is changed in the vertical position by the addition of the force of gravity.

direction of increasing x is given by

$$F_x = k(a-x-a_o)-k(a+x-a_o) = -2kx$$

From Newton's law, the equation of motion as

$$m\frac{d^2x}{dt^2} = -2kx$$

or

$$\frac{d^2x}{dt^2} = -\frac{2k}{m}x$$

which describes SHM at angular frequency

$$\omega = \sqrt{\frac{2k}{m}}$$

and period

$$T = 2\pi\sqrt{\frac{m}{2k}}$$

Transverse Oscillations

Consider a mass m suspended between two identical springs, each of spring constant k and relaxed length a_o (Fig. 1.12). In the equilibrium position each spring is stretched to length a and exerts a force given by

$$T_o = k(a-a_o) \tag{1.19}$$

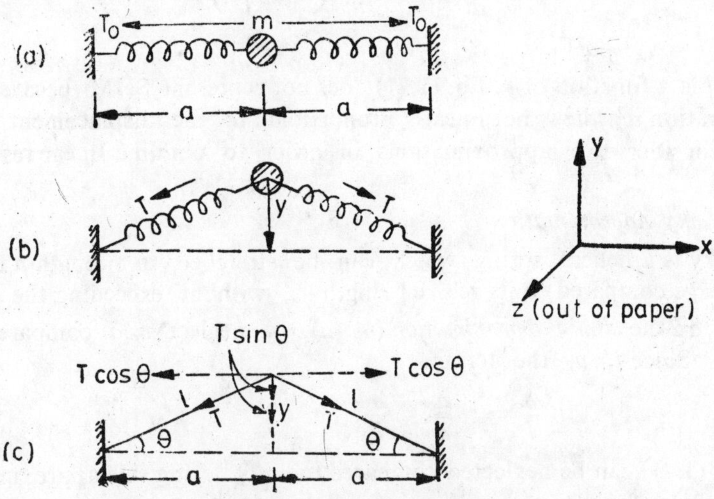

Fig. 1.12 Transverse oscillations

For simplicity, we neglect the effect of gravity which produces a little 'sag'. The system has three *degrees of freedom*. The mass can move along x, y or z direction. Its motion along x-axis results in longitudinal oscillations already discussed above. The system is symmetrical with respect to x, y and z axes. If the mass is constrained to move along one direction, it will have no tendency to move along any other direction. Here we will consider one of the two transverse motions, namely, the one along the y direction.

Let the mass be pulled in the $+y$ direction and released. It will execute vertical oscillations. Let y be the displacement of the mass from the equilibrium position at any instant of time and let l be the length of each spring at this time. In the displaced position Fig. 1.12(b) the tension in each spring is given by

$$T = k\ (l - a_o) \tag{1.20}$$

The tension is exerted along the axis of each spring. Taking the x and y components of this force [Fig. 1.12(c)] we see that the mass experiences no net force along the x direction but along the y direction each spring contributes a force $T \sin \theta$. Thus the restoring force is

$$F_y = -2T \sin \theta = -2T \frac{y}{l}$$

From Newton's law, we have

$$m \frac{d^2y}{dt^2} = -\frac{2T}{l} y$$

Using Eq. (1.20) we have

$$\frac{d^2y}{dt^2} = -\frac{2k}{m} \left(1 - \frac{a_o}{l} \right) y \tag{1.21}$$

where $l = (y^2 + a^2)^{\frac{1}{2}}$

Since l is a function of y, Eq. (1.21) does not represent SHM because the acceleration d^2y/dt^2 is not linearly proportional to the displacement y. We shall consider two approximations in order to obtain a linear restoring force.

(a) Slinky Approximation

A slinky is a helical spring which can be stretched to a length a that is very large compared to its relaxed length a_o, without exceeding the elastic limit. In the *slinky approximation* ($a_o \ll a$) we neglect a_o/a compared to unity. Since $l > a$, the term

$$\frac{a_o}{l} = \frac{a_o}{a} \cdot \frac{a}{l}$$

in Eq. (1.21) can be neglected compared to unity. Using this approximation Eq. (1.21) reduces to

$$\frac{d^2y}{dt^2} = -\frac{2k}{m} y$$

which represents SHM with angular frequency

$$\omega = \sqrt{\frac{2k}{m}}$$

and time period

$$T = 2\pi \sqrt{\frac{m}{2k}} \tag{1.22}$$

Notice that we can have large oscillations which are perfectly harmonic. Further, the frequency of transverse oscillations (under slinky approximation) is the same as that for longitudinal oscillations.

(b) *Small Oscillation Approximation*

The slinky approximation does not apply to springs for which a_o cannot be neglected compared to a. In such cases we use the *small oscillation approximation.* We assume that the oscillator always stays close to its equilibrium position, so that its displacement is small compared to any characteristic length of the system. In other words, we assume that

$$y \ll a \text{ or } l$$

so that

$$l = a\left(1+\frac{y^2}{a^2}\right)^{1/2} \simeq a$$

where we have neglected terms of the order $\frac{y^2}{a^2}$ and higher in the Binomial expansion of $(1+y^2/a^2)^{1/2}$.

Using this approximation Eq. (1.21) becomes

$$\frac{d^2y}{dt^2} = -\frac{2k}{m}\left(1-\frac{a_o}{a}\right)y$$

which represents SHM at angular frequency

$$\omega = \sqrt{\frac{2k}{m}\left(1-\frac{a_o}{a}\right)}$$

and time period

$$T = 2\pi\sqrt{\frac{m}{2k\left(1-\frac{a_o}{a}\right)}} \tag{1.23}$$

Notice that if the slinky approximation does not apply the period of longitudinal oscillations [Eq. (1.18)] is not the same as that of transverse oscillations [Eq. (1.23)]. The ratio of the time periods in the two cases is

$$\frac{T_{\text{long}}}{T_{\text{trans}}} = \left(1-\frac{a_o}{a}\right)^{1/2}$$

Thus, the longitudinal oscillations are more rapid that transverse oscillations This conclusion can be verified by replacing the springs by rubber ropes for which the slinky approximation does not hold.

The Simple Pendulum

A simple pendulum is an idealized system consisting of a massless inextensible string, fixed rigidly at one end, having a point mass at the other (Fig. 1.13). When the mass is displaced from the equilibrium position O and released, it oscillates. Suppose the mass is at P at any instant of time during oscillation. Let α be the angle subtended by the string with the vertical at that instant. The force acting vertically downwards at P is $m\,g$ where g is the acceleration due to gravity. This force has a component $m\,g\cos\alpha$ which acts along the string and is balanced by the tension in the string. The tangential component is $m\,g\sin\alpha$ and is directed opposite to increasing α. Thus, the restoring force is given by

$$F = -m\,g\sin\alpha$$

$$= -m\,g\,\alpha\left(1-\frac{\alpha^2}{3!}+\frac{\alpha^4}{5!}-\ldots\ldots\right)$$

where α is expressed in radians.

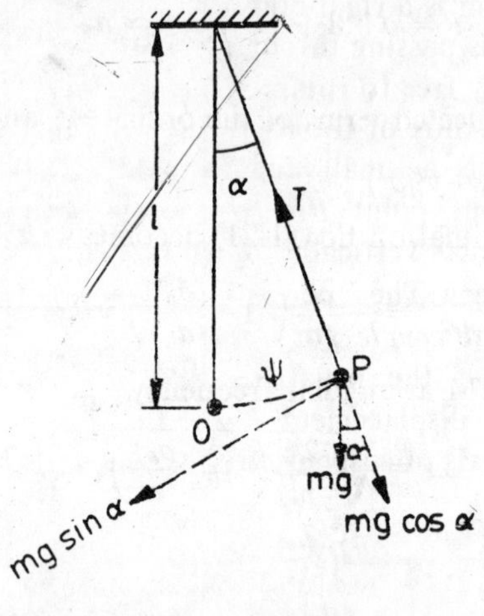

Fig. 1.13 The simple pendulum

The restoring force is not proportional to α, the angular displacement, hence the oscillation is not harmonic. However, if α is so small that terms of order α^2 and higher can be neglected compared to unity* we have

$$F = -mg\,\alpha = -mg\,\frac{\psi}{l}$$

since $\psi = l\,\alpha$, where ψ is the displacement OP along the arc. From Newton's law, the equation of motion is

$$m\,\frac{d^2\psi}{dt^2} = -\frac{mg}{l}\,\psi$$

or
$$\frac{d^2\psi}{dt^2} = -\frac{g}{l}\,\psi$$

*The error in replacing $\sin\alpha$ by α is nearly 1% for α = 0·25 radian (or 14°). For smaller angles the error is even less.

which represents SHM at angular frequency given by

$$\omega = \sqrt{\frac{g}{l}} \tag{1.24}$$

and time period given by

$$T = 2\pi\sqrt{\frac{l}{g}} \tag{1.25}$$

Notice that the period of oscillations is independent of the mass m of the pendulum.

The Compound Pendulum

A compound pendulum is a rigid body, of any shape, capable of oscillating about a horizontal axis passing through it. Figure 1.14 shows a vertical section of a rigid body free to rotate about a point P. The distance l between point P and the centre of gravity G is called the *length* of the pendulum. The pendulum is given a small angular displacement θ and released. It begins to oscillate about point P. In the displaced position, the weight $m\,g$ of the pendulum acts vertically downwards at G', the new position of the centre of gravity. The pendulum tends to return under the influence of a *reactive couple* (or *torque*). The moment of the restoring couple $= -m\,g\,l \sin\theta$, the negative sign indicating that the couple is directed opposite to displacement. If I is the moment of inertia of the pendulum about the axis of suspension through P, this couple is also equal

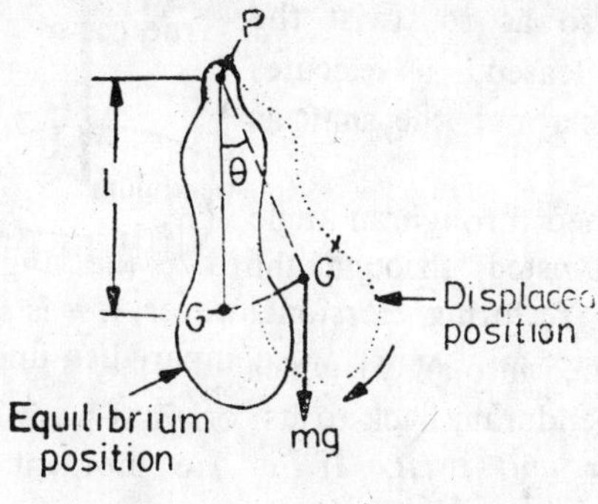

Fig. 1.14 The Compound pendulum

to $I\dfrac{d^2\theta}{dt^2}$ where $\dfrac{d^2\theta}{dt^2}$ is the angular acceleration. Thus,

$$I\frac{d^2\theta}{dt^2} = -m\,g\,l\sin\theta$$

If θ (expressed in radians) is small, as before, we replace $\sin\theta \cong \theta$, so that

$$\frac{d^2\theta}{dt^2} = -\frac{mgl}{I}\,\theta$$

Thus, the pendulum executes SHM and its time period is given by

$$T = 2\pi \sqrt{\frac{I}{mgl}}$$

It is sometimes more useful and convenient to introduce the moment of inertia about a parallel axis through G, the centre of gravity. If this is written as $I_o = mk^2$, where k is the *radius of gyration* of the pendulum, then we have, from the theorem of parallel axes,

$$I = I_o + ml^2 = m(k^2 + l^2)$$

$$\therefore \qquad T = 2\pi \sqrt{\frac{\left(\frac{k^2}{l} + l\right)}{g}}$$

The time period is the same as that of a simple pendulum of length $L = \frac{k^2}{l} + l$. This length L is called the *length of the equivalent simple pendulum.*

The Torsional Pendulum

A torsional pendulum consists of a massive body, such as a disc or cylinder, attached at its mid-point to a shaft or a wire suspended from a rigid support (Fig. 1.15). If the disc is turned in the horizontal plane (so as to twist the shaft) and then released, it executes torsional oscillations about the shaft as the axis.

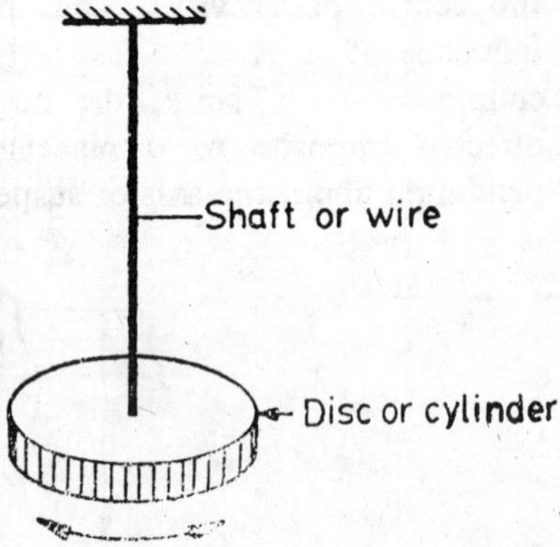

Fig. 1.15 The torsional pendulum

If the disc is turned through an angle θ, the shaft is also twisted through the same angle θ. A restoring torsional couple $= -\tau\theta$ is called into play, which tends to bring the pendulum back to its original position. Here τ is the *torsional couple per unit twist.* If I is the moment of inertia of the disc about the shaft as the axis and $\frac{d^2\theta}{dt^2}$, its angular acceleration, the couple due to the acceleration is given by $I\frac{d^2\theta}{dt^2}$. In the dynamic equilibrium, both these couples must balance, giving

$$I\frac{d^2\theta}{dt^2} = -\tau\theta$$

or

$$\frac{d^2\theta}{dt^2} = -\frac{\tau}{I}\theta$$

Thus the disc executes SHM and its time period is given by

$$T = 2\pi\sqrt{\frac{I}{\tau}}$$

From mechanics of materials, τ, in case of shaft is given by

$$\tau = \frac{\pi\eta d^4}{32l}$$

where d is the diameter of the shaft, l its length and η is the modulus of rigidity of its material. In the case of a wire of radius r, length l and modulus of rigidity η, we have

$$\tau = \frac{\pi\eta r^4}{2l}$$

Liquid Column in a U-tube

A mass oscillating in SHM need not be a particle or a rigid body The body of a fluid as a whole can also oscillate. One such example is the oscillation of the liquid column in a U-tube (Fig 1.16). The column of

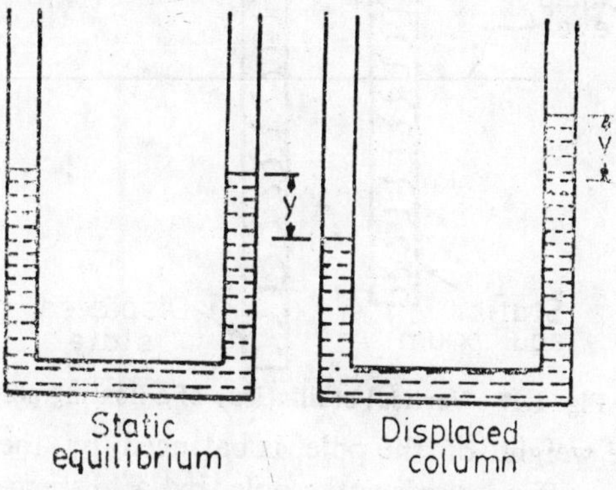

Fig. 1.16 Oscillations of a liquid column

the liquid is displaced through y by gently blowing into the tube. The columns exhibit vertical oscillations. Let L, A and ρ be respectively, the length of the liquid column, area of cross-section of the tube and density of the liquid We shall neglect viscous effects Since the right-hand side column is higher by $2y$, with respect to the column on the left-hand side, the mass of this column of liquid is $m = 2A\rho y$. The restoring force (which is a gravitational force) is given by

$$F = -mg = -2A\rho gy = -Ky$$

where the force constant $K = 2A\rho g$. The angular frequency of the harmonic oscillation is

$$\omega = \sqrt{\frac{K}{M}}$$

where $M = \rho AL$ is the total mass of the liquid in oscillation. Thus

$$\omega = \sqrt{\frac{2A\rho g}{\rho AL}} = \sqrt{\frac{2g}{L}}$$

The time period of oscillation is

$$T = 2\pi \sqrt{\frac{L}{2g}}$$

It is interesting to note that the period of oscillation does not depend on the density of the liquid or the area of cross-section of the tube.

Floating Objects

Consider a pole of cross-sectional area A and mass M floating in a vertical position in a liquid of density ρ (Fig. 1.17). This is the static equilibrium

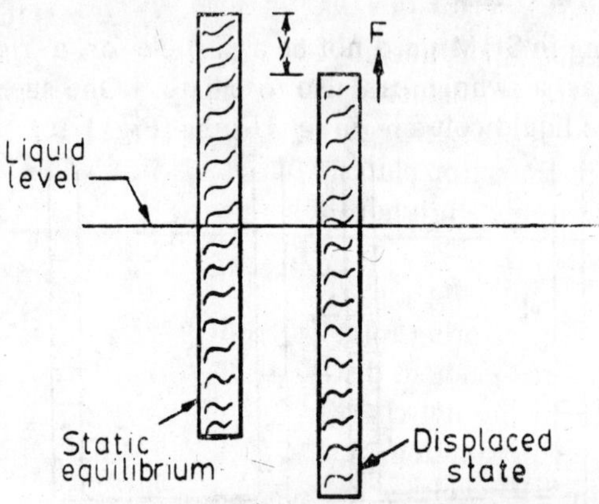

Fig. 1.17 Vertical oscillations of a floating pole

state, because the weight of the pole is balanced by the weight of the liquid it displaces If we displace the pole by a distance y (by dipping it further in the liquid), the *buoyant force* on the pole increases by ρAgy because ρAy is the mass of the liquid displaced by this further dipping ; g being the acceleration due to gravity. We have neglected viscous effects. The restoring force F on the pole is given by

$$F = -\rho Agy = -Ky$$

where $K = \rho Ag$ is the force constant. The angular frequency of the resulting harmonic oscillations is given by

$$\omega = \sqrt{\frac{\rho Ag}{M}}$$

and period

$$T = 2\pi \sqrt{\frac{M}{\rho Ag}}$$

The Electrical System

We have so far restricted ourselves only to mechanical systems. We shall now discuss harmonic oscillations of an electrical circuit consisting of a capacitor C and an inductor L (Fig. 1.18). The equilibrium state is the

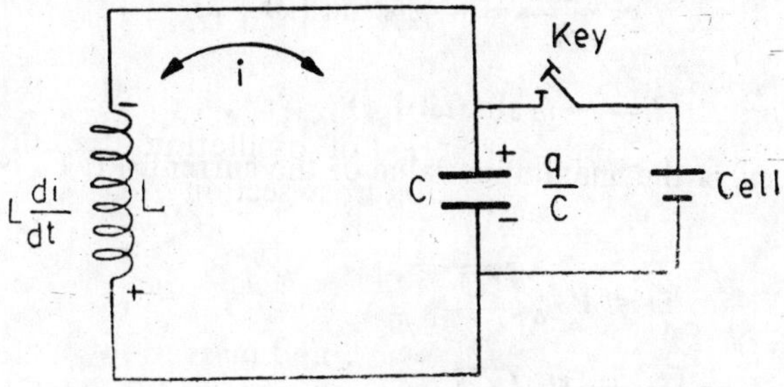

Fig. 1.18 The oscillatory circuit

state when the capacitor is uncharged and no current is flowing in the circuit. This state is disturbed by pressing the key, thus charging the capacitor. Let q be the charge on the capacitor so that $V = q/C$ is the voltage across the capacitor plates. When the key is released, the capacitor starts discharging through the inductor, i.e. the charge changes with time and a current $i = \frac{dq}{dt}$ is established in the inductor.

In this circuit the restoring force is due to the force of repulsion between electrons. This force tends to distribute electrons equally on the capacitor plates so that there is no net charge. Inductance, on the other hand, tends to oppose this redistribution, i.e. it opposes the increase of current. At any instant of time, the voltage across the inductor is

$$V = -L\frac{di}{dt} = -L\frac{d^2q}{dt^2}$$

The minus sign indicates that the voltage opposes the increase of current. From Kirchhoff's law this voltage must equal the voltage q/C across the capacitor plates, giving,

$$-L\frac{d^2q}{dt^2} = \frac{q}{C}$$

or

$$\frac{d^2q}{dt^2} = -\omega^2 q$$

with

$$\omega = \frac{1}{\sqrt{LC}}$$

Thus, in an electrical circuit consisting of an inductance L and a capacitance C, the charge oscillates harmonically with an angular frequency $\omega = 1/\sqrt{LC}$ and period $T = 2\pi\sqrt{LC}$. At any instant of time, the charge q is given by

$$q = q_o \cos(\omega t+\phi)$$

where q_o is the maximum value of the charge and ϕ is the phase of electron oscillations. The current in the circuit is given by

$$i = \frac{dq}{dt} = -\omega q_o \sin(\omega t+\phi)$$

or
$$i = -i_o \sin(\omega t+\phi)$$

where $i_o = \omega q_o$, is the maximum value of the current. If V_o is the applied voltage,

$$i_o = V_o\sqrt{\frac{C}{L}}$$

since $q_o = CV_o$ and $\omega = 1/\sqrt{LC}$

Energy Considerations

The LC circuit resembles a mass-spring system in the sense that each has a characteristic frequency. To understand how an LC circuit oscillates, let us assume that initially the capacitor C carries a charge q and the current in inductor L is zero. At this instant, the electrostatic energy stored in the capacitor is

$$F_e = \frac{1}{2}\,\frac{q^2}{C}$$

and that in the inductance is zero, since $i = 0$ initially. As time passes, the capacitor starts discharging through the inductance and a current $i = dq/dt$ is established in the inductor. As q decreases, E_e decreases and i increases, so that the energy now appears around inductance as the current is building up. When the capacitor is completely discharged, the magnetic energy,

$$E_m = \frac{1}{2}\,Li^2$$

associated with inductance is maximum because the current is maximum and $E_e = 0$ since $q = 0$. Thus, although at this time $q = 0$, $\dfrac{dq}{dt}$ is not zero; it is, in fact, maximum. The large current in the inductor starts transporting charge across the capacitor plates and the capacitor is charged again. It starts discharging again and the current now flows in the opposite direction. Eventually the current returns to its initial value and the process continues. The energy exchange occurs between the electric field of the capacitance and the magnetic field of the inductance. The total energy of the system is conserved, since the system considered here does not contain any resistive component, so that there is no dissipation of energy. Thus

$$E = E_e + E_m$$

$$= \frac{1}{2}\frac{q^2}{C} + \frac{1}{2} L i^2 = \text{constant}$$

Differentiating and setting $\frac{dE}{dt} = 0$, yields $\left(\text{using } i = \frac{dq}{dt}\right)$

$$L\frac{d^2q}{dt^2} + \frac{q}{C} = 0 \tag{1.26}$$

which is the force equation obtained earlier.

A comparison between Eqs (1.2) and (1.26) indicates that *mass in mechanical systems and magnetic field inertia in electrical systems play analogous role.* The mass controls the velocity change for a given force and the magnetic field controls the rate of change of current for a given voltage.

1.11 INITIAL CONDITIONS

In Sec. 1.10 we have computed the frequency (and period) of the harmonic oscillations of a few systems having one degree of freedom. We noticed that the frequency was determined solely by the elastic and inertial properties of the system under study and is not dependent on the way the system was set into oscillation initially. The system, once disturbed (in any arbitrary way) and released, oscillates with its own characteristic frequency called the *natural frequency.* Let us now learn what determines the other two constants, namely amplitude A and phase constant ϕ, of the oscillator. These constants are determined from the way the system is started. Consider the example of a simple pendulum. The pendulum can be set into oscillation by a variety of ways. We can displace it (say, to the right) and let go, so that it has a finite initial displacement and no initial velocity or we can give it a kick when it is at its equilibrium position, in which case, it has a finite initial velocity but no initial displacement or we can start with some initial displacement and then push it when we let it go, in which case, it has a finite initial displacement and a finite initial velocity. The way a system is set into oscillation is called the *initial conditions.*

Suppose at time $t = 0$ (i.e. when the system is released) we give an initial displacement ψ_o and an initial velocity V_o which are both specified. This is the most general way in which a motion can be started. The values of ψ_o and V_o will determine the constants A and ϕ. The displacement and velocity of a system in SHM at any time t are respectively given by

$$\psi(t) = A\cos(\omega t + \phi)$$

and

$$V(t) = \frac{d\psi}{dt} = -A\,\omega\sin(\omega t + \phi)$$

Setting $t = 0$ in these equations, we have

$$\psi_o = A \cos \phi$$

and $$V_o = -A\omega \sin \phi$$

which give

$$\tan \phi = -\frac{\omega \psi_o}{V_o}$$

and $$A = \left(\psi_0^2 + \frac{V_0^2}{\omega^2} \right)^{1/2}$$

Knowing ω, ψ_o and V_o ; A and ϕ can be determined. Thus only the initial values of ψ and $\frac{d\psi}{dt}$ are required to completely predict the motion of the oscillator at any later time t.

In fact, once the initial conditions are specified, it is not even necessary to solve Eq. (1.2) of motion in order to determine the future motion of the system. This can, quite generally, be shown as follows :

We know that any function $\psi(t)$ is completely specified by its Taylor's series expansion

$$\psi(t) = \psi(0) + t \left(\frac{d\psi}{dt} \right)_{t=0} + \frac{t^2}{2!} \left(\frac{d^2\psi}{dt^2} \right)_{t=0} + \frac{t^3}{3!} \left(\frac{d^3\psi}{dt^3} \right)_{t=0} + \ldots \quad (1.27)$$

where $\psi(t)$ satisfies the equation

$$\frac{d^2\psi}{dt^2} = -\omega^2 \psi(t) \quad (1.28)$$

In expansion (1.27) $\psi(0) = \psi_o$ and $\left(\frac{d\psi}{dt} \right)_{t=0} = V_o$ are specified by the initial conditions stated above. Eq. (1.28) then determines $\left(\frac{d^2\psi}{dt^2} \right)_{t=0}$, which is given by

$$\left(\frac{d^2\psi}{dt^2} \right)_{t=0} = -\omega^2 \psi_o$$

The next time derivative of Eq. (1.28) determines $\left(\frac{d^3\psi}{dt^3} \right)_{t=0}$, which is give by

$$\left(\frac{d^3\psi}{dt^3} \right)_{t=0} = -\omega^2 \left(\frac{d\psi}{dt} \right)_{t=0} = -\omega^2 V_o$$

Substituting for $\psi(o)$, $\left(\frac{d\psi}{dt} \right)_{t=0}$, $\left(\frac{d^2\psi}{dt^2} \right)_{t=0}$, etc. in Eq. (1.27) we have

$$\psi(t) = \psi_o + tV_o - \frac{t^2}{2!} \omega^2 \psi_o - \frac{t^3}{3!} \omega^2 V_o - \ldots\ldots$$

$$= \psi_o\left(1 - \frac{\omega^2 t^2}{2!} + \ldots\right) + \frac{V_o}{\omega}\left(\omega t - \frac{\omega^3 t^3}{3!} + \ldots\right)$$

$$= \psi_o \cos \omega t + \frac{V_o}{\omega} \sin \omega t$$

which indeed is the solution [see Eq. (1.4)] we have obtained earlier with $\psi_o = a$ and $\frac{V_o}{\omega} = b$. This equation determines $\psi(t)$ at any later time t, as ω, ψ_o and V_o are known.

Thus, we conclude that once the initial conditions, namely ψ_o and $d\psi/dt$ at $t=0$ are specified, the values of all the higher derivatives at $t = 0$ are determined and hence the future motion of the oscillator is known. One can generalize and state that the motion of an oscillator, acted upon by any force that depends upon ψ, will be determined completely just by giving specific values to its initial displacement and its initial velocity. In mathematical language, this implies that the solution of any second-order differential equation has just two arbitrary constants.

SOLVED EXAMPLES

Example 1.1 A particle oscillates with SHM of amplitude 4 cm and a frequency of 5 Hz. At time $t = 0$ the particle is at its equilibrium position ($\psi = 0$).

(a) Write down the equation describing the position of the particle as a function of time in the form $\psi = A \cos (\omega t + \phi)$, giving the numerical values of A, ω and ϕ.

(b) What are the values of ψ, $\frac{d\psi}{dt}$, and $\frac{d^2\psi}{dt^2}$ at $t = \frac{10}{3}$ s?

Solution

(a) The equation of motion is $\psi = A \cos (\omega t + \phi)$, where

amplitude $A = 4$ cm

frequency $\nu = 5$ Hz

$\therefore$ angular frequency $\omega = 2\pi\nu = 2\pi \times 5 = 10\pi$ rad s^{-1}

Now, at $t = 0$, $\psi = 0$. Substituting in $\psi = A \cos (\omega t + \phi)$ we have

$$\cos \phi = 0, \text{ giving } \phi = \pm \pi/2$$

Thus $A = 4$ cm, $\omega = 10\pi$ rad s^{-1} and $\phi = \pm \pi/2$

(b) For $\phi = +\frac{\pi}{2}$ we have

$$\psi = A \cos\left(\omega t + \frac{\pi}{2}\right) = -A \sin \omega t$$

$$\therefore \qquad \frac{d\psi}{dt} = -A\omega \cos \omega t$$

and
$$\frac{d^2\psi}{dt^2} = A\omega^2 \sin \omega t$$

At $t = \frac{10}{3}$ s, we have

$$\psi = -4 \sin \left(10\pi \times \frac{10}{3}\right) = -4 \sin \left(\frac{100\pi}{3}\right) = -4 \sin \left(33\pi + \frac{\pi}{3}\right)$$

$$= +4 \sin \frac{\pi}{3} = +2\sqrt{3} \text{ cm}$$

$$\frac{d\psi}{dt} = 20\pi \text{ cm s}^{-1}$$

and
$$\frac{d^2\psi}{dt^2} = -200\,\pi\sqrt{3} \text{ cm s}^{-2}$$

For $\phi = -\frac{\pi}{2}$; ψ, $\frac{d\psi}{dt}$ and $\frac{d^2\psi}{dt^2}$ are respectively $-2\sqrt{3}$ cm, -20π cm s^{-1} and $200\,\pi\sqrt{3}$ cm s^{-2}.

Example 1.2 A particle vibrates with SHM of amplitude 5 cm and a period of 6 s. How long does it take to move from one end of its path to a position 2.5 cm from the equilibrium position on the same side ?

Solution

The equation for the displacement of the particle in SHM is

$$\psi = A \cos (\omega t + \phi)$$

where $A = 5$ cm and $\omega = \frac{2\pi}{T} = \frac{2\pi}{6} = \frac{\pi}{3}$ rad s^{-1}

Let us suppose that at $t = 0$, the particle is at one end of its path, i.e. at $t = 0$, $\psi = A$ so that

$$A = A \cos (0 + \phi) \quad \text{or} \quad \cos \phi = 1 \quad \text{or} \quad \phi = 0$$

Setting $\phi = 0$ we have

$$\psi = A \cos \omega t$$

The value of t for which $\psi = 2.5$ cm is given by

$$2.5 = 5 \cos \omega t$$

or
$$\cos \omega t = \frac{1}{2}$$

or
$$\omega t = \frac{\pi}{3}$$

$$\therefore \qquad t = \frac{\pi}{3\omega} = 1 \text{ s} \left(\because \omega = \frac{\pi}{3} \text{ rad s}^{-1}\right)$$

Example 1.3 Show that the values of T, the period of the three simple harmonic oscillations (a), (b) and (c) in Fig. 1.19 are in the ratio of $1 : \sqrt{2} : \frac{1}{\sqrt{2}}$. All springs are identical, each of spring constant k and of mass that is negligible compared to mass m.

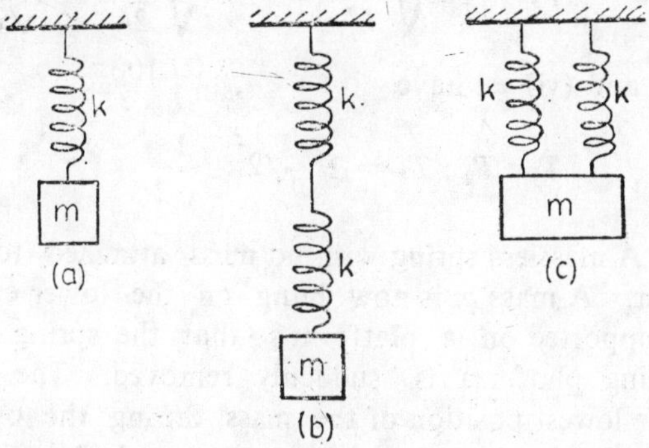

Fig. 1.19

Solution

Case (a) : Let us suppose that an elongation y is produced in the spring when a force mg is applied to it. The equilibrium position is given by

$$F = mg = ky \tag{i}$$

If the system is displaced and released, it will execute SHM of period given by

$$T_1 = 2\pi \sqrt{\frac{m}{k}} \tag{ii}$$

Case (b) : As the length of the spring is doubled, a given force mg will produce double the extension in this case. Let y' be the extension produced and k', the spring constant of the combined system. In the equilibrium position ($\because y' = 2y$)

$$F = mg = k'y' = 2\,k'y \tag{iii}$$

Comparing (i) and (iii) we get

$$k = 2k' \qquad \text{or } k' = k/2$$

The period T_2 of SHM in this case is

$$T_2 = 2\pi \sqrt{\frac{m}{k'}} = 2\pi \sqrt{\frac{2m}{k}} = \sqrt{2}\,T_1 \tag{iv}$$

Case (c) : In this case, the extension y'' produced in each spring by a force mg is half that produced in case (a), i.e. $y'' = y/2$.

If k'' is the spring constant of the combined system, we have

$$F = mg = k''\,y'' = \frac{k''}{2}\,y \tag{v}$$

Comparing (i) and (v) we get

$$k'' = 2k$$

The period T_3 of SHM in this case is

$$T_3 = 2\pi\sqrt{\frac{m}{k''}} = 2\pi\sqrt{\frac{m}{2k}} = \frac{T_1}{\sqrt{2}} \qquad \text{(vi)}$$

From (ii), (iv) and (vi) we have

$$T_1 : T_2 : T_3 = 1 : \sqrt{2} : \frac{1}{\sqrt{2}}$$

Example 1.4 A massless spring with no mass attached to it hangs from a rigid support. A mass m is now hung on the lower end of the spring. The mass is supported on a platform so that the spring remains relaxed. The supporting platform is suddenly removed. The mass begins to oscillate. The lowest position of the mass during the oscillation is 5 cm below the place where it was resting on the platform. (a) What is the frequency of oscillation? (b) What is the velocity when the mass is 2.5 cm below its original resting place? Take $g = 10$ m s^{-2}.

Solution

It is clear from Fig. 1.20 that the separation between the two extreme positions of the oscillating mass is 5 cm. Therefore, the equilibrium position is 2.5 cm below the supporting platform. In other words, the force mg produces an extension of $y = 2.5$ cm in the spring. If k is the constant of the spring, we have

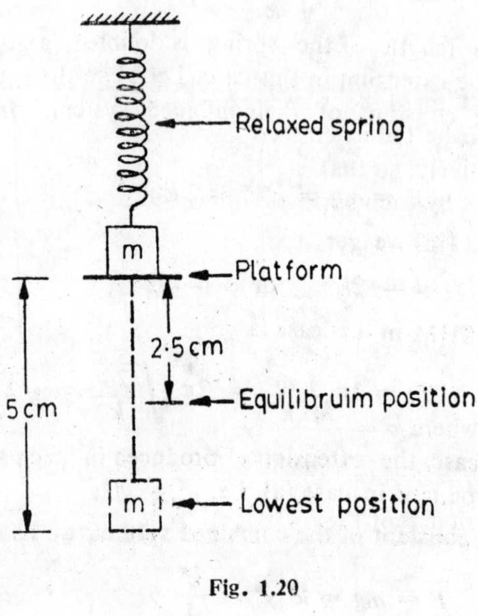

Fig. 1.20

$$mg = ky$$

Substituting for y and g we get

$$\frac{k}{m} = \frac{g}{y} = \frac{1000 \text{ cm s}^{-2}}{2.5 \text{ cm}} = 400 \text{ s}^{-2}$$

The angular frequency of oscillation is

$$\omega = \sqrt{\frac{k}{m}} = \sqrt{400} = 20 \text{ rad s}^{-1}$$

The frequency of oscillation is

$$\nu = \frac{\omega}{2\pi} = 3.18 \text{ Hz.}$$

When the mass is 2.5 cm below the platform, it is passing through the equilibrium position and hence has maximum velocity given by

$$V_{max} = A\omega = 2.5 \times 20 = 50 \text{ cm s}^{-1}$$

since A, the amplitude of oscillation is 2.5 cm.

Example 1.5 Two massless springs of force constants k_1 and k_2 are connected to mass m placed on a horizontal frictionless surface as shown in Fig. 1.21 (a) and (b). Obtain the expression for the time period of horizontal oscillations in each case.

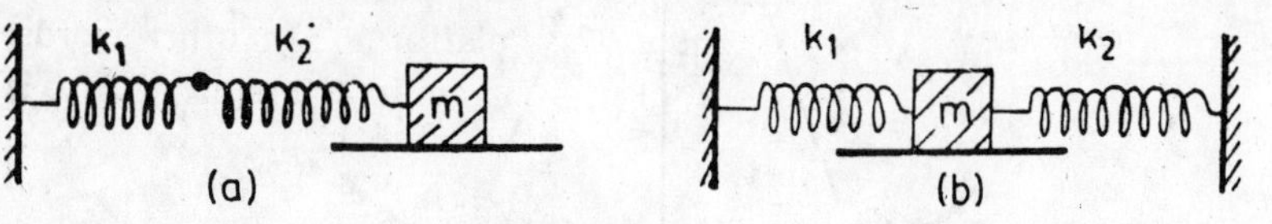

Fig. 1.21

Solution

Case (a): Let the mass be displaced through a distance x, say, to the right. Let x_1 and x_2 be the extensions produced in the springs of constants k_1 and k_2 respectively, so that $x = x_1+x_2$. Since, the restoring force F exerted on the mass by each spring is the same, we have

$$F = -k_1x_1 = -k_2x_2$$

giving $\quad x_1 = -F/k_1$ and $x_2 = -F/k_2$

$$\therefore \quad x = x_1+x_2 = -F\left(\frac{1}{k_1}+\frac{1}{k_2}\right)$$

whence $F = -kx$, where $k = \dfrac{k_1k_2}{k_1+k_2}$ is the effective force constant of the combination.

Hence the time period T of the oscillation is

$$T = 2\pi\sqrt{\frac{m}{k}} = 2\pi\sqrt{\frac{m(k_1+k_2)}{k_1\,k_2}}$$

Case (b): In this case, if the mass is displaced through a distance x, say to the right, the spring k_1 is extended by x and spring k_2 is compressed by x; so that the restoring force exerted by each spring on mass m is in same direction, tending to bring it to its original position. If F_1 and F_2 are the restoring forces due to k_1 and k_2 respectively, we have

Total restoring force $F = F_1+F_2 = -k_1x-k_2x = -(k_1+k_2)\,x = -kx$

where $k = k_1+k_2$ is the effective force constant of the combined system.

Hence

$$T = 2\pi\sqrt{\frac{m}{k}} = 2\pi\sqrt{\frac{m}{k_1+k_2}}$$

Example 1.6 A small spherical steel ball is placed a little away from the centre of a concave mirror whose radius of curvature is 2.5 m. When the ball is released, it begins to oscillate about the centre. What is the period of the oscillations? Neglect friction and take $g = 10\ ms^{-2}$.

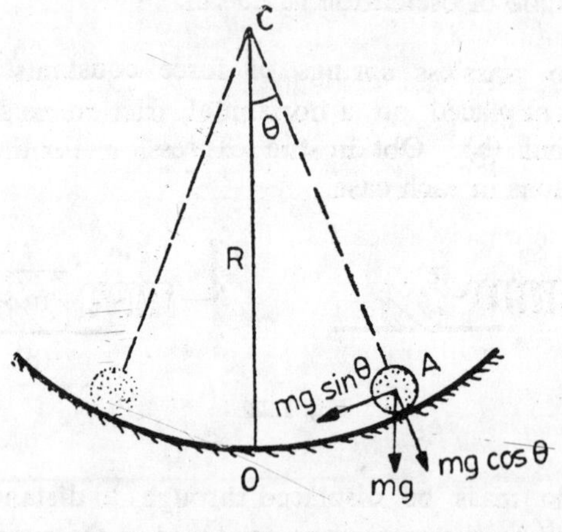

Fig. 1.22

Solution

Place a small steel ball at A, a little away from the centre O of a concave mirror of radius of curvature $R\ (= OC = AC)$ as shown in Fig. 1.22. Let $\angle AOC = \theta$. If m is the mass of the ball, its weight mg acts vertically downwards at A. This force is resolved into two rectangular components: $mg \cos\theta$ (which is balanced by the reaction of the mirror) and $mg \sin\theta$ (which provides the restoring force F). Thus

$$F = -mg\sin\theta$$

$$= -mg\,\theta \qquad \text{(since } \theta \text{ is small, } R \text{ being very large)}$$

$$= -\frac{mg\psi}{R} \qquad (\because\ \psi = R\theta\ ;\ \psi \text{ being the arc } OA)$$

$$= -K\psi$$

where force constant $K = mg/R$. Thus the motion is harmonic and the angular frequency is given by

$$\omega = \sqrt{\frac{k}{m}} = \sqrt{\frac{g}{R}}$$

$$\therefore \quad T = 2\pi \sqrt{\frac{R}{g}}$$

$$= 2 \times 3.142 \times \sqrt{\frac{2.5}{10}} = 3.14 \text{ s}$$

Example 1.7 A simple pendulum is displaced from its mean position O to a position P until the height of P above O is 0.05 m (see Fig. 1.23). It is then released. Calculate its velocity when it passes the mean position O. Take $g = 10 \text{ ms}^{-2}$ and neglect the friction offered by the medium.

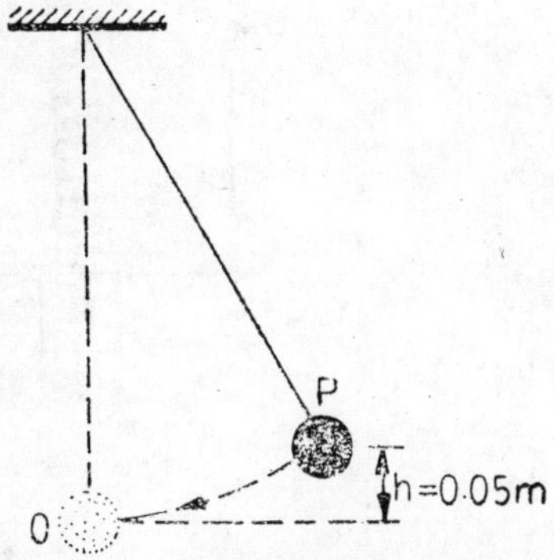

Fig. 1 23

Solution

The potential energy of the pendulum at P = work done (against gravity) in raising the bob of mass m to a vertical height $h = mgh$.

When the pendulum is released, it oscillates. The energy when it passes O is entirely kinetic and is equal to $\frac{1}{2}mV^2$ where V is the velocity at the instant of time the pendulum passes O. Since frictional losses are neglected, the entire potential energy at P must be converted into kinetic energy at O, giving

$$\tfrac{1}{2}mV^2 = mgh$$

or

$$V = \sqrt{2gh}$$

$$= \sqrt{2 \times 10 \times 0.05}$$

$$= 1 \text{ ms}^{-1}$$

Notice that the velocity depends only on g and h and is independent of the length and mass of the pendulum.

Example 1.8 A uniform spring of constant k and a finite mass m is loaded with a mass M. If m is not negligible compared to M, show that the period of vertical oscillations is

$$T = 2\pi \sqrt{\frac{(M + m/3)}{k}}$$

Solution

Consider a uniform spring of length l and mass m suspended from a rigid support, carrying a mass M at its lower free end, as shown in

Fig. 1.24. Consider an element (of the spring) of length ds at a distance s from the fixed end. Since mass per unit length of the spring is m/l, the mass dm of this element is

$$dm = \frac{m}{l}\, ds \tag{i}$$

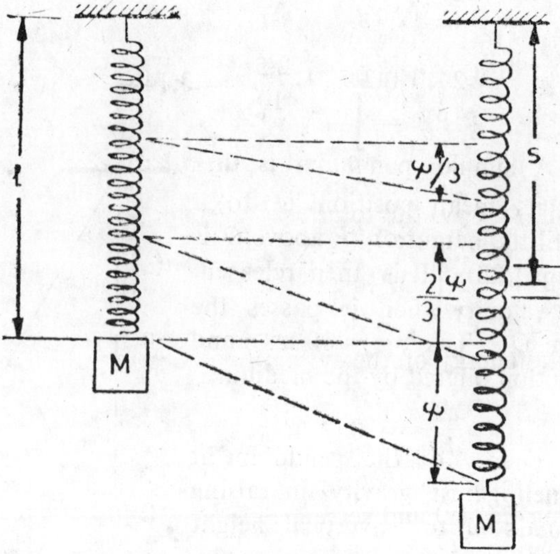

Fig. 1.24

Let us assume that the various parts of the spring undergo displacements proportional to their distance from the fixed end, as indicated in the diagram. We can now compute the total kinetic energy of the spring at an instant when the mass M has a displacement ψ. Now the displacement per unit length of the spring is ψ/l. Therefore, the displacement of the element located at s is $\frac{\psi s}{l}$ and its velocity v is

$$v = \frac{s}{l}\frac{d\psi}{dt} \tag{ii}$$

From Eqs. (i) and (ii) the kinetic energy of the spring at any instant of time can be obtained by integrating the above expression, treating $d\psi/dt$ as constant. Hence, we have

$$(\text{KE})_{\text{spring}} = \frac{1}{2}\frac{m}{l^3}\left(\frac{d\psi}{dt}\right)^2 \int_0^l s^2\, ds$$

$$= \frac{1}{6}\, m \left(\frac{d\psi}{dt}\right)^2$$

The kinetic energy of the mass M is

$$(\text{KE})_{\text{mass}} = \frac{1}{2}\, M \left(\frac{d\psi}{dt}\right)^2$$

The total KE of the spring-mass system

$$=\frac{1}{2}\left(M+\frac{m}{3}\right)\left(\frac{d\psi}{dt}\right)^2 \tag{iii}$$

If k is the spring constant, the restoring force F when a displacement ψ is given to the system is $-k\psi$. Therefore, the potential energy of the system is given by

$$\text{PE} = \int_0^{\psi} k\psi\, d\psi = \frac{1}{2}k\psi^2 \tag{iv}$$

From Eqs. (iii) and (iv) the total energy E of the system is

$$E = \frac{1}{2}\left(M+\frac{m}{3}\right)\left(\frac{d\psi}{dt}\right)^2+\frac{1}{2}k\psi^2 \tag{v}$$

Since, the total energy of the system must remain constant, we have

$$\frac{dE}{dt} = 0$$

Differentiating Eq. (v) and setting $\frac{dE}{dt} = 0$, we get

$$\frac{d^2\psi}{dt^2} = -\frac{k\psi}{(M+m/3)}$$

indicating that the system executes SHM of period

$$T = 2\pi\sqrt{\frac{(M+m/3)}{k}}$$

The above calculation is not exact because we have assumed that the extension of an element of the spring is proportional to the distance from the fixed end and that the velocity $d\psi/dt$ is the same for all elements of the spring. In fact, different elements undergo different accelerations. The expression for time period will hold if $m \ll M$, in which case, the stretching force does not vary appreciably with distance along the spring and can be treated as roughly constant. Thus, the expression for time period T is only approximate and the factor $m/3$ can, at best, be regarded as a small correction factor that increases the period slightly. Hence, if $m < M$ but not negligible compared to M, the finite mass of the spring effectively slows down the oscillations.

Example 1.9 Compute the frequency of an electrical circuit consisting of a coil of inductance 0.1 mH and a capacitor of 1.0 μF. What is the maximum current in the circuit if the capacitor is charged to 5 V? Neglect the resistance of the coil.

Solution

$$L = 0.1 \text{ mH} = 10^{-4} \text{ H}$$

$$C = 1.0 \ \mu\text{F} = 10^{-6} \text{ F}$$

$$\text{Frequency} \quad \nu = \frac{1}{2\pi\sqrt{LC}}$$

$$= \frac{1}{2\times3.142\times(10^{-4}\times10^{-6})^{1/2}}$$

$$= 1.59\times10^{5} \text{ Hz}$$

$$\text{Maximum current} = V\sqrt{\frac{C}{L}}$$

$$= 5\times\sqrt{\frac{10^{-6}}{10^{-4}}}$$

$$= 0.5 \ A$$

Example 1.10 A string of length $L = 100$ cm is stretched with a tension $T = 10$ N between two fixed points A and B as shown in Fig. 1.25 (a). A mass $m = 10$ g is fixed at a distance $a = 25$ cm from point A. Determine the frequency of the vertical oscillations of the mass, assuming that the tension remains constant for small displacements.

Solution

Figure 1.25 (b) shows the displaced position of the mass. The net restoring force acting on the mass is

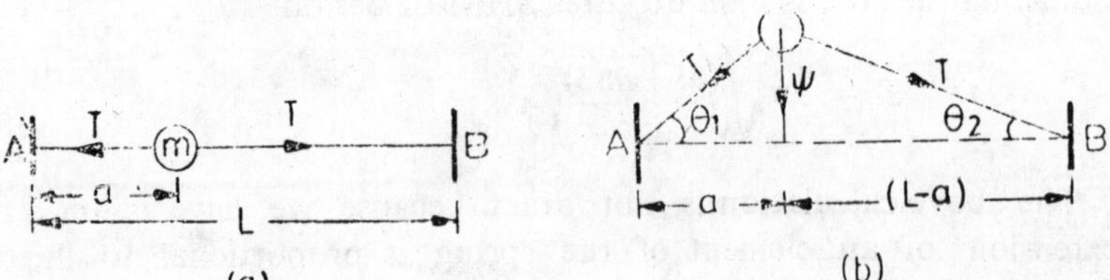

Fig. 1.25

$$F = -T\sin\theta_1 - T\sin\theta_2$$

If the displacement ψ is small

$$\sin\theta_1 \simeq \tan\theta_1 = \frac{\psi}{a}$$

$$\sin\theta_2 \simeq \tan\theta_2 = \frac{\psi}{L-a}$$

$$\therefore \quad F = -T\left(\frac{1}{a} + \frac{1}{L-a}\right)\psi = -\frac{TL\psi}{a(L-a)} = -K\psi$$

where $\quad K = \dfrac{TL}{a(L-a)}$

Hence
$$\omega = \sqrt{\frac{TL}{ma\ (L-a)}}$$

and
$$\nu = \frac{1}{2\pi}\sqrt{\frac{TL}{ma(L-a)}}$$

Substituting for $T = 10$ N, $m = 10\text{g} = 10^{-2}$ kg, $L = 1$ m and $a = 0.25$ m we get

$$\nu = 11.6 \text{ Hz}$$

Example 1.11 A mass m is attached to a spring of spring constant k via a friction-less pulley of radius r and mass M as shown in Fig. 1.26. Determine the frequency of vertical oscillations of the mass using (a) Newton's laws of motion and (b) energy considerations.

Solution

Figure 1.26(c) shows the displaced state of the system.

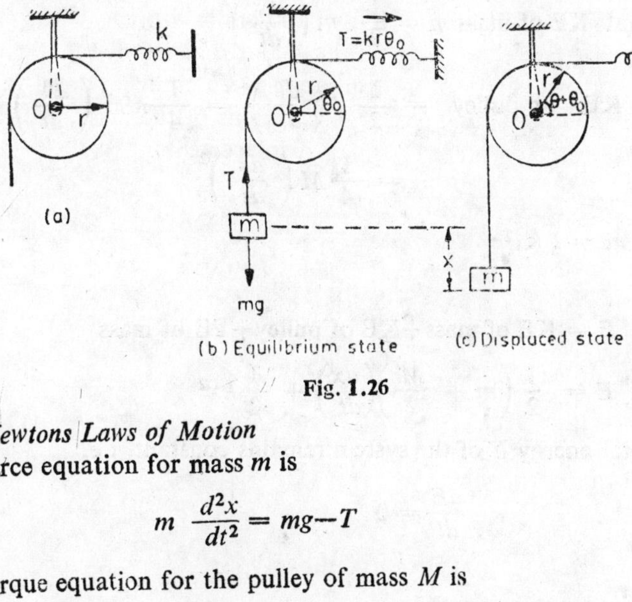

Fig. 1.26

(a) *Newtons Laws of Motion*

The force equation for mass m is

$$m\ \frac{d^2x}{dt^2} = mg - T \qquad \text{(i)}$$

The torque equation for the pulley of mass M is

$$I\ \frac{d^2\theta}{dt^2} = Tr - kr^2\ (\theta + \theta_0) \qquad \text{(ii)}$$

where $I = \frac{1}{2}\ Mr^2$ is the moment of inertia of the pulley about an axis passing through the centre O and perpendicular to the plane of the pulley. But, in the static equilibrium, we have [see Fig. 1.26(b)]

$$mg = kr\theta$$

Eliminating T between (i) and (ii) and using $mg = kr\theta_0$, we get

$$m\ \frac{d^2x}{dt^2} + \frac{1}{2}\ Mr\frac{d^2\theta}{dt^2} = -kr\theta$$

But $x = r\theta$ and $\frac{d^2x}{dt^2} = r\frac{d^2\theta}{dt^2}$.

Hence we get

$$(m+\tfrac{1}{2} M)\frac{d^2x}{dt^2} = -kx$$

or

$$\frac{d^2x}{dt^2} = -\omega^2 x$$

with

$$\omega = \left(\frac{k}{m+M/2}\right)^{1/2}$$

(b) *Energy Considerations*

The total energy E of the system is the sum of the kinetic energy of the mass, the rotational kinetic energy of the pulley and the potential energy of the mass.

Translational KE of mass $m = \frac{1}{2} m\left(\frac{dx}{dt}\right)^2$

Rotational KE of pulley $= \frac{1}{2} I\left(\frac{d\theta}{dt}\right)^2 = \frac{1}{4} Mr^2\left(\frac{d\theta}{dt}\right)^2$

$$= \frac{1}{4} M\left(\frac{dx}{dt}\right)^2$$

PE of mass $m = \frac{1}{2} kx^2$

Hence

$$E = \text{KE of mass} + \text{KE of pulley} + \text{PE of mass}$$

or

$$E = \frac{1}{2}\left(m + \frac{M}{2}\right)\left(\frac{dx}{dt}\right)^2 + \frac{1}{2}kx^2$$

Now the total energy E of the system remains constant, i.e.

$$\frac{dE}{dt} = 0$$

This gives

$$\left(m+\frac{M}{2}\right)\frac{dx}{dt}\cdot\frac{d^2x}{dt^2} + kx\cdot\frac{dx}{dt} = 0$$

Since $\frac{dx}{dt}$ is not always zero, we have

$$\frac{d^2x}{dt^2} = -\frac{kx}{(m+M/2)}$$

giving

$$\omega = \left(\frac{k}{m+M/2}\right)^{\frac{1}{2}}$$

Example 1.12 A simple pendulum consists of a rod of mass m and length l which is pivoted at O and carries a mass M at the other end as shown in Fig. 1.27. Using energy considerations, determine the frequency of the pendulum, if (a) $m \ll M$ and (b) m is comparable with M.

Solution

Let ψ be the displacement of the pendulum at any instant of time when its velocity is $\frac{d\psi}{dt}$ and α is the angle subtended with the vertical.

(a) $$\text{KE} = \tfrac{1}{2} M\left(\frac{d\psi}{dt}\right)^2$$

$$\text{PE} = Mgh$$

But in triangle $O\,C\,B$,

$$\frac{l-h}{l} = \cos\alpha \qquad \text{or} \qquad h = l\,(1-\cos\alpha)$$

$$\therefore \quad \text{PE} = 2\,Mgl\,(1-\cos\alpha)$$

$$= 2\,Mgl\,\sin^2\frac{\alpha}{2}$$

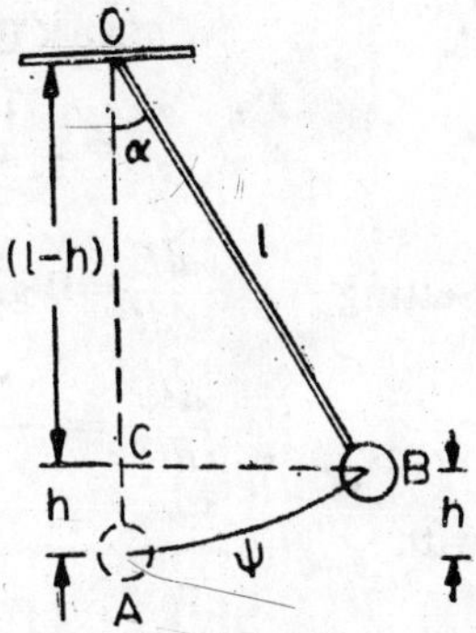

Fig. 1.27

Now, for small displacements

$$\sin\alpha = \frac{\psi}{l}$$

$$\therefore \quad \text{PE} = \frac{1}{2}\frac{Mg\psi^2}{l}$$

Total energy E is given by

$$E = \text{KE}+\text{PE}$$

$$= \frac{1}{2}\,M\left(\frac{d\psi}{dt}\right)^2 + \frac{1}{2}\frac{Mg\psi^2}{l}$$

Setting $\frac{dE}{dt} = 0$ gives

$$\frac{d^2\psi}{dt^2} = -\frac{g\psi}{l}$$

$$\therefore \quad \omega = \sqrt{\frac{g}{l}} \qquad \text{or } \nu = \frac{1}{2\pi}\sqrt{\frac{g}{l}}$$

(b) If the mass m of the rod is not negligible, then as in Eq. (iii) of Example 1.8, we have

$$\text{KE} = \text{KE of mass } M + \text{KE of rod}$$

$$= \frac{1}{2}M\left(\frac{d\psi}{dt}\right)^2 + \frac{1}{2}\left(\frac{m}{3}\right)\left(\frac{d\psi}{dt}\right)^2$$

$$= \frac{1}{2}\left(M+\frac{m}{3}\right)\left(\frac{d\psi}{dt}\right)^2$$

Since the rod is uniform, its weight $m\,g$ acts through the centre of the rod. Hence we have

$$PE = \text{PE of mass } m + \text{PE of rod}$$

$$= Mgl\,(1-\cos\alpha) + mg\left(\frac{l}{2}\right)\left(1-\cos\alpha\right)$$

$$= \left(M+\frac{m}{2}\right)gl\,(1-\cos\alpha)$$

$$= \frac{1}{2}\left(M+\frac{m}{2}\right)\frac{g}{l}\,\psi^2$$

$\therefore$
$$E = KE + PE$$

$$= \frac{1}{2}\left(M+\frac{m}{3}\right)\left(\frac{d\psi}{dt}\right)^2 + \frac{1}{2}\left(M+\frac{m}{2}\right)\frac{g}{l}\,\psi^2$$

Setting $\frac{dE}{dt} = 0$ gives

$$\frac{d^2\psi}{dt^2} = -\,\omega^2\psi$$

with

$$\omega = \left\{\frac{(M+m/2)}{(M+m/3)}\,\frac{g}{l}\right\}^{1/2}$$

If $m \ll M$, we have $\omega = \sqrt{\frac{g}{l}}$ as in case (a) above.

Example 1.13 A circular solid cylinder of radius r and mass m is connected to a spring of spring constant k as shown in Fig. 1.28. Determine the frequency of horizontal oscillations of the system if the cylinder (a) slips on the surface without rolling and (b) rolls on the surface without slipping. Neglect friction.

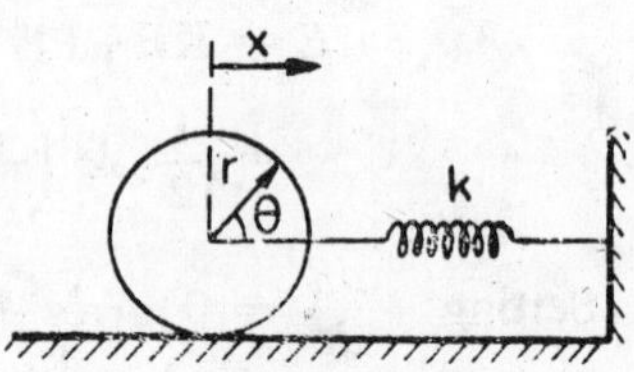

Fig 1.28

Solution

(a) In this case, the total energy E of the system consists of translational kinetic energy and potential energy of the cylinder. If the instantaneous displacement is x, we have

$$\text{Translational KE} = \frac{1}{2}\,m\left(\frac{dx}{dt}\right)^2$$

$$PE = \frac{1}{2}\,kx^2$$

$$E = KE + PE$$

$$=\frac{1}{2}m\left(\frac{dx}{dt}\right)^2+\frac{1}{2}kx^2$$

setting $\frac{dE}{dt}=0$ gives

$$\frac{d^2x}{dt^2}=-\frac{k}{m}x$$

Therefore

$$\omega=\sqrt{\frac{k}{m}}$$

(b) In this case, the total energy E of the system consists of translational and rotational kinetic energy and potential energy.

$$\text{Translation KE}=\frac{1}{2}m\left(\frac{dx}{dt}\right)^2$$

$$\text{Rotational KE}=\frac{1}{2}I\left(\frac{d\theta}{dt}\right)^2$$

$$=\frac{1}{4}m\left(\frac{dx}{dt}\right)^2$$

Since $r\theta=x$ and $I=1/2mr^2$ is the moment of inertia of the cylinder about an axis passing through the centre and prependicular to the plane of the paper. Therefore

$$\text{Total KE}=\frac{3}{4}m\left(\frac{dx}{dt}\right)^2$$

and

$$\text{PE}=\tfrac{1}{2}kx^2$$

$\therefore$

$$E=\frac{3}{4}m\left(\frac{dx}{dt}\right)^2+\frac{1}{2}kx^2$$

In friction is neglected, $\frac{dE}{dt}=0$ must be zero which gives

$$\frac{d^2x}{dt^2}=-\frac{2k}{3m}x$$

Thus

$$\omega=\sqrt{\frac{2k}{3m}}$$

Example 1.13 A rectangular pan of base length 2 L is partly filled with water up to a height h as shown in Fig. 1.29 (a). When the pan is pushed a little, the water begins to slosh. Assuming that the water surface remains practically flat during sloshing, show that the time period of the sloshing mode is given by

$$T=2\pi\sqrt{\frac{L^2}{3gh}}$$

where g is the acceleration due to gravity.

Solution

Figure 1.29 (a) shows the static equilibrium and Fig. 1.29 (b) shows one extreme position of the surface (assumed flat) of the sloshing liquid. Let us obtain the equation of motion of the centre of mass of the liquid as it sloshes. The motion is obviously two dimensional. Let $(\bar{x}, \bar{y})$ be the coordinates of the centre of mass of the liquid while it is sloshing and let $(\bar{x}_0, \bar{y}_0)$ be the coordinates of the centre of mass of the liquid when it is in equilibrium. It is clear from Fig. 1.29 (a) that

$$(\bar{x}_0, \bar{y}_0) = \left(L, \frac{h}{2}\right)$$

Figure 1.29 (b) consists of two portions of the liquid; rectangular portion $ABCD$ and triangular portion ADE. The centre of mass $(\bar{x}, \bar{y})$ of the entire portion $ABCE$ can be determined as follows : Let $(\bar{x}_1, \bar{y}_1)$ be the coordinates of the centre of mass of the portion $ABCD$ and $(\bar{x}_2, \bar{y}_2)$ be the coordinates of the centre of mass of portion ADE. Then

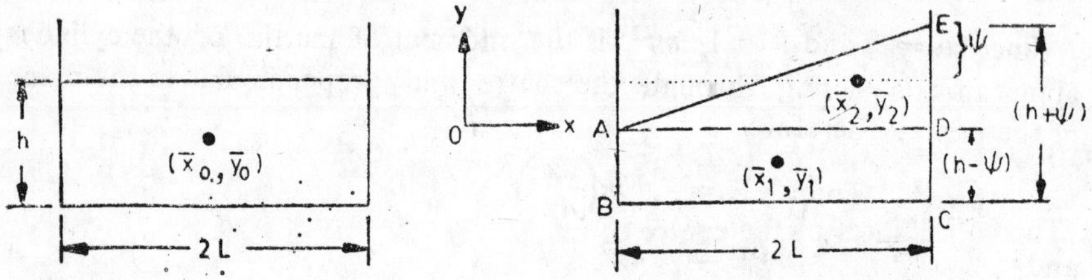

(a) Equilibrium state (b) Displaced state

Fig. 1.29

$$(\bar{x}_1, \bar{y}_1) = \left(L, \frac{h-\psi}{2}\right)$$

where ψ is the vertical displacement of liquid surface. The centre of mass of triangle ADE is at the centroid. Therefore

$$(\bar{x}_2, \bar{y}_2) = \left[\frac{2L+2L+0}{3}, \frac{(h+\psi)+(h-\psi)+(h-\psi)}{3}\right]$$

$$= \left(\frac{4L}{3}, \frac{3h-\psi}{3}\right)$$

If b is the width of the pan and ρ the density of liquid, then

M_1 (= mass of rectangular portion $ABCD$ of the liquid)

$$= \rho b(2L)(h-\psi)$$

$$= 2\rho bL(h-\psi)$$

M_2 (= mass of triangular portion *ADE* of the liquid)

$$= \tfrac{1}{2}\rho b \times 2L \times 2\psi$$

$$= 2\rho bL\psi$$

The coordinates $(\bar{x}, \bar{y})$ of the sloshing liquid are given by

$$\bar{x} = \frac{M_1\bar{x}_1 + M_2\bar{x}_2}{M_1 + M_2}$$

$$= \frac{2\rho bL(h-\psi)L + 2\rho bL\psi 4L/3}{2\rho bL(h-\psi) + 2\rho bL\psi}$$

or

$$\bar{x} = L + \tfrac{1}{3}L\frac{\psi}{h}$$

Similarly

$$\bar{y} = \frac{M_1\bar{y}_1 + M_2\bar{y}_2}{M_1 + M_2}$$

$$= \frac{2\rho bL(h-\psi)\frac{1}{2}(h-\psi) + 2\rho bL\psi(3h-\psi)/3}{2\rho bL(h-\psi) + 2\rho bL\psi}$$

or

$$\bar{y} = \tfrac{1}{2}h + \frac{\psi^2}{6h}$$

$$\therefore \quad (\bar{x}, \bar{y}) = \left(L + \frac{L}{3}\frac{\psi}{h},\ \frac{h}{2} + \frac{\psi^2}{6h} \right)$$

The coordinates of the centre of mass at equilibrium are

$$(\bar{x}_0, \bar{y}_0) = \left(L, \frac{h}{2} \right)$$

If the entire mass is assumed to be concentrated at the centre of mass of the system, then the increase in potential energy is $Mg\,(\bar{y} - \bar{y}_0)$, where M is the mass of the liquid. Now

$$\bar{y} - \bar{y}_0 = \frac{h}{2} + \frac{\psi^2}{6h} - \frac{h}{2} = \frac{\psi^2}{6h}$$

and

$$\bar{x} - \bar{x}_0 = L + \frac{L}{3}\frac{\psi}{h} - L = \frac{L\psi}{3h}$$

we find that

$$\bar{y} - \bar{y}_0 = \frac{3h}{2L^2}(\bar{x} - \bar{x}_0)^2$$

$\therefore$ Increase of potential energy $= Mg\left(\dfrac{3h}{2L^2}\right)(\bar{x} - \bar{x}_0)^2$

If the origin is shifted to $(\bar{x}_0, \bar{y}_0)$ and the potential energy at $(\bar{x}_0, \bar{y}_0)$ is taken to be zero, then potential energy at $X = (\bar{x} - \bar{x}_0)$ is given by

$$\text{PE} = \frac{3}{2}\ \frac{Mgh}{L^2}X^2$$

In other words, the potential energy of the centre of mass is proportional to the square of the displacement $X = (\bar{x} - \bar{x}_0)$ of the centre of mass. Hence the motion of the centre of mass is simple harmonic. For a harmonic oscillator with $\omega = \sqrt{\dfrac{K}{M}}$

we have

$$\text{PE} = \tfrac{1}{2} KX^2$$

Thus, in our case,

$$K = \frac{3Mgh}{L^2}$$

giving

$$\omega^2 = \frac{3gh}{L^2}$$

or

$$T = 2\pi \sqrt{\frac{L^2}{3\,gh}}$$

The sloshing water in lakes is called a *seiche.* The kinetic energy is almost entirely due to horizontal flow (along x) and potential energy is due to a change in the vertical level of water. In a lake of depth 150 m and length 60 km, the seiches of period of 1 hour have been observed.

QUESTIONS

1. Define SHM. Give two examples of SHM not mentioned in the text. Why is SHM of central importance in the study of oscillations and waves ?
2. Name the three physical parameters that characterize a SHM. Give the meaning of each.
3. 'All simple harmonic motions are periodic, but all periodic motions are not simple harmonic.' Comment.
4. Show, by the appropriate choice of constants a and b in equation

$$\psi(t) = a \sin \omega t + b \cos \omega t$$

that equally valid solutions for $\psi(t)$ are

$$\begin{aligned}\psi(t) &= A \cos(\omega t + \phi)\\ &= A \sin(\omega t + \delta)\\ &= A \cos(\omega t - \phi)\\ &= A \sin(\omega t - \delta)\end{aligned}$$

and check that they satisfy the equation

$$\frac{d^2\psi}{dt^2} = -\omega^2\psi$$

5. Show that the following solutions of equation $\frac{d^2\psi}{dt^2}+\omega^2\psi = 0$ are equivalent.

$$\psi(t) = a \sin \omega t + b \cos \omega t$$
$$= A \sin (\omega t+\delta)$$
$$= A \cos (\omega t-\phi)$$
$$= C_1 \exp (i\omega t)+C_2 \exp (-i\omega t)$$

6. Show that the equation $\psi(t) = A \sin (2\pi t/T+\delta)$ of a particle executing SHM implies that the energy of the oscillating particle does not change with time.
7. Show that the equation $\psi(t) = A \cos (\omega t-\phi)$ of a particle executing SHM indicates that the time period of motion is $2\pi/\omega$.
8. Deduce an expression for the energy of a harmonic oscillator of mass m, amplitude A and frequency ν. At what displacement is the energy half-kinetic and half-potential ?
9. Explain clearly how a harmonically varying quantity can be represented by (i) a rotating vector and (ii) a complex exponential.
10. A mass m is attached to two identical rubber bands, each of stiffness constant k and of mass that is negligible compared to m. Obtain the expression for the frequency of the small amplitude transverse and longitudinal oscillations of the mass. Hence, show, that the longitudinal oscillations are more rapid than the transverse oscillations.
11. One end of a uniform spring of constant k and a finite mass m is attached to a rigid wall. The other end of the spring is attached to a body of mass $M (>m)$ placed on a horizontal frictionless surface. Show that the angular frequency of the horizontal harmonic oscillations is $[k/(M+m/3)]^{1/2}$

    (*Hint. See Example 1.8.*)
12. Using energy considerations, deduce the expression for the frequency of the small-amplitude oscillations of a simple pendulum and show that it is independent of the mass of the pendulum bob.
13. A hollow metal sphere is filled with water and hung from a rigid support by a long thread. When the sphere is made to oscillate in a plane, the period of its oscillations is found to be 2 s. A small hole is then made at the bottom of the sphere. As the water flows out, it is observed that the period of oscillations first increases and then decreases ; eventually regaining the value of 2 seconds, when the sphere has become empty. Explain.

    (Hint : As water flows out, the centre of gravity of the sphere is lowered till the sphere is half empty, resulting in an increase in the effective length of the pendulum. Beyond this state, the centre of gravity starts rising.)
14. Deduce an expression for the time period of vertical oscillations of a liquid column in a uniform U-tube. Neglect viscous effects Check the equation for dimensional consistency.
15. Using energy considerations, deduce the expression for the frequency of an oscillatory electrical circuit consisting of an inductance and a capacitance. Neglect the resistance of the coil.

PROBLEMS

1. A simple harmonic motion of amplitude 0.01 m has a time period of 2 s. Calculate the velocity and acceleration when the displacement is half the amplitude.
2. A particle executing SHM has an amplitude of 5 cm and a period of 2 s. Find the velocity of the particle at a point where its acceleration is half the maximum value.
3. A body moves with a SHM of frequency 0.5 Hz and amplitude 4.0 cm. Starting at a time when the displacement is +4.0 cm, find its displacement, velocity and acceleration 1.25 s later.
4. A particle vibrates with SHM of amplitude 5 cm and period 6 s. How long will it take to move from one end of its path on one side of the equilibrium position to a position 2.5 cm on the other side of the equilibrium position? What is the magnitude of its velocity at this point ?
5. A particle executes SHM with amplitude A. If its starting point from rest is (a) $\psi = +A$, (b) $\psi = -A$, (c) $\psi = A/2$ and (d) $\psi = 0$, find the different values of the phase constant (ϕ or δ) for the solutions

$$\psi(t) = A \cos(\omega t + \phi)$$

$$\psi(t) = A \sin(\omega t + \delta)$$

6. A particle executes SHM with amplitude A. If its starting point from rest is (a) $\psi = +A$, (b) $\psi = -A$, (c) $\psi = A/2$ and (d) $\psi = A/\sqrt{2}$, find the different values of the phase constant (ϕ or δ) for the solutions

$$\psi(t) = A \cos(\omega t - \phi)$$

$$\psi(t) = A \sin(\omega t - \delta)$$

7. The amplitude of the SHM of a particle is 5.0 cm. When its displacement is +3.0 cm, its velocity has a magnitude of 0.5 ms^{-1}. Find the period of its motion. If the mass of the particle is 10 g, what will be its total energy ?
8. A particle is vibrating in SHM with an amplitude of 10 cm. What fraction of the total energy is kinetic when the displacement of the particle from the mean position is 5 cm? At what displacement is the energy half kinetic and half potential?
9. Compare the displacement-time graphs of the two motions given by

$$\psi_1 = 0.01 \sin \pi t$$

and $$\psi_2 = 0.01 \sin(\pi t + \pi)$$

where the physical quantities are measured in SI system of units. State the amplitude and period of each motion and the phase difference between them.

10. A load of mass 0.5 kg hangs from a spring of force constant 10 Nm^{-1}. The mass is pulled down 0.05 m from its equilibrium position and then released. (a) What is the distance between the two most widely separated positions of the mass ? (b) How long does it take to traverse this distance ?

(In the following problems, take $g = 9.8\ ms^{-2}$, wherever necessary)

11. When a mass is hung from the lower end of a spring of negligible mass, an extension of 10 cm is produced in the spring. The mass is set into vertical oscillations. Calculate the period of oscillation.
12. A heavy object, placed on a shock absorber such as a rubber pad, compresses it by 1 cm. If the object is given a vertical tap, it will oscillate Compute the frequency of oscillation. Neglect damping.

(Hint : Assume that the pad acts like an elastic spring.)

13. An uniform spring has certain mass suspended from it. The spring is cut into two equal halves, and the same mass is suspended from one of the halves. Will the frequency of vertical oscillations be the same as before? How is the frequency in the second case related to the first ?
14. A uniform spring of force constant k is cut into two pieces whose lengths are in the ratio of 1 : 2. What is the force constant of each piece in terms of k ?
15. Two identical springs, each of force constant k, are connected as shown in Fig. 1.30. In each case, find the value of the effective force constant of the system in terms of k, for the oscillation of body A.

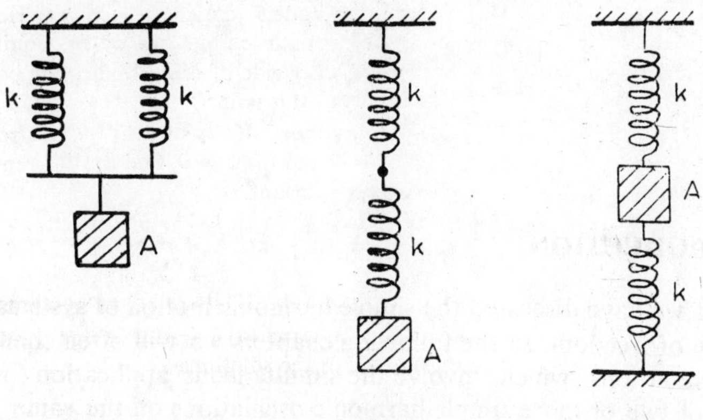

Fig. 1.30

16. A pendulum clock shows accurate time. If the length of its pendulum increases by 0.1 per cent, what will be the error in time per day.
17. By what percentage should the length of the pendulum of a clock be changed so that it keeps correct time when moved from a place where $g = 9.80\ \text{ms}^{-2}$ to a place where $g = 9.82\ \text{ms}^{-2}$?
18. A person normally weighing 60 kg stands on a platform which is oscillating up and down harmonically with a time period of 1.0 s and an amplitude of 10 cm. If a weighing machine on the platform gives the person's weight against time, what will be the maximum and minimum readings shown by it?
19. A platform is executing SHM in a vertical direction with an amplitude of 5 cm and an angular frequency of 20 rad s^{-1}. A block is placed on the platform at the lowest point of its path. (a) At what point will the block leave the platform ? (b) How far will the block rise above the highest point reached by the platform ? Take $g = 10\ \text{ms}^{-2}$.
20. Suppose a tunnel is dug through the earth (assumed to be a homogeneous sphere) from one side to the other along a diameter. A body dropped into the tunnel takes a time t_1 to reach the centre of the earth. If the acceleration due to gravity had remained unchanged, the body would have taken a time t_2 to traverse this distance. What is the ratio $t_1 : t_2$? Neglect friction offered to the motion.

(In the first case, the motion of the body is simple harmonic with a period $T = 2\pi\sqrt{R/g}$, where R is the radius of the earth and g is the acceleration due to gravity at its surface.)

2

Superposition of Harmonic Oscillations

2.1 INTRODUCTION

In Chap. 1 we have discussed the simple harmonic motion of systems having one degree of freedom. In the following chapters we will often come across physical situations which involve the simultaneous application (or superposition) of two or more simple harmonic oscillations on the same system. Such situations are particularly common in acoustics. A microphone diaphragm or a human eardrum is subjected simultaneously to various vibrations. The actual motion or the diaphragm or the eardrum corresponds to the resultant effect of the superposition of the various vibrations. In this chapter, we shall consider some specific cases of this superposition process. In order to obtain the resultant of two or more harmonic oscillations, we shall make use of a very important principle called the *superposition principle* which states that "*The resultant of two or more harmonic displacements is simply the algebraic sum of the individual displacements*". In the following section, we shall discuss the validity of this principle.

2.2 THE SUPERPOSITION PRINCIPLE AND LINEARITY

In Sec. 1 10 we have analysed the oscillations of systems with one degree of freedom, in which the moving part always stays close to the equilibrium position For small oscillations, the restoring force F is proportional to displacement ψ ($F = -K\psi$) and the resulting oscillations are harmonic. The equation of motion for the oscillations [see Eq. (1.2)] contains terms that depend only on ψ or $\frac{d^2\psi}{dt^2}$ with no terms that depend on higher powers of ψ or $\frac{d^2\psi}{dt^2}$ Such an equation is said to be linear. Thus, *a linear differential equation is the one that contains terms that depend only on the first powers of the variable and its derivatives.* On the other hand, if the moving part of the system oscillates violently, i.e. with a large amplitude, as, for example, in the case of a simple pendulum when sin α cannot be replaced by α (see page 24) the restoring force would contain terms that depend on ψ, ψ^2, ψ^3, etc. In other words, the restoring force would not be linear in ψ and the equation of motion will then be *nonlinear*.

Nonlinear equations are generally difficult to solve. Often one has to take recourse to numerical methods. In this book, we shall deal only with linear differential equations. Fortunately, we come across many interesting physical situations which, to a very good approximation, can be described by linear equations. Further, an equation is said to be *homogeneous* if it contains no terms independent of the variable ψ.

Linear homogeneous differential equations have a very interesting and important property : *The sum of any two solutions is itself a solution. This is also the statement of the superposition principle.*[1] The superposition principle holds only for linear differential equations. It does not apply if the equations are nonlinear. In other words, the sum of two solutions of a nonlinear equation is not itself a solution of the equation.

We shall prove these statements by considering a particularly simple nonlinear differential equation. Let us say that the oscillations of a system with one degree of freedom are governed by an equation of the form

$$\frac{d^2\psi}{dt^2} = -\omega^2\psi+\alpha\psi^2+\beta\psi^3+\ldots\ldots \tag{2.1}$$

This is the equation of motion of a simple pendulum when sin α cannot be replaced by α. Equation (2.1) is nonlinear because it contains terms like ψ^2, ψ^3, etc. If the constants α, β, etc. in Eq. (2.1) are all zero (or can be taken to be zero as a sufficiently good approximation), then Eq. (2.1) is linear and homogeneous.

Equation (2.1) is not easy to solve[2]. We do not need to solve it either. Let us suppose that we have somehow solved it and that $\psi_1(t)$ and $\psi_2(t)$ are the two solutions of Eq. (2.1). By hypothesis, ψ_1 and ψ_2 must each satisfy Eq. (2.1). Thus we have

$$\frac{d^2\psi_1}{dt^2} = -\omega^2\psi_1+\alpha\psi_1^2+\beta\psi_1^3+\ldots\ldots\ldots \tag{2.2}$$

and $$\frac{d^2\psi_2}{dt^2} = -\omega^2\psi_2+\alpha\psi_2^2+\beta\psi_2^3+\ldots\ldots\ldots \tag{2.3}$$

It is interesting to find out whether or not the superposition of ψ_1 and ψ_2 defined by the sum $\psi = \psi_1+\psi_2$, satisfies the same equation of motion

[1]We have already used this principle in Chap. 1 without explicitly stating it. While solving Eq. (1.2) we found that $\psi(t) = a \sin \omega t$ and $\psi(t) = b \cos \omega t$ both satisfy Eq. (1.2). We went a step further and showed that $\psi(t) = a \sin \omega t + b \cos \omega t$ also satisfies the equation and hence was also a solution of the equation.

[2]The nonlinear simple pendulum can be solved exactly. See Berkeley Physics Course Vol .1 p. 225 McGraw-Hill Book Company, 1965 or H.J. Pain, The Physics of Vibrations and Waves p. 300, John Wiley and Sons Ltd., 1976.

(2.1). We shall assume that it does and then arrive at a contradiction. If $\psi = \psi_1+\psi_2$ is also a solution of Eq. (2.1) we must have (replacing ψ by $\psi_1+\psi_2$)

$$\frac{d^2}{dt^2}(\psi_1+\psi_2) = -\omega^2(\psi_1+\psi_2)+\alpha(\psi_1+\psi_2)^2+\beta(\psi_1+\psi_2)^3+\ldots \tag{2.4}$$

Adding Eqs. (2.2) and (2.3), we get

$$\frac{d^2\psi_1}{dt^2}+\frac{d^2\psi_2}{dt^2} = -\omega\psi_1-\omega\psi_2+\alpha\psi_1^2+\alpha\psi_2^2+\beta\psi_1^3+\beta\psi_2^3+\ldots \tag{2.5}$$

If our assumption that $\psi_1+\psi_2$ is also a solution of Eq. (2.1), then Eqs. (2.4) and (2.5) must both hold. These equations agree if and only if all the following conditions are satisfied :

$$\frac{d^2}{dt^2}(\psi_1+\psi_2) = \frac{d^2\psi_1}{dt^2}+\frac{d^2\psi_2}{dt^2}$$

$$-\omega^2(\psi_1+\psi_2) = -\omega^2\psi_1-\omega^2\psi_2$$

$$\alpha(\psi_1+\psi_2)^2 = \alpha(\psi_1^2+\psi_2^2)$$

$$\beta(\psi_1+\psi_2)^3 = \beta(\psi_1^3+\psi_2^3)$$

The first two conditions are true, but the last two conditions are not true, unless α and β are zero. In other words, our assumption that $\psi_1+\psi_2$ is a solution of Eq. (2.1) can be realized if and only if the nonlinear terms (i.e. those with constants α, β, etc.) are absent. Thus we conclude that the superposition of two solutions is itself a solution, if and only if the equation is linear. The superposition principle does not hold for nonlinear equations.

The Importance of the Superposition Principle

To illustrate the application of the superposition principle, let us consider the example of the motion of a simple pendulum. For small oscillation the equation of motion, namely,

$$\frac{d^2\psi}{dt^2} = -\omega^2\psi$$

is linear. Under a given set of initial conditions (i.e. displacement and velocity at $t=0$), let the solution be ψ_1, given by

$$\psi_1 = A_1 \cos(\omega t+\phi_1)$$

where A_1 and ϕ_1 are determined from the given initial conditions (see Sec. 1.11). Remember that angular frequency ω does not depend upon the initial conditions. Under another set of initial conditions, the displacement ψ_2 is given by

$$\psi_2 = A_2 \cos(\omega t+\phi_2)$$

where A_2 and ϕ_2 are determined from the new set of initial conditions. Now suppose we prescribe a third set of initial conditions as follows. We superpose the initial conditions corresponding to ψ_1, and ψ_2. In other words, we give an initial displacement equal to the sum of the initial displacement corresponding to ψ_1 and that corresponding to ψ_2 and give an initial velocity equal to the sum of the initial velocities corresponding to ψ_1 and ψ_2. Then, from the superposition principle, the new motion described by ψ_3 is simply given by the superposition of ψ_1 and ψ_2, i.e.

$$\psi_3 = \psi_1 + \psi_2$$

There is no need to solve the equation of motion to find the new motion. Thus, *the resultant of two or more harmonic displacements is given by the algebraic sum of the individual displacements*. Remember, this result holds only if the equation of motion is linear. We shall now apply this principle to deduce the resultant of various kinds of superpositions of two or more harmonic displacements.

2.3 SUPERPOSITION OF TWO COLLINEAR HARMONIC OSCILLATIONS

Oscillations Having Equal Frequencies

Suppose we have two SHMs of equal frequencies but of different amplitudes and phase constants acting on a particle (or a system) in the x direction. The displacements x_1 and x_2 of the two harmonic motions, of the same angular frequency ω, are given by

$$x_1 = A_1 \cos(\omega t + \phi_1) \tag{2.6}$$

and

$$x_2 = A_2 \cos(\omega t + \phi_2) \tag{2.7}$$

where A_1 and A_2 are the amplitudes and ϕ_1 and ϕ_2 are the phase constants of the two motions. The resultant motion of the system, which moves in the x direction under the simultaneous effect of the two harmonic oscillations, can be found by the following methods.

(a) *Analytical Method*

We use the superposition principle which states that the resultant displacement x is equal to the sum of the individual displacements x_1 and x_2, i.e.

$$x = x_1 + x_2$$
$$= A_1 \cos(\omega t + \phi_1) + A_2 \cos(\omega t + \phi_2)$$

Using the trignometric identity $\cos(\alpha + \beta) = \cos\alpha \cos\beta - \sin\alpha \sin\beta$, this equation can be rewritten as

$$x = (A_1 \cos\phi_1 + A_2 \cos\phi_2) \cos\omega t - (A_1 \sin\phi_1 + A_2 \sin\phi_2) \sin\omega t \tag{2.8}$$

Now let (see also Fig. 2.1)

$$A_1 \sin\phi_1 + A_2 \sin\phi_2 = A \sin\delta \tag{2.9}$$

and

$$A_1 \cos\phi_1 + A_2 \cos\phi_2 = A \cos\delta \tag{2.10}$$

where A and δ are constants to be determined. Using the transformations (2.9) and (2.10) in Eq. (2.8) we have

$$x = A \cos (\omega t + \delta) \tag{2.11}$$

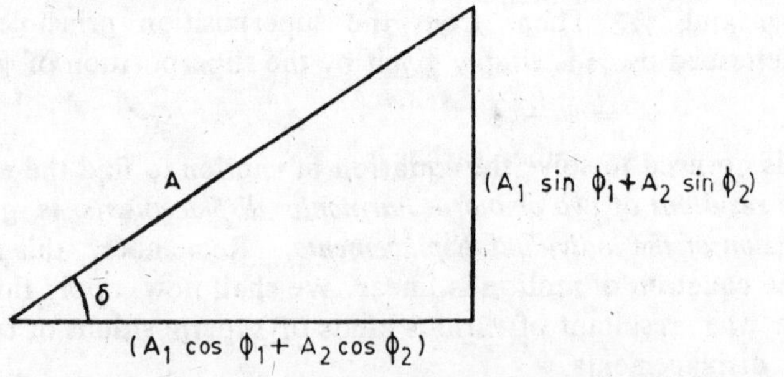

Fig. 2.1 Transformation into A and δ.

Equation (2.11) shows that the resulting motion is simple harmonic with an angular frequency ω, the same as that of the individual SHMs. The resulting motion has an amplitude A and a phase constant δ. A and δ can be evaluated from Eqs. (2.9) and (2.10). Squaring these equations and adding, we find that the resultant amplitude A is given by

$$A^2 = A_1^2 + A_2^2 + 2A_1A_2 \cos (\phi_2 - \phi_1) \tag{2.12}$$

Dividing Eqs. (2.9) and (2.10) we find that the phase constant of the resulting motion is given by

$$\tan \delta = \frac{A_1 \sin \phi_1 + A_2 \sin \phi_2}{A_1 \cos \phi_1 + A_2 \cos \phi_2} \tag{2.13}$$

Thus we conclude that the resultant effect of two collinear SHMs of equal frequencies is a simple harmonic motion of the same frequency but having amplitude and phase constant given respectively by Eqs. (2.12) and (2.13).

(b) *Vector Addition of Amplitudes* (*Geometrical Method*)

The rotating vector representation of SHM (see Sec. 1.8) provides a very simple geometrical method of obtaining the resultant of two SHMs of equal frequencies. In Fig. 2.2 (a) let OP_1 be a rotating vector of length A_1 (the amplitude of the first oscillation) making an angle $(\omega t + \phi_1)$ with the x-axis at time t, where ω is the angular frequency of the oscillation and ϕ_1 its phase constant. The projection ON_1 of OP_1 on the x-axis is the displacement x_1 of this motion at time t. Let OP_2 be a rotating vector of lenth A_2 at an angle $(\omega t + \phi_2)$. Its projection ON_2 on the x-axis is the second SHM of the same frequency ω, amplitude A_2 and phase constant ϕ_2. The superposition (*veotor sum*) of these two motions is then represented by the vector OP as defined by the parallelogram law of vector addition. As OP_1 and OP_2 rotate at the same angular frequency, we can imagine the parallelogram $OP_1 PP_2$ rotating at the same angular frequency. The resultant vector OP can be obtained as the vector sum of OP_1 and P_1P as shown in Fig. 2.2 (b), the vector P_1P being equal to vector OP_2. Since $\angle P_1ON_1 = \omega t + \phi_1$ and $\angle P_2ON_2 = \omega t + \phi_2$ the angle between OP_1 and P_1P is just $\phi_2 - \phi_1 = \phi$. Hence we have

$$A^2 = (A_1 + A_2 \cos \phi)^2 + (A_2 \sin \phi)^2$$

or

$$A^2 = A_1^2 + A_2^2 + 2A_1A_2 \cos \phi$$

which is the same as obtained earlier [see Eq. (2 12)].

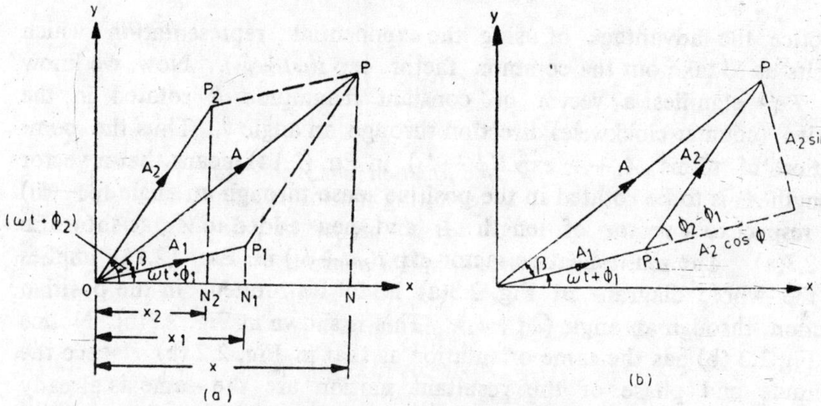

Fig. 2.2 (a) Superposition of two rotating vectors of the same frequency
(b) Vector triangle for obtaining the resultant vector

The total phase of the resultant motion is given by $\angle PON$. Let this be $(\omega t + \delta)$, where δ is the phase constant of the resultant motion. It is evident from Fig. 2.2 (b) that

$$\delta = \beta + \phi_1$$

$$\therefore \qquad \tan \delta = \tan (\beta + \phi_1) = \frac{\tan \beta + \tan \phi_1}{1 + \tan \beta \tan \phi_1}$$

Now $\tan \beta = A_2 \sin \phi / (A_1 + A_2 \cos \phi)$
with $\phi = \phi_2 - \phi_1$
Substituting for $\tan \beta$ in the above equation and simplifying, we get

$$\tan \delta = \frac{A_1 \sin \phi_1 + A_2 \sin \phi_2}{A_1 \cos \phi_1 + A_2 \cos \phi_2}$$

which agrees with Eq. (2.13) obtained earlier.

(c) *Use of Complex Exponential Representation*
The use of complex exponential representation of SHM (see Sec. 1.9 Chap. 1) leads us directly to the results obtained above. In the complex exponential representation, the rotating vectors OP_1 and OP_2 are described by the following equations :

$$Z_1 = A_1 \exp i(\omega t + \phi_1)$$

and $$Z_2 = A_2 \exp i(\omega t + \phi_2)$$

The displacements x_1 and x_2 are respectively the real parts of Z_1 and Z_2. From the superposition principle, the resultant complex displacement Z is given by

$$
\begin{aligned}
Z &= Z_1+Z_2 \\
&= A_1 \exp i(\omega t+\phi_1)+A_2 \exp i(\omega t+\phi_2)
\end{aligned}
$$

This equation may be recast in the form

$$Z = \{A_1+A_2) \exp i(\phi_2-\phi_1)\} \exp i(\omega t+\phi_1) \tag{2.14}$$

Notice the advantage of using the exponential representation which permits us to take out the common factor $\exp i(\omega t+\phi_1)$. Now, we know that $Re^{i\theta}$ signifies a vector of constant magnitude R rotated in the positive (counter-clockwise) direction through an angle θ. Thus the combination of terms $A_1+A_2 \exp i(\phi_2-\phi_1)$ in Eq. (2.14) means that a vector of length A_2 is to be rotated in the positive sense through an angle $(\phi_2-\phi_1)$ with respect to a vector of length A_1 and then added to A_1, as shown in Fig. 2.3(a). The multiplicative factor $\exp i(\omega t+\phi_1)$ in Eq. (2.14) implies that the whole diagram in Fig. 2.3(a) is to be rotated, in the positive direction, through an angle $(\omega t+\phi_1)$. This is shown in Fig. 2.3(b). Notice that Fig 2.3 (b) has the same orientation as that in Fig. 2.2 (b). Hence the amplitude and phase of the resultant motion are the same as already obtained above.

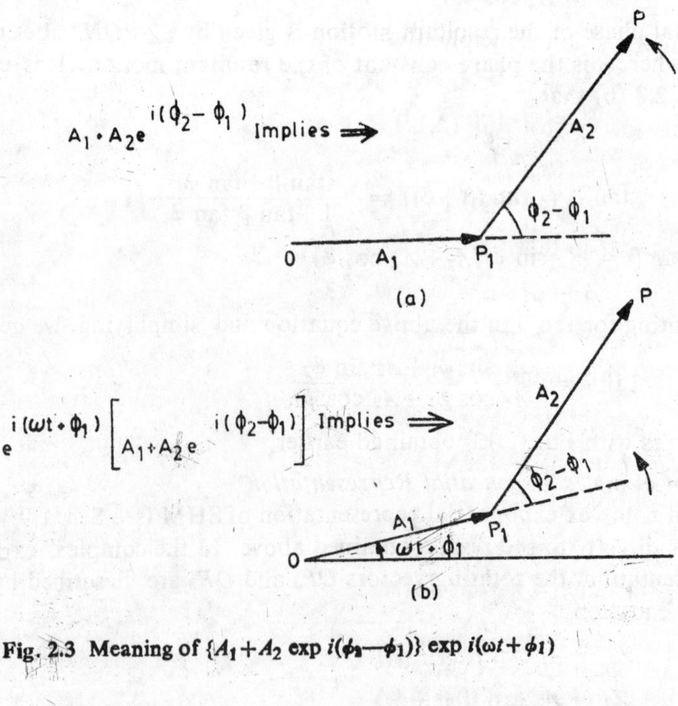

Fig. 2.3 Meaning of $\{A_1+A_2 \exp i(\phi_2-\phi_1)\} \exp i(\omega t+\phi_1)$

It is evident from Eq. (2.12) that the amplitude of the resulting oscillations is maximum given by

$$A_{max} = A_1 + A_2$$

if $\cos(\phi_2 - \phi_1) = +1$ or $\phi_2 - \phi_1 = 2m\pi$, where m is an integer with values $m = 0, 1, 2, 3, \ldots$. On the other hand, the resultant amplitude is minimum given by

$$A_{min} = A_1 - A_2$$

if $\cos(\phi_2 - \phi_1) = -1$ or $\phi_2 - \phi_1 = (2m+1)\pi$. For other values of the phase difference $(\phi_2 - \phi_1)$ the resultant amplitude A lies between A_{max} and A_{min}.

Oscillations Having Different Frequencies : Beats

In the subsequent chapters we shall come across many physical phenomena in which the moving part of a system is subjected simultaneously to two harmonic oscillations of different frequencies. To analyse the resulting motion of the system, let us consider two harmonic oscillations of different amplitudes A_1 and A_2 and different angular frequencies ω_1 and ω_2. For simplicity, we assume that the two oscillations have the same phase constant which we take to be zero[3]. The two harmonic oscillations can be written as

$$x_1 = A_1 \cos \omega_1 t \tag{2.15}$$

$$x_2 = A_2 \cos \omega_2 t \tag{2.16}$$

From superposition principle, the resulting oscillation is given by

$$x = x_1 + x_2 = A_1 \cos \omega_1 t + A_2 \cos \omega_2 t \tag{2.17}$$

We shall now recast Eq. (2.17) into a particularly simple form. Let us define on *average frequency* ω_a and a *modulation frequency* ω_m as :

$$\omega_a = \tfrac{1}{2}(\omega_1 + \omega_2) \qquad \text{and} \qquad \omega_m = \tfrac{1}{2}(\omega_2 - \omega_1)$$

where $\omega_2 \gg \omega_1$, so that

$$\omega_1 = \omega_a - \omega_m$$

$$\omega_2 = \omega_a + \omega_m$$

Substituting for ω_1 and ω_2 in Eq. (2.17) we get

$$x = A_1 \cos(\omega_a - \omega_m)\, t + A_2 \cos(\omega_a + \omega_m)\, t$$

or $$x = (A_1 + A_2) \cos \omega_m t \cos \omega_a t - (A_1 - A_2) \sin \omega_m t \sin \omega_a t \tag{2.18}$$

Now, as before, let

$$(A_1 + A_2) \cos \omega_m t = A_m \cos \delta_m \tag{2.19}$$

and $$(A_1 - A_2) \sin \omega_m t = A_m \sin \delta_m \tag{2.20}$$

---

[3]We could take the initial phases to be different and non-zero. It turns out that a finite initial phase difference only adds to the mathematical complexity and is, in general, not of major significance in this case.

Using these transformations in Eq. (2.18) gives

$$x = A_m \cos(\omega_a t + \delta_m) \tag{2.21}$$

where A_m and δ_m are given by [use Eqs. (2.19) and (2.20)]

and
$$A_m^2 = A_1^2 + A_2^2 + 2A_1A_2 \cos(2\omega_m t) \tag{2.22}$$

$$\tan \delta_m = \frac{(A_1 - A_2)\sin \omega_m t}{(A_1 + A_2)\cos \omega_m t} \tag{2.23}$$

The formal resemblance of Eq. (2.21) with the equation of SHM is misleading. In fact, the oscillation described by Eq. (2.21) is not harmonic since its amplitude A_m and phase constant δ_m both vary with time according to Eqs. (2.22) and (2.23) respectively. This oscillation can, at best, be described as periodic with an angular frequency ω_a, the average of the two component frequencies.

Beats. Recasting of the superposition (2.17) in the form of Eq. (2.21) becomes useful if ω_1 and ω_2 are nearly equal, i.e.

$$\omega_2 \simeq \omega_1$$

so that
$$\omega_m \ll \omega_a$$

In that case, the 'modulated' amplitude A_m and 'modulated' phase δ_m vary only slightly with time and may be treated as sensibly constant during the time scale of interest, which in our case, is the period $(2\pi/\omega_a)$ of the fast oscillation. Therefore, Eq. (2.21) will represent an 'almost' harmonic oscillation at an angular frequency ω_a. The resulting oscillation, in the case when the two frequencies of the SHMs are nearly equal, exhibits what are called *beats*.

The amplitude A_m of the resulting motion is maximum ($= A_1 + A_2$) when [see Eq. (2.22)]

$$\cos(2\omega_m t) = +1$$

or
$$2\omega_m t = 0, 2\pi, 4\pi, \ldots$$

or
$$(\omega_2 - \omega_1)\, t = 0, 2\pi, 4\pi, \ldots$$

or
$$2\pi(\nu_2 - \nu_1)\, t = 0, 2\pi, 4\pi, \ldots$$

or when
$$t = 0, \frac{1}{\nu_2 - \nu_1}, \frac{2}{\nu_2 - \nu_1}, \ldots$$

Here $\nu_1\ (= \omega_1/2\pi)$ and $\nu_2\ (= \omega_2/2\pi)$ are the frequencies of the two SHMs expressed in hertz.

Hence the time interval t_b between two consecutive maxima $= \dfrac{1}{\nu_2 - \nu_1}$. The frequency ν_b of the maxima $= \nu_2 - \nu_1$.

The amplitude A_m of the resulting motion is minimum ($= A_2 - A_1$) when

$$\cos(2\omega_m t) = -1$$

or when $t = \dfrac{1}{2(\nu_2 - \nu_1)}, \dfrac{3}{2(\nu_2 - \nu_1)}, \dfrac{5}{2(\nu_2 - \nu_1)} \ldots$ etc. Hence the frequency of the minima is also $\nu_b = \nu_2 - \nu_1$. Between any two maxima, there is a minimum. The periodic variation of the amplitude of the motion, resulting from the superposition of SHMs of slightly different frequencies, is known as the *phenomenon of beats*. One maximum of amplitude followed by a minimum is technically called a *beat*. The time period t_b between the successive beats is called the *beat period* given by

$$t_b = \frac{1}{\nu_2 - \nu_1}$$

and the *beat frequency* ν_b is given by

$$\nu_b = \frac{1}{t_b} = \nu_2 - \nu_1$$

Hence the beat frequency is equal to the difference between the frequencies of the component oscillations.

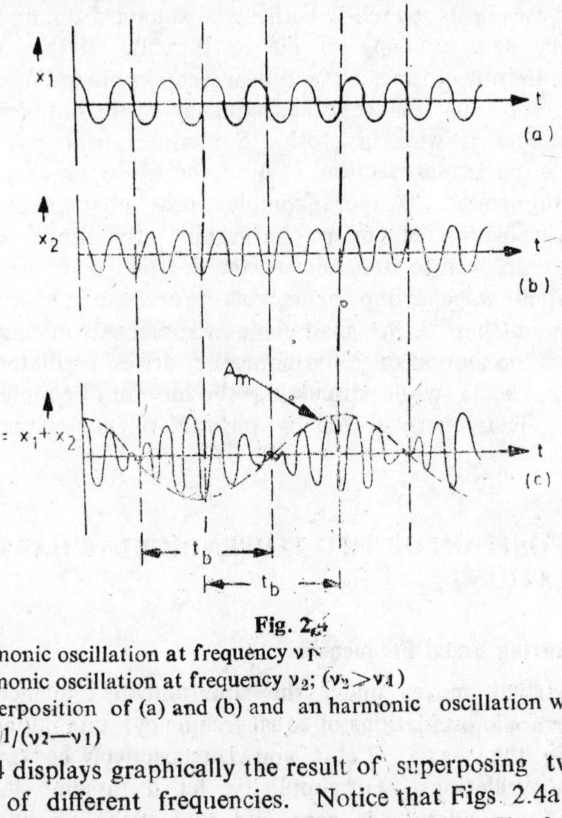

Fig. 2.4

(a) Harmonic oscillation at frequency ν_1

(b) Harmonic oscillation at frequency ν_2; ($\nu_2 > \nu_1$)

(c) Superposition of (a) and (b) and an harmonic oscillation with period $t_b = 1/(\nu_2 - \nu_1)$

Figure 2.4 displays graphically the result of superposing two harmonic oscillations of different frequencies. Notice that Figs. 2.4a and 2.4b are harmonic oscillations but their superposition shown in Fig. 2.4c is periodic but not harmonic.

Beats in Sound. Place two tuning forks of the same frequency on a resonance box and sound them. A continuous sound will be heard. The intensity of sound does not increase or decrease with time. Now stick a little wax to the prong of one of them so as to reduce its frequency. We will now actually hear beats. The intensity of the resulting sound will increase and decrease periodically with time. By counting the number of beats heard in a given interval of time, we can calculate the beat frequency and hence determine the difference between the frequencies of the two forks. To take a specific case, if two tuning forks are vibrating side by side at 480 and 486 Hz, their combined effect would be a vibration at 483 Hz (average of 480 and 486 Hz) passing through a maximum of amplitude 6 times (= 486−480) every second.

Applications of Beats

The phenomenon of beats is of great practical importance. Beats can be used to determine the small difference between frequencies of two sources of sound. Musicians often make use of beats in tuning their instruments. A piano tuner uses beats to tell whether his standard tuning fork has the same frequency as the string of his instrument. If the two differ in frequency, i.e. are out of tune, he will hear beats. He adjusts the tension in the string and thus changes the frequency of the note emitted by the string and matches it with his fork. Sometimes beats are deliberately produced in a particular section of an orchestra to give a pleasing tone to the resulting sound. A more complex beat phenomenon, resulting from the superposition of many harmonic oscillations of different frequencies, is employed to transmit a signal from one place to another. The beats called wave groups or packets propagate in space. We shall deal with them in Chap. 9. We shall come across beats in chapter 4 where we shall discuss the motion of a harmonically driven oscillator. We shall again encounter beats while discussing the motion of coupled oscillators (see Chap. 5). Thus, there are many physical phenomena which involve beats.

2.4 SUPERPOSITION OF TWO PERPENDICULAR HARMONIC OSCILLATIONS

Oscillations Having Equal Frequencies

Suppose a particle moves under the simultaneous influence of two perpendicular harmonic oscillations of equal frequency, one along the x-axis, the other along the y-axis. Let A_1 and A_2 respectively be the amplitudes of the x and y oscillations. For simplicity, let us assume that the phase constant of the x oscillation is zero and that of the y oscillation is δ, so that δ is the phase difference between them.[4] There is no loss of generality in doing so. Thus, the two rectangular SHMs can be written as

[4]It may be remarked that the knowledge of the phase constant of a simple harmonic motion does not furnish any useful information about that motion, for the simple reason that the phase constant can have any value depending on how we start the oscillation at time $t = 0$. In the case of two or more SHMs, the relevant and much more meaningful quantity is the phase difference δ between them. We small see that the subsequent motion of the particle depends on δ and not on the individual phase constants.

$$x = A_1 \cos \omega t \tag{2.24}$$

$$y = A_2 \cos (\omega t + \delta) \tag{2.25}$$

where x and y are the displacements along two mutually perpendicular directions. The resulting motion of the particle can be obtained as follows:

(a) Analytical Method

The path followed by the particle can be traced by eliminating time t from Eqs (2.24) and (2.25) so that we are left with an expression involving only x and y and the constant δ. Expanding the argument of the cosine in Eq. (2.24), we have

$$\frac{y}{A_2} = \cos \omega t \cos \delta - \sin \omega t \sin \delta$$

But from Eq. (2.24), $\cos \omega t = x/A_1$ and, therefore, $\sin \omega t = \left(1 - \dfrac{x^2}{A_1^2}\right)^{1/2}$

Therefore,

$$\frac{y}{A_2} = \frac{x}{A_1} \cos \delta - \left(1 - \frac{x^2}{A_1^2}\right)^{1/2} \sin \delta$$

or

$$\left(\frac{x}{A_1} \cos \delta - \frac{y}{A_2}\right) = \left(1 - \frac{x^2}{A_1^2}\right)^{1/2} \sin \delta$$

Squaring both sides we have

$$\frac{x^2}{A_1^2} + \frac{y^2}{A_2^2} - \frac{2xy}{A_1 A_2} \cos \delta = \sin^2 \delta \tag{2.26}$$

This is the general equation of an ellipse whose axes are inclined to the coordinate axes. Hence, *the path followed by the particle, which is subjected to two rectangular SHMs of equal frequencies, is, in general, an ellipse.*

Let us consider a few special cases :

(i) $\delta = 0$. In this case, Eq. (2.26) reduces to

$$\frac{x^2}{A_1^2} + \frac{y^2}{A_2^2} - \frac{2xy}{A_1 A_2} = 0$$

or

$$\left(y - \frac{A_2}{A_1} x\right)^2 = 0$$

This represents a pair of coincident straight lines, $y = \dfrac{A_2}{A_1} x$, having a positive slope A_2/A_1 and passing through the origin. The resultant motion is rectilinear and takes place along a diagonal of a rectangle of sides $2A_1$ and $2A_2$ such that x and y always have the same sign, both positive or both negative (Fig 2.5). The direction of motion can be easily determined from the defining Eqs. (2.24) and (2.25) by setting $\delta = 0$,

$$x = A_1 \cos \omega t$$

$$y = A_2 \cos \omega t$$

which immediately give $y = \dfrac{A_2}{A_1} x$, the equation of the straight line of slope $\frac{A_2}{A_1}$. At time $t = 0$, we have, $x = A_1$, $y = A_2$ so that the particle is at P at $t = 0$ (see Fig. 2.5). As time passes the cosines begin to

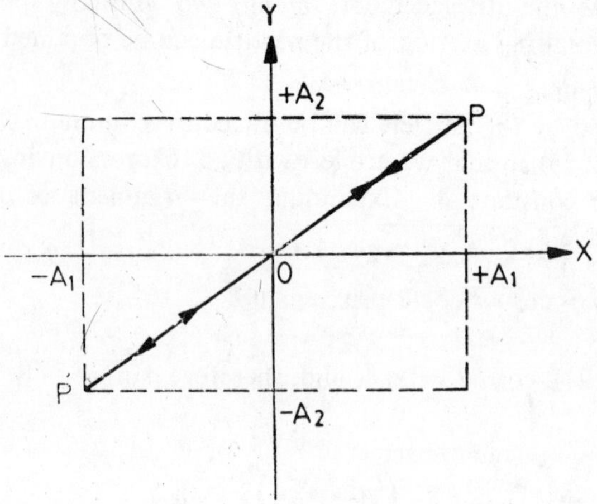

Fig. 2.5 Superposition of two perpendicular SHMs of the same frequency for phase difference $\delta = 0$

decrease until x and y become zero when $\omega t = \pi/2$. The particle moves from P to O. After this time, x and y become negative and at time when $\omega t = \pi$, x becomes $-A_1$ and y is $-A_2$. The particle moves from O to P'. After this the particle retraces its path. The particle continues to vibrate along the straight line $P\,O\,P'$. This represents what in optics is called a *linearly polarized vibration.*

(ii) $\delta = \frac{\pi}{2}$. In this case, Eq. (2.26) reduces to

$$\frac{x^2}{A_1^2} + \frac{y^2}{A_2^2} = 1$$

which is the equation of an ellipse whose principal axes lie along the x and y axes, as shown in Fig. 2.6. The particle moves in an elliptical path. The direction of its motion can be determined from the defining equations (2.24) and (2.25) with $\delta = \pi/2$.

$$x = A_1 \cos \omega t$$

$$y = A_2 \cos\left(\omega t + \frac{\pi}{2}\right) = -A_2 \sin \omega t$$

From these equations, the equation of the ellipse, obtained above, immediately follows. At time $t=0$, $x= A_1$ and $y=0$, so that the particle is at point P at $t = 0$ (see Fig. 2.6). As time t begins to increase from zero, x begins to decrease from its maximum positive value A_1 and y immediately begins to go negative. At a time when $\omega t = \pi/2$, x becomes zero and y equals A_2. The particle moves from P to Q during this time.

The subsequent motion of the particle is indicated by arrows in the diagram. The particle traces out an ellipse in the clockwise sense. This

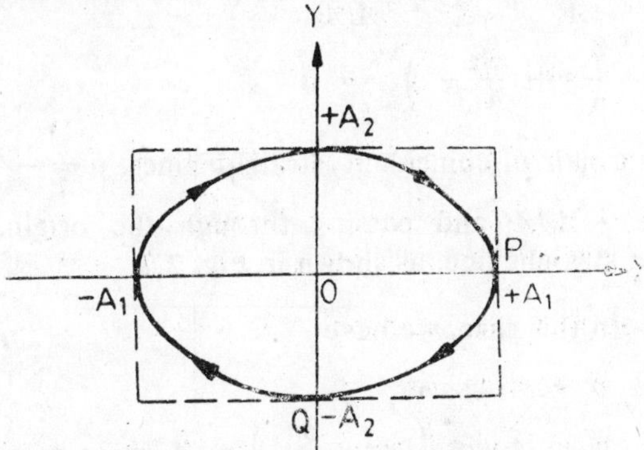

Fig. 2.6 Superposition of two perpendicular SHMs of the same frequency and phase differences $\delta = \pi/2$

represents what in optics is called the *right-handed elliptically polarized vibration*. The rotating electric field vector of the electromagnetic wave is always confined in one plane, with its tip tracing out an ellipse in the clockwise direction.

If, in addition, $A_1 = A_2 = A$, the ellipse degenerates into a circle

$$x^2+y^2 = A^2$$

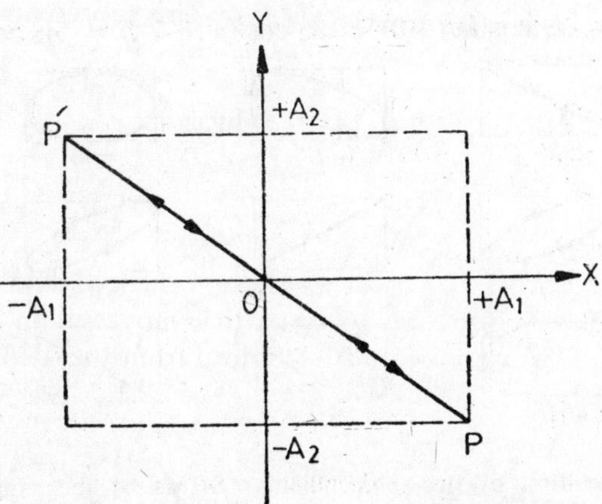

Fig. 2.7 Superposition of two perpendicular SHMs of the same frequency and phase difference $\delta = \pi$

Thus, *two harmonic oscillations, at right angles to each other, of equal amplitudes and equal frequencies but with phases differing by* $\pi/2$ *are equivalent to a uniform circular motion, the radius of the circle being equal to the amplitude of either oscillation.* Conversely, *a uniform circular motion can be resolved into two SHMs, at right angles to each other, their amplitudes being equal while their phases differ by* $\pi/2$ (see also Sec. 1.9, Ch. 1).

(iii) $\delta = \pi$. In this case, Eq. (2.26) becomes

$$\frac{x^2}{A_1^2} + \frac{y^2}{A_2^2} + \frac{2xy}{A_1 A_2} = 0$$

or

$$\left(y + \frac{A_2}{A_1} x \right)^2 = 0$$

This represents a pair of coincident straight lines, $y = -\frac{A_2}{A_1} x$, having a negative slope $-A_2/A_1$ and passing through the origin. The ellipse degenerates into a straight line, as shown in Fig. 2.7.

(iv) $\delta = \frac{3\pi}{2}$. In this case. we have

$$x = A_1 \cos \omega t$$

$$y = A_2 \cos \left(\omega t + \frac{3\pi}{2} \right) = A_2 \sin \omega t$$

which give

$$\frac{x^2}{A_1^2} + \frac{y^2}{A_2^2} = 1$$

We have an ellipse of the same form as in case (ii), but the motion is now *counter-clockwise*. In optics, such a vibration is called the *left-handed elliptically polarized vibration.*

The sequence of motions for a few values of δ in the range 0 to 2π is illustrated in Fig. 2.8. Notice that the resulting motion is the same for $\delta = 0$ or 2π. This is expected since for $\delta = 0$ or 2π

$$y = A_2 \cos (\omega t + 0) = A_2 \cos (\omega t + 2\pi) = A_2 \cos \omega t$$

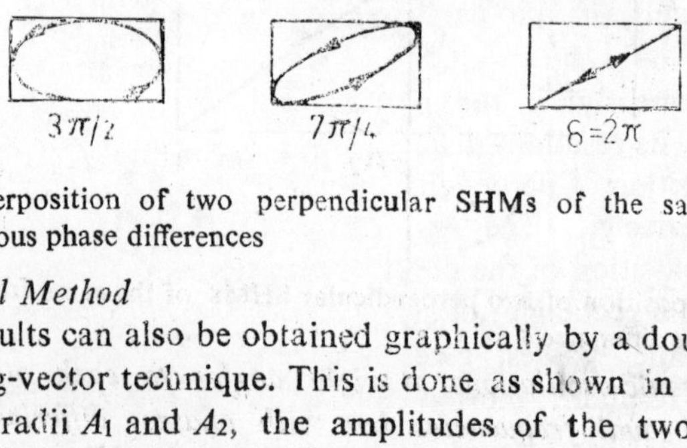

Fig. 2.8 Superposition of two perpendicular SHMs of the same frequency for various phase differences

(b) *Graphical Method*

The above results can also be obtained graphically by a double application of the rotating-vector technique. This is done as shown in Fig 2.9. Draw two circles of radii A_1 and A_2, the amplitudes of the two perpendicular SHMs. The circle of radius A_1 defines the SHM along the x-axis. Let O_1P_1 be the position of the rotating vector at a certain instant of time t. The projection of O_1P_1 on the x-axis ($O_1N_1 = OX = x$) gives the instantaneous displacement,

$$x = A_1 \cos \omega t$$

The circle of radius A_2 defines the SHM along the y-axis. Let O_2P_2 be the position of the rotating vector at time t. The projection of O_2P_2 on the y-axis ($O_2N_2 = OY = y$) gives the instantaneous perpendicular displacement

$$y = A_2 \cos (\omega t + \delta)$$

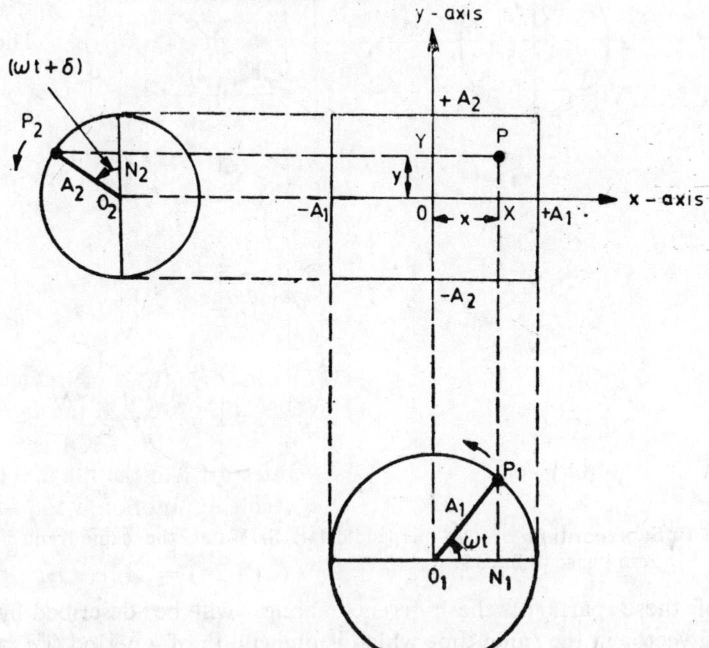

Fig. 2.9 Geometrical representation of the superposition of two SHMs at right angles to each other

If the particle has SHM only along the x-axis, its displacement at time t would be $x = OX$ where X is the projection of P_1 on the x-axis. On the other hand, if the particle has SHM only along the y-axis, its displacement would be $y = OY$, where Y is the projection of P_2 on the y-axis. Consequently, if the particle was subjected to both the SHMs simultaneously, its resultant displacement at time t would be OP. Point P is the intersection of perpendiculars drawn from P_1 and P_2 on the x and y axes respectively. The two displacements together describe the instantaneous position of the point P with respect to the origin O that lies at the centre of a rectangle of sides $2A_1$ and $2A_2$. The path followed by point P, as time passes, gives the resultant motion. We shall now construct the resultant motion for a few special values of the phase difference δ.

(i) $\delta = 0$: In this case, the two perpendicular motions are

$$x = A_1 \cos \omega t$$

$$y = A_2 \cos \omega t$$

The application of the above method to this particular case is shown in Fig. 2.10. Each *reference* circle is divided into the *same* number of equal parts, say, eight. Since the frequency of the two SHMs is the same,

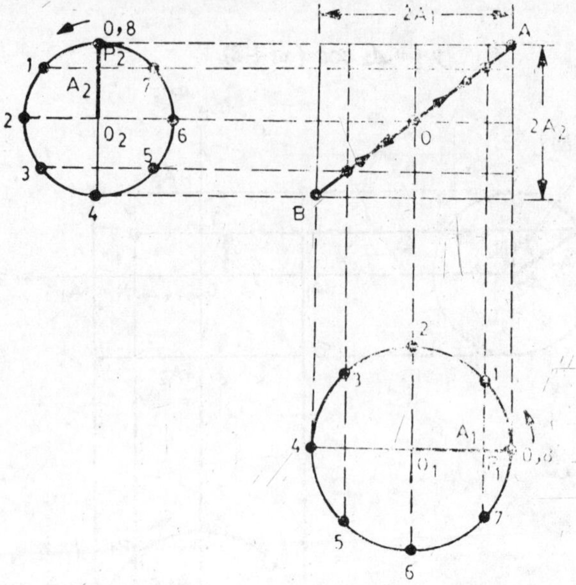

Fig. 2.10 Superposition of two perpendicular SHMs of the same frequency and zero phase difference

each of these parts of the reference circles will be described by each rotating vector in the same time which is one eighth of a period (i.e. $\pi/4\omega$). The positions of the points P_1, P_2, on the reference circles, are shown at a number of instants separated by one-eighth of a period. The points are numbered 0, 1, 2 . 8, beginning with $t = 0$, when O_1P_1 (see Fig. 2.9) is parallel to the x-axis and O_2P_2 parallel to the y-axis, so that the phase difference δ is zero. The projections from these corresponding positions of P_1 and P_2 then give us a set of intersections, as shown in Fig. 2.10, representing the instanteneous positions of the point P (see Fig. 2.9) as it moves within the rectangle. The locus defined by these points is a straight line *AOB* with a positive slope.

(ii) $\delta = \dfrac{\pi}{4}$ The application of the rotating vector method to this particular case is shown in Fig. 2.11. The positions of the points P_1 and P_2, on the two reference circles, are shown at a number of instants separated by one-eighth of the period of each component motion. The points are numbered 0, 1, 2, ..., 8 in sequence, starting with $t = 0$ when O_1P_1 (see Fig. 2.9) is parallel to the x-axis, and O_2P_2 is at angle $\delta = \dfrac{\pi}{4}$ or 45° measured in counterclockwise sense, from the y-axis, so that the phase difference δ is $\pi/4$. The projections of these corresponding positions of

P_1 and P_2 give us a set of points of intersection, as shown in Fig. 2.11. These intersections represent the instantaneous positions of the point P as it moves within the rectangle of sides $2A_1$ and $2A_2$. The locus of these points is an inclined ellipse, described in the clockwise sense as shown. The exact shape of the curve can be ascertained by dividing the reference circles into 16, 32 ... etc. parts instead of 8.

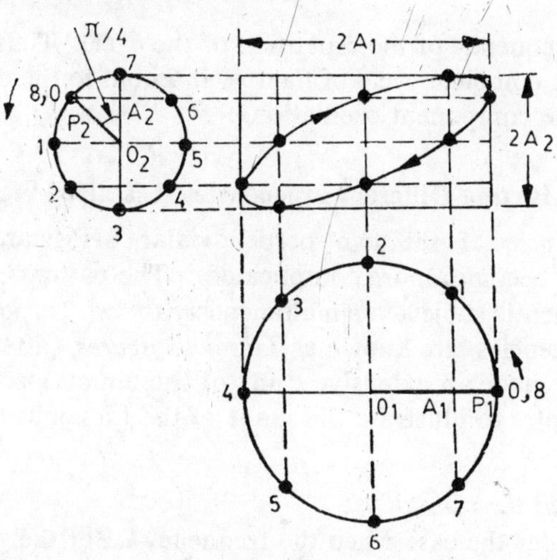

Fig. 2.11 Superposition of two perpendicular SMSs of the same frequency and a phase difference of $\pi/4$.

The resulting motion for other values of the phase difference can be similarly constructed. The sequence of motions is shown in Fig. 2.8.

Oscillations Having nearly Equal Frequencies. When the frequencies of the perpendicular oscillations are exactly equal, the curve traced out by the particle will remain perfectly steady. But, if the frequencies are nearly (but not quite) equal, the shape of the curve will change slowly on account of a gradual change in the phase difference between the two component oscillations and all forms of curves corresponding to the various values of the phase difference will be traced out.

For instance, when $\delta = 0$, the path traced out is a pair of coincident straight lines. As δ increases from 0 to $\pi/2$, the straight lines open out into an oblique ellipse, which, passing through its various intermediate forms, becomes a symmetrical ellipse when $\delta = \frac{\pi}{2}$ (see Fig. 2.8). As δ increases further from $\pi/2$ to π, the symmetrical ellipse changes to an oblique ellipse which degenerates into coincident straight lines when $\delta = \pi$. As δ changes from π to 2π, the reverse process takes place until at $\delta = 2\pi$, the original coincident straight lines are recovered. The whole cycle of figures is then repeated.

It is evident that the cycle of figures will be repeated only after one component motion has gained (or lost) *one* complete vibration over the other, Hence, if ν_1 and ν_2 are the frequencies of the two rectangular oscillations and t_c is the time for one complete cycle of figures, we have

$$\nu_1 t_c - \nu_2 t_c = \pm 1 \quad \text{or} \quad \nu_1 - \nu_2 = \pm \frac{1}{t_c} = \pm \nu_c$$

where ν_c is the frequency of the repetition of the cycle. Thus, the frequency of repetition of a complete cycle of figures is equal to the difference of the frequencies of the component oscillations.

2. Oscillations Having Different Frequencies (Lissajous Figures)

When the frequencies of the two perpendicular SHMs are not equal, the resulting motion becomes more complicated. The patterns, that are traced by a particle which is subjected simultaneously to two perpendicular SHMs of different frequencies, are known as *Lissajous figures,* after J.A. Lissajous (1822-1880) who made an extensive study of these motions. We shall now give a few examples to illustrate the shape of the Lissajous figure for some special cases.

(i) *Frequencies in the ratio of 1 : 2*

Let us first consider the case when the frequency ω_2 of the y oscillation is twice the frequency ω_1 the x oscillation, i.e. $\omega_1 = \omega$ and $\omega_2 = 2\omega$. The two SHMs are then given by

$$x = A_1 \cos \omega t \tag{2.27}$$

$$y = A_2 \cos (2\omega t + \delta) \tag{2.28}$$

where A_1 and A_2 are their respective amplitudes and δ is the phase difference between them.

The shape of the Lissajous figure can be obtained either by analytical or graphical method. In the analytical method, we find the locus of the instantaneous particle positions by eliminating time t from the above equations. Expanding the argument of the cosine in Eq. (2.28), we have

$$\frac{y}{A_2} = \cos 2\omega t \cos \delta - \sin 2\,\omega t \sin \delta$$

$$= (2\cos^2 \omega t - 1) \cos \delta - 2 \sin \omega t \cos \omega t \sin \delta$$

But from Eq. (2.27), $\cos\omega t = \dfrac{x}{A_1}$ and $\sin \omega t = \left(1 - \dfrac{x^2}{A_1^2}\right)^{1/2}$. Therefore,

$$\frac{y}{A_2} = \left(\frac{2x^2}{A_1^2} - 1\right) \cos \delta - \frac{2x}{A_1}\left(1 - \frac{x^2}{A_1^2}\right)^{1/2} \sin \delta$$

Rearranging we have

$$\left(\frac{y}{A_2} + \cos \delta\right) - \frac{2x^2}{A_1^2} \cos \delta = - \frac{2x}{A_1}\left(1 - \frac{x^2}{A_1^2}\right)^{1/2} \sin \delta$$

which, on squaring and upon simplification, reduces to

$$\left(\frac{y}{A_2}+\cos\delta\right)^2+\frac{4x^2}{A_1^2}\left(\frac{x^2}{A_1^2}-1-\frac{y}{A_2}\cos\delta\right)=0 \quad (2.29)$$

This is an equation of the fourth degree which, in general, represents a closed curve having two loops. For a given value of δ, the curve corresponding to Eq. (2.29) can be traced using the knowledge of coordinate geometry. Equation (2.29) reduces to a particularly simple form for $\delta = 0$. Setting $\cos\delta = 1$ in this equation, we have

$$\left(\frac{y}{A_2}+1\right)^2+\frac{4x^2}{A_1^2}\left(\frac{x^2}{A_1^2}-1-\frac{y}{A_2}\right)=0$$

or
$$\left(\frac{y}{A_2}+1-\frac{2x^2}{A_1^2}\right)^2=0$$

This represents two coincident parabolas with their vertices at $(0, -A_2)$ as shown in Fig. 2.12, the equation of each parabola being

$$\frac{y}{A_2}+1-\frac{2x^2}{A_1^2}=0$$

or
$$x^2=\frac{A_1^2}{2A_2}(y+A_2)$$

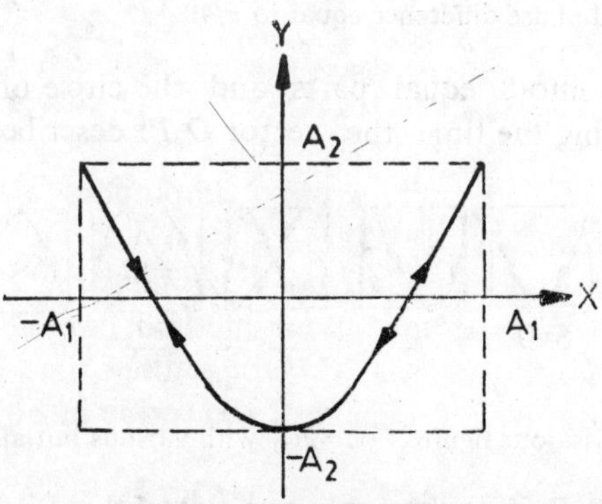

Fig. 2.12 Superposition of two perpendicular SHMs with frequencies in the ratio 1 : 2 and phase difference equal to zero

The analytical method becomes very cumbersome for values of δ other than zero. In such cases, the resultant motion can be constructed quite conveniently by the graphical method. Figure 2.13 shows how the rotating vector technique is used to obtain the shape of the Lissajous figure when $\delta = \pi/4$ and $\omega_2 = 2\omega_1$. The rotating vector O_2P_2 subtends an angle $\pi/4$ at time $t = 0$ with the y-axis so that the y oscillation has an initial phase of $\pi/4$, but the rotating vector O_1P_1 is along the x-axis at this instant of time, so that the x oscillation has no initial phase ; the phase difference between them is thus $\pi/4$. The y oscillation is twice as fast as the x oscillation. Therefore, we choose to divide the

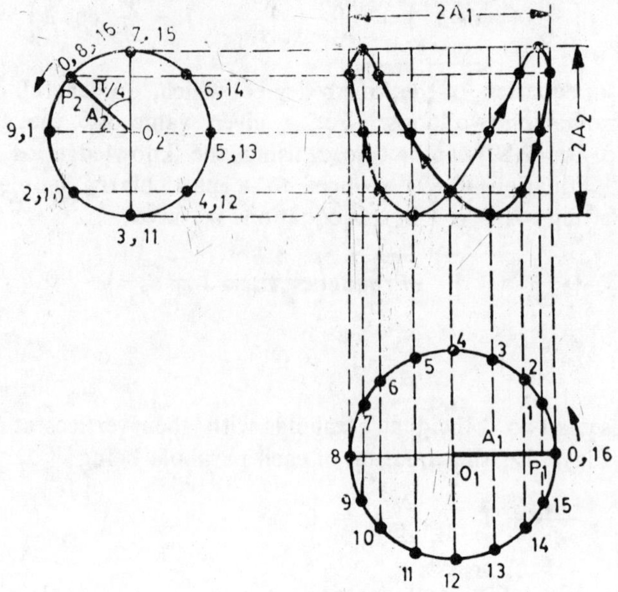

Fig. 2.13 Superposition of two perpendicular SHMs with frequencies in the ratio 1 : 2 and phase difference equal to $\pi/4$.

circle of radius A_2 into 8 equal parts and the circle of radius A_1 into 16 equal parts. During the time the vector O_2P_2 describes one-eighth of its

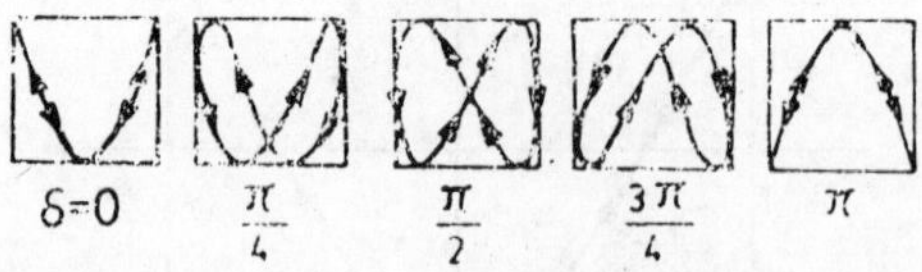

Fig. 2.14 Lissajous figures : $\omega_2=2\omega_1$ with various initial phase differences

circle, the vector O_1P_1 describes only one-sixteenth of its circle. During one complete cycle of ω_2 we go through only half a cycle of ω_1 and the points on the reference circles are marked accordingly. One must, of course, go through a complete cycle of ω_2 in order to obtain one complete period of the combined motion.

The combined motion corresponding to other phase differences can be similarly constructed. Figure 2.14 shows the sequence of these motions for values of δ in the range 0 to π.

Frequencies in any commensurate ratio

The analytical and graphical methods described above are, in general, applicable to perpendicular oscillations with frequencies in any commensurate ratio $m : n$, *reduced to lowest terms*. The analytical method becomes very cumbersome when the frequency ratio exceeds 3. In such cases, the

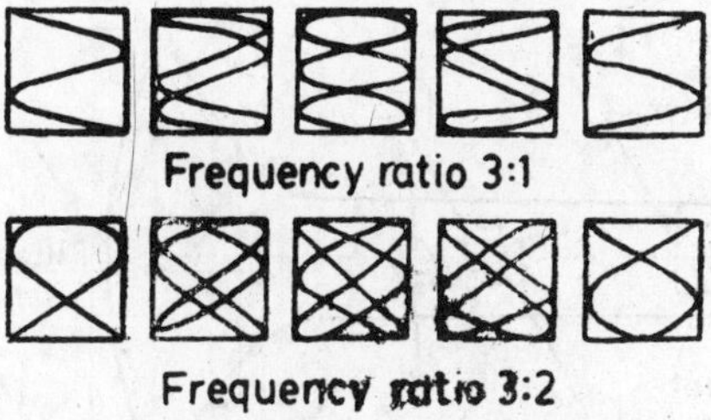

Fig. 2.15 Lissajous figures for various initial phase differences; frequency ratios 3 : 1 and 3 : 2

graphical method is more suitable. Figure 2.15 shows the shapes of Lissajous figures for various values of initial phase difference when the frequencies are in the ratio 3 : 1 and 3 : 2. If the frequencies of the two perpendicular oscillations are not commensurate, the resulting motion is not periodic. It does not repeat itself. In such cases, the particle will describe an endless curve.

A study of Figs. 2.8, 2.14 and 2.15 reveals some very interesting features of Lissajous figures when the frequencies of the two perpendicular oscillations are in a commensurate ratio.

(a) The resultant curve is inscribed in a rectangle of sides $2A_1$ and $2A_2$, where A_1 and A_2 are the amplitudes of the component oscillations.

(b) The resulting motion is periodic since the curve returns to itself.

(c) The sides of the rectangle are tangential to the curve at a number of points and the ratio of the numbers of these tangential points along the x-axis to those along the y-axis is the inverse of the ratio of the corresponding frequencies.

Demonstration of Lissajous Figures. A visual record of Lissajous figures can be obtained be means of a cathode-ray oscillograph (see Fig. 2.16). Two rectangular oscillations are simultaneously imposed upon a beam of cathode rays by connecting two sources of electrical oscillations to horizontal plates XX and vertical plates YY of the oscillograph. Thus the beam of cathode rays is subjected simultaneously to two perpendicular deflections. The beam falls on a fluorescent screen on which the Lissajous figure corresponding to the resultant motion can be seen. If the frequencies of the electrical oscillations are not exactly in a simple ratio, the figure will be seen to change its form slowly. For more complicated frequency ratios, very beautiful patterns are obtained.

Uses of Lissajous Figures. Lissajous figures can be used to determine the ratio of two exactly commensurate frequencies. The Lissajous figure is steady and, by inspection, we can find the ratio of the frequencies of the component oscillations. Let ν_1 and ν_2 be the frequencies of the oscillations along x and y axes respectively and let t_c be the time during which a complete cycle of the figures is described. Then, during one cycle, the number of oscillations made by the particle parallel to the x-axis will be $\nu_1 t_c$ and that of the oscillations parallel to the y-axis will be $\nu_2 t_c$. Hence

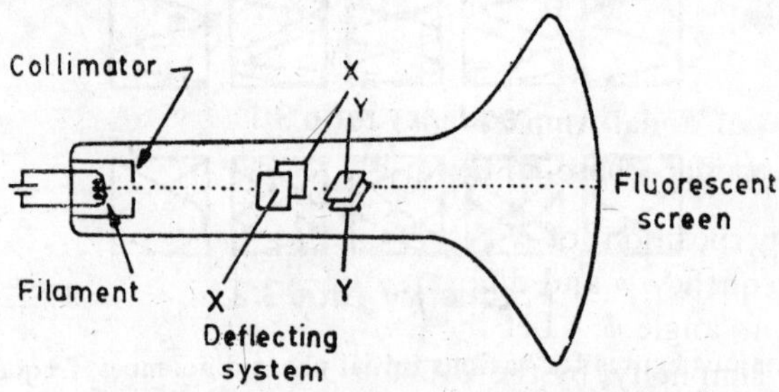

Fig. 2.16 The cathode-ray oscillograph

$\frac{\nu_1}{\nu_2} = \frac{\nu_1 t_c}{\nu_2 t_c}$. In other words, the ratio of the frequencies of the x and y oscillations will be equal to the inverse ratio of the maximum number of intersections of the Lissajous figure on the two lines parallel to the x and y axes respectively.

Lissajous figures may also be used to compare two nearly equal frequenecies. If the frequencies of the two component oscillations are not exactly equal, the Lissajous figure will change gradually, as discussed earlier. We have seen that, if ν_1 and ν_2 are nearly equal frequencies and t_c is the time for a complete cycle of change of Lissajous figure,

$$\nu_1 - \nu_2 = \pm \frac{1}{t_c}$$

The sign may be determined by observing the direction of change of the pattern to find out which of the two oscillations gains over the other.

2.5 SUPERPOSITION OF MANY HARMONIC OSCILLATIONS

The methods of combining two harmonic oscillations, outlined in Sec. 2.3, can readily be extended to an arbitrarily large number of oscillations. The general case, in which the amplitudes, frequencies and initial phases of the component oscillations are all different, is of no great importance, as it does not find applications in physics. Two situations, in particular, are of great interest and wide application. These are as follows :

(a) Superposition of a number of harmonic oscillations, all of the same frequency and amplitude and with equal successive initial phase differences.

(b) Superposition of a number of harmonic oscillations, all of the same amplitude and initial phase difference and with equal successive frequency differences.

The former finds application in the analysis of multiple-source interference effects in optics (such as multiple-beam interferometry and diffraction), while the latter has special relevance to the problem of wave groups or packets (as discussed in Chap. 9). We shall deal with the two cases separately.

1. Oscillations of Equal Amplitudes, Equal Frequencies and Equal Successive Initial Phase Differences

Consider a superposition of N harmonic oscillations each of amplitude A_0, angular frequency ω and differing in initial phase from its neighbouring oscillation by an angle ϕ. Let the first of these component oscillations be described, for simplicity, by the equation

$$x_1 = A_0 \cos \omega t$$

The other oscillations are then given by

$$x_2 = A_0 \cos (\omega t + \phi)$$

$$x_3 = A_0 \cos (\omega t + 2\phi)$$

$$\vdots$$

$$x_N = A_0 \cos \{\omega t + (N-1)\phi\}$$

From the superposition principle, the resultant motion is given by

$$x = A_0 \cos \omega t + A_0 \cos (\omega t + 2\phi) + \ldots\ldots + A_0 \cos\{\omega t + (N-1)\phi\} \qquad (2.30)$$

The resultant motion can be obtained by either of the following methods :

(a) *Vector Addition of Amplitudes*

Figure 2.17 is the geometrical representation of the mathematical expression (2.30). Vectors OP_1, P_1P_2......$P_{N-1}P_N$ respectively represent the first, the second,......, the Nth harmonic oscillation. Vector OP_N represents the resultant vector and its length A is the resultant amplitude. The combining vectors form successive sides of a regular polygon. We know that any regular polygon can be inscribed in a circle (not shown in the diagram) having some radius R and with its centre at a point C. All the corners O, P_1, P_2,...,P_N lie on the circle and the angle subtended at C by any individual vector (such as OP_1) is equal to the angle ϕ between adjacent vectors. Hence, the total angle OCP_N subtended at C by the resultant vector is equal to $N\phi$. Now, in a circle, the perpendicular (not shown in the diagram) drawn from its centre on any chord (such as OP_N or OP_1) bisects the chord as well as the angle the chord subtends at the centre. Hence,

$$A = 2\,R \sin\left(\frac{N\phi}{2}\right)$$

and

$$A_0 = 2\,R \sin\left(\frac{\phi}{2}\right)$$

Therefore

$$A = A_0\,\frac{\sin\,(N\phi/2)}{\sin\,(\phi/2)} \tag{2.31}$$

It is evident from Fig. 2.17 that the projection of OP_N on the x-axis gives the resultant displacement, i.e.

$$x = A\cos\,(\omega t+\delta) \tag{2.32}$$

where A [given by Eq. (2.31)] is the amplitude and δ is the phase constant of the resulting oscillation. From the geometry of Fig. 2.17 we have

$$\delta = \angle COP_1 - \angle COP_N$$

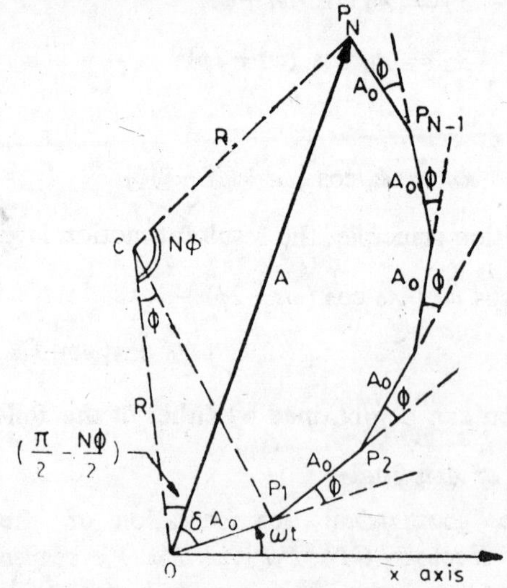

Fig. 2.17 Superposition of N harmonic oscillations of equal amplitude A_0, equal angular frequency ω and constant incremental phase difference ϕ.

Now $\angle COP_1 = \dfrac{\pi}{2} - \dfrac{\phi}{2}$ and $\angle COP_N = \dfrac{\pi}{2} - \dfrac{N\phi}{2}$. Therefore,

$$\delta = (N-1)\,\frac{\phi}{2} \tag{2.33}$$

Hence, the resultant oscillation is described by the equation

$$x = A_0\,\frac{\sin\,(N\phi/2)}{\sin\,(\phi/2)}\cos\left\{\omega t+(N-1)\,\frac{\phi}{2}\right\} \tag{2.34}$$

(b) *Use of Complex Exponential Representation*

The use of complex exponentials simplifies the algebra considerably. The superposition Eq. (2.30) is the real part of the complex function $f(t)$, where

$$
\begin{aligned}
f(t) &= A_0\,[e^{i\omega t} + e^{i(\omega t+\phi)} + e^{i(\omega t+2\phi)} + \ldots + e^{i\{\omega t+(N-1)\phi\}}] \\
&= A_0\, e^{i\omega t}[1 + e^{i\phi} + e^{2i\phi} + \ldots + e^{i(N-1)\phi}] \\
&= A_0\, e^{i\omega t}\,(1 + a + a^2 + \ldots + a^{N-1}) \\
&= A_0\, e^{i\omega t} S
\end{aligned}
$$

where $a = e^{i\phi}$ and S is the geometric series stated below

$$S = 1 + a + a^2 + \ldots + a^{N-1}$$

Then $aS = a + a^2 + \ldots + a^{N-1} + a^N$

$\therefore\ (a-1)S = a^N - 1$

or

$$
\begin{aligned}
S &= \frac{a^N - 1}{a - 1} \\
&= \frac{e^{iN\phi} - 1}{e^{i\phi} - 1} \qquad (\because a = e^{i\phi}) \\
&= \frac{e^{iN\phi/2}}{e^{i\phi/2}}\left(\frac{e^{iN\phi/2} - e^{-iN\phi/2}}{e^{i\phi/2} - e^{-i\phi/2}}\right)
\end{aligned}
$$

or

$$S = e^{i(N-1)\phi/2}\,\frac{\sin (N\phi/2)}{\sin (\phi/2)}$$

Thus

$$
\begin{aligned}
f(t) &= A_0 e^{i\omega t} e^{i(N-1)\phi/2}\,\frac{\sin(N\phi/2)}{\sin (\phi/2)} \\
&= A_0\,\frac{\sin (N\phi/2)}{\sin (\phi/2)} \exp i\{\omega t + \tfrac{1}{2}(N-1)\phi\}
\end{aligned}
$$

Finally, x in Eq. (2.30) is the real part of $f(t)$. Hence

$$x = A_0\,\frac{\sin(N\phi/2)}{\sin (\phi/2)}\cdot \cos\,\{\omega t + \tfrac{1}{2}(N-1)\,\phi\}$$

which is Eq. (2.34) obtained earlier. The amplitude of the resultant oscillation is given by

$$A = A_0\,\frac{\sin (N\phi/2)}{(\sin (\phi/2)}$$

This expression is used in the analysis of the diffraction pattern by a plane optical diffraction grating. When N becomes very large, ϕ becomes very small and the polygon (see Fig. 2.17) becomes an arc of the circle with centre C, the length of the arc then becomes $N A_0$, with A as the chord. For large N, $N-1 \simeq N$, so that, from Eq. (2.33) we have

$$\delta = (N-1)\,\frac{\phi}{2} \simeq \frac{N\phi}{2}$$

Now, for small ϕ, $\sin \phi \simeq \phi$

Hence, in this limit,

$$A = A_0 N \frac{\sin \delta}{\delta}$$

In the limit $\delta \to 0$, $\dfrac{\sin \delta}{\delta} \to 1$, so that $A_0 N$ is the value of A when $\delta \to 0$. Let us call it $A(0)$. Thus

$$A = A(0) \frac{\sin \delta}{\delta}$$

This expression is used in the study of diffraction of light by a single slit.

Now, let us check to see whether Eq. (2.34) reduces to the expression we have already obtained when only two terms are present in Eq. (2.30). Setting $N = 2$ in Eq. (2.34) we have

$$x = 2A_0 \cos\left(\frac{\phi}{2}\right) \cos\left(\omega t + \frac{\phi}{2}\right)$$

or

$$x = A \cos(\omega t + \delta)$$

where

$$A = 2A_0 \cos\left(\frac{\phi}{2}\right) \text{ and } \delta = \frac{\phi}{2}$$

This agrees with Eq. (2.11) with A and δ given by Eqs. (2.12) and (2.13) if we set $A_1 = A_2 = A_0$ and $\phi_1 = 0$, $\phi_2 = \phi$. Equation (2.13) reduces to

$$\tan \delta = \frac{A_0 \sin \phi}{A_0(1 + \cos \phi)} = \frac{2 \sin(\phi/2) \cos(\phi/2)}{2 \cos^2(\phi/2)} = \tan\left(\frac{\phi}{2}\right)$$

giving

$$\delta = \frac{\phi}{2}$$

Equation (2.12) becomes

$$A^2 = 2A_0^2 (1 + \cos \phi) = 4A_0^2 \cos^2\left(\frac{\phi}{2}\right)$$

or

$$A = 2A_0 \cos\left(\frac{\phi}{2}\right)$$

2. Oscillations of Equal Amplitudes, Equal Phase Constants, and Equal Successive frequency differences.

Consider a superposition of N different harmonic oscillations having equal amplitudes A_0, equal phase constants (assumed zero, for simplicity) and angular frequencies distributed uniformly between the lowest frequency, ω_1 and the highest frequency, ω_2. Such a situation finds application in the propagation of pulses and wave packets. We shall deal with this problem, in details, in Chap. 9.

For the moment, we shall only obtain the resultant motion corresponding to the above superposition, which is given by

$$x = A_0 \cos \omega_1 t + A_0 \cos (\omega_1 + \delta\omega)t + A_0 \cos (\omega_1 + 2\delta\omega)t + \ldots$$
$$+ A_0 \cos \omega_2 t \tag{2.35}$$

where $\delta\omega$ is the frequency spacing between neighbouring components, i.e.

$$\delta\omega = \frac{\omega_2 - \omega_1}{N-1} = \frac{\Delta\omega}{N-1} \tag{2.36}$$

where $\Delta\omega = \omega_2 - \omega_1$ is called the bandwidth. As before, the superposition (2.35) is the real part of the function $f(t)$ where

$$f(t) = A_0 \,[e^{i\omega_1 t} + e^{i(\omega_1 + \delta\omega)t} + e^{i(\omega_1 + 2\delta\omega)t} + \ldots\ldots + e^{i\{\omega_1 t + (N-1)\delta\omega\}t}]$$
$$= A_0 \, e^{i\omega_1 t} \,\{1 + a + a^2 + \ldots + a^{N-1}\}$$
$$= A_0 \, e^{i\omega_1 t} S$$

where $\quad a = e^{i\delta\omega t}$

and $\quad S = 1 + a + a^2 + \ldots + a^{N-1}$

Multiplication by a gives

$$aS = a + a^2 + \ldots \quad + a^{N-1} + a^N$$

Subtracting, we have

$$(a-1)\, S = a^N - 1$$

or

$$S = \frac{a^N - 1}{a - 1}$$
$$= \frac{e^{iN\delta\omega t} - 1}{e^{i\delta\omega t} - 1}$$
$$= \frac{e^{iN\delta\omega t/2}}{e^{i\delta\omega t/2}} \left(\frac{e^{iN\delta\omega t/2} - e^{-iN\delta\omega t/2}}{e^{i\delta\omega t/2} - e^{-i\delta\omega t/2}} \right)$$

or

$$= \exp \left\{ \frac{i}{2}(N-1)\delta\omega t \right\} \frac{\sin (\frac{1}{2} N \,\delta\omega t)}{\sin (\frac{1}{2} \,\delta\omega t)}$$

Thus $f(t)$ is given by

$$f(t) = A_0 \;\exp i \left\{ \omega_1 t + \tfrac{1}{2}\,(N-1)\delta\omega \right\} \frac{\sin (\frac{1}{2} N \delta\omega t)}{\sin (\frac{1}{2} \delta\omega t)}$$

From Eq. (2.36)

$$\omega_1 + \tfrac{1}{2}(N-1)\; \delta\omega = \omega_1 + \tfrac{1}{2}(\omega_2 - \omega_1)$$
$$= \tfrac{1}{2}\,(\omega_1 + \omega_2)$$
$$= \omega_a$$

where ω_a is the average of the two extreme frequencies. Thus we have

$$f(t) = A_0 \; e^{i\omega_a t} \; \frac{\sin (\frac{1}{2} \; N \; \delta\omega t)}{\sin (\frac{1}{2} \; \delta\omega t)}$$

Now x in Eq (2·35) is the real part of $f(t)$. Hence x is given by

$$x = A_o \frac{\sin(\frac{1}{2} N \delta\omega t)}{\sin(\frac{1}{2} \delta\omega t)} \cdot \cos \omega_a t \tag{2.37}$$

or

$$x = A_m \cos \omega_a t \tag{2.38}$$

where A_m, the modulation amplitude, is given by

$$A_m = A_o \frac{\sin(\frac{1}{2} N \delta\omega t)}{\sin(\frac{1}{2}\delta\omega t)} \tag{2.39}$$

Since A_m is time-dependent, the resulting oscillation is not harmonic. Let us check to see whether it reduces to the familiar form for beats when there are just two terms present. Setting $N = 2$ in Eq. (2.39), we have

$$A_m = A_0 \frac{\sin(\delta\omega t)}{\sin(\frac{1}{2} \delta\omega t)}$$

$$= 2 A_0 \cos(\tfrac{1}{2} \delta\omega t)$$

This agrees with Eq. (2.22) if we set $A_1 = A_2 = A_0$ and $\omega_m = \frac{1}{2}(\omega_2 - \omega_1) = \frac{\delta\omega}{2}$; since Eq. (2.22) then reduces to

$$A_m^2 = 2 A_0^2 (1 + 2 \cos \delta\omega t) = 4 A_0^2 \cos^2(\tfrac{1}{2} \delta\omega t)$$

or

$$A_m = 2 A_0 \cos(\tfrac{1}{2} \delta\omega t)$$

Equation (2.23) reduces to

$$\tan \delta_m = 0$$

or

$$\delta_m = 0$$

Thus. Eq. (2.38) reduces to the familiar Eq. (2.21) for beats when N is set equal to 2.

We shall use Eq. (2.37) in Chap. 9 while discussing the propagation of pulses and wave packets.

SOLVED EXAMPLES

Example 2.1 A particle is simultaneously subjected to two simple harmonic motions in the same direction, each of frequency 5 Hz. If the amplitudes are 0.005 m and 0.002 m respectively, and phase difference between them is 45°, find the amplitude of the resultant displacement and its phase relative to the first component. Write down the expression for the resultant displacement as a function of time.

Solution

Let the phase constant ϕ_1 of the first component be zero. then the phase constant ϕ_2 of the second component is 45°, i.e.

$$\phi_1 = 0, \ \phi_2 = 45°$$

Now $\quad A_1$ = amplitude of first motion = 0.005 m

and $\quad A_2$ = amplitude of second motion = 0.002 m

The amplitude A of the resultant motion is given by Eq. (2.12)

$$A^2 = A_1^2 + A_2^2 + 2A_1A_2 \cos(\phi_2 - \phi_1)$$
$$= (0.005)^2 + (0.002)^2 + 2 \times 0.005 \times 0.002 \times \cos 45^\circ$$
$$= 43.14 \times 10^{-6} \text{ m}^2$$

Thus $\quad A = 6.57 \times 10^{-3}$ m

The phase constant δ of the resultant motion is given by Eq. (2.13), which gives,

$$\tan \delta = \frac{0.005 \sin 0^\circ + 0.002 \sin 45^\circ}{0.005 \cos 0^\circ + 0.002 \cos 45^\circ}$$
$$= 0.2205$$

$$\therefore \quad \delta = 12.5^\circ = \frac{\pi}{14.4}$$

Now frequency of each motion is, $\nu = 5$ Hz

Angular frequency $\quad \omega = 2\pi\nu = 10\,\pi$ rad s^{-1}

With these values of A, δ and ω, the expression (2.11) for the resultant displacement becomes

$$x = 6.57 \times 10^{-3} \cos\left(10\pi t + \frac{\pi}{14.4}\right)$$

where x is expressed in metres and t in seconds.

Example 2.2 Two vibrations along the same line are described by the equations

$$x_1 = 0.03 \cos 10\,\pi t$$
$$x_2 = 0.03 \cos 12\,\pi t$$

where x_1, x_2 are measured in metres and t in seconds. Obtain the equation describing the resultant motion and hence find the beat period.

Solution

Using the superposition principle, the resultant motion is given by

$$x = x_1 + x_2$$
$$= 0.03\,(\cos 10\,\pi t + \cos 12\,\pi t)$$

Using the trignometric identity,

$$\cos \alpha + \cos \beta = 2 \cos\left(\frac{\alpha+\beta}{2}\right) . \cos\left(\frac{\alpha-\beta}{2}\right),$$ we have

$$x = 0.06 \cos (\pi t) \cos (11 \pi t)$$

which is of the form

$$x = A_m \cos \omega_a t$$

with $$A_m = 0.06 \cos (\pi t) \text{ and } \omega_a = 11 \pi$$

It is clear that A_m is a slowly varying modulated amplitude and $\cos \omega_a t$ is the fast oscillation of angular frequency $\omega_a = 11 \pi$ rad s^{-1} which is the average of the two component frequencies. Now A_m is maximum, if

$$\cos \pi t = \pm 1$$

or $$\pi t = 0, \pi, 2\pi, \ldots$$

or $$t = 0, 1, 2, \ldots$$

Hence the beat period is $t_b = 1$ s. This is the time interval between two consecutive maximum values of A_m^2.

Example 2.3 Two tuning forks A and B of nearly equal frequencies are employed in an optical experiment to produce Lissajous figures. On slightly loading fork A, it is observed that the cycle of change of figure slows down from 10 to 20 seconds. If the frequency of fork B is 256 Hz, determine the frequency of fork A before and after loading.

Solution :

Let ν_1 be the frequency of fork A and t_c the time for a complete cycle of change of a Lissajous figure before the fork is loaded. Then, if ν_2 is the frequency of fork B, we have

$$\nu_1 = \nu_2 \pm \frac{1}{t_c}$$

$$= 256 \pm \frac{1}{10}$$

$$= 256.1 \text{ Hz} \quad \text{or} \quad 255.9 \text{ Hz}$$

Thus the frequency ν_1 of fork A is either 255.9 Hz or 256.1 Hz. We know that on loading the frequency of the fork decreases. If ν_1 is 255.9 Hz, then, on loading it, ν_1 decreases; hence the difference $\nu_2 - \nu_1$ must increase, which would decrease the value of t_c, the time for a complete cycle of change of figure with the loaded fork. But t_c is observed to increase from 10 to 20 s. Hence the frequency of fork A cannot be 255.9 Hz. It must be 256.1 Hz before loading. After loading, the frequency of fork A becomes

$$\nu_1' = 256 + \frac{1}{20}$$

$$= 256.05 \text{ Hz}$$

Example 2.4 A particle is simultaneously subjected to three simple harmonic motions, all of the same frequency and in the same direction. If the amplitudes are 0.5 mm, 0.4 mm and 0.3 mm respectively, and the phase difference between the first and second is 45°, and between the second and the third is 30°, find the amplitude of the resultant displacement and its phase relative to the first motion of amplitude 0.5 mm.

Solution

We shall obtain the resultant amplitude by the method of vector addition of amplitudes. Let us assume that the phase constant of the first component is zero. The vector diagram can then be drawn as shown in Fig. 2.18.

Vector OP_2 is the resultant of vectors OP_1 and P_1P_2 of amplitudes $A_1 = 0.5$ mm and $A_2 = 0.4$ mm respectively. The phase difference between them is $\phi = \phi_2 - \phi_1 = 45° - 0° = 45°$. The resultant amplitude A is given by

$$A^2 = A_1^2 + A_2^2 = 2A_1A_2 \cos\phi$$

$$= (0.5)^2 + (0.4)^2 + 2 \times 0.5 \times 0.4 \times \cos 45°$$

$$= 0.6928 \text{ mm}^2$$

$$\therefore \quad A = 0.8324 \text{ mm}$$

The phase of the resultant of OP_1 and P_1P_2 relative to the first component OP_1 is α. In $\Delta\, OP_1P_2$ we have

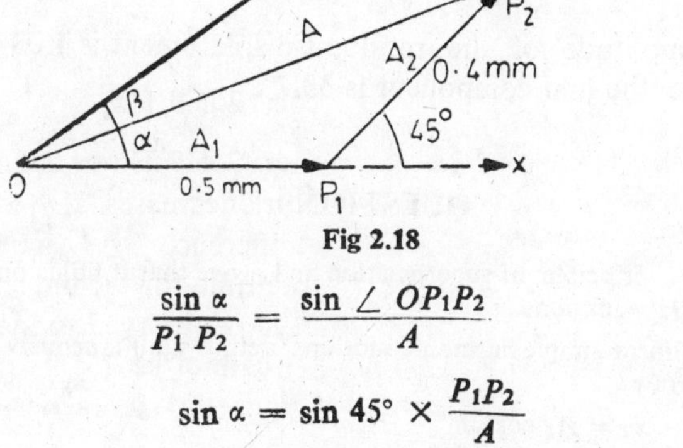

Fig 2.18

$$\frac{\sin\alpha}{P_1P_2} = \frac{\sin \angle OP_1P_2}{A}$$

$$\sin\alpha = \sin 45° \times \frac{P_1P_2}{A}$$

$$= \sin 45° \times \frac{0.4}{0.8324}$$

$$= 0.3398$$

$$\alpha = 19° 52' = 19.87°$$

Now the phase difference between

$$OP_2 \text{ and } P_2P_3 = \phi' = 30° + (45° - 19.87°)$$
$$= 55.13°$$

Vector OP_3 represents, in magnitude and direction, the resultant of vectors OP_2 and P_2P_3. Hence R is the resultant of the three vectors OP_1, P_1P_2, and P_2P_3. R is given by

$$R^2 = A^2 + A_3^2 + 2AA_3 \cos \phi'$$
$$= (0.8324)^2 + (0.3)^2 + 2 \times 0.8324 \times 0.3 \times \cos 55.13°$$
$$= 1.068 \text{ mm}^2$$
$$\therefore \quad R = 1.03 \text{ mm}$$

The phase δ of the resultant vector OP_3 relative to the first component OP_1 is, $\delta = \alpha + \beta$. Now β can be obtained as follows :

In ΔOP_2P_3, we have

$$\frac{\sin \beta}{P_2P_3} = \frac{\sin \angle OP_2P_3}{OP_3}$$

or

$$\sin \beta = \sin (180° - 55.13°) \frac{P_2P_3}{OP_3}$$
$$= \sin 124.87° \times \frac{0.3}{1.03}$$
$$= 0.2390$$
$$\therefore \quad \beta = 13° \, 50' = 13.83°$$

Hence

$$\delta = \alpha + \beta = 33.7°$$

Thus, the amplitude of the resultant displacement is 1.03 mm and its phase relative to the first component is 33.7°.

QUESTIONS

1. State the principle of superposition and prove that it holds only for linear differential equations.
2. Two collinear simple harmonic motions acting simultaneously on a particle are given by

$$x_1 = A_1 \cos \omega t$$
$$x_2 = A_2 \cos (\omega t + \phi)$$

Show that the resultant motion of the particle is simple harmonic. Also obtain the expression for the amplitude and phase constant of the resultant motion in terms of A_1, A_2 and ϕ.

3. Using the rotating vector representation, obtain the resultant motion of a particle subjected simultaneously to two simple harmonic motions in the same direction having equal amplitudes and equal frequencies and differing in phase by $\pi/4$.
4. What are beats ? Give an analytical description of the phenomenon of beats and show that the beat frequency is equal to the difference between the frequencies of the component oscillations.
5. Two simple harmonic motions, of equal amplitudes and slightly different frequencies, are in phase at time $t = 0$. Draw the displacement—time curves of them and construct their superposition to show the existence of beats. What are the uses of beats ?
6. Trace graphically and analytically the motion of a particle that is subjected to two perpendicular simple harmonic motions of equal frequencies, different amplitudes and phases differing by (a) zero and (b) $\pi/2$.
7. Show that two harmonic oscillations, at right angles to each other, of equal amplitudes and equal frequencies but with phases differing by $\pi/2$, are equivalent to a uniform circular motion, the radius of the circle being equal to the amplitude of either oscillation.
8. What are Lissajous figures ? Trace graphically the form of the Lissajous figure traced out by a particle subjected to two perpendicular simple harmonic motions of unequal amplitudes, time periods in the ratio of 1 : 2 and phases differing by (a) zero, (b) $\pi/4$ and (c) $\pi/2$.
9. A particle is subjected to two simple harmonic motions of slightly different frequencies. Explain how the shape of the curve traced out by the particle changes with time and show that the frequency of repetition of a particular curve is equal to the difference of the frequencies of the two component oscillations.
10. What are Lissajous figures ? How are they experimentally demonstrated ? Explain how these figures are used to determine the difference between two nearly equal frequencies.
11. A particle is subjected simultaneously to N simple harmonic motions of the same frequency. If the amplitude of each oscillation is A_0 and ϕ is the phase difference between successive oscillations, show that amplitude of the resultant oscillation is given by

$$A = A_0 \frac{\sin (N \phi/2)}{\sin (\phi/2)}$$

12. A particle is subjected simultaneously to N simple harmonic oscillations having frequencies distributed uniformly between ν_1 and ν_2. If the amplitude of each oscillation is A_0, initial phase of each is zero and $\delta\nu$ is the frequency difference between successive components, show that the resultant displacement of the particle is given by

$$x(t) = A_0 \frac{\sin (\pi N \delta\nu t)}{\sin (\pi \delta \nu t)} \cdot \cos [\, \pi(\nu_1+\nu_2)\, t\,]$$

PROBLEMS

1, A particle is subjected simultaneously to two simple harmonic motions of the same frequency and in the same direction. If their amplitudes are 5 mm and 3 mm respectively and the phase of the second component relative to the first is 30°, find the amplitude of the resultant displacement and its phase relative to the first component.

2. Two collinear simple harmonic motions, acting simultaneously on a particle are given by

$$x_1 = 0.3 \cos 2\pi t$$
$$x_2 = 0.2 \sin (2\pi t - \pi/3)$$

where x is expressed in cm and t in seconds. Write down the expression for the resultant displacement as a function of time.

3. Two vibrations along the same line are described by the equations

$$x_1 = 0.05 \cos 8\pi t$$
$$x_2 = 0.03 \cos 10\pi t$$

where x is expressed in metres and t in seconds. Obtain the equation describing the resultant motion and hence find the beat period. Draw a careful sketch of the resultant displacement over one beat period.

4. Two collinear simple harmonic motions, each of amplitude 1.0 cm and frequencies 8 Hz and 6 Hz respectively, act simultaneously on a particle. Assuming that they are initially in phase, draw the displacement—time curves for the two motions. Draw a careful sketch of the resultant displacement over one beat period. What is the beat frequency ?

5. Two parts of a sonometer wire, divided by a movable knife-edge, differ in length by 2 mm and produce 2 beats per second when sounded together. If the total length of the wire is 1 metre, find the frequencies of the two parts of the wire.

6. Find the frequency of the combined motion of each of the following :
(a) $\sin (2\pi t - \pi/4) + \cos (2\pi t)$
(b) $\sin (10\pi t) + \cos (11\pi t + \pi/4)$
(c) $\cos(3t) - \sin (\pi t)$
(d) $a \cos (2\pi \nu t) + b \sin (2\pi \nu t + \pi/3)$

7. Two vibrations, at right angles to each other, are described by the equations

$$x = 3 \cos 5\pi t$$
$$y = 2 \cos (5\pi t + \pi/3)$$

where x and y are expressed in centimetres and t in seconds. Construct the curve for the combined motion.

8. Two vibrations, at right angles to each other, are described by the equations

$$x = 3 \cos 4\pi t$$
$$y = 3 \cos (8\pi t + \pi/3)$$

where x and y are expressed in centimetres and t in seconds. Construct the Lissajous figure of the combined motion.

9. Construct the Lissajous figures for the following motions :
(a) $x = a \cos \omega t,\ y = b \sin \omega t$
(b) $x = a \cos 2\omega t,\ y = a \cos [2\omega t - (\pi/4)]$
(c) $x = a \cos 2\omega t,\ y = b \cos \omega t$
(d) $x = a \cos 2\omega t,\ y = a \sin 2\omega t$

10. A particle is simultaneously subjected to three simple harmonic motions, all of the same frequency and in the same direction. If the amplitudes are 1.0 cm, 0.5 cm 0.25 cm and respectively and the phase of the second relative to the first is 30° and that of tne third relative to the second is 60°, find the amplitude of the resultant displacement and its phase relative to the first component.

3

Free Damped Oscillations

3.1 INTRODUCTION

In the preceding chapters we were concerned entirely with free oscillations of undamped physical systems. We have seen that the total energy of a harmonic oscillator remains constant. Once started, the oscillations continue for ever with a constant amplitude (which is determined from the initial conditions) and a constant frequency (which is determined by the inertial and elastic properties of the system). Simple harmonic motions which persist indefinitely without loss of amplitude are called *free* or *undamped*. However, observation of the free oscillations of a real physical system reveals that the energy of the oscillator gradually decreases with time and the oscillator eventually comes to rest. For example, the amplitude of a pendulum oscillating in air decreases with time and it ultimately stops. The vibrations of a tuning fork die away with the passage of time. This happens because, in actual physical systems, the friction (or damping) is always present. Friction resists motion.

The presence of resistance to motion implies that frictional or damping force acts on the system. The damping force acts in opposition to the motion, doing negative work on the system, leading to a dissipation of energy. When a body moves through a medium such as air, water, etc. its energy is dissipated due to friction ane appears as heat either in the body itself or in the surrounding medium or both. There is another mechanism by which an oscillator losses energy. The energy of an oscillator may decrease not only due to friction in the system, but also due to *radiation*. The oscillating body imparts periodic motion to the particles of the medium in which it oscillates, thus producing *waves*. For example, a tuning fork produces sound waves in the medium which results in a decrease in its energy. All sounding bodies are subject to dissipative forces, or otherwise, there would be no loss of energy by the body and consequently no emission of sound energy could occur. Thus sound waves are produced by radiation from machanical oscillatory systems. We shall learn later that the electromagnetic waves are produced by radiations from oscillating electric and magnetic fields.

The effect of radiation by an oscillating system and of the friction present in the system is that the amplitude of oscillations gradually diminishes with time. The reduction in amplitude (or energy) of an oscillator is called *damping* and the oscillations are said to be *damped.*

3.2 DAMPING FORCES

The damping of a real system is a complex phenomenon involving several kinds of damping forces. The damping force of a fluid (liquid or gas) to a moving object is some function of the velocity of the object. The damping force that depends on velocity is referred to as *viscous damping force.* The magnitude of this force is well described by the equation

$$F = p_1 v + p_2 v^2$$

where v is the magnitude of the velocity of the object. The direction of this resistive force is opposite to that of the velocity. If v is small compared to the ratio p_1/p_2, the damping force will be proportional to the first power of v. Thus, for small velocities,

$$F = -pv \tag{3.1}$$

where p, the viscous damping coefficient, represents the damping force per unit velocity. The negative sign indicates that the force opposes motion, tending to reduce velocity. In other words, the viscous damping force is a retarding force. Since, the velocity of most oscillating systems is usually small, the damping force exerted by the fluid in contact with the system is likely to be viscous. Viscous forces are generally much smaller than inertial and elastic forces in a system. However, damping devices called dampers are sometimes deliberately introduced in a system for vibration control. The damping force exerted by such devices may be comparable in magnitude to the inertial and elastic forces.

In real systems, it is likely that the moving part is in contact with an unlubricated surface, as in the case of horizontal oscillations of a body attached to a spring (see Fig. 1.9). The oscillating body is always in contact with the horizontal surface. The resulting frictional force opposes the motion and can often be idealized as a force of constant magnitude. Such a force is usually referred to as a *Coulomb friction force*.

In a solid, some part of energy may be lost due to imperfect elasticity or internal friction of the material. It is very difficult to estimate this type of damping. Experiments suggest that a resistive force proportional to the amplitude and independent of the frequency may serve as a satisfactory approximation. This kind of damping in solids is referred to as *structural damping.*

Thus, the damping of a real system is a complex phenomenon involving several kinds of damping forces such as viscous damping, Coulomb friction and structural damping. Because it is generally very difficult to predict the magnitude of the damping forces, one usually has to rely on experience and experiment so as to make a reasonably good estimate. It is a common practice to approximate the damping of a system by an equivalent viscous damping, for the simple reason that viscous damping is the most convenient to handle mathematically. Thus, according to this approximation, the magnitude of the viscous force to be used in a particular problem is chosen to be the one that would produce the same rate of energy dissipation as the actual damping forces. This usually provides a good estimate.

The inclusion of damping forces complicates the analysis considerably. Fortunately, in actual systems, the damping forces are usually small and can often be ignored. In situations, where they are not negligibly small, the viscous damping model is the most convenient mathematically. We shall use this model, under a simplifying assumption, that the velocity of the moving part of the system is small, so that the damping force is linear in velocity as in Eq. (3.1). If the velocity is not small, the damping force exerted on the system may be represented more closely by a force proportional to the square of the velocity. We shall not deal with such forces. The effect of the linear viscous damping force on the free oscillations of simple systems, with one degree of freedom, is considered in the next section.

3.3 DAMPED OSCILLATIONS OF A SYSTEM HAVING ONE DEGREE OF FREEDOM

We shall investigate the effect of damping on the harmonic oscillations of a simple system having one degree of freedom. One such system is shown in Fig. 3.1. When the system is displaced from its equilibrium state and released, it begins to move. The forces acting on the system are :—

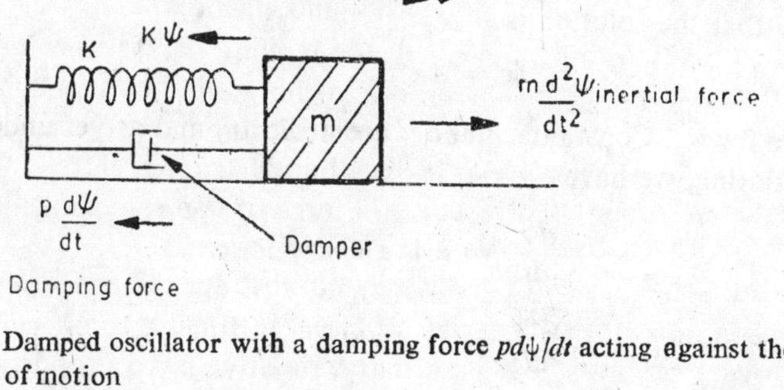

Fig. 3.1 Damped oscillator with a damping force $pd\psi/dt$ acting against the direction of motion

(i) a restoring force $-K\psi$, where K is the coefficient of the restoring force and ψ is the displacement, and

(ii) a damping force $-p\dfrac{d\psi}{dt}$, where p is the coefficient of the damping force and $\dfrac{d\psi}{dt}$ is the velocity of the moving part of the system. From Newton's law for a rigid body in translation, these forces must balance with Newton's force $m\dfrac{d^2\psi}{dt^2}$, where m is the mass of the oscillator and $\dfrac{d^2\psi}{dt^2}$, its acceleration. Since, the restoring force and the damping force act in a direction opposite to Newton's force, we have

$$m\frac{d^2\psi}{dt^2} = -K\psi - p\frac{d\psi}{dt}$$

Remember, this equation holds only for small displacements and small velocities. This equation can be rewritten as :

$$\frac{d^2\psi}{dt^2} + \gamma\frac{d\psi}{dt} + \omega_0^2\,\psi = 0 \tag{3.2}$$

with
$$\gamma = p/m \tag{3.3}$$
and
$$\omega_0^2 = K/m \tag{3.4}$$

Notice that dimensionally $\gamma = \dfrac{p}{m} = \dfrac{\text{force}}{\text{velocity}\times\text{mass}} = \dfrac{MLT^{-2}}{LT^{-1}M} = T^{-1}$. the same as the dimension of frequency.

It is easy to see that in Eq. (3.2) the damping is characterized by the quantity γ, having the dimension of frequency, and the constant ω_0 represents the angular frequency of the system in the absence of damping and is called the *natural frequency* of the oscillator. Equation (3.2) is the differential equation of the damped oscillator. To find out how the displacement varies with time, we need to solve Eq. (3.2) with constants γ and ω_0 given respectively by Eqs. (3.3) and (3.4).

The General Solution

To solve Eq. (3.2) we make use of the exponential function again. Let us assume that the solution is

$$\psi = A\,e^{\alpha t}$$

and solve for α. Constants A and α are arbitrary and as yet undetermined. Differentiating, we have

$$\frac{d\psi}{dt} = \alpha A\ e^{\alpha t}$$

$$\frac{d^2\psi}{dt^2} = \alpha^2\,A\,e^{\alpha t}$$

Substitution in Eq. (3.2) yields

$$(\alpha^2+\gamma\alpha+\omega_0^2)\,A\,e^{\alpha t} = 0$$

For this equation to hold for all values of t, the term in the brackets must vanish, i.e.

$$\alpha^2+\gamma\alpha+\omega_0^2 = 0$$

The two roots of this quadratic equation are

$$\alpha_1 = -\frac{\gamma}{2}+\frac{1}{2}(\gamma^2-4\omega_0^2)^{1/2}$$

and
$$\alpha_2 = -\frac{\gamma}{2}-\frac{1}{2}(\gamma^2-4\omega_0^2)^{1/2}$$

Thus the two possible solutions of Eq. (3.2) are

$$\psi_1 = A_1\, e^{\alpha_1 t} = A_1 \exp\left[-\frac{\gamma}{2}+\frac{1}{2}\left(\gamma^2-4\omega_0^2\right)^{1/2}\right]t$$

and
$$\psi_2 = A_2\, e^{\alpha_2 t} = A_2 \exp\left[-\frac{\gamma}{2}-\frac{1}{2}\left(\gamma^2-4\omega_0^2\right)^{1/2}\right]t$$

Since Eq. (3.2) is linear, the superposition principle is applicable. Hence, the general solution is given by the superposition of the two solutions, i.e.

$$\psi = \psi_1+\psi_2$$

or
$$\psi = A_1 \exp\left[-\frac{\gamma}{2}+\left(\frac{\gamma^2}{4}-\omega_0^2\right)^{1/2}\right]t$$

$$+A_2 \exp\left[-\frac{\gamma}{2}-\left(\frac{\gamma^2}{4}-\omega_0^2\right)^{1/2}\right]t \tag{3.5}$$

Here A_1 and A_2 are arbitrary constants to be determined from the initial conditions, namely, the initial displacement and the initial velocity.

The nature of the motion depends on the character of the roots α_1 and α_2. The roots may be real or complex depending on whether $\gamma > 2\omega_0$ or $\gamma < 2\,\omega_0$ respectively. In fact, three different kinds of motion are possible, depending on whether $\gamma > 2\omega_0$, $\gamma = 2\omega_0$ and $\gamma < 2\,\omega_0$. Each condition describes a particular kind of behaviour of the system. We shall now treat each case separately.

Case I : $\gamma > 2\,\omega_0$ (*Large Damping*) In this case, the damping term $\gamma/2$ dominates the stiffness term ω_0 and the term $(\gamma^2/4-\omega_0^2)^{1/2}$ in Eq. (3.5) is a real quantity with a positive value, say, q, i.e.

$$\left(\frac{\gamma^2}{4}-\omega_0^2\right)^{1/2} = q$$

so that displacement ψ as a function of time is given by

$$\psi = A_1 \exp\left(-\frac{\gamma}{2}+q\right)t+A_2 \exp\left(-\frac{\gamma}{2}+q\right)t \tag{3.6}$$

The velocity is given by

$$\frac{d\psi}{dt} = \left(-\frac{\gamma}{2}+q\right)A_1 \exp\left(-\frac{\gamma}{2}+q\right)t$$

$$-\left(\frac{\gamma}{2}+q\right)A_2 \exp\left(-\frac{\gamma}{2}+q\right)t \tag{3.7}$$

These equations describe the behaviour of a heavily damped oscillator, as for example, a pendulum in a viscous medium such as a dense oil. As stated earlier, the constants A_1 and A_2 are determined from the initial conditions. Let us assume that the oscillator is at its equilibrium position) ($\psi = 0$) at time $t = 0$. At this instant it is given a kick so that it has a finite velocity, say, V_0 at this time, i.e. at $t = 0$.

$$\psi = 0$$

$$\frac{d\psi}{dt} = V_0.$$

Equations (3.6) and (3.7) then give (setting $t = 0$)

$$0 = A_1 + A_2$$

$$V_0 = \left(-\frac{\gamma}{2} + q\right)A_1 - \left(\frac{\gamma}{2} + q\right)A_2$$

giving $$A_1 = -A_2 = \frac{V_0}{2q}$$

Thus, under the above initial conditions, Eqs. (3.6) and (3.7) become

$$\psi = \frac{V_0}{2q}\, e^{-\gamma t/2}\,(e^{qt} - e^{-qt})$$

or $$\psi = \frac{V_0}{q}\, e^{-\gamma t/2}\sinh(qt) \tag{3.8}$$

and $$\frac{d\psi}{dt} = \frac{V_0}{2}\, e^{-\gamma t/2}\left\{(e^{qt} + e^{-qt}) - \frac{\gamma}{2q}(e^{qt} - e^{-qt})\right\}$$

or $$\frac{d\psi}{dt} = V_0\, e^{-\gamma t/2}\left\{\cosh(qt) - \frac{\gamma}{2q}\sinh(qt)\right\} \tag{3.9}$$

Figure 3.2 illustrates the behaviour of a heavily damped system when it is disturbed from equilibrium by a sudden impulse at $t = 0$. It is the displacement—time graph of Eq. (3.8). For small values of time t, the term $e^{-\gamma t/2}$ is very nearly unity, the displacement increases with time since $\sinh(qt)$ increases as t increases. Very soon, however, the term $e^{-\gamma t/2}$ starts contributing and the displacement decays exponentially with time, eventually becoming zero. The turning point occurs at a time $t = t_0$ when $d\psi/dt = 0$. Equation (3.9) tells us that this happens at a time $t = t_0$ satisfying

$$\tanh(qt_0) = \frac{2q}{\gamma}$$

Thus, the displacement increases until time $t = t_0$, after which it slowly returns to zero. Since, displacement ψ never becomes negative, there is no oscillation at all. Such a motion is called *dead beat*. We come across such a motion in the case of a dead beat galvanometer (see Sec. 3.6).

Case II : $\gamma \simeq 2\,\omega_0$ *(Critical Damping)*. This is a special case of a heavily damped motion. Using the notation $q = (\gamma^2/4 - \omega_0^2)^{1/2}$ of case I, we see that, in this case, $q = 0$ and Eq. (3.6) becomes

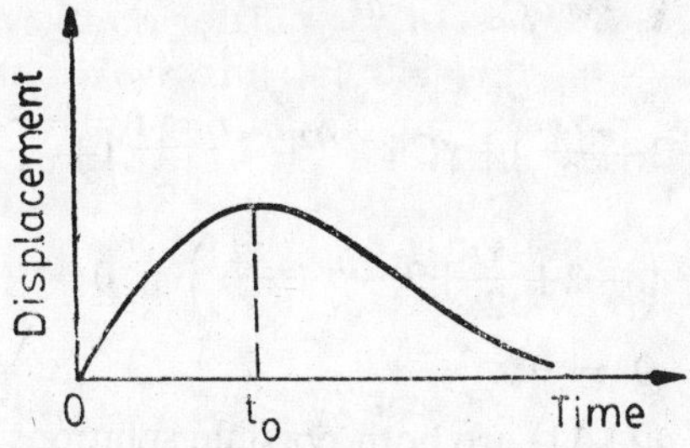

Fig. 3.2 Non-oscillatory behaviour of a heavily damped oscillator

$$\psi = (A_1 + A_2)\, e^{-\gamma t/2}$$

or
$$\psi = B\, e^{-\gamma t/2} \tag{3.10}$$

where $B = A_1 + A_2$, is a constant. In other words, Eq. (3.10) is the solution of Eq (3.2) for $\gamma = 2\,\omega_0$. In this case, the two roots α_1 and α_2 become identical. Notice that the solution (3.10) contains only one adjustable constant B. This solution is only a partial solution, since the solution of any second—order differential equation must contain two adjustable constants. This can be understood as follows. If Eq. (3.10) were a complete solution of Eq. (3.2), then the velocity of the oscillator would be given by

$$\frac{d\psi}{dt} = -B\frac{\gamma}{2}\, e^{-\gamma t/2}$$

When the system is disturbed from equilibrium ($\psi = 0$) by giving an impulse (i.e. by imparting a velocity V_0) at $t = 0$, we have, from the above two equations

$$B = 0$$

$$V_0 = -\frac{\gamma}{2} B$$

implying, thereby, that V_0 is also zero, which is not our initial condition. Hence our trial solution yields only a partial solution in the case when $q = 0$.

We can verify that a second solution is represented by the trial solution

$$\psi = Ct\, e^{-\gamma t/2} \tag{3.11}$$

giving
$$\frac{d\psi}{dt} = C\, e^{-\gamma t/2}\left(1 - \frac{\gamma t}{2}\right)$$

and
$$\frac{d^2\psi}{dt^2} = C\frac{\gamma}{2}\, e^{-\gamma t/2}\left(-2 + \frac{\gamma t}{2}\right)$$

Substituting for ψ, $\frac{d\psi}{dt}$ and $\frac{d^2\psi}{dt^2}$ in Eq. (3.2) with ω_0^2 replaced by $\frac{\gamma^2}{4}$ i.e.

$$\frac{d^2\psi}{dt^2}+\gamma\frac{d\psi}{dt}+\frac{\gamma^2}{4}\psi=0$$

We have,

$$C\frac{\gamma}{2}e^{-\gamma t/2}\left(-2+\frac{\gamma t}{2}\right)+\gamma C\,e^{-\gamma t/2}\left(1-\frac{\gamma t}{2}\right)+\frac{\gamma^2}{4}Ct\,e^{-\gamma t/2}=0$$

or $$\frac{\gamma}{2}C\,e^{-\gamma t/2}\left(-2+\frac{\gamma t}{2}+2-\gamma t+\frac{\gamma t}{2}\right)=0$$

or $$0=0$$

Thus, Eqs. (3.10) and (3.11) are both possible solutions of Eq. (3.2) in the special case when $\gamma = 2\,\omega_0$. From the superposition principle, the general solution is given by

$$\psi = B\,e^{-\gamma t/2}+Ct\,e^{-\gamma t/2}=(B+Ct)\,e^{-\gamma t/2} \tag{3.12}$$

and $$\frac{d\psi}{dt}=\left\{C-\frac{\gamma}{2}(B+Ct)\right\}e^{-\gamma t/2}$$

The constants B and C can be determined from the initial conditions. If at $t = 0$, $\psi = 0$ and $\frac{d\psi}{dt} = V_0$, we have, from the above equations,

$$B = 0$$

$$C = V_0$$

Thus, under these initial conditions, the displacement ψ in Eq, (3.12) is given by

$$\psi = V_0\,t\,e^{-\gamma t/2} \tag{3.13}$$

and $$\frac{d\psi}{dt}=C\left(1-\frac{\gamma t}{2}\right)e^{-\gamma t/2} \tag{3.14}$$

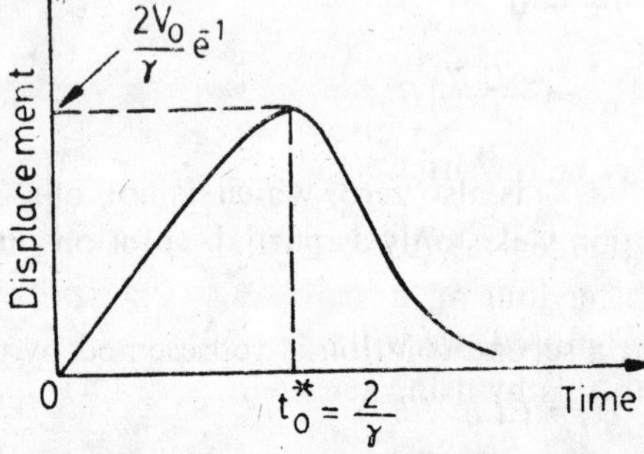

Fig. 3.3 Displacement-time behaviour of a weekly damped oscillaton

Figure 3.3 is a graph of ψ against t in Eq. (3.13). It illustrates the displacement—time behaviour of a damped system with $\gamma = 2\omega_0$, when it is disturbed from equilibrium by a sudden impulse For small values of t, the term $e^{-\gamma t/2}$ is very nearly unity and displacement [Eq. (3.13)] increases linearly with time t. After sometime, $e^{-\gamma t/2}$ starts changing and the displacement decays exponentially with time, eventually becoming zero. The turning point occurs at a time t_0, when $\frac{d\psi}{dt} = 0$. From Eq. (3.14) this happens at $t = t_0$ given by

$$1 - \frac{\gamma t_0}{2} = 0$$

or
$$t_0 = \frac{2}{\gamma} = \frac{1}{\omega_0}$$

The displacement increases until time $t = t_0$, after which it decays to zero. A comparison of Eqs. (3.8) and (3.13) reveals that the decay rate is much faster when $\gamma = 2\,\omega_0$ than when $\gamma > 2\,\omega_0$. In both cases, there is no oscillation at all, since ψ never becomes negative.

The motion described by Eq. (3.13) is called *critically damped*. The necessary condition for critical damping is $\gamma = 2\,\omega_0$. Suppose we are faced with a problem in which we desire a high rate of decay without oscillation. Evidently, the optimum choice is critical damping. We come across such a problem in pointer-type galvanometers, where we would want the pointer to move immediately to the correct position and stay there without annoying oscillation (see Sec. 3.6).

Case III $\gamma < 2\omega_0$ *(Small Damping)*. When $\gamma < 2\omega_0$, the damping is small and this gives the most important kind of behaviour, namely, *oscillatory damped harmonic motion*, for then, the expression $\left(\frac{\gamma^2}{4} - \omega_0^2\right)^{1/2}$ in the exponentials in Eq. (3.5) is an imaginary quantity. Writing this as

$$\left(\frac{\gamma^2}{4} - \omega_0^2\right)^{1/2} = \sqrt{-1}\left(\omega_0^2 - \frac{\gamma^2}{4}\right)$$

$$= i\omega^*$$

where, $\omega^* = \left(\omega_0{}^2 - \frac{\gamma^2}{4}\right)^{1/2}$ is a real positive quantity, the displacement Eq. (3.5) may be rewritten as

$$\psi = e^{-\gamma t/2}\{(A_1 \exp(i\omega^* t) + A_2 \exp(-i\omega^* t)\} \tag{3.15}$$

To compare the behaviour of a damped oscillator with the ideal case in which damping is ignored, we will recast Eq. (3.15) into a more familiar form. We can do this by using the identities,

$$e^{i\theta} = \cos\theta + i\sin\theta$$
$$e^{-i\theta} = \cos\theta - i\sin\theta$$

so that Eq. (3.15) can be written as

$$\psi = e^{-\gamma t/2}\{(A_1 + A_2)\cos\omega^* t + i(A_1 - A_2)\sin\omega^* t\}$$

If we choose

$$A_1 + A_2 = A \cos \delta$$

$$i\,(A_1 - A_2) = A \sin \delta$$

where A and δ are constants which depend upon the initial conditions, we find, after substitution,

$$\psi = A\, e^{-\gamma t/2} \cos(\omega^* t - \delta) \qquad (3.16)$$

with $$\omega^* = \omega_0 \left(1 - \frac{\gamma^2}{4\omega_0^2} \right)^{1/2} \qquad (3.17)$$

Differentiating Eq. (3.16), we obtain the expression for the velocity of the oscillator, which reads

$$\frac{d\psi}{dt} = -A\, e^{-\gamma t/2} \left\{ \omega^* \sin(\omega^* t - \delta) + \frac{\gamma}{2} \cos(\omega^* t - \delta) \right\} \qquad (3.18)$$

Equation (3.16) shows that the motion is oscillatory. The oscillation is not simple harmonic, since its, 'amplitude' $A\, e^{-\gamma t/2}$ is not constant but decreases with time. The motion is not even periodic, since it never repeats itself, each swing being of smaller amplitude than the preceding one. However, if γ is very small compared to ω_0, the amplitude will remain sensibly constant over a large number of cycles of the harmonic term $\cos(\omega^* t - \delta)$ in which case, the motion is nearly periodic and simple harmonic. The angular frequency of the oscillation is ω^* given by Eq. (3.17) which is less than the natural angular frequency of free undamped oscillations. Strictly speaking, we are really not justified in using the terms 'amplitude' and 'frequency' for a motion which is not periodic. But, when damping is small, the motion is nearly periodic, we may use these terms with some reservations.

To illustrate the behaviour of a weakly damped oscillator, let us choose the initial conditions, namely, that at $t = 0$, $\psi = 0$ and $\frac{d\psi}{dt} = V_0$. Using these conditions in Eqs. (3.16) and (3.18) we get

$$0 = A \cos \delta$$

and $$V_0 = -A \left(\frac{\gamma}{2} \cos \delta - \omega^* \sin \delta \right)$$

yielding $$\delta = \frac{\pi}{2} \qquad (A = 0\,;\ \text{being a trivial case})$$

and $$A = \frac{V_0}{\omega^*}$$

Using these values of A and δ in Eqs (3.16) and (3.18) we find that, under the above initial conditions, the displacement and velocity of the oscillator are, respectively, given by

$$\psi = \frac{V_0}{\omega^*} e^{-\gamma t/2} \sin \omega^* t = A(t) \sin \omega^* t \tag{3.19}$$

with $A(t) = \frac{V_0}{\omega^*} e^{-\gamma t/2} = A_0\, e^{-\gamma t/2}$

A_0 being the value of $A(t)$ when $\gamma = 0$

and $$\frac{d\psi}{dt} = V_0\, e^{-\gamma t/2} \left(\cos \omega^* t - \frac{\gamma}{2\omega^*} \sin \omega^* t \right) \tag{3.20}$$

Figure 3.4 depicts the behaviour of a weakly damped oscillator. It is a graph of ψ against t of the motion described by Eq. (3.19). The constant A_0 is the value of $A(t) = \frac{V_0}{\omega^*} e^{-\gamma t/2}$ in the absense of damping ($\gamma = 0$), i.e. $A_0 = V_0/\omega^*$. Since the maximum values of sin $(\omega^* t)$ are $+1$ and -1 alternately, the displacement-time graph of oscillation is bounded by the dotted curves $A_0 e^{-\gamma t/2}$ and $-A_0 e^{-\gamma t/2}$.

Thus, although, the amplitude decreases exponentially with time, the weekly damped oscillator executes some sort of oscillatory motion. The motion does not repeat itself and is, therefore, not periodic in the usual sense of the term. However, it still has a time period $T^* = 2\pi/\omega^*$, which is the time interval between two alternate zeros of displacement. The time period between two successive zeros of displacement is $T^*/2$. This is also the time interval between a maximum and the next minimum value of the displacement, but the maxima and minima are not exactly halfway between the zeros. Thfs is obvious from Eq. (3.20), because at a maximum or a minimum of displacement, the velocity is zero, giving

$$\cos \omega^* t - \frac{\gamma}{2\omega^*} \sin \omega^* t = 0$$

or $$\tan \omega^* t = \frac{2\omega^*}{\gamma}$$

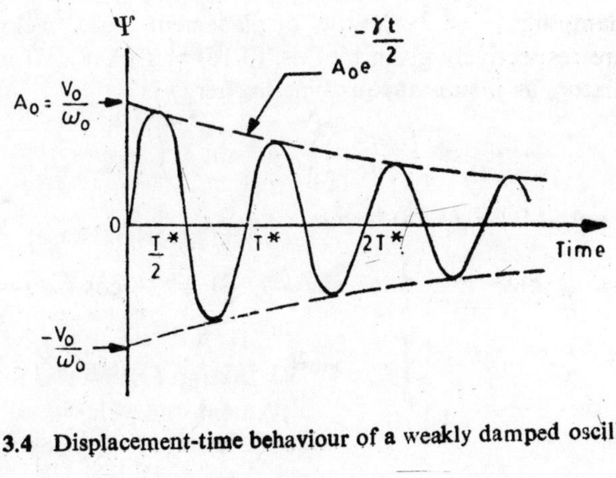

Fig. 3.4 Displacement-time behaviour of a weakly damped oscillator

The values of t satisfying this equation are the instants at which ψ is either a positive maximum or a negative maximum. In the case when $\gamma \ll 2\omega_0$, $\frac{2\omega_0}{\gamma} \to \infty$, so that

$$\omega^* t \to \frac{\pi}{2}, \frac{3\pi}{2}, \frac{5\pi}{2}, \ldots,$$

The first maximum of ψ occurs at a time $t = t_1$ given by

$$\omega^* t_1 = \frac{\pi}{2}$$

or

$$t_1 = \frac{\pi}{2\omega^*} = \frac{T^*}{4}$$

i.e. the maximum is exactly midway between the two zeros of ψ. Thus, only in the case of negligibly small damping, are the maxima and minima halfway between the zeros of displacement as in the case of simple harmonic motion.

Effect Of Damping. The effect of damping is two-fold: (a) The amplitude of oscillation decreases exponentially with time as

$$A(t) = A_0 e^{-\gamma t/2}$$

where A_0 is the amplitude in the absence of damping and (b) the angular frequency ω^* of the damped oscillator is less than ω_0, the frequency of the undamped oscillation. The relation between them is

$$\omega^* = \omega_0 \left(1 - \frac{\gamma^2}{4\omega_0^2}\right)^{1/2}$$

3.4 ENERGY OF A WEAKLY DAMPED OSCILLATOR

We shall now develop an expression for the average energy of a weakly damped oscillator at any instant of time. We have seen that, in the case of weak damping ($\gamma < 2\omega_0$), the displacement and velocity of the oscillator are respectively given by Eqs. (3.16) and (3.18). If m is the mass of the oscillator, its instantaneous kinetic energy is

$$\frac{1}{2} m \left(\frac{d\psi}{dt}\right)^2$$

which with help of Eq. (3.18) becomes

$$\text{KE} = \frac{1}{2} mA^2 e^{-\gamma t} \left\{ \omega^* \sin(\omega^* t - \delta) + \frac{\gamma}{2} \cos(\omega^* t - \delta) \right\}^2$$

$$= \frac{1}{2} mA^2 e^{-\gamma t} \left\{ \omega^{*2} \sin^2(\omega^* t - \delta) + \omega^* \gamma \sin(\omega^* t - \delta) \cos(\omega^* t - \delta) + \frac{\gamma^2}{4} \cos^2(\omega^* t - \delta) \right\}$$

The instantaneous potential energy of the oscillator is given by

$$\mathrm{PE} = \int_0^{\psi} K\psi\, d\psi = \frac{1}{2} K\psi^2$$

Using Eq. (3.16) we have, since $K = m\omega_0^2$,

$$\mathrm{PE} = \frac{1}{2} m\omega_0^2 A^2 e^{-\gamma t} \cos^2(\omega^* t - \delta)$$

The total energy of the oscillator at any instant of time is then given by

$$\begin{aligned} E(t) &= \mathrm{KE} + \mathrm{PE} \\ &= \frac{1}{2} m A^2 e^{-\gamma t} \left\{ \omega^{*2} \sin^2(\omega^* t - \delta) + \frac{\omega^* \gamma}{2} \sin 2(\omega^* t - \delta) \right. \\ &\quad \left. + \left(\frac{\gamma^2}{4} + \omega_0^2 \right) \cos^2(\omega^* t - \delta) \right\} \end{aligned} \tag{3.21}$$

If damping is very small ($\gamma \ll 2\omega_0$), as is usually the case, the term $e^{-\gamma t}$ in Eq. (3.21) does not change appreciably during one time period $T^* = 2\pi/\omega^*$ of the oscillation. Thus, assuming that $e^{-\gamma t}$ is sensibly constant during period T^* of the oscillations, the time-averaged energy of the oscillator is given by

$$\begin{aligned} \langle E(t) \rangle &= \frac{1}{2} m A^2 e^{-\gamma t} \left\{ \omega^{*2} \langle \sin^2(\omega^* t - \delta) \rangle \right. \\ &\quad + \frac{\omega^* \gamma}{2} \langle \sin 2(\omega^* t - \delta) \rangle \\ &\quad \left. + \left(\frac{\gamma^2}{4} + \omega_0^2 \right) \langle \cos^2(\omega^* t - \delta) \rangle \right\} \end{aligned} \tag{3.22}$$

where notation $\langle \quad \rangle$ implies averaging over one time period T^*. A function $f(t)$ averaged over T, is by definition, given by

$$\langle f(t) \rangle = \frac{\int_0^T f(t)\,dt}{\int_0^T dt} = \frac{1}{T} \int_0^T f(t)\,dt$$

Thus

$$\langle \sin^2(\omega^* t - \delta) \rangle = \frac{1}{T^*} \int_0^{T^*} \sin^2\left(\frac{2\pi t}{T^*} - \delta \right)^2 dt$$

To integrate, let us use the transformation

$$\frac{2\pi t}{T^*} - \delta = \alpha$$

so that $$dt = \frac{T^*}{2\pi} d\alpha.$$

Then

$$\langle \sin^2 (\omega^* t - \delta) \rangle = \frac{1}{2\pi} \int_{-\delta}^{2\pi-\delta} \sin^2 \alpha \, d\alpha = \frac{1}{4\pi} \int_0^{2\pi} (1 - \cos 2\alpha) \, d\alpha = \frac{1}{2}$$

Similarly

$$\langle \cos^2 (\omega^* t - \delta) \rangle = \frac{1}{2}$$

and $$\langle \sin 2 (\omega^* t - \delta) \rangle = 0$$

Substituting for these time-averaged values in Eq. (3.22), we get

$$\langle E(t) \rangle = \frac{1}{4} m A^2 e^{-\gamma t} \left(\omega^{*2} + \frac{\gamma^2}{4} + \omega_0^2 \right)$$

Now, since $\omega^{*2} = \omega_0^2 - \frac{\gamma^2}{4}$, we have

$$\langle E(t) \rangle = \frac{1}{2} m A^2 \omega_0^2 e^{-\gamma t}$$

or $$\langle E(t) \rangle = E_0 e^{-\gamma t} \tag{3.23}$$

where $E_0 = \frac{1}{2} m A^2 \omega_0^2$, is the total energy of an undamped oscillator (see Ch. 1). Hence the energy of a weakly damped oscillator diminishes exponentially with time. The decay of the total energy is illustrated in Fig. 3.5.

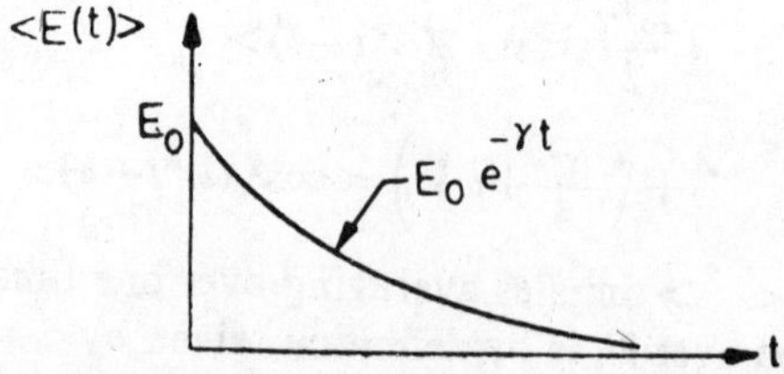

Fig. 3.5 Exponential decay of total energy during damping of harmonic oscillations

The average power dissipation during one time period is given by

$$\begin{aligned} \langle P(t) \rangle &= \text{rate of loss of energy} \\ &= -\frac{d}{dt} \langle E(t) \rangle \\ &= \gamma E_0 e^{-\gamma t} \\ &= \gamma \langle E(t) \rangle \end{aligned}$$

This expression may also be obtained as follows : Since the loss of energy is due to the work done by the oscillator to overcome the force of friction $F = -p \frac{d\psi}{dt}$, the instantaneous power $P(t)$ is given by

$$P(t) = \frac{\text{work}}{\text{time}} = \frac{F \cdot \delta\psi}{\delta t} = F \frac{d\psi}{dt}$$

where $\delta\psi$ is the change in displacement in time δt. Thus

$$P(t) = p\left(\frac{d\psi}{dt}\right)^2 = m\gamma\left(\frac{d\psi}{dt}\right)^2 \tag{3.24}$$

Now using Eq. (3.18) we have

$$P(t) = m\gamma A^2 e^{-\gamma t}\left\{\omega^{*2}\sin^2(\omega^* t-\delta) + \frac{\omega^*\gamma}{2}\sin 2(\omega^* t-\delta) + \frac{\gamma^2}{4}\cos^2(\omega^* t-\delta)\right\}$$

Hence, the power dissipation during one time period of oscillation is given by

$$\begin{aligned}\langle P(t)\rangle &= m\gamma A^2 e^{-\gamma t}\left\{\omega^{*2}\langle\sin^2(\omega^* t-\delta)\rangle + \frac{\omega^*\gamma}{2}\langle\sin 2(\omega^* t-\delta)\rangle + \frac{\gamma^2}{4}\langle\cos^2(\omega^* t-\delta)\rangle\right\}\\ &= \tfrac{1}{2}\gamma m A^2 e^{-\gamma t}\left(\omega^{*2}+\frac{\gamma^2}{4}\right)\\ &= \tfrac{1}{2}\gamma m A^2\omega_0^2 e^{-\gamma t}\\ &= \gamma\langle E(t)\rangle\end{aligned} \tag{3.25}$$

As mentioned earlier, this loss of energy is due to the friction in the system (leading to heating) and the emission of radiation from the system (resulting in waves).

3.5 METHODS OF DESCRIBING THE DAMPING OF AN OSCILLATOR

From the foregoing analysis, it is clear that the damped oscillator is characterized by two parameters. ω_0 and γ. The constant ω_0 is the angular frequency of undamped oscillations and γ is the measure of daming present in the system. Quantities ω_0 and γ are of the same dimension (s^{-1}).

It may be recalled that the foregoing analysis of the effect of damping has been based on an assumption that the damping force is proportional to the velocity of the moving part of the system (i.e. $F = -p\dfrac{d\psi}{dt}$). In other words, the damping is assumed to be viscous. From Eq. (3.24) it is clear the rate of energy dissipation is proportional to the square of the velocity i.e. $P(t) \propto \left(\dfrac{d\psi}{dt}\right)^2$.

It may be pointed out that the exponential decay of energy as described by Eq. (3.23) may arise from different kinds of dissipative processes. For example, in an oscillatory electrical circuit (considered in the next section), the rate of energy dissipation in a resistor is proportional to the square of the current. This situation is, therefore, closely analogous to the mechanical oscillator with viscous damping. The exponential decay of energy has also been observed in some atomic and nuclear processes. The above viscous damping model explains the behaviour of these processes quite satisfactorily. Thus, the special case of a mechanical oscillator with viscous damping, gives us some useful information about the behaviour of diverse kinds of physical situations involving dissipation of energy due to damping.

For convenience in applying our results to diverse kinds of physical systems, three different methods of describing damping have been used. Each method is based on a parameter defined in terms of ω_0 and γ, the two constants that characterize a damped system.

Logarithmic Decrement

This method measures the rate at which the amplitude decreases with time. Consider a weakly damded oscillator initially at rest at its equilibrium position, when it is given an impulse. The future motion of the oscillator is described by Eq. (3.19).

$$\psi = A_0 e^{-\gamma t/2} \sin \omega^* t$$

$$= A(t) \sin \frac{2\pi t}{T^*}$$

where $A(t) = A_0 e^{-\gamma t/2}$, A_0 being the amplitude in the absence of damping ($\gamma = 0$) and T^* $\left(= \frac{2\pi}{\omega^*} \right)$ is the time period of the oscillations. Fig. 3.6 shows the displacement-time curve of this motion.

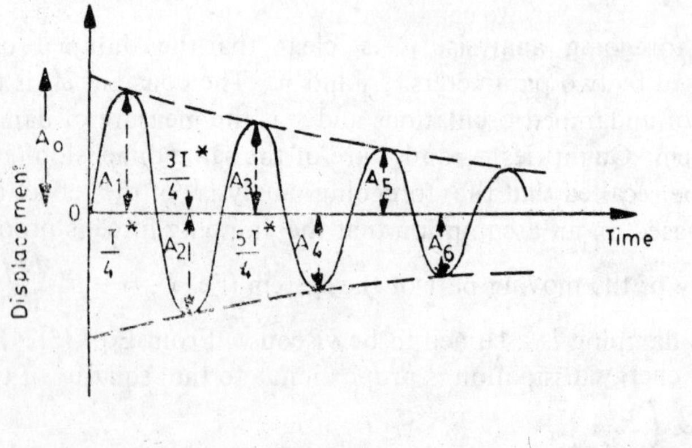

Fig. 3.6 Logarithmic decrement of a damped oscillator. Displacement ψ is zero at time $t = 0$, $T^*/2, T^*$,etc. and maximum (positive or negative) at $t = T^*/4$, $3T^*/4$, $5T^*/4$.

At time $t = \dfrac{T^*}{4}$, the displacement given by

$$A = A_0 \exp\left(-\frac{\gamma}{2} \cdot \frac{T^*}{4}\right) \sin\left(\frac{2\pi}{T^*} \frac{T^*}{4}\right)$$
$$= A_0 \exp(-\gamma T^*/8)$$

rises to its first maximum value, say, A_1. The amplitude A_1 is given by

$$A_1 = A_0 \exp(-\gamma T^*/8)$$

At time $t = \dfrac{3T^*}{4}$, the displacement becomes maximum again (this time it is $-$ve). The amplitude A_2 [see Fig. 3.6] is given by

$$A_2 = A_0 \exp(-3\gamma T^*/8)$$

The next maximum of ψ occurs at $t = 5T^*/4$, so that

$$A_3 = A_0 \exp(-5\gamma T^*/8)$$

and so on.

Notice that

$$\frac{A_1}{A_2} = \frac{A_2}{A_3} = \dots = \frac{A_{n-1}}{A_n} = \exp\left(\frac{\gamma}{2} \cdot \frac{T^*}{2}\right) = d \text{ (a constant)}$$

The constant d which is the ratio of two successive amplitudes of the damped oscillation, is called the *decrement* of the motion. The logarithm of decrement d is called *logarithmic decrement* of the motion and is usually denoted by the symbol λ.

$$\lambda = \log_e d = \log_e \exp\left(\frac{\gamma}{2} \cdot \frac{T^*}{2}\right) = \frac{\gamma}{2} \cdot \frac{T^*}{2}$$

Logarithmic decrement is the natural logarithm of the ratio of two successive amplitudes that are separated by half a period, $T^*/2$; the larger amplitude being the numerator.

It is clear that

$$\frac{A_1}{A_2} = d = e^{\lambda}$$
$$\frac{A_1}{A_3} = \frac{A_1}{A_2} \cdot \frac{A_2}{A_3} = d^2 = e^{2\lambda}$$
$$\vdots$$
$$\frac{A_1}{A_7} = e^{6\lambda}$$
$$\vdots$$
$$\frac{A_1}{A_n} = e^{(n-1)\lambda}$$

or $$\lambda = \frac{1}{(n-1)} \log_e \left(\frac{A_1}{A_n} \right)$$

Thus, logarithmic decrement, λ, of the motion can be measured by observing several successive amplitudes on both sides of the equilibrium position For example, if $n = 7$ (seven successive throws)

$$\lambda = \frac{1}{6} \log_e \left(\frac{A_1}{A_7} \right) = \frac{2.303}{6} \log_{10} \left(\frac{A_1}{A_7} \right)$$

This method of describing damping by the parameter, λ, is used in applying the necessary correction to the first deflection or the first throw θ_1 of a ballistic galvanometer, when a certain quantity or charge is passed through its coil. The relation between the correct throw θ_0 (i.e. the throw if damping were absent) and the first throw θ_1, is,

$$\theta_1 = \theta_0 \exp(-\gamma T^*/8)$$

where T^* is the period of oscillation of the coil. Thus,

$$\theta_0 = \theta_1 \exp(\gamma T^*/8) = \theta_1 e^{\lambda/2}$$

If $\lambda \ll 1$ (as is usually the case), we have, since $e^{\lambda/2} \simeq 1 + \frac{\lambda}{2}$,

$$\theta_0 = \theta_1 \left(1 + \frac{\lambda}{2} \right)$$

Thus, knowing the logarithmic decrement (λ) for a given galvanometer, we can easily correct the first throw (θ_1) for damping.

Relaxation Time

Another method of expressing the damping effect on the motion is in terms of the time taken by the amplitude to decrease by a factor of $(1/e)$ of its original value ; $e = 2.718$.

This time is called the *relaxation time* or *modulus of decay* and is usually denoted by the symbol τ. From the expression

$$A(t) = A_0 e^{-\gamma t/2}$$

It is clear that the amplitude $A(t)$ decreases to

$$e^{-1} = \frac{1}{2.718} = 0.368$$

of its original value A_0 in time $t = \tau$ given by

$$\frac{\gamma \tau}{2} = 1$$

or $$\tau = \frac{2}{\gamma}$$

The relaxation time is a measure of how rapidly the motion is damped out by friction. The higher the value of γ, the shorter is the relaxation time. In many problems in physics involving decay of energy, the magnitude of the dissipative process is estimated by measuring the relaxation time.

The relaxation time and the logarithmic decrement of the motion are obviously related to each other. They are related as

$$\lambda = \frac{\gamma}{2} \cdot \frac{T^*}{2} = \frac{T^*}{2\tau}$$

i.e. Logarithmic decrement is the ratio between half the period of oscillations and the relaxation time. Thus λ is a measure of the fraction of the decrease in amplitude which occurs in one-half cycle

Quality Factor or Q-Value

The third method of expressing damping in an oscillatory system measures the rate at which energy decays. In this method, we define a parameter called the *quality factor* or Q-value of the system as

$$Q = \frac{\omega^*}{\gamma}$$

where $\omega^* = \omega_0 \left(1 - \frac{\gamma^2}{4\omega_0^2}\right)^{1/2}$, is the angular frequency of the damped oscillation. That the Q-value of the system measures the rate of decay of energy, can be understood as follows :

We have seen [see Eq. (3.23)] that the average energy stored in the oscillator is $\langle E(t) \rangle$ and that the average rate of loss of enerpy is [see Eq. (3.25)]

$$\frac{d}{dt} \langle E(t) \rangle = \langle P(t) \rangle = \gamma \langle E(t) \rangle$$

The average energy dissipated in time period T^* $(= 2\pi/\omega^*)$ is

$$\gamma T^* \langle E(t) \rangle = \frac{2\pi\gamma}{\omega^*} \langle E(t) \rangle$$

$$= \frac{2\pi}{Q} \langle E(t) \rangle = \frac{2\pi}{Q} \times (\text{average stored energy})$$

or

$$Q = 2\pi \times \frac{\text{average energy stored in one period}}{\text{average energy lost in one period}}$$

Thus, the quality factor of a damped harmonic oscillator may be defined as *2π times the ratio between average energy stored and average energy lost per period.*

We shall mostly be concerned with situations in which $\gamma \ll 2\omega_0$. For such situations

$$\omega^* \simeq \omega_0$$

Thus, to a very close approximation, we may write

$$Q = \frac{\omega_0}{\gamma}$$

which is a constant of the damped system. Q is a pure number ; large compared to unity for oscillating systems with small rates of dissipation of energy. The lower the damping, the higher the value of Q. As $\gamma \to 0$, $Q \to \infty$. In such cases the motion of the oscillator [Eq (3.16)] is given very nearly by

$$\psi = A \exp\left(-\frac{\omega_0 t}{2Q}\right) \cos(\omega_0 t - \delta) \tag{3.26}$$

The average energy of the oscillator is given by [see Eq. (3.23)]

$$\langle E(t) \rangle = E_0 \exp\left(-\frac{\omega_0 t}{Q}\right) \tag{3.27}$$

It may be noted that Q is closely related to the number of oscillations (or cycles) over which the energy fall to $\left(\frac{1}{e}\right)$ of its original value E_0. From Eq. (3.27), this happens in time $t = \Gamma$ where Γ is given by

$$\frac{\omega_0 \Gamma}{Q} = 1$$

or

$$\Gamma = \frac{Q}{\omega_0} = \frac{T_0 Q}{2\pi}$$

where T_0 is the period of oscillations. From Eq. (3.26) it is clear that during this time Γ, the number n of complete oscillations executed is given by

$$n = \frac{\omega_0}{2\pi} \Gamma = \frac{Q}{2\pi}$$

Thus, the average energy falls to $\left(\frac{1}{e}\right)$ of its original value in $Q/2\pi$ cycles of free oscillations. The energy decay time Γ and amplitude decay time τ are related as

$$\Gamma = \frac{\tau}{2}$$

Thus Γ is the *mean* decay time of free damped oscillations.

In terms of ω_0 and Q, Eq. (3.2), for a damped system, can be rewritten in the form

$$\frac{d^2\psi}{dt^2} + \frac{\omega_0}{Q} \frac{d\psi}{dt} + \omega_0^2 \psi = 0 \tag{3.28}$$

This form is, in many cases, a very convenient one in analysing the effect of damping in a variety of physical systems, both mechanical and non-mechanical.

Relation between the logarithmic decrement, relaxation time and quality factor

In the case $\gamma \ll 2\omega_0$, the three parameters are given by

$$\lambda = \frac{\gamma T_0}{4}$$

$$\tau = \frac{2}{\gamma}$$

$$Q = \frac{\omega_0}{\gamma}$$

In terms of Q, the parameters λ and τ are given by

$$\lambda = \frac{\pi}{2Q}$$

$$\tau = \frac{2Q}{\omega_0}$$

The smaller the damping, the larger are Q and τ, indicating that it takes a longer time for the oscillations to damp out. Also the smaller the damping, the smaller is λ, indicating that the reduction in the amplitude in a half-cycle is smaller. It may be remarked that these conclusions are independent of the way the system is set into motion (i e. initial conditions).

3.6 IMPORTANT EXAMPLES OF DAMPED HARMONIC OSCILLATORS

Moving Coil Galvanometer

A galvanometer consists of a rectangular frame containing a number of turns of fine insulated coil of copper wire suspended (or pivoted) between the poles of a permanent horse-shoe magnet. When an impulse is given to the coil, it deflects. The *deflecting couple* is opposed by a *restoring couple*, $C\theta$, where θ is the deflection of the coil and C (in the case of a suspended coil) is the torsional couple per unit twist in the suspension fibre and is given by

$$C = \frac{\pi \eta r^4}{2l}$$

where l and r are respectively the length and radius of the suspension fibre and η is the modulus of rigidity of its material.

There are two kinds of damping processes, namely, *mechanical* damping *electromagnetic* damping. The former is due to the friction of the air and the latter is due to the induced current set up in the coil as it moves in a magnetic field, thus producing a field which opposes the motion of the coil. Both these dampings are proportional to the angular velocity $\frac{d\theta}{dt}$ of the coil. The mechanical damping is $-p\frac{d\theta}{dt}$, where p is the coefficient of friction and the eletcromagnetic damping is $-q\frac{d\theta}{dt}$, with

$$q = \frac{n^2A^2H^2}{R}$$

where n is number of turns in the coil, A its face area, H the strength of the magnetic field and R is the total resistance in the circuit.

If I is the moment of inertia of the coil about the axis of suspension, the equation of motion of the coil is

$$I\frac{d^2\theta}{dt^2} = -C\,\theta - p\frac{d\theta}{dt} - q\frac{d\theta}{dt}$$

or

$$\frac{d^2\theta}{dt^2} + \gamma\frac{d\theta}{dt} + \omega_0^2\theta = 0$$

where $\gamma = (p+q)/I$

and $\omega_0^2 = C/I$

This equation is the same as that of a damped oscillator [Eq. (3.2)], except that the displacement is angular instead of a linear one.

Dead Beat Galvanometer

In the case of large damping $\gamma > 2\omega_0$ or $(p+q)^2 > 4CI$, the motion of the coil is non-oscillatory or *dead beat*. The parameter p that measures air resistance is usually smaller than q that measures electromagnetic damping. The values of these parameters can be controlled. For a given coil in a given magnetic field, q is inversely proportional to R. The higher the resistance in series with the galvanometer, the smaller is the damping. In a dead beat galvanometer R is very small, at most equal to the resistance of the coil. When a current is passed in the coil, it slowly deflects to a new equilibrium position without any oscillation.

Critically Damped Galvanometer

In a critically damped galvanometer $\gamma = 2\omega_0$ or

$$(p+q) = 2\sqrt{CI}$$

This condition is realized by introducing a resistance R which is necessary to just stop the oscillations of the coil. This very special condition corresponds to what is called critical damping and R is called the *critical damping resistance*. Under conditions of critical damping, the coil smoothly and quickly approaches its new equilibrium position when a steady current is passed through it. Such a behaviour is highly advantageous in electrical meters (ammeters, voltmeters, etc.) where one would like to take a steady reading immediately after the meter is connected to the circuit.

Ballistic or Under-damped Galvanometer

Sometimes, instead of measuring a steady current, it is necessary to measure a transient current (or charge). The current produced in the coil, while the charge is passing, flows only for a short time. The time period of oscillations of the coil must be greater than the time for which the charge flows in the coil. In other words, we should make the moving system such that the current, due to flow of charge, will have ceased before the moving part has had time to move appreciably from its position of rest and it is deflected only after the charge has passed through the coil. A galvanometer having a long period of mechanical oscillations of its coil will meet this requirement. Such a galvanometer is called *ballistic*. Since the time period T is given by

$$T = 2\pi \sqrt{\frac{I}{C}}$$

a ballistic galvanometer should heve a large moment of inertia, I. If I is large, $\gamma = (p+q)/I$ is small. If $\gamma < 2\omega_0$, the motion of the coil is oscillatory. Thus, for ballistic work, a galvanometer with long time period and low damping is preferred. The damping can be reduced further by using a resistance R much larger than the critical damping resistance. The logarithmic decrement of the galvanometer is measured by observing the successive throws, and the first throw is corrected for damping, as explained in Sec 3.5.

The LCR Circuit

The electrical circuit illustrated in Fig. 3.7, involving inductance L, capacitance C and a resistance R, is an exellcent example of damped harmonic oscillations. When $R = 0$, the oscillations of the circuit are undamped with angular frequency $\omega_0 = 1/\sqrt{LC}$ (see Chap. 1). We shall see that resistance R plays the part of a resistive or dissipative force analogous to that of friction or viscosity in the case of mechanical oscillations.

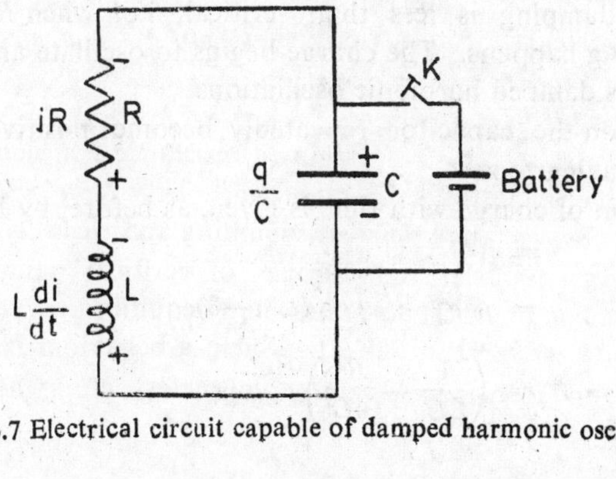

Fig. 3.7 Electrical circuit capable of damped harmonic oscillations

On pressing key K, the capacitor gets charged by the battery. When the key is released, the battery is thrown out of the circuit and the capacitor begins to discharge through the inductance and the resistance. Suppose that, at a given instant of time t, the charge on the capacitor is q and the current in the circuit is i, so that $V = q/C$ is the voltage across the capacitor plates at time t. The induced *emf* across the inductane is $L\frac{di}{dt}$ and the potential difference across the resistance is iR. Since there is no external emf (battery being out of the circuit) we have, from Kirchhoff's law,

$$\frac{q}{C} = -L\frac{di}{dt} - iR$$

or
$$L\frac{di}{dt} + iR + \frac{q}{C} = 0$$

Now, since, $i = dq/dt$, we have

$$L\frac{d^2q}{dt^2} + R\frac{dq}{dt} + \frac{q}{C} = 0$$

or
$$\frac{d^2q}{dt^2} + \gamma\frac{dq}{dt} + \omega_0^2 = 0$$

where $\gamma = R/L$ and $\omega_0^2 = 1/LC$

This equation is of the same form as that of a damped mechanical oscillator (Eq. 3.2) except that the variable ψ is the charge q in this case. It is interesting to note that the inductance is analogous to mass m, the resistance to the viscous damping factor p and the inverse of the capacitance to the stiffness constant K.

When the damping is large, i e. when $\gamma > 2\omega_0$ or $R > 2\sqrt{\frac{L}{C}}$ the charge decays gradually until the capacitor is discharged. The discharge is non-oscillatory or dead beat. When the damping is critical, i.e. when $R = 2\sqrt{L/C}$, the discharge becomes just non-oscillatory and dies out quickly.

When the damping is less than critical, i.e, when $R < 2\sqrt{L/C}$, an interesting thing happens. The charge begins to oscillate and the electrical system exhibits damped harmonic oscillations.

The charge on the capacitor repeatedly becomes positive and negative, eventually decaying to zero.

The variation of charge with time is given, as before, by Eq.(3.16) which now reads

$$q = q_0\, e^{-Rt/2L} \cos(\omega^* t - \delta)$$

with
$$\omega^* = \left(\frac{1}{LC} - \frac{R^2}{4L^2}\right)^{1/2}$$

This expression gives the angular frequency of oscillations. Since $\omega_0 = 1/\sqrt{LC}$ is the natural frequency of undamped oscillations, it is evident that the damping in this case is due to a finite value of resistance R, the energy dissipated appearing as heat.

The quality factor Q of the circuit is

$$Q = \frac{\omega^*}{\gamma} = \frac{L\omega^*}{R}$$

The Q-value is larger (i.e, damping is smaller) if R is smaller. If R is small $\omega^* \simeq \omega_0$, so that

$$Q = \frac{L\omega_0}{R} = \frac{1}{R}\sqrt{\frac{L}{C}}$$

For a purely inductive circuit, i.e. when $R = 0$, the quality factor Q will by infinite.

SOLVED EXAMPLES

Example 3.1 A massless spring, suspended from a rigid support, carries a flat disc of mass 100g at its lower end. It is observed that the system oscillates with a frequency of 10 Hz and the amplitude of the damped oscillations reduces to half its undamped value in one minute. Calculate (a) The resistive force constants (b) the relaxation time of the system, (c) its quality factor and (d) the force constant of the spring.

Solution :

(a) The amplitude of the damped oscillator at an instant t is given by

$$A = A_0 e^{-\gamma t/2}$$

Since $\dfrac{A}{A_0} = \frac{1}{2}$ when $t = 1$ minute $= 60$ s, we have,

$$e^{-\gamma/2\times 60} = \tfrac{1}{2}$$

or $$e^{+30\gamma} = 2$$

or $$30\,\gamma = \log_e 2$$

or $$\gamma = \frac{2.303\times 0.301}{30} = 0.023\ \text{s}^{-1}$$

The resistive force constant p is related to γ as

$p/m = \gamma$, where m = mass of the oscillator $= 100\ \text{g} = 0.1\ \text{kg}$

Thus $$p = m\gamma = 0.1\times 0.023 = 0.0023\ \text{Nsm}^{-1}$$

(b) The relaxation time τ of the system is

$$\tau = \frac{2}{\gamma} = 86.96 \text{ s}$$

(c) The quality factor of the system is

$$Q = \frac{\omega^*}{\gamma}$$

where $\omega^* = 2\pi\nu^* = 2\pi \times 10 = 20\pi$ rad s^{-1} ; ν^*, being the frequency of the damped oscillations, which is given to be 10 Hz. Hence

$$Q = \frac{20 \times 3.142}{0.023}$$

$$= 2732$$

(d) The force constant K of the spring is related to the frequency ω_0 of undamped oscillations, as

$$\omega_0 = \sqrt{\frac{K}{m}}$$

or $\qquad K = m\omega_0^2$

Now, ω_0^2 is determined from the relation

$$\omega^{*2} = \omega_0^2 - \frac{\gamma^2}{4} \text{ which gives}$$

$$\omega_0^2 = 3949$$

Thus $\qquad K = m\omega_0^2$

$$= 0.1 \times 3949$$

$$= 394.9 \text{ Nm}^{-1}$$

Example 3.2 A massless spring of spring constant 10 Nm^{-1} is suspended from a rigid support and carries a mass of 0.1 kg at its lower end. The system is subjected to a resistive force $-pv$, where p is a constant and v is the velocity. It is observed that the system performs damped oscillatory motion and its energy decays to $1/e$ of its initial value in 50 s.

(a) What is the value of p ?

(b) What is the Q value of the osillator ?

(c) Show that the fractional change in the frequency of the damped oscillations is $\approx (8\,Q^2)^{-1}$.

What is the percentage change in frequency due to damping ? What conclusion will you draw from it ?

Solution

(a) $\qquad m = 0.1$ kg.

The decay of energy of the damped oscillator is given by

$$E(t) = E_0\, e^{-\gamma t}$$

where E_0 is the initial energy and $\gamma = p/m$, p being the constant of the resistive force. It is given that when $t = 50$ s

$$\frac{E(t)}{E_0} = \frac{1}{e} = e^{-1}$$

or $$e^{-50\gamma} = e^{-1}$$

or $$\gamma = \frac{1}{50} = 2\times10^{-2}\ \text{s}^{-1}$$

Hence $$p = m\gamma = 0.1\times2\times10^{-2} = 2\times10^{-3}\ \text{kg s}^{-1} \text{ or } \text{Nsm}^{-1}$$

Since dimensionally $p = \dfrac{\text{force}}{\text{velocity}}$, the unit of p is newton second (metre)$^{-1}$ or kg s^{-1}.

(b) Since $K = 10$ Nm^{-1}, the angular frequency ω_0 in the absence of damping is

$$\omega_0 = \sqrt{\frac{K}{m}} = \sqrt{\frac{10}{0.1}} = 10\ \text{rad s}^{-1}$$

The angular frequency of damped oscillations is

$$\omega^{*2} = \omega_0^2 - \frac{\gamma^2}{4}$$
$$= 10^2 - 10^{-4}$$
$$\simeq 10^2$$

$\therefore$ $$\omega^* \simeq 10\ \text{rad s}^{-1}$$

The quality factor Q is

$$Q = \frac{\omega^*}{\gamma} = 500$$

(c) The fractional change in the frequency is given by $\dfrac{\omega_0 - \omega^*}{\omega_0}$

Now $$\omega^* = \omega_0\left(1 - \frac{\gamma^2}{4\omega_0^2}\right)^{1/2}$$
$$= \omega_0\left(1 - \frac{1}{4Q^2}\right)^{1/2}$$
$$\simeq \omega_0\left(1 - \frac{1}{8Q^2}\right)$$

where higher order terms are neglected since $\dfrac{1}{4Q^2} \ll 1$.

Then $$\frac{\omega_0 - \omega^*}{\omega_0} \simeq \frac{1}{8Q^2}$$

The percentage change in frequency is

$$\left(\frac{\omega_0 - \omega^*}{\omega_0}\right)\times 100 = \frac{100}{8Q^2} = \frac{100}{8\times(500)^2} = 0.5\times10^{-5}$$

From this we conclude that, for oscillations with a high Q value (which is usually the case), the change in frequency due to damping is negligibly small.

Example 3.3 A underdamped ballistic galvanometer has a time period of 2 s. When a transient current is passed, the first deflection of the spot (with a lamp and scale arrangement) is observed to be 20 cm. After 10 complete oscillations, the deflection reduces to 2 cm. What is the logarithmic decrement of the galvanometer ? What would be the first deflection if the damping were absent ?

Solution :

The amplitude of the damped oscillations is given by

$$A(t) = A_0 e^{-\gamma t/2}$$

Since, the time period T^* of the damped oscillations is 2 s., the time taken for 10 complete oscillations = 20 s. Referring to Fig. 3.6, on page 104 the first throw occurs at $t = T^*/4$, the second at $t = 3T^*/4$, the third at $t = 5\,T^*/4$, and so on. Since the successive throws (maximum deflections) occur at intervals of $T^*/2$, the 21st throw occurs after 20s, i.e. at $t = \dfrac{T^*}{4} + 20$,

Therefore
$$A_1 = A_0 \exp\left(-\frac{\gamma}{2}\cdot\frac{T^*}{4}\right) = A_0 e^{-\gamma/4} \quad (\because T^* = 2\text{s})$$

$$A_{21} = A_0 \exp\left\{-\frac{\gamma}{2}\left(\frac{T^*}{4} + 20\right)\right\} = A_0 e^{-\gamma/4 - 10\gamma}$$

Now $A_1 = 20$ cm and $A_{21} = 2$ cm. Therefore, we have

$$\frac{A_1}{A_{21}} = \frac{20}{2} = e^{10\gamma}$$

or
$$\gamma = 0.23 \text{ s}^{-1}$$

The logarithmic decrement λ is given by

$$\lambda = \frac{\gamma}{2}\cdot\frac{T^*}{2}$$
$$= 0.115$$

The relation between the correct deflection A_0 (i.e. the deflection in the absence of damping) and the first deflection is

$$A_1 = A_0\, e^{-\lambda/2}$$
$$A_0 = A_1 e^{\lambda/2}$$
$$= 20 \times e^{0.115/2}$$
$$= 21.2 \text{ cm}$$

Example 3.4 Calculate the frequency, the relaxation time and the quality factor of an LCR circuit with $L = 1$ mH, $C = 5$ μF and $R = 0.5\ \Omega$.

Solution :

Given

$$L = 1 \text{ mH} = 10^{-3} \text{ H} \quad C = 5\ \mu\text{F} = 5\times10^{-6} \text{ F} \quad R = 0.5\ \Omega$$

The angular frequency ω^* of the damped oscillations is

$$\omega^* = \left(\frac{1}{LC} - \frac{R^2}{4L^2}\right)^{1/2}$$
$$= 1.414\times10^4 \text{ rad s}^{-1}$$

Frequency ν^* is $\nu^* = \omega^*/2\pi = 0.225\times10^4$ Hz

The relaxation time τ of oscillations is

$$\tau = \frac{2}{\gamma} = \frac{2L}{R} = 4\times10^{-3}\ \text{s}$$

The quality factor Q is

$$Q = \frac{\omega^*}{\gamma} = \frac{L\omega^*}{R} = 28.3$$

Example 3.5 According to classical electromagnetic theory, an electron in an atom executes harmonic oscillations in a straight line and hence undergoes acceleration. An accelerated electron radiates energy and therefore, behaves as a damped harmonic oscillator, the damping being due to the radiation it emits. According to the theory, the electron radiates energy at the rate of $\frac{Kq^2\,\omega_0^4 A^2}{3c^3}$ watts where $K = 9\times10^9\ \text{N m}^2\ C^{-2}$, $q = 1.6\times10^{-19}C$ is the electric charge, $c=3\times10^8\ \text{ms}^{-1}$, is velocity of light, ω_0 is angular frequency and A is the amplitude of oscillations. If the emitted radiation has a wavelength 6000Å, calculate (a) the Q-value of the oscillator and (b) the radiation life-time (i.e. the time for the energy to fall to e^{-1} of its energy originally).

Given, electronic mass $=9\times10^{-31}$ kg.

Solution

(a) We know that [see Eq. (3.25)] the average power radiated $<P(t)>$ is related to average energy radiated $<E(t)>$ as

$$<P(t)> = \gamma <E(t)>$$

where

$$= \frac{\omega_0}{Q} <E(t)>$$

$$<E(t)> = \frac{1}{2} m A^2 \omega_0^2$$

It is given that

$$<P(t)> = \frac{Kq^2\,\omega_0^4 A^2}{3c^3}$$

$$\therefore \qquad \frac{<P(t)>}{<E(t)>} = \frac{2}{3}\frac{Kq^2\,\omega_0^2}{mc^3}$$

or

$$\frac{\omega_0}{Q} = \frac{2}{3}\frac{Kq^2\,\omega_0^2}{mc^3}$$

or

$$Q = \frac{3mc^3}{2\,Kq^2\omega_0}$$

Since $\lambda_0 = 6000$ Å and $\nu_0\lambda_0 = c$ we have

$$\omega_0 = 2\pi\nu_0 = \frac{2\pi c}{\lambda_0}$$

$$\therefore \qquad Q = \frac{3\, mc^2 \lambda_0}{4\, \pi K q^2}$$

Substituting for the given values of m, c, λ_0, K and q, we get

$$Q \simeq 6 \times 10^7$$

(b) From Eq. (3.27) we have

$$<E(t)> = E_0 \exp\left(-\frac{\omega_0 t}{Q}\right)$$

Thus, the energy decays to e^{-1} of its initial value E_0 in time τ_0 where

$$\tau_0 = \frac{Q}{\omega_0} = \frac{Q\,\lambda_0}{2\pi c}$$

$$\simeq 2 \times 10^{-8}\ \text{s}$$

Hence the radiation life-time of the atom, according to the classical electromagnetic theory, is of the order of 10^{-8} s.

Example 3.6 The system shown in Fig. 3.8 is subjected to a resistive force $F = -pv$ where p is a constant and v is the velocity. The system is at rest initially when a velocity of 6.8 cm s^{-1} is given to it. If $k = 10$ Nm^{-1}, $m = 10$ kg and $p = 8$ Ns m^{-1}, determine the subsequent displacement and velocity of the mass.

Solution :

The equation of motion is

$$m\frac{d^2\psi}{dt^2} + p\frac{d\psi}{dt} + 2k\psi = 0$$

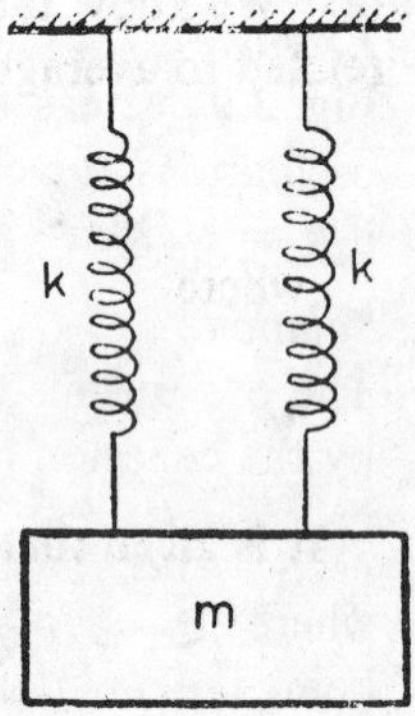

Fig. 3.8

Since the springs are connected in parallel, the effective spring constant of the system is $k+k = 2k$. The solution of the above equation is

$$\psi = e^{-\gamma t/2}(A \cos \omega^* t + B \sin \omega^* t)$$

where $\qquad \gamma = p/m,\ \omega^{*2} = \omega_0^2 - \frac{\gamma^2}{4}$: with

$\omega_0^2 = 2\,k/m$. Substituting for the given numerical values we have

$$\omega_0 = 1.41\ \text{rad s}^{-1}$$

$$\gamma = 0.8\ \text{s}^{-1}$$

$$\omega^* = 1.36\ \text{rad s}^{-1}$$

Substituting these values in the above equation we have

$$\psi = e^{-0.4t}(A \cos 1.36t + B \sin 1.36t) \qquad \text{(i)}$$

and

$$\frac{d\psi}{dt} = -0.4\, e^{-0.4t}(A \cos 1.36t + B \sin 1.36t)$$
$$+1.36\, e^{-0.4t}(B \cos 1.36t - A \sin 1.36t) \qquad \text{(ii)}$$

The initial conditions are: At $t = 0$, $\psi = 0$ and $\frac{d\psi}{dt} = 0.068 \text{ ms}^{-1}$. Using these conditions in Eqs. (i) and (ii) we have

$$A = 0$$

$$B = \frac{0.068}{1.36} = 0.05 \text{ m}$$

Substituting for A and B in Eq. (i) we have

$$\psi = 0.05\, e^{-0.4t} \sin 1.36t \tag{iii}$$

which determines the subsequent motion of the system. From Eq. (ii) the subsequent velocity is given by

$$\frac{d\psi}{dt} = 0.05e^{-0.4t}\,(1.36 \cos 1.36t - 0.4 \sin 1.36t) \tag{iv}$$

Notice from Eq (iii) that at $t=0$, $\psi=0$ and from Eq. (iv) the velocity at $t = 0$ is

$$\left(\frac{d\psi}{dt}\right)_{t=0} = 0.05 \times 1.36 = 0.068 \text{ ms}^{-1}$$

which are our initial conditions which confirms the correctness of Eqs. (iii) and (iv).

Example 3.7 The amplitude of vertical oscillations of the system shown in Fig. 3.9 decreases to 20% of the initial value after five consecutive cycles of oscillations. Determine the damping coefficient p (frictional force $= -pv$) if $k = 80 \text{ Nm}^{-1}$ and $m = 2.5$ kg.

Solution :

The oscillations of the damped system are described by the equation [see Eq. (3.16)]

$$\psi = A_0\, e^{-\gamma t/2} \cos(\omega^* t - \delta)$$

where $\gamma = p/m$, $\omega^* = \omega_0\,(1 - \gamma^2/4\omega_0^2)^{1/2}$; ω_0 being the angular frequency if damping were absent. Since the springs are connected in series, the effective spring constant K is given by

$$\frac{1}{K} = \frac{1}{k} + \frac{1}{k} = \frac{2}{k}$$

or

$$K = \frac{k}{2}$$

Hence

$$\omega_0 = \sqrt{\frac{K}{m}} = \sqrt{\frac{k}{2m}} = \sqrt{\frac{80}{2 \times 2.5}}$$

$$= 4 \text{ rad s}^{-1}$$

Fig. 3.9

The maximum amplitude occurs at value of t satisfying $\cos(\omega^* t-\delta) = +1$. The maximum amplitudes are

$$A_1 = A_0 \exp\left(-\frac{\gamma t_1}{2}\right)$$

$$A_2 = A_0 \exp\left(-\frac{\gamma t_2}{2}\right)$$

$$A_3 = A_0 \exp\left(-\frac{\gamma t_3}{2}\right), \text{ etc.}$$

It is given that

$$\frac{A_1}{A_6} = \frac{1}{0.20} = 5$$

The logarithmic decrement is given by

$$\lambda = \frac{1}{(n-1)} \ln\left(\frac{A_1}{A_n}\right) = \frac{1}{5} \ln 5 = 0.32$$

But $$\lambda = \frac{\gamma}{2} \cdot \frac{T^*}{2} = \frac{\pi\gamma}{2\omega^*} = \frac{\pi\gamma}{2\omega_0\left(1-\frac{\gamma^2}{4\omega_0^2}\right)^{1/2}}$$

or $$(\pi^2+\lambda^2)\,\gamma^2 = 4\,\lambda^2\,\omega_0^2$$

or $$\gamma = \frac{2\,\lambda\,\omega_0}{(\pi^2+\lambda^2)^{1/2}}$$

Substituting for λ and ω_0, we get

$$\gamma = 0.81 \text{ s}^{-1}$$

and $$p = m\gamma \simeq 2 \text{ N s m}^{-1}$$

Example 3.8 A block is placed on a horizontal plane surface and held between two springs as shown in Fig. 3.10. The coefficient of kinetic friction between the block and the surface is μ which has a constant value. The block is given an initial displacement x_0 from the equilibrium position and released. Investigate the subsequent motion of the block.

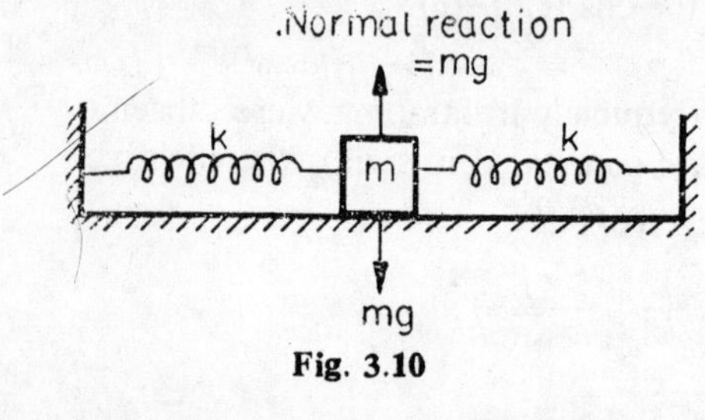

Fig. 3.10

Solution

Suppose the block is displaced to the right. Let x be the displacement at any instant of time t. The forces acting on the block are (i) the restoring force $-2\,kx$ and (ii) the frictional force = (coefficient of friction)×(normal reaction) = $+\mu mg$. The constant frictional force always opposes motion. The equation of motion is

$$m\frac{d^2x}{dt^2} = -2kx + \mu mg$$

$$= -2k\left(x - \frac{\mu mg}{2k}\right)$$

If we transform to a new variable

$$x' = x - \frac{\mu mg}{2k} \tag{i}$$

We have

$$\frac{d^2x'}{dt^2} = -\frac{2k}{m}x' \tag{ii}$$

(since μ, m, g and k are constants)

which gives $\omega = \sqrt{\dfrac{2k}{m}}$

or $T = 2\pi\sqrt{\dfrac{m}{2k}}$

Thus we find that the frequency of damped oscillations is the same as that of the undamped natural oscillations. In other words, a constant frictional force has no effect on the frequency of vibration of a system.

The solution of Eq. (*ii*) is

$$x' = A\cos\omega t + B\sin\omega t \tag{iii}$$

The motion of the mass is described by variable x. Using Eq. (*i*) in Eq. (*iii*) we have

$$x = (A\cos\omega t + B\sin\omega t) + \frac{\mu mg}{2k} \tag{iv}$$

Therefore, the velocity is given by

$$\frac{dx}{dt} = \omega(B\cos\omega t - A\sin\omega t) \tag{v}$$

Now the initial conditions are : At $t = 0$, $x = x_0$ and $dx/dt = 0$. Equations (*iv*) and (*v*) will satisfy these conditions if

$$B = 0$$

$$A = x_0 - \frac{\mu mg}{2k}$$

Hence the motion of the system is given by

$$x = \left(x_0 - \frac{\mu mg}{2k}\right)\cos\omega t + \frac{\mu mg}{2k}$$

or $$x = \left(x_0 - \frac{\mu mg}{2k}\right)\cos\frac{2\pi t}{T} + \frac{\mu mg}{2k}$$

At time $t = 0$; $x = x_0$, the maximum value of x. At the end of half a cycle, i.e. at $t = T/2$, the value of x is given by

$$x = -\left(x_0 - \frac{\mu mg}{2k} \right) + \frac{\mu mg}{2k} = -\left(x_0 - \frac{\mu mg}{k} \right)$$

At time $t = T/2$, the mass m is at the extreme left position. Thus we find that the amplitude decreases from x_0 to $(x_0 - \mu mg/k)$ during the first half cycle. Let us say that the maximum displacement to the left is x_1, i.e.

$$x_1 = x_0 - \frac{\mu mg}{k}$$

This is the displacement at $t = T/2$. Applying the same reasoning, the displacement at $t = T$ must decrease to $(x_1 - \mu mg/k) = (x_0 - 2\mu mg/k)$. Hence due to damping, the amplitude of oscillation decreases by $\mu mg/k$ during each half-cycle. Thus we conclude that the motion of the system is not simple harmonic (because its amplitude is not constant) but the motion is periodic with period,

$$T = 2\pi \sqrt{\frac{m}{2k}}$$

QUESTIONS

1. A massless spring of spring constant K is suspended from a rigid support and carries a mass m at its lower end. The system is subjected to a resistive force $F = -bv$, where b is a constant and v is the velocity of the mass.
   (a) Set up the differential equation of motion for free oscillations of the system.
   (b) Show that $\psi = A_1 \exp(\alpha_1 t) + A_2 \exp(\alpha_2 t)$ is a solution of this equation, ψ being the instantaneous displacement and A_1 and A_2 being two arbitrary constants.
   (c) Find α_1 and α_2 in terms of m, K, and b.
2. Discuss the nature of the motion of the system in Q. 1 in the case when $b > 4$ mK, under the following initial conditions :
   (a) Initial displacement is zero and initial velocity is v_0.
   (b) Initial velocity is zero and initial displacement is ψ_0.
3. Verify that $\psi = (A + Bt) e^{\alpha t}$ is a solution of the differential equation of motion of the system in Q. 1 in the case when $b = 4$ m K and find α in terms of m, K and b. Evaluate the constants A and B under the condition that, initially, the mass has a finite displacement ψ_0 and zero velocity. Discuss the nature of motion.
4. Establish the equation of motion of a damped harmonic oscillator subjected to a resistive force that is proportional to the first power of its velocity. If the damping is less than critical, show that the motion of the system is oscillatory with its amplitude decaying exponentially with time.

5. Verify that $\psi = A\, e^{-\alpha t} \cos \omega t$ is a possible solution of the equation

$$\frac{d^2\psi}{dt^2} + \gamma \frac{d\psi}{dt} + \omega_0^2\, \psi = 0$$

and find α and ω in terms of γ and ω_0.

6. Show that the average energy of a weakly damped harmonic oscillator decays exponentially with time.

7. What do you understand by 'logarithmic decrement', 'relaxation time' and 'quality factor' of a weakly damped harmonic oscillator. What are the relationships between them.

8. A harmonic oscillator of mass m and stiffness constant K is subjected to a resistive force $F = -pv$, where v is the velocity of the oscillator and p a constant. The oscillator is displaced by ψ_0 from its equilibrium position and released at time $t = 0$. Assuming that $p \ll 4$ mK, discuss the subsequent motion of the oscillator and draw the displacement-time graph for it.

9. Show that average power dissipation $\langle P \rangle$ from the oscillator of Q. 8 is given by

$$\langle P \rangle = p/m \langle E \rangle$$

where $\langle E \rangle$ is the average stored energy.

10. Set out the differential equation of the motion of the coil of a moving coil galvanometer and discuss the conditions under which its motion is (a) dead beat (b) critically damped and (c) oscillatory. What do you understand by the term 'logarithmic decrement' of the instrument ? What is its significance and use ?

11. In an LCR circuit, show that it is the resistance alone that is responsible for the damping of oscillations. Discuss the conditions under which the discharge of the capacitor is oscillatory. What is the quality factor of such a circuit ? What does a high Q value signify ?

12. Show that the damping has practically no effect on the frequency of a harmonic oscillator having a very high Q value, say, of the order of 10^6.

13. Show that the amplitude of a weakly damped oscillator reduces to half its initial value in time $t = \tau \ln 2$, where τ is the relaxation time of the oscillator.

PROBLEMS

1. A massless spring, suspended from a rigid support, carries a mass of 200 g at its lower end. It is observed that the system oscillates with time period of 0.2 s and the amplitude of oscillations reduces to half its initial undamped value in 30 seconds. Calculate (a) the relaxation time of the system, (b) its quality factor and (c) the spring constant.

2. An object of mass 0.1 kg is hung from a spring whose spring constant is 40 Nm^{-1}. The object is subjected to a resistive force $F = -pv$, where p is a constant and v is the velocity of the mass. It is observed that the frequency of the damped oscillations is 99/100 of the undamped value. (a) What is the value of p ? (b) What is the Q-value of the oscillator ? (c) By what factor is the amplitude of oscillations reduced after 10 complete oscillations ?

3. A massless spring of spring constant 9 Nm^{-1} is suspended from a rigid support and carries a mass of 0.1 kg at its lower end. The system oscillates in a liquid that exerts a viscous force $F = -pv$, where p is a constant v is the velocity. It is observed that the energy decays to half its original value in

5 seconds. (a) What is the value of p ? (b) What is the Q-value of the oscillator ? (c) What is the change in the frequency due to damping ?

4. The energy of a piano string of frequency 256 Hz reduces to half its initial value in 2 s. What is the Q-value of the string ?
5. The quality factor of a sonometer wire of frequency 500 Hz is 5000. In what time will its energy reduce to $1/e$ of its value in the absence of damping ?
6. The Q-value of an underdamped harmonic oscillator of frequency 480 Hz is 80000. Calculate the time in which its amplitude reduces to $1/e$ of its initial value. How many oscillations does it make in this time ?
7. A linear harmonic oscillator of mass 10 g is at rest at its equilibrium position in a viscous medium that exerts a resistive force $F = -pv$, where p is a constant and v is the velocity. The oscillator is given a kick at time $t = 0$ and the amplitude of the first oscillation is observed to be 2 cm. After 10 complete oscillations the amplitude falls to 2 mm. If the period of the damped oscillations is 1 second, find (a) the value of p, (b) the relaxation time of the oscillator (c), its Q-value, (d) its logarithmic decrement, (e) its frequency if damping were absent, (f) the amplitude of the first oscillation if damping were absent and (g) the energy stored in the system initially.
8. A capacitor of capacity $2\mu F$ is discharged through a series combination of a resistance of 2Ω and an inductor of inductance 100 millihenry. Is the discharge oscillatory ? If so, what is its frequency ?
9. In an LCR circuit $L = 0.5$ henry and $C = 5\mu F$. What should be the maximum value of the resistance R for the discharge to be oscillatory ?
10. The Q-value of an LCR circuit with $L = 0.2$ henry, and $C = 2\mu F$ is 100 and its frequency is 8000 Hz. What is the value of R ? How will the frequency of the discharge change if the resistance is reduced to zero ?

4

Forced Oscillations and Resonance

4.1 INTRODUCTION

In Chap. 1 we studied the free undamped oscillations of various simple systems having one degree of freedom. In Chap. 3 we investigated the effect of damping on the free oscillations of these systems and found that, due to the friction present in the system, the amplitude of oscillations decreases exponentially with time and the frequency of the natural oscillations is slightly diminished. Usually, the change in frequency is too small to be of any significance. In this chapter we shall investigate the behaviour of a weakly damped harmonic oscillator, when an external time-dependent force is applied to the system so as to maintain the amplitude of oscillations. Without loss of genetality, we shall specialize to a harmonically varying driving force of frequency not necessarily the same as that of the oscillator and investigate how the system responds to the driving force as its frequency is gradually changed.

The problem of a damped harmonic oscillator driven by an externally applied harmonic force is of profound importance in physics. Very often a system is set into oscillation by linking (or coupling) it in some way with another oscillating system (which we shall call the driver). For example, in a resonance tube, the air column vibrates because it is linked (by sound waves) to a vibrating tuning fork. The diaphragm of a microphone vibrates because it is linked (by sound waves) to the vibrations of, say, a musical instrument. The diaphragm of a loud-speaker vibrates because it is linked (by current oscillations) to the output circuit of an amplifier. The electrical circuit in a radio receiver oscillates because it is linked (by radio waves) to the oscillatory system (i e. the transmitter) in a broadcasting station. In all these examples, the driven oscillator picks up energy from the driving system and oscillates. A harmonic oscillator driven by an externally applied harmonic force is said to execute *forced oscillations*.

In the examples mentioned above, the driven system extracts energy from the driving system without any appreciable feed-back of energy from the former to the latter. The transfer of energy is essentially a one-way process. This is because either the coupling between the two is very weak (as in the case of a radio receiver and the transmitter) or the driving system has such a large reservoir of energy that the energy fed back into it is negligible (as in the case of an amplifier coupled to a loud speaker). The driving system thus remains practically unaffected by the forced oscillations of the driven system. The driving system only serves as the supplier of a periodic force. In this chapter we shall analyse the behaviour of only those systems which satisfy this condition. A much more complicated case where the feedback of energy cannot be neglected and two oscillatory systems have to be treated on an equal footing, without labelling one as the driver and the other as the driven, will be dealt with in the next chapter.

4.2 FORCED OSCILLATIONS OF A ONE-DIMENSIONAL DAMPED HARMONIC OSCILLATOR

The general behaviour of the systems in the examples mentioned above is adequately described by the simple mechanical oscillator illustrated in Fig. 4.1. When the mass m is displaced from its equilibrium position and released, it experiences a restoring force $-K\psi$, where K is the spring constant and ψ is the instantaneous displacement. If there were no other force, the mass would oscillate harmonically with its natural angular frequency $\omega_0 = \sqrt{K/m}$. Under the influence of the restoring force, the mass begins to move. If the instantaneous velocity is $d\psi/dt$, the mass also experiences a frictional force $-p\, d\psi/dt$ (assumed viscous) where p is the coefficient of

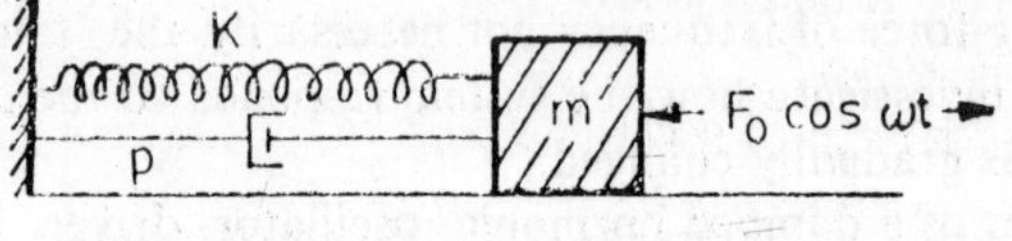

Fig. 4.1 Mechanical forced oscillators with force $F_0 \cos \omega t$ applied to damped oscillator.

the damping force. Finally the mass is subjected to a time-dependent force $F(t)$ which opposes the restoring force as well as the frictional force and helps motion. The equation of motion of the forced oscillator is given by

$$m\frac{d^2\psi}{dt^2} = -K\psi - p\frac{d\psi}{dt} + F(t)$$

This is an inhomogeneous second-order linear differential equation. In order to investigate the effect of $F(t)$ on the motion of the oscillator, we shall choose, for $F(t)$, a harmonic force of the type $F(t) = F_0 \cos \omega t$, where F_0 is the magnitude of the force and ω is its angular frequency. The equation of motion of a damped oscillator driven by an externally applied harmonic force then reads

$$m \frac{d^2\psi}{dt^2} + p \frac{d\psi}{dt} + K\psi = F_0 \cos \omega t$$

or
$$\frac{d^2\psi}{dt^2} + \gamma \frac{d\psi}{dt} + \omega_0^2\psi = f_0 \cos \omega t \tag{4.1}$$

where $\gamma = p/m$, $\omega_0 = \sqrt{K/m}$ and $f_0 = F_0/m$.

The General Solution

Before we obtain the complete solution of Eq. (4.1), let us analyse the situation qualitatively. In Chap. 3 we discussed the motion of the oscillator described by Eq. (4.1) with the right-hand side equal to zero, i,e. when the external force was absent. We found that a weakly damped oscillator ($\gamma < 2\omega_0$) oscillates harmonically with an effective angular frequency $\omega^* = \left(\omega_0^2 - \frac{\gamma^2}{4}\right)^{1/2}$. If the oscillator is displaced from its equilibrium position and left to itself, it will oscillate with a frequency $\nu^* = \omega^*/2\pi$. However, when the external periodic force is applied, it will try to impose its own frequency $\nu = \omega/2\pi$ on the oscillator. We must, therefore, expect that the actual motion in this case is some sort of superposition of two oscillations ; one at frequency ν^* of the free damped oscillations and the other at frequency ν of the driving force. That the mathematically complete solution of Eq. (4.1) is indeed a sum of two motions can be shown as follows :

Suppose we have found a solution-call it ψ_1—to Eq. (4.1) so that

$$\frac{d^2\psi_1}{dt^2} + \gamma \frac{d\psi_1}{dt} + \omega_0^2\psi_1 = f_0 \cos \omega t$$

Now suppose ψ_2 is the solution of equation

$$\frac{d^2\psi}{dt^2} + \gamma \frac{d\psi}{dt} + \omega_0^2\psi = 0$$

so that

$$\frac{d^2\psi_2}{dt^2} + \gamma \frac{d\psi_2}{dt} + \omega_0^2\psi_2 = 0$$

Then by simple addition of these two equations, we have

$$\frac{d^2}{dt^2}(\psi_1 + \psi_2) + \gamma \frac{d}{dt}(\psi_1 + \psi_2) + \omega_0^2(\psi_1 + \psi_2) = f_0 \cos \omega t$$

Thus the superposition $\psi_1+\psi_2$ is as much a solution of Eq. (4.1) as is ψ_1 alone. There is no mathematical reason to exclude ψ_2. In fact, we shall show later, that the solution ψ_1 by itself does not tell complete history of the forced oscillator and that we are obliged to include the contribution of ψ_2 Thus $\psi_1+\psi_2$ is the complete solution of Eq (4.1) In mathematics, solution ψ_1 is called the *particular solution* and solution ψ_2 is called the *complementary function.*

In Chap. 3, we have already found ψ_2. In the case of low damping ($\gamma < 2\omega_0$) we have seen that ψ_2 is given by [see Eq. (3.16)]

$$\psi_2 = A\, e^{-\gamma t/2} \cos(\omega^* t - \delta)$$

where $\omega^* = (\omega_0^2 - \gamma^2/4)^{1/2}$ is the angular frequency of free damped oscillation and δ is the phase constant. The amplitude $Ae^{-\gamma t/2}$ of the oscillation decays exponentially with time. The motion corresponding to solution ψ_2 dies out with time. The motion corresponding to solution ψ_1 stays all the time. Thus, the situation is somewhat like this : Suppose that the oscillator is at its equilibrium position at time $t = 0$ when the external periodic force of frequency ν is switched on. A tussle ensues between the damping force tending to retard the motion and the driving force tending to help the motion. The damped oscillator 'likes' to oscillate at its own frequency ν^* and the driving force forcing the oscillator to 'obey' and oscillate at the frequency ν of the driving force. After a sufficiently long time (longer than the relaxation time $\tau = 2/\gamma$ of the damped oscillations) the oscillator succumbs, as it were, to the driving force and settles down to a harmonic oscillation with the frequency ν of the driviug force. Thus for sometime after the driving force is switched on, the motion of the oscillator is given by solutions $\psi_1+\psi_2$. This is called the *transient state*. After the damped oscillations corresponding to ψ_2 have died out, the oscillator executes harmonic oscillations at the frequency of the driving force. This is called the *steady state*. Thus the displacement ψ_1 of the oscillator in the steady state is given by

$$\psi_1 = B \cos(\omega t - \phi)$$

where B is the amplitude and ϕ is the phase constant of the oscillations relative to that of the driving force. For simplicity, the phase constant of the driving force is taken to be zero. The complete solution of Eq. (4.1) is written as

$$\psi = \psi_1+\psi_2 = Ae^{-\gamma t/2} \cos(\omega^* t - \delta) + B\cos(\omega t - \phi) \tag{4.2}$$

In the transient state dealt with in Sec. 4.4, both the terms on the right-hand side of Eq. (4.2) are operative, resulting in a motion that is a superposition of two harmonic oscillations of angular frequencies ω^* and ω giving rise to beats called *transient beats*. In the steady state, only the second term in Eq. (4 2) contributes, the oscillation corresponding to the first term having died out. The motion of the forced oscillator is harmonic at an angular frequency ω of the driving force. These observations can be justified by the following simple experiment.

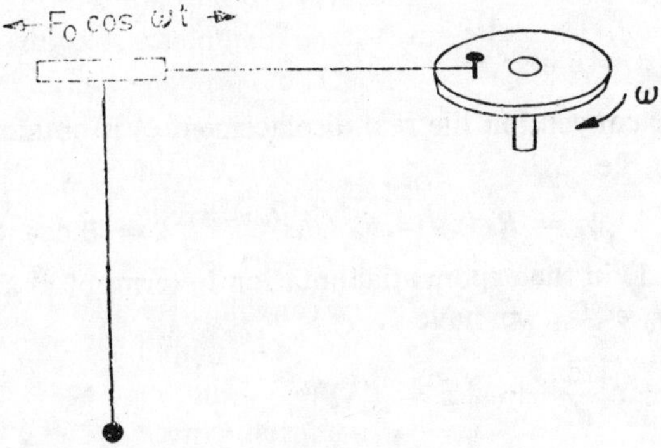

Fig. 4.2

Consider a simple pendulum hanging from a support which can execute horizontal oscillations as shown in Fig. 4.2. Let us suppose that at time $t = 0$ the pendulum is at its equilibrium position when the support starts oscillating harmonically. The pendulum begins to ocillate under the influence of the periodic impulses imparted to it by the oscillating support. Initially we find that the motion of the pendulum is quite erratic. This transient behaviour persists for a fairly long time, since due to very little friction offered by the air, the damped oscillations take a long time to decay to zero. But eventually the pendulum is seen to execute SHM of constant amplitude at the same frequency as that of the support. The phase of the pendulum oscillations is, however, not the same as that of the driving force. This is the steady state of the forced oscillator. We shall first analyse the steady state behaviour which obviously is much more important from the point of view of practical applications. Later, for the sake of completeness we shall analyse the transient behaviour.

4.3 STEADY STATE BEHAVIOUR OF A FORCED OSCILLATOR

We are interested in analysing the steady state behaviour of a forced oscillator. In other words, we wish to know the motion of a damped oscillator after the force has been acting for a sufficiently long time. We are also interested in knowing how this behaviour depends on the frequency of the driving force. Equation (4.1) describes the forced oscillations. In the steady state, we expect that the displacement of the oscillator is given by

$$\psi_s = B \cos (\omega t - \phi) \tag{4.3}$$

where the constants B and ϕ are yet undetermined. To determine these constants it is convenient to use the exponential notation introduced in Chapter 1. In terms of this notation, displacement ψ_s can be represented by a vector z as

$$z = B\, e^{i(\omega t-\phi)}$$

According to our convention the real displacement ψ_s is obtained by taking the real part of z, i.e.

$$\psi_s = R_e(z) = R_e\{Be^{i(\omega t-\phi)}\} = B\cos(\omega t-\phi)$$

Rewriting Eq. (4.1) in the exponential notation in terms of z and representing $f_0 \cos \omega t$ by $f_0\, e^{i\omega t}$, we have

$$\frac{d^2z}{dt^2} + \gamma\frac{dz}{dt} + \omega_0^2 z = f_0\, e^{i\omega t}$$

$$= f_0\, e^{i(\omega t-\phi+\phi)}$$

$$= f_0\, e^{i(\omega t-\phi)}\, e^{i\phi} \tag{4.4}$$

with $z = B\, e^{i(\omega t-\phi)}$, where B is the amplitude of the forced oscillator and ϕ its phase constant relative to that of the driving force. Differentiating z with respect to t we have

$$\frac{dz}{dt} = i\omega B\, e^{i(\omega t-\phi)}$$

and

$$\frac{d^2z}{dt^2} = -\omega^2 B\, e^{i(\omega t-\phi)}$$

Substituting for z, $\frac{dz}{dt}$ and $\frac{d^2z}{dt^2}$ in Eq. (4.4) we have

$$\{B(\omega_0^2-\omega^2+i\omega\gamma)-f_0\, e^{i\phi}\}\, e^{i(\omega t-\phi)}=0$$

which can be satisfied for all t if

$$B(\omega_0^2-\omega^2+i\omega\gamma)-f_0\, e^{i\phi} = 0$$

or

$$B = \frac{f_0\, e^{i\phi}}{\omega_0^2-\omega^2+i\omega\gamma}$$

or

$$Be^{-i\phi} = \frac{f_0}{\omega_0^2-\omega^2+i\omega\gamma}$$

or

$$B(\cos\phi - i\sin\phi) = \frac{f_0}{\omega_0^2-\omega^2+i\omega\gamma}$$

The complex conjugate of this equation is

$$B(\cos\phi + i\sin\phi) = \frac{f_0}{\omega_0^2-\omega^2-i\omega\gamma}$$

These two equations, on addition and subtraction, give

$$B\cos\phi = \frac{f_0(\omega_0^2-\omega^2)}{(\omega_0^2-\omega^2)^2+\omega^2\gamma^2} \tag{4.5}$$

and
$$B\sin\phi = \frac{\omega\gamma f_0}{(\omega_0^2-\omega^2)^2+\omega^2\gamma^2} \tag{4.6}$$

The constants B and ϕ can now be obtained from Eqs. (4.5) and (4.6). Squaring and adding these equations we have

$$B^2 = \frac{f_0^2}{(\omega_0^2-\omega^2)^2+\omega^2\gamma^2}$$

so that
$$B = \frac{f_0}{\{(\omega_0^2-\omega^2)^2+\omega^2\gamma^2\}^{1/2}} = \frac{F_0/m}{\{(\omega_0^2-\omega^2)^2+\omega^2\gamma^2\}^{1/2}} \tag{4.7}$$

Dividing Eq. (4.6) by Eq. (4.5) we have

$$\tan\phi = \frac{\omega\gamma}{(\omega_0^2-\omega^2)} \tag{4.8}$$

or
$$\phi = \tan^{-1}\left\{\frac{\omega\gamma}{(\omega_0^2-\omega^2)}\right\} \tag{4.9}$$

Hence z is given by

$$z = B\,e^{i(\omega t-\phi)} \tag{4.10}$$

where B is given by Eq. (4.7) and ϕ by Eq. (4.9). Taking the real part of z we get the displacement ψ_s which reads

$$\psi_s = B\cos(\omega t-\phi) \tag{4.11}$$

Equations (4.7), (4.9) and (4.11) give us three pieces of information about the steady state behaviour of the forced oscillator. These are :

(1) That a phase difference ϕ exists between the displacement and the driving force. From Eq. (4.9) we see that the value of ϕ depends upon the angular frequency ω of the driving force and the constants γ and ω_0 of the oscillator.

(2) The amplitude B of the forced oscillations depends on the constants F_0 and ω of the driving force and the constants ω_0 and γ of the oscillator.

(3) The motion of the oscillator is completely independent of the initial conditions, i.e. the way in which the oscillator is set into motion; the amplitude, phase and frequency of the oscillator depending only on the constants F_0 and ω of the driving force and on the oscillator constants m, ω_0 and γ. No matter how we start the oscillator, its motion will eventually settle down into that represented by Eq. (4.11). *Steady state motion is the motion of a system that has forgotten how it started.*

Mechanical Impedance The mechanical impedance Z_m is defined as the force required to produce unit velocity in the oscillator, i.e. $Z_m = F/V$. The velocity of the oscillator is $V = dz/dt$, which from Eq. (4.10) reads

$$V = i\,\omega B\, e^{i(\omega t - \phi)}$$

The force F is $F_0\, e^{i\omega t}$. Therefore, the mechanical impedance of a driven oscillator is

$$Z_m = \frac{F_0\, e^{i\omega t}}{i\omega B e^{i(\omega t - \phi)}} = \frac{F_0}{i\omega B e^{-i\phi}} = \frac{m f_0}{i\omega B e^{-i\phi}}$$

Using Eq. (4.5), we have

$$Z_m = \frac{m}{i\omega}\,(\omega_0^2 - \omega^2 + i\omega\gamma)$$

or

$$Z_m = m\gamma - \frac{im}{\omega}\,(\omega_0^2 - \omega^2)$$

Thus, the impedance of a mechanical driven oscillator is a complex quantity. It has a mechanical part $m\gamma = p$, where p is the coefficient of the damping force and a reactive part $m(\omega^2 - \omega_0^2)/\omega$ which results from the reaction of the oscillator to the driving force. The complex conjugate Z_m^* of Z_m is

$$Z_m^* = m\gamma + \frac{im}{\omega}\,(\omega_0^2 - \omega^2)$$

The magnitude $|Z_m|$ of the impedance is given by

$$|Z_m|^2 = Z_m Z_m^* = m^2\gamma^2 + \frac{m^2}{\omega^2}\,(\omega_0^2 - \omega^2)^2$$

or

$$|Z_m| = m\left\{\gamma^2 + \frac{1}{\omega^2}\,(\omega_0^2 - \omega^2)^2\right\}^{1/2} \tag{4.12}$$

In terms of $|Z_m|$, the amplitude B of the forced oscillations can be expressed as [see Eq. (4.7)].

$$B = \frac{F_0}{\omega\,|Z_m|} \tag{4.13}$$

Dependence of the Amplitude of the Forced Oscillator on the Frequency of the Driving Force

We shall now study the behaviour of the forced oscillator as the frequency of the driving force is varied from a low value ($\omega \ll \omega_0$) to a high value ($\omega \gg \omega_0$). It is clear from Eq. (4.7) that amplitude B of the forced oscillations depends upon the constants (m, ω_0, γ) of the oscillator itself and upon the magnitude F_0 and angular frequency ω of the driving force.

For given values of m, ω_0, γ and F_0 the amplitude becomes maximum when the denominator in Eq. (4.7) is minimum i.e. when

$$D \equiv (\omega_0^2 - \omega^2)^2 + \omega^2\gamma^2$$

is minimum. This happens at angular frequency $\omega = \omega_r$ satisfying the following two conditions

$$\frac{dD}{d\omega} = 0 \quad \text{and} \quad \frac{d^2D}{d\omega^2} \geqslant 0$$

Differentiating D w.r.t. ω and setting it equal to zero, we obtain

$$\omega_r = \left(\omega_0^2 - \frac{\gamma^2}{2}\right)^{1/2} \tag{4.14}$$

It is easy to verify that $d^2D/d\omega^2$ is indeed positive at this value of ω, provided $\gamma < \sqrt{2}\,\omega_0$, which is usually the case. In fact γ is always much smaller than ω_0, so that $\omega_r \simeq \omega_0$ in cases where damping is negligible. Thus, the amplitude of the forced oscillator becomes maximum if the frequency of the driving force is very nearly equal to the frequency of natural oscillations. This is called the condition of resonance and ω_r satisfying Eq. (4.14) is called the *resonant* frequency. It may be noted that in cases where damping is negligible $\omega_r \simeq \omega_0$, otherwise ω_r is slightly less than ω_0.

Substituting ω_r for ω_0 in Eq. (4.7) we have

$$B_{max} = \frac{F_0/m}{\gamma\left(\omega_0^2 - \dfrac{\gamma^2}{4}\right)^{1/2}} \tag{4.15}$$

showing that the smaller the value of γ, the greater the value of B_{max} (see Fig. 4.3). If $\gamma \ll \omega_0$; $B_{max} = F_0/m\gamma\omega_0 = F_0/p\omega_0$. Thus near resonance $\omega_r \simeq \omega_0$, the amplitude is controlled by the resistive term p. The forced oscillator is said to be ***resistance controlled.*** It may be noticed from Eq. (4.12) that near resonance the impedance Z_m of the oscillator is minimum, resulting in a large amplitude. The large response of the oscillator to the driving force near $\omega \simeq \omega_0$ is the most striking feature of the motion. The repeated application of a relatively small force (F_0) can help build up a large amplitude near resonance.

If the driving frequency is very small compared to the natural frequency of free oscillations i.e. if $\omega \ll \omega_0$, we find from Eq. (4.7) that the amplitude of oscillations becomes

$$\frac{F_0/m}{\omega_0^2\left(1 - \dfrac{\omega^2}{\omega_0^2} \cdot \dfrac{\gamma^2}{\omega_0^2}\right)^{1/2}}$$

which for weak damping ($\gamma < \omega_0$) is approximately equal to $F_0/m\omega_0^2 = F_0/K$, where K is the stiffness constant. But this is the displacement which a constant force F_0, would produce. Remember that when $\omega \to 0$, $F(t) \to F_0$. This is equivalent to saying that the term $m\, d^2\psi/dt^2 = -\, m\omega^2\psi$ in Eq. (4.1) plays a relatively small role compared to the term $K\psi$ at very low frequencies. In other words, the response of the oscillator is controlled by the stiffness constant K at very low frequency of the driving force. The oscillator is said to be *stiffness controlled*. Notice that, when $\omega \ll \omega_0$, the impedance Z_m becomes very large. Hence, at very low frequencies, the amplitude becomes very small. As ω is increased, the amplitude increases, until at resonance ($\omega \simeq \omega_0$), the amplitude assumes a very large value.

On the other hand, if the driving frequency is very high compared to the natural frequency of free oscillations, i.e. if $\omega \gg \omega_0$, the amplitude of oscillations becomes

$$\frac{F_0/m}{\omega^2\left(1+\frac{\gamma}{\omega_0^2}\cdot\frac{\omega_0^2}{\omega^2}\right)^{1/2}}$$

which, for weak damping ($\gamma < \omega_0$), is nearly equal to $F_0/m\omega^2$, indicating that the amplitude falls as ω increases. Notice that when $\omega \gg \omega_0$, the impedance again becomes very large, hence the amplitude becomes small. The stiffness term $K\psi$ in Eq. (4.1) becomes very small compared to the inertia term $m\frac{d^2\psi}{dt^2} = -m\omega^2\psi$ (when ω is very high), because of the large acceleration associated with high frequencies, so that the response of the oscillator is controlled by the *inertia*. The oscillator is said to be *mass (or inertia) controlled*.

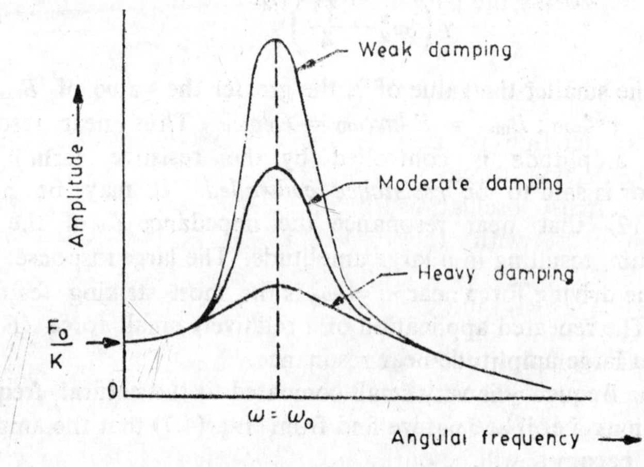

Fig. 4.3 Dependence of the amplitude of a forced oscillator on the frequency of the driving force

Figure 4.3 illustrates the response of the forced oscillator as the frequency of the driving force is increased from a very low value to a very high value. The amplitude of the forced oscillator is plotted against the frequency of the driving force for different cases of damping ; other parameters such as m, K and F_0 being kept constant. If damping is weak, the peak value of the amplitude B_{max} occurs at $\omega = \omega_0$, which represents the condition of *amplitude resonance*. It is evident that the peak value B_{max} is smaller for higher damping coefficient p, but it always occurs at or near $\omega = \omega_0$, provided damping is not too large. Only for heavy damping does the peak value occur at frequencies less than ω_0 $\left[\text{being equal to } \left(\omega_0^2 - \frac{\gamma^2}{2}\right)^{1/2}\right]$. Notice that the fall in amplitude is steeper on either side of $\omega = \omega_0$ for weak damping than for heavy damping. Figure 4.3 also shows that the steady state motion of the oscillator is not very sensitive to the value of the damping coefficient p except in the range of frequencies near the resonant frequency.

Dependence of the Phase of the Forced Oscillator on the Frequency of the Driving Force

The phase constant of the forced oscillator in the steady state is given by Eq. (4.8) and the displacement is given by Eq. (4.11). Since the driving force has been taken to be

$$F(t) = F_0 \cos \omega t$$

it is clear that ϕ in Eq. (4.11) is the angle by which the driving force leads the displacement or by which the displacement lags behind the driving force. As can be seen from Eq. (4.8), the phase angle ϕ depends on damping characterized by the parameter γ and the relative values of ω and ω_0.

If $\omega \ll \omega_0$, Eq. (4.8) gives $\tan \phi \simeq \frac{\gamma}{\omega_0} \cdot \frac{\omega}{\omega_0}$, so that $\tan \phi$ is a small positive quantity tending to zero in the limit $\omega \to 0$. Thus, if the frequency of the driving force is very small compared to the natural frequency of free oscillations, the oscillator displacement is very nearly in phase with the driving force. On the other hand,

$$\text{when } \omega \gg \omega_0, \ \tan \phi \simeq -\frac{\gamma}{\omega} \simeq -\frac{\gamma}{\omega_0} \frac{\omega_0}{\omega},$$

which, for weak damping ($\gamma < \omega_0$), has a very small negative value, indicating that $\phi \simeq \pi$. Thus, if the frequency of the driving force is very high compared to the natural frequency of free oscillations, the oscillator displacement will be out of phase with the driving force. The oscillator acceleration (which is 180° out of phase with the displacement) will, however, be in phase with the driving force in the case $\omega \gg \omega_0$. But at resonance ($\omega \simeq \omega_0$), $\tan \phi = \infty$ i.e. $\phi = \pi/2$ or 90°. Thus at resonance, the displacement will differ in phase from the driving force by $\pi/2$. In other words, the displacement will be maximum when the driving force is zero and vice versa.

It is clear that the phase angle ϕ increases continuously from 0 ($\omega \ll \omega_0$) to 180° ($\omega \gg \omega_0$), passing through 90° at resonance ($\omega \simeq \omega_0$). Figure 4.4 shows the dependence of phase angle ϕ on angular frequency of the driving force, for different cases of damping.

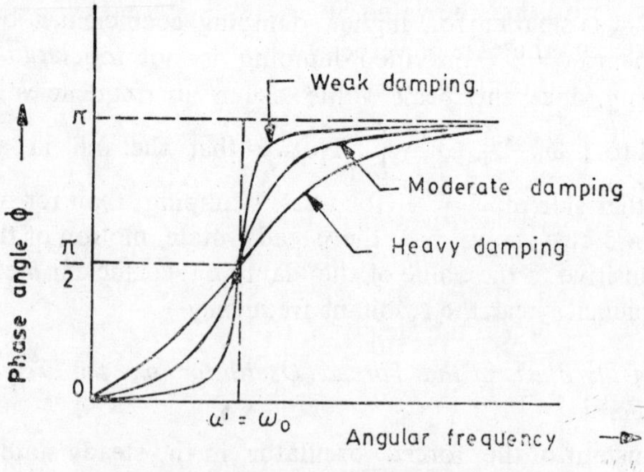

Fig. 4.4 Dependence of the phase of a forced oscillator on the frequency of driving force. Displacement ψ lags the force F by ϕ radians

Velocity Resonance

In the steady state, the displacement ψ_s of a harmonically driven damped oscillator is given by Eq. (4.11). Therefore, the velocity $V = \frac{d\psi_s}{dt}$ of the forced oscillator at an instant of time t is given by

$$V = \frac{d\psi_s}{dt} = -B\omega \sin(\omega t - \phi)$$

$$= B\omega \cos\left(\omega t - \phi + \frac{\pi}{2}\right)$$

or $$V = V_0 \cos(\omega t - \delta) \tag{4.16}$$

where B and ϕ are respectively given by Eqs (4.7) and (4.8) and $V_0 = B\omega$ is the amplitude of velocity given by [using Eq. (4.7)]

$$V_0 = \frac{F_0/m}{\left\{\frac{(\omega_0^2 - \omega^2)^2}{\omega^2} + \gamma^2\right\}^{1/2}} \tag{4.17}$$

and $$\delta = \phi - \pi/2 \tag{4.18}$$

A comparison of Eqs (4.11) and (4.16) reveals that the velocity leads the displacement in phase by $\pi/2$. The velocity amplitude V_0 varies with ω, the angular frequency of the driving force. It is clear from Eq. (4.17) that $V_0 = 0$ when $\omega = 0$. As ω is increased, V_0 increases reaching a maximum value $V_{0\max}$ at a value of ω for which the denominator is minimum i.e.

$$\frac{d}{d\omega}\left[\frac{(\omega_0^2 - \omega^2)^2}{\omega^2} + \gamma^2\right] = 0$$

It is easy to verify that the value of ω satisfying this condition is $\omega = \omega_0$, which represents the case of velocity resonance. At resonance, the maximum value of V_0 is [setting $\omega = \omega_0$ in Eq. 4.17).]

$$V_{0\max} = \frac{F_0}{m\gamma} = \frac{F_0}{p}$$

indicating that $V_{0\max}$ decreases as p increases. As ω is increased to a very high value $\omega \gg \omega_0$, $V_0 \simeq F_0/m\omega^2$, if damping is not too heavy. In the limit $\omega \to \infty$, $V_0 \to 0$. Figure 4.5 shows the dependence of the velocity of the forced oscillator (in the steady state) on the frequency of the driving force.

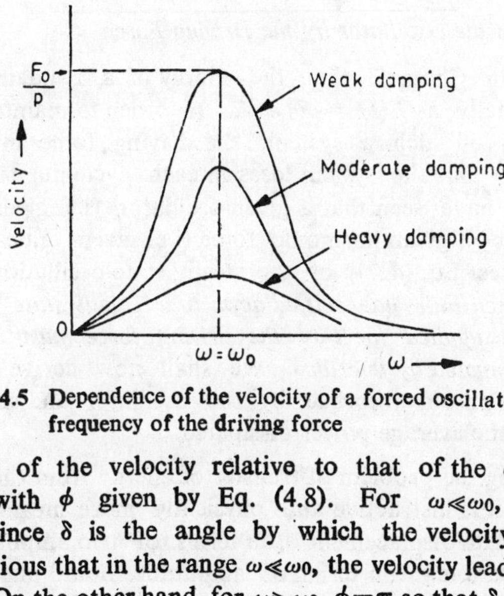

Fig. 4.5 Dependence of the velocity of a forced oscillator on the frequency of the driving force

The phase of the velocity relative to that of the driving force is $\delta = \phi - \pi/2$, with ϕ given by Eq. (4.8). For $\omega \ll \omega_0$, $\phi = 0$, so that $\delta = -\pi/2$. Since δ is the angle by which the velocity lags behind the force it is obvious that in the range $\omega \ll \omega_0$, the velocity leads the force by an angle of $\pi/2$. On the other hand, for $\omega \gg \omega_0$, $\phi = \pi$ so that $\delta = \pi - (\pi/2) = (\pi/2)$

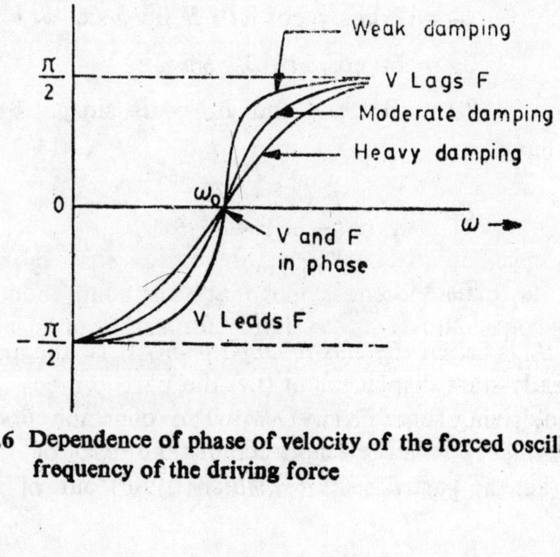

Fig. 4.6 Dependence of phase of velocity of the forced oscillator on the frequency of the driving force

implying that, in the case of very high frequencies of the driving force, the velocity of the oscillator lags behind the force by an angle of $\pi/2$. However, at resonance ($\omega = \omega_0$), $\phi = \pi/2$; so that $\delta = \frac{\pi}{2} - \frac{\pi}{2} = 0$. *Thus at resonance, the velocity of the driven oscillator is in phase with the driving force.* This is, therefore, the most favourable circumstance for the transference of energy from the driving force to the oscillator, for both V and F are in phase and V has its maximum value (see the next subsection). Figure 4.6 shows the dependence of the phase of the velocity of the oscillator on the frequency of the driving force.

Power Supplied to the Oscillator by the Driving Force

We have seen in Chap. 3 that the energy of a free damped oscillator decays exponentially as $E(t) = E_0 e^{-\gamma t}$. In order to maintain the steady state oscillations of such a system, the driving force must supply the necessary energy that the system loses in each cycle due to the presence of friction. We have seen that a given oscillator (i.e. given values of m, K and p) driven by a given harmonic force (i.e. given values of F_0 and ω), the amplitude [see Eq. (4.7)] of the steady state oscillations is constant. Thus, *when the oscillator has settled down to a steady state of oscillations, the average power supplied to it by the driving force must be equal to the average power dissipated by friction.* We shall now derive the expression for the average power absorbed by the oscillator and show that it is, indeed, equal to the average power dissipated.

While discussing the problem of transfer of energy from the driving force to the oscillator, it is instructive and physically more meaningful to describe the steady state displacement ψ_s in terms of two amplitudes B_{el} and B_{ab} instead of describing it in terms of amplitude B and phase constant ϕ as in Eq. (4.7). which can be rewritten as

$$\psi_s = B \cos \phi \cos \omega t + B \sin \phi \sin \omega t$$

or

$$\psi_s = B_{el} \cos \omega t + B_{ab} \sin \omega t \tag{4.19}$$

where $B_{el} = B \cos \phi$ and $B_{ab} = B \sin \phi$. From Eqs. (4.5) and (4.6) we have

$$B_{el} = \frac{f_0 (\omega_0^2 - \omega^2)}{(\omega_0^2 - \omega^2)^2 + \omega^2 \gamma^2} \tag{4.20}$$

$$B_{ab} = \frac{f_0 \omega \gamma}{(\omega_0^2 - \omega^2)^2 + \omega^2 \gamma^2} \tag{4.21}$$

The constant B_{el} is called the *elastic amplitude.* It is the amplitude of that part of the steady-state displacement (i.e. the part $B_{el} \cos \omega t$) which is in phase with the driving force $F_0 \cos \omega t$. The constant B_{ab} is called the *absorptive amplitude.* It is the amplitude of that part of the steady-state displacement (i.e. the part $B_{ab} \sin \omega t$) which is 90° out of phase with the driving force $F_0 \cos \omega t$. These names are chosen for reasons explained in the sequence. In terms of B_{el} and B_{ab} the velocity of the oscillator is

$$\frac{d\psi_s}{dt} = \omega\,(B_{ab} \cos \omega t - B_{el} \sin \omega t) \tag{4.22}$$

If the driving force $F(t) = F_0 \cos \omega t$ produces an infinitesimal displacement $d\psi_s$ in time dt, then the work done or energy supplied during time $dt = F(t)\, d\psi_s$. Hence the instantaneous power (i.e. rate at which work is done) supplied is given by

$$P(t) = F(t)\frac{d\psi_s}{dt}$$

Substituting for $\frac{d\psi_s}{dt}$ from Eq. (4.22), we have

$$P(t) = F_0\, \omega \cos \omega t\,(B_{ab} \cos \omega t - B_{el} \sin \omega t) \tag{4.23}$$

Denoting the time average over one cycle by the brackets $< >$, we find that the average power supplied in one cycle is given by

$$P = \langle P(t) \rangle = F_0\, \omega B_{ab} \langle \cos^2 \omega t \rangle - F_0 \omega B_{el} \langle \cos \omega t \sin \omega t \rangle$$

But $$\langle \cos^2 \omega t \rangle = \frac{1}{T}\int_0^T \cos^2 \omega t\, dt = \frac{1}{T}\int_0^T \cos^2\left(\frac{2\pi t}{T}\right)dt = \frac{1}{2}$$

where $T = \frac{2\pi}{\omega}$ is the time period of the steady-state oscillations.

Similarly,

$$\langle \cos \omega t \sin \omega t \rangle = \tfrac{1}{2} \langle \sin 2\omega t \rangle$$

$$= \frac{1}{2T}\int_0^T \sin\left(\frac{4\pi t}{T}\right)dt$$

$$= 0$$

Thus, we obtain the average power supplied over one cycle,

$$P_{in} = \tfrac{1}{2} F_0\, \omega B_{ab} \tag{4.24}$$

Equation (4.24) shows that the time-averaged input power is proportional to the amplitude B_{ab} of that part of the steady-state displacement ψ_s, which is 90° out of phase with the driving force. The time-averaged input power is entirely due to the term $B_{ab} \sin \omega t$ in Eq. (4.19) It is for this reason that we gave the name absorptive amplitude to B_{ab}. The term B_{el} does contribute to the instantaneous power absorption $P(t)$, as in Eq. (4.23), but averages to zero over one cycle of steady-state oscillation.

The power supplied by the driving force is not stored in the system, but is dissipated as work done in moving the system against the force of friction. The instantaneous power dissipated through friction is given by

$$P(t) = \text{instantaneous frictional force} \times \text{instantaneous velocity}$$

$$= p \frac{d\psi_s}{dt} \cdot \frac{d\psi_s}{dt} = p \left(\frac{d\psi_s}{dt} \right)^2$$

Substituting for $\frac{d\psi_s}{dt}$ from Eq. (4.22), we have

$$P(t) = p\omega^2 (B_{ab}^2 \cos^2 \omega t + B_{el}^2 \sin^2 \omega t - 2 B_{ab} B_{el} \cos \omega t \sin \omega t) \quad (4.25)$$

Averaging over one cycle of oscillations, as before, we find that the average power dissipated through friction is given by

$$P_{\text{dis}} = \frac{p\omega^2}{2} \left(B_{ab}^2 + B_{el}^2 \right) \quad (4.26)$$

Use of Eqs. (4.20) and (4.21) yields

$$P_{\text{dis}} = \frac{p\omega^2 f_0^2}{2\{(\omega_0^2 - \omega^2)^2 + \omega^2 \gamma^2\}} = \tfrac{1}{2} F_0 \, \omega B_{ab}$$

which is the same as Eq. (4.24). Thus

$$P_{\text{input}} = P_{\text{dissipated}}$$

i.e. in the steady state the average power supplied equals the average power dissipated. Notice that the instantaneous input power is not equal to the instantaneous power dissipated. Therefore, at any instant of time, the power stored in the oscillator is not constant. However, when an averaging is carried out over one cycle, the time-averaged input power and the time-averaged power dissipated become equal. It is for this reason that the forced oscillator settles down to a steady state of oscillations with a constant amplitude.

Significance of Elastic and Absorptive Amplitudes. It is evident from Eq. (4.19) that the steady-state displacement ψ_s is a superposition of two terms: the elastic term $B_{el} \cos \omega t$ (which is in phase with the driving force) and the absorptive term $B_{ab} \sin \omega t$ (which is 90° out of phase with the driving force). As shown above, the elastic term does not contribute to the time-averaged power absorption which is governed entirely by the absorptive term. Furthermore, at resonance (i.e. at $\omega \simeq \omega_0$), B_{el} is zero [see Eq. (4.20)]. This does not mean that B_{el} is unimportant. In fact, if the driving frequency is far from resonance, the elastic terms is a dominant one. This can be easily seen from Eqs. (4.20) and (4.21) which give

$$\frac{B_{el}}{B_{ab}} = \frac{\omega_0^2 - \omega^2}{\gamma \omega}$$

For $\omega < \omega_0$, the ratio B_{el}/B_{ab} is positive and can assume a large value if ω is sufficiently small. For $\omega \gg \omega_0$, the ratio B_{el}/B_{ab} is negative and its magnitude can be made as large as we please by choosing ω sufficiently large. For both these cases, we have $\gamma\omega \ll |\omega_0^2-\omega^2|$ and we may neglect the small contribution $B_{ab} \sin \omega t$ to ψ_s. We have seen above, that far from resonance, the power absorption is negligible compared to that near resonance. Thus, far from resonance the steady-state solution is just given by $B_{el} \cos \omega t$ with B_{el} given by

$$B_{el} = \frac{f_0}{\omega_0^2-\omega^2}$$

Here we have neglected the term $\gamma^2\omega^2$ in the denominator of Eq. (4.20). Thus, far from resonance,

$$\psi_s \cong \frac{f_0 \cos \omega t}{(\omega_0^2-\omega^2)} \tag{4.27}$$

An interesting observation is that the damping constant γ has completely disappeared from the resulting Eq. (4.27). In fact, by direct substitution, one can immediately check that Eq. (4.27) gives the exact steady-state solution of the equation of motion [Eq. (4.1)] if we set $\gamma = 0$ in this equation.

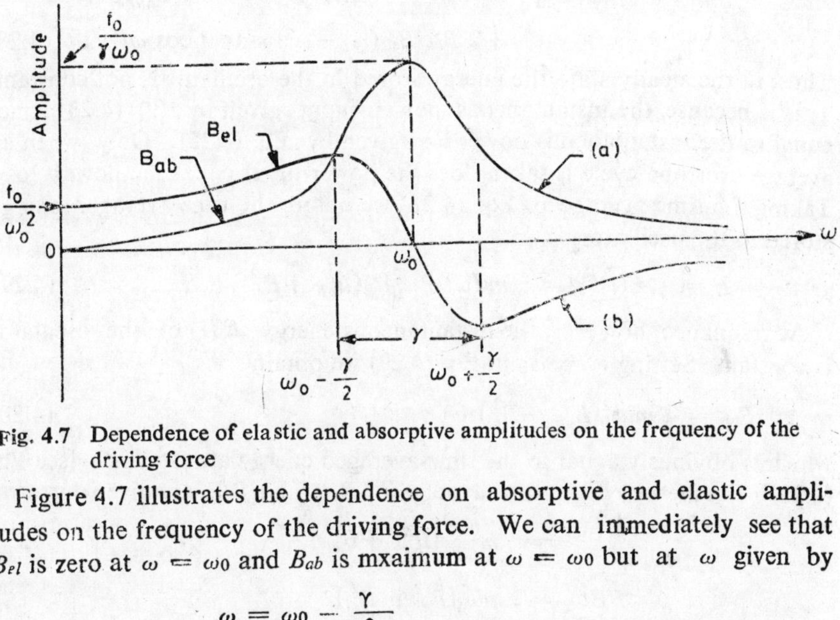

Fig. 4.7 Dependence of elastic and absorptive amplitudes on the frequency of the driving force

Figure 4.7 illustrates the dependence on absorptive and elastic amplitudes on the frequency of the driving force. We can immediately see that B_{el} is zero at $\omega = \omega_0$ and B_{ab} is mxaimum at $\omega = \omega_0$ but at ω given by

$$\omega = \omega_0 - \frac{\gamma}{2}$$

they are equal. We come across these curves in the study of optical dispersion where an electron in an atom is the forced oscillator and the oscillating electric vector of the electromagnetic wave provides the driving force. If the frequency of the wave is nearly equal to the frequency of electron oscillations in the atom, the atom absorbs a large amount of energy from the wave and the characteristic absorption curve is the one shown in Fig. 4.7 *a*. On the other hand, curve (*b*) in Fig 4.7 represents the variation of refractive index of the medium with frequency of the wave in the region of anomalous dispersion.

Energy of the Driven Oscillator. The instantaneous kinetic energy of the oscillator in the steady state is given by [use Eq. (4.22)].

$$KE = \frac{1}{2} m \left(\frac{d\psi_s}{dt}\right)^2$$

$$= \tfrac{1}{2} m\, \omega^2 (B_{ab}^2 \cos^2 \omega t + B_{el}^2 \sin^2 \omega t - 2B_{ab} B_{el} \sin \omega t \cos \omega t)$$

The instantaneous potential energy is given by [use Eq. (4.19) and $\omega_0^2 = K/m$]

$$PE = \tfrac{1}{2} K \psi_s^2$$

$$= \tfrac{1}{2} m\, \omega_0^2 (B_{ab}^2 \sin^2 \omega t + B_{el}^2 \cos^2 \omega t + 2B_{ab} B_{el} \sin \omega t \cos \omega t)$$

The total energy of the oscillator at any instant of time is

$$E(t) = KE + PE$$

$$= \tfrac{1}{2} m \{(\omega^2 B_{ab}^2 + \omega_0^2 B_{el}^2) \cos^2 \omega t + (\omega^2 B_{el}^2 + \omega_0^2 B_{ab}^2) \sin^2 \omega t$$

$$+ 2 B_{ab} B_{el} (\omega_0^2 - \omega^2) \sin \omega t \cos \omega t\} \qquad (4.28)$$

Thus, in the steady state the energy stored in the oscillator is not constant. This is because the instantaneous power input given by Eq. (4.23) is not equal to the instantaneous power loss given by Eq. (4.25). Only when an average over one cycle is taken does the power input equal the power loss. Taking the time-average of Eq. (4.28) we obtain the time-averaged energy stored in the oscillator.

$$E = \langle E(t) \rangle = \tfrac{1}{4} m (\omega^2 + \omega_0^2)\ (B_{ab}^2 + B_{el}^2) \qquad (4.29)$$

At resonance, however, the instantaneous energy $E(t)$ of the oscillator is constant. Setting $\omega = \omega_0$ in Eq. (4.29) we obtain

$$E_{res} = \tfrac{1}{2} m\, \omega_0^2 (B_{ab}^2 + B_{el}^2) \qquad (4.30)$$

which is obviously equal to the time-averaged energy at resonance [see Eq. (4.29)]. Furthermore, the time-average of *K.E.* and *P.E.* are given by

$$\langle KE \rangle = \tfrac{1}{4} m\omega^2 (B_{ab}^2 + B_{el}^2)$$

$$\langle PE \rangle = \tfrac{1}{4} m\omega_0^2 (B_{ab}^2 + B_{el}^2)$$

Notice that at resonance ($\omega = \omega_0$) the time-averaged kinetic energy equals the time-averaged potential energy. In Eq. (4.28) the term with ω_0^2 is the time averaged kinetic energy and that with ω^2 is the time-averaged potential energy. They are equal only when $\omega = \omega_0$. If ω is large compared with ω_0, the time-averaged energy is predominently kinetic and when ω is small compared with ω_0 it is predominantly potential. This can be understood qualitatively as follows : If $\omega \gg \omega_0$, the frequency of steady-state oscillations is so large that the oscillator velocity reverses on a time scale which is so short that it fails to follow the driving force and hence does not succeed in acquiring a substantial displacement and consequently, a substantial potential energy. On the other hand when $\omega \ll \omega_0$, the oscillator frequency is small and it acquires sufficient displacement during the long time scale but its velocity never becomes too large. Consequently its kinetic energy never becomes large. The time-averaged energy is therefore predominently potential.

We shall now make a very important observation. From Eqs. (4.26) and (4.30) we notice that, at resonance ($\omega = \omega_0$).

$$E = \frac{P_{\text{dis}}}{\gamma} = \Gamma \times P_{\text{dis}} \tag{4.31}$$

where $\Gamma = 1/\gamma$ is the decay time of free damped oscillations (See Chap. 3). In other words, *at resonance the time-averaged energy stored in the oscillator equals the average power dissipated through friction times the mean decay time of the free damped oscillations.* This can be qualitatively understood as follows : If the driving force were not applied, the energy of the oscillator would decay exponentially and most of it would be lost during the decay Γ. But when the external force is applied at a frequency ω equal to the frequency ω_0 of free oscillations, the amplitude starts increasing until a steady state is reached when the average power supplied (or put in) by the driving force equals the average power dissipated by friction. Since friction dissipates most of the energy in the decay time Γ, the steady state energy must be equal to the energy supplied during time Γ. Thus, in the steady state we expect that the stored energy is nearly equal to the average input power times the decay time Γ. Since the average input power equals the average power dissipated, the result in Eq. (4.31) immediately follows.

Response : Sharpness of Resonance and Quality Factor

We shall now consider how the oscillator responds to the driving force when the driving frequency is slowly varied. The response of the oscillator obviously depends upon the magnitude of power it extracts from the driving force during each cycle of its oscillation. We know that the time-averaged input power is given by [see Eqs. (4.21) and (4.24)].

$$P = \tfrac{1}{2} F_0 \, \omega \, B_{ab} = \frac{F_0^2 \gamma}{2m} \left\{ \frac{\omega^2}{(\omega_0^2 - \omega^2)^2 + \gamma^2 \omega^2} \right\} \tag{4.32}$$

The maximum value of P occurs at a frequency ω satisfying the following two conditions

$$\frac{d}{d\omega} \left[\frac{\omega^2}{(\omega_0^2 - \omega^2)^2 + \gamma^2 \omega^2} \right] = 0$$

and
$$\frac{d^2}{d\omega^2}\left[\frac{\omega^2}{(\omega_0^2-\omega^2)^2+\gamma^2\omega^2}\right] \leqslant 0$$

It is straight forward to see that these conditions are satisfied for $\omega = \omega_0$, in other words the maximum value of P (denoted by P_{max}) occurs at resonance. Setting $\omega = \omega_0$ in Eq. (4.32) we have

$$P_{max} = \frac{F_0^2}{2m\gamma}$$

In terms of P_{max} Eq. (4.32) can be rewritten as

$$P = P_{max}\frac{\gamma^2\omega^2}{(\omega_0^2-\omega^2)^2+\gamma^2\omega^2} \tag{4.33}$$

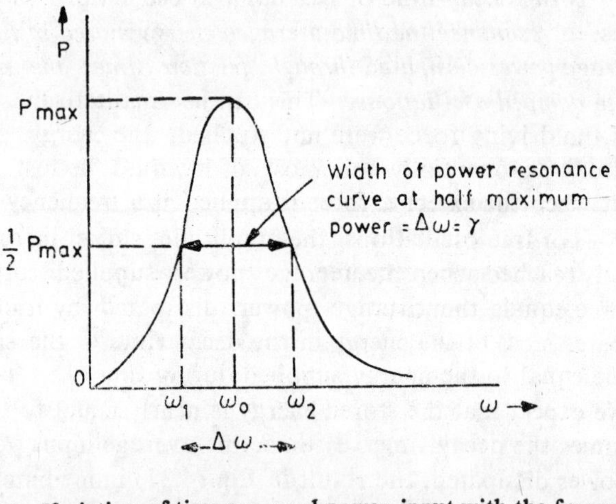

Fig. 4.8 Variation of time-averaged power input with the frequency of the driving force

Figure (4.8) depicts the variation of P as a function of ω. It is a plot of P against ω in Eq. (4.33) for given values of γ and ω_0. The values of ω at which P is half its maximum value are called *half power points*. From Eq. (4.33) the half power points are the values of ω satisfying the equation

$$\frac{1}{2} = \frac{\gamma^2\omega^2}{(\omega_0^2-\omega^2)^2+\gamma^2\omega^2}$$

or
$$\omega^2 = \omega_0^2 \pm \gamma\omega$$

These are two quadratic equations in ω, namely, $\omega^2+\gamma\omega-\omega_0^2 = 0$ and $\omega^2-\gamma\omega-\omega_0^2 = 0$. Each of these equations has one positive root and one negative root. Since angular frequency ω cannot be negative, we retain only the positive roots, which are

$$\omega_1 = -\frac{\gamma}{2} + \left(\omega_0^2 + \frac{\gamma^2}{4}\right)^{\frac{1}{2}}$$

$$\omega_2 = +\frac{\gamma}{2} + \left(\omega_0^2 + \frac{\gamma^2}{4}\right)^{\frac{1}{2}}$$

The frequency interval between the two half-power points is

$$\Delta\omega = \omega_2 - \omega_1 = \gamma$$

The frequency interval $\Delta\omega$ between the two half-power points is called *full frequency width at half-maximum power* or simply *band width.* Since $\gamma = 1/\Gamma$ where Γ is the energy decay time, or mean decay time of free damped oscillations [see Chap. 3], we have

$$\Delta\omega \cdot \Gamma = 1 \tag{4.33a}$$

In other words, *the frequency width of the resonance curve for driven oscillators is equal to the inverse of the mean decay lifetime for free damped oscillations.* Equation [4.33(a)] is of profound importance from the point of view of experimental observation. Experimentally it is difficult to observe the free decay of oscillations, i.e. it is extremely difficult to make a reasonable estimate of the decay time Γ. But it is easier to observe the resonant response of a driven system, i.e. it is easy to estimate $\Delta\omega$. One can, therefore, obtain the decay time Γ for free oscillations by studying instead the resonance response of the system and computing $\Delta\omega$. The decay time Γ is then immediately obtained from Eq. (4.33a).

We shall next discuss the sharpness of resonance which is measured in terms of what is called *quality* or *Q factor* defined as

$$Q = \frac{\omega_0}{\omega_2 - \omega_1} = \frac{\omega_0}{\Delta\omega} = \frac{\text{resonant frequency}}{\text{bandwidth}}$$

Since $\Delta\omega = \gamma = \frac{1}{\Gamma}$, we have

$$Q = \frac{\omega_0}{\gamma} = \Gamma\,\omega_0 \tag{4.34}$$

This definition, however, may also be obtained directly as follows [see Ch. 3). By definition,

$$Q = 2\pi\,\frac{\text{average energy stored in one period}}{\text{average energy lost in one period}}$$

$$= 2\pi\,\frac{E}{P_{\text{dis}}\,T}$$

where P_{dis} is the average power dissipated and $T = (2\pi/\omega)$ is the time period of oscillations. Thus $P_{\text{dis}} \times T$ is the average energy lost in one period. Using Eqs. (4.26) and (4.29) we have [since $\gamma = p/m$]

$$Q = \frac{\omega^2 + \omega_0^2}{2\gamma\omega}$$

Near resonance $\omega \simeq \omega_0$, we have

$$Q = \frac{\omega_0}{\gamma}$$

which agrees with Eq. (4.34). For low damping, γ is very small compared with ω_0 and Q will be very large, making the resonance very sharp. Thus, *quality factor Q is a measure of the sharpness of resonance in the case of driven harmonic oscillators.* This can be understood if we rewrite Eq. (4.7) for the amplitude of the steady-state oscillations in terms of Q as defined by Eq. (4.34). Using Eq. (4.34) in Eq. (4.7) we have

$$B = \frac{f_0\, Q/\omega\omega_0}{\left\{1+Q^2\left(\frac{\omega_0}{\omega}-\frac{\omega}{\omega_0}\right)^2\right\}^{\frac{1}{2}}}$$

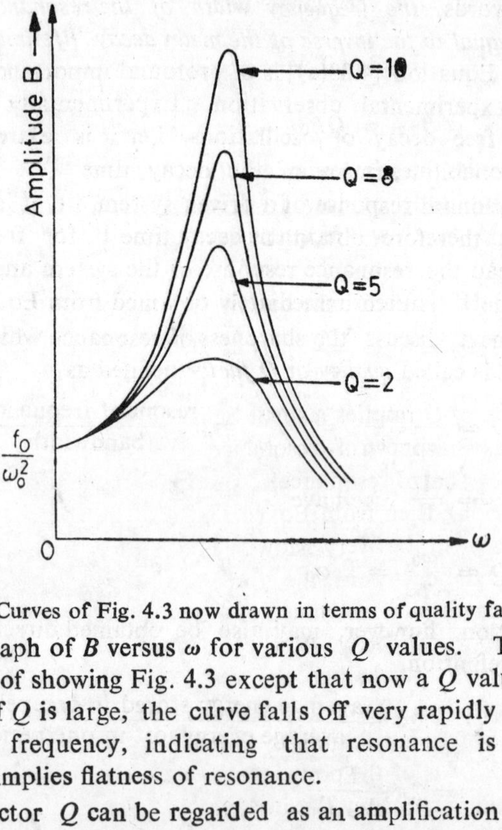

Fig. 4.9 Curves of Fig. 4.3 now drawn in terms of quality factor Q.

Figure 4.9 is a graph of B versus ω for various Q values. This figure is just another way of showing Fig. 4.3 except that now a Q value is attached to each curve. If Q is large, the curve falls off very rapidly on both sides of the resonant frequency, indicating that resonance is very sharp. Smaller Q value implies flatness of resonance.

The quality factor Q can be regarded as an amplification factor. This can be understood as follows: It is obvious from Eq. (4.7) that at low driving frequencies ($\omega \to 0$), the amplitude of oscillations has a value given by

$$B_0 = \frac{f_0}{\omega_0^2}$$

We have already seen that the maximum value of B is given by [see Eq. (4.15)]

$$B_{\max} = \frac{f_0}{\gamma\left(\omega_0^2 - \frac{\gamma^2}{4}\right)^{\frac{1}{2}}}$$

So that

$$\frac{B_{\max}}{B_0} = \frac{\omega_0^2}{\gamma\left(\omega_0^2 - \frac{\gamma^2}{4}\right)^{\frac{1}{2}}} = \frac{\omega_0}{\gamma\left(1 - \frac{\gamma^2}{4\omega_0^2}\right)^{\frac{1}{2}}}$$

$$= \frac{Q}{\left(1 - \frac{1}{4Q^2}\right)^{\frac{1}{2}}}$$

For low damping, Q is very large, hence

$$\frac{B_{\max}}{B_0} = Q\left(1 - \frac{1}{4Q^2}\right)^{-\frac{1}{2}} \simeq Q\left(1 + \frac{1}{8Q^2}\right) \simeq Q$$

Thus $\qquad B_{\max} = QB_0$

In other words, the maximum amplitude (which occurs at resonance) is Q times the amplitude at low frequencies of the driving force. Thus, the amplitude at low frequencies is amplified by a factor Q at resonance.

In terms of the Q-factor, the bandwidth $\Delta\omega$ can be written as

$$\Delta\omega = \gamma = \frac{\gamma}{\omega_0} \cdot \omega_0 = \frac{\omega_0}{Q}$$

A high value of Q implies a small bandwidth. Extremely narrow bandwidth indicates sharpness of resonance. If a system is observed to have a very narrow (or sharp) resonance response (as measured by amplitude or power absorption), then from Eq. (4.33a) it is clear that the decay of its free oscillations will be very slow. Flat (or broad) resonance response indicates that the decay of the free oscillations will be fast. Thus the sharpness or flatness of resonance of a driven system depends on whether the free oscillations decay slowly or quickly. A measurement of the decay time Γ tells us whether the resonance is sharp or flat. What is our criterion of sharpness or flatness of resonance or slowness or quickness of the decay of free oscillations ? We can say that the resonance is sharp if the bandwidth $\Delta\omega$ of the power resonance curve is very small compared with the resonant frequency ω_0, i.e.

$$\frac{\Delta\omega}{\omega_0} \ll 1 \qquad (4.35\text{ a})$$

and we can say that the decay of free oscillations is slow if the oscillator loses only a small fraction of its energy in one cycle of oscillation. From equation

$$E = E_0 \, e^{-\gamma t}$$

it is obvious that the loss of energy δE in time δt is

$$\delta E = -\gamma \, F_0 \, e^{-\gamma t} \, . \, \delta t$$

Hence the fraction of energy lost in time δt is given by

$$\frac{\delta E}{E} = -\gamma \delta t$$

If δt is put equal to time period $2\pi/\omega_0$ of free oscillations, we have

$$\frac{\delta E}{E} = -\frac{2\pi\gamma}{\omega_0}$$

Then a slow decay means

$$\frac{2\pi\gamma}{\omega_0} \ll 1 \tag{4.35 b}$$

Since $\gamma = \Delta\omega = \omega_0/Q$, the conditions described in Eqs. (4.35 a) and (4.35 b) can both be met by saying that the dimensionless factor Q must be large. Thus a large Q-value implies sharpness of resonance which, in turn, implies slowness of the decay of free oscillations.

This relation between the resonance width of forced oscillations and the decay of free oscillations is characteristic of a wide variety of oscillatory physical systems (we have used a mechanical system as an example). A few such systems are mentioned below. Whenever the free oscillations of such a system show an exponential decay of energy with time, it also shows resonance characteristics if driven by a harmonically varying force.

4.4 DRIVEN LCR CIRCUIT

In Sec. 4.3 we have discussed the resonant behaviour of a simple mechanical system consisting of a mass and a spring that is driven harmonically by an external periodic force. We shall now discuss resonance in electrical systems. One of the most familiar and important resonant systems is an electrical circuit consisting of a capacitance (C), an inductance (L) and a resistance (R). In Chap. 3 we have discussed the free damped oscillations in such a system. We shall now analyse the resonant response of such a system when it is driven by an external source of alternating *emf* to supply the necessary energy to maintain the oscillations. Let us first discuss resonance in a series *LCR* circuit.

Series Resonance Circuit

A series *LCR* circuit is shown in Fig. 4.10. The circuit is driven by an alternating applied voltage $V = V_0 \cos \omega t$ of angular frequency ω. Let I

be the current in the circuit at a given instant of time with q as the charge on the capacitor at that instant. The potential difference across the capacitor plates is given by

$$V_c = \frac{q}{C}$$

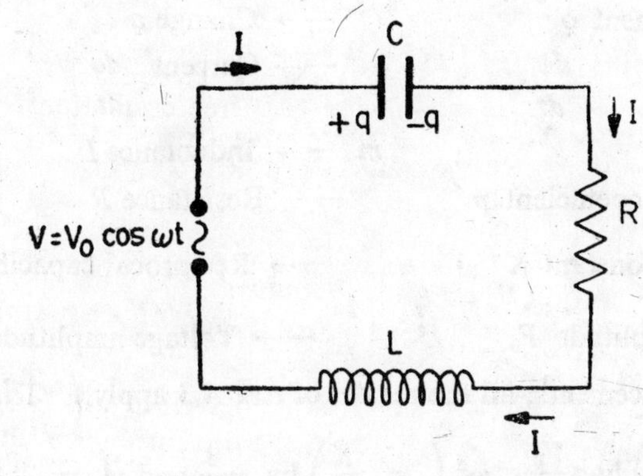

Fig. 4.10 Series LCR circuit driven by an alternating voltage

Since the current is changing with time, there is a potential differenee between the ends of the inductance L. This voltage is proportional to the rate of change of current I and is given by

$$V_L = L\frac{dI}{dt}$$

The instantaneous value of V_L is positive or negative depending on whether (at that instant) the current is increasing or decreasing with time. Finally the potential difference V_R across the resistance R is given by

$$V_R = IR$$

If $V = V_0 \cos \omega t$ is the applied voltage, it is distributed among the three components C, L and R. In other words,

$$V = V_C + V_L + V_R$$

or

$$L\frac{dI}{dt} + IR + \frac{q}{C} = V_0 \cos \omega t$$

Since $I = \dfrac{dq}{dt}$, we have

$$L\frac{d^2q}{dt^2} + R\frac{dq}{dt} + \frac{q}{C} = V_0 \cos \omega t$$

or

$$\frac{d^2q}{dt^2} + \frac{R}{L}\frac{dq}{dt} + \frac{q}{LC} = \frac{V_0}{L}\cos \omega t \tag{4.36}$$

Compare this equation with Eq. (4.1) of a mechanical oscillator which reads

$$\frac{d^2\psi}{dt^2}+\frac{p}{m}\frac{d\psi}{dt}+\frac{K}{m}\psi=\frac{F_o}{m}\cos\omega t \qquad (4.37)$$

Equation (4.36) becomes identical with Eq. (4.37) if the following replacements are made :

Displacement ψ	$\longrightarrow$ Change q
Velocity $\frac{d\psi}{dt}$	$\longrightarrow$ Current $\frac{dq}{dt}$
Mass m	$\longrightarrow$ Inductance L
Damping coefficient p	$\longrightarrow$ Resistance R
Stiffness constant K	$\longrightarrow$ Reciprocal capacitancc $\frac{1}{C}$
Force amplitude F_0	$\longrightarrow$ Voltage amplitude V_0

with these replacements, all the results of Art. 4.3 apply. Thus, replacing $f_0\left(=\frac{F_0}{m}\right)$ by $\frac{V_0}{L}$, $\omega_0^2\left(=\frac{K}{m}\right)$ by $\frac{1}{LC}$ and $\gamma\left(=\frac{p}{m}\right)$ by R/L and representing, as before, $V=V_0\cos\omega t$ by $V=V_0e^{i\omega t}$, we find that the current in the circuit is given by the real part of the expression [see the discussion after Eq. (4.4)]

$$I=\frac{dq}{dt}=i\,\omega q_0\,e^{i(\omega t-\phi)}=i\omega q_0 e^{-i\phi}\,e^{i\omega t}$$

where q_0 is the amplitude of charge q which is given by the real part of

$$q=q_0\,e^{i(\omega t-\phi)}$$

As in Eq. (4.5), q_0 is given by

$$q_0\,e^{-i\phi}=\frac{V_0/L}{\left(\frac{1}{LC}-\omega^2\right)+i\omega\frac{R}{L}}$$

Thus I is given by

$$I=\frac{i\omega\frac{V_0}{L}e^{i\omega t}}{\left(\frac{1}{LC}-\omega^2\right)+i\omega\frac{R}{L}}=I_0\,e^{i\omega t} \qquad (4.38)$$

where $I_0=\dfrac{V_0}{R+i\left(\omega L-\frac{1}{\omega C}\right)}$ is the amplitude of current.

Since $V = V_0\, e^{-i\omega t}$, we have

$$I = \frac{V}{Z_e} \tag{4.39}$$

where
$$Z_e = R + i\left(\omega L - \frac{1}{\omega C}\right) \tag{4.40}$$

Equation (4.38) is the expression for Ohm's law for a circuit involving L and C in addition to R. For a circuit involving only a resistance R, Ohm's law takes the scalar from $I = V/R$. The denominator Z_e in Eq. (4.39) is, therefore, the effective resistance in the circuit. It is called the *electrical impedance* (or simply impedance) in the circuit and is measured in ohms. Clearly, the impedance Z_e of the a.c. circuit is made up of two parts ; the ohmic resistance R and the quantity $\left(\omega L - \dfrac{1}{\omega C}\right)$ called the *reactance,* usually denoted by the letter X_e. The reactance is also made up of two parts ; $L\omega$, the reactance due to inductance (which is the value of the effective resistance presented by the inductance to the current of frequency ω) and $1/\omega C$, the reactance due to capacitance (which is the value of the effective resistance presented by the capacitance to the current of frequency ω). Equation (4.39) tells us that Z_e is the vector sum of R and $X_e = \left(\omega L - \dfrac{1}{\omega C}\right)$. This is illustrated in Fig. 4.11.

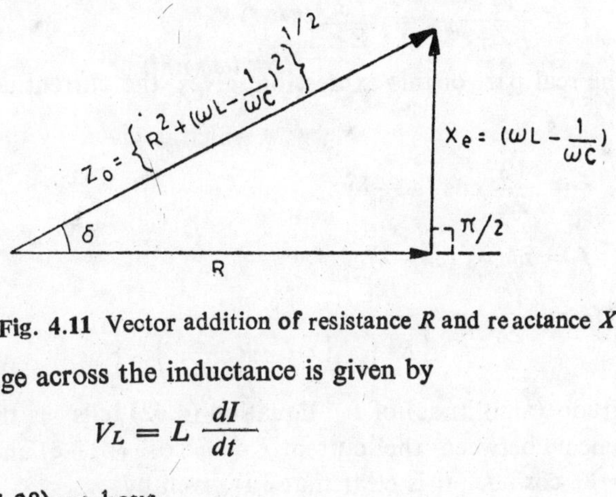

Fig. 4.11 Vector addition of resistance R and reactance X_e

The voltage across the inductance is given by

$$V_L = L\,\frac{dI}{dt}$$

From Eq. (4.38) we have

$$V_L = i\omega L\, I_0\, e^{i\omega t} = i\omega L\, I$$

The product ωLI has the dimensions of ohms times current in amperes and is, therefore, measured in volts. From the above expression and from the fact that $i = e^{i\pi/2}$ it is clear that the *phase of the voltage V_L across the inductance is $\dfrac{\pi}{2}$ (or 90°) ahead of that of the current in the circuit.*

Similarly, the voltage across the capacitance is

$$V_C = \frac{q}{C} = \frac{1}{C}\int I\,dt = \frac{I_0}{C}\int e^{i\omega t}\,dt = \frac{I_0\,e^{i\omega t}}{i\omega C} = -\frac{iI}{\omega C}$$

This relation tells us that (since $-i = e^{-i\pi/2}$), *the phase of the voltage V_C across the capacitance lags the current by 90°*). The voltage across the resistance R is in phase with the current. The quantities ωL and $1/\omega C$ are called reactances because they introduce a phase relationship as well as an effective resistance. This phase relationship is illustrated in Fig. 4.11.

Separating the magnitude Z_0 of Z_e and its phase part δ, we can rewrite Eq. (4.40) as

$$Z_e = Z_0\,e^{i\delta} = R + i\left(\omega L - \frac{1}{\omega C}\right)$$

Its complex conjugate is

$$Z_e^* = Z_0\,e^{-i\delta} = R - i\left(\omega L - \frac{1}{\omega C}\right)$$

From these two equations, the magnitude Z_0 of the impedance is given by

$$Z_0 = \sqrt{Z_e Z_e^*} = \left[R^2 + \left(\omega L - \frac{1}{\omega C}\right)^2\right]^{1/2} \tag{4.41}$$

Therefore, the current in the circuit is given by [see Eq. (4.38)]

$$I = \frac{V_0\,e^{i\omega t}}{Z_0\,e^{i\delta}} = \frac{V_0}{Z_0}\,e^{i(\omega t-\delta)}$$

Extracting the real part of this expression, gives the current actually in the circuit

$$I = \frac{V_0}{Z_0}\cos(\omega t - \delta)$$

or

$$I = I_0\cos(\omega t - \delta) \tag{4.42}$$

where

$$I_0 = \frac{V_0}{Z_0} = \frac{V_0}{\left\{R^2 + \left(\omega L - \frac{1}{\omega C}\right)^2\right\}^{1/2}}$$

is the magnitude (amplitude) of I Equation (4.42) tells us that there is a phase difference δ between the current $I = I_0\cos(\omega t - \delta)$ and the applied voltage $V = V_0\cos\omega t$. It is clear that δ is given by

$$\delta = \tan^{-1}\left\{\frac{\left(\omega L - \frac{1}{\omega C}\right)}{R}\right\} \tag{4.43}$$

The phase difference δ between the current and the applied voltage depends upon the relative values of ωL and $1/\omega C$, which, for given values of L and C, depend upon the angular frequency ω of the applied voltage. The following three cases arise.

Case 1 : Applied frequency ω less than resonant frequency ω_0. In this case ωL is less than $1/\omega C$. The reactance in the circuit is predominantly *capacitive* and tan δ is a negative quantity. In the limit $\omega \to 0$, $\omega L \to 0$ and $1/\omega C \to \infty$ and $\delta \to -\frac{\pi}{2}$. Hence, *for very small driving frequencies, the current leads the applied voltage by $\pi/2$.* (see Fig. 4.12b). The impedance of the circuit is large and hence the current is very small. (see Fig. 4.12a).

Case 2 : Applied frequency ω greater than resonant frequency ω_0. In this case, the reactance in the circuit is predominantly *inductive*. Since tan δ is a positive quantity, the applied voltage leads the current. For very high frequencies of the applied voltage $1/\omega C \to 0$ and ωL tends to infinity. So that $\delta \to +\frac{\pi}{2}$ and impedance becomes very large. Thus for very high frequencies of the driving voltage, the current in the circuit becomes very small and lags behind the voltage by $\pi/2$ (see Figs. 4.12a and 4.12b).

Case 3 : Applied frequency ω equal to the resonant frequency ω_0. In this case the reactance in the circuit is zero, therefore an impedance in the circuit is equal to the resistance R and is, hence, the least. The peak value current is maximum given by

$$I_0 = \frac{V_0}{R}$$

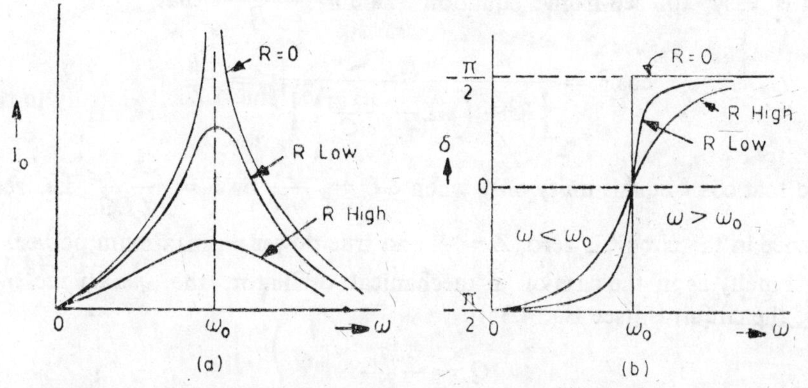

Fig. 4.12 Dependence of the peak value of the current and its phase on the frequency of the applied voltage. The resonant frequency is ω_0.

which approaches infinity as $R \to 0$. This is, therefore, the case of resonance. At resonance ($\omega = \omega_0$) the current and the applied voltage are in phase. This is analogous to the velocity resonance in the case of mechanical oscillators. Figures 4.12a and 4.12b show how the current I_0 and phase angle δ vary with the frequency of the driving voltage. Notice that at resonance I_0 is maximum and $\delta = 0$.

The power delivered to the $L-C-R$ circuit by the applied voltage can be easily computed. The instantaneous power $P(t)$ is given by

$$P(t) = VI$$

$$= V_0 \cos \omega t \times I_0 \cos (\omega t - \delta)$$

$$= V_0 I_0 (\cos^2 \omega t \cos \delta + \tfrac{1}{2} \sin 2\omega t \sin \delta)$$

Taking the time-average of this equation over one full cycle, we find that the time-average power delivered to the circuit is given by

$$P = \langle P(t) \rangle = \tfrac{1}{2} V_0 I_0 \cos \delta$$

(since the average value of sin 2 ωt over one cycle is zero). Far from resonance ($\omega \ll \omega_0$ or $\omega \gg \omega_0$), δ tends to $-\frac{\pi}{2}$ or $+\frac{\pi}{2}$ and $\cos \delta \to 0$. Hence, the average power delivered to the circuit is negligibly small (tending to zero) if the frequency of the applied voltage is either too small or too large compared to the resonance frequency. At resonance, however, $\omega = \omega_0$ and $\delta = 0$, hence $\cos \phi = 1$. Hence, at resonance, the power delivered is the maximum

$$P_{max} = \tfrac{1}{2} V_0 I_0$$

Thus,

$$\text{true power} = \text{maximum power} \times \cos \delta$$

It is for this reason that cos δ is called the *power factor* of the a. c. circuit.

It is easy to see from equation $\tan \delta = \dfrac{X_L - X_C}{R}$ that

$$\cos \delta = \frac{R}{\left\{R^2 + \left(\omega L - \dfrac{1}{\omega C}\right)^2\right\}^{1/2}} = \frac{R}{Z_0}$$

so that cos δ equals unity only when $\omega L = \dfrac{1}{\omega C}$ or $\omega = \dfrac{1}{\sqrt{LC}}$, i.e. reactance in the circuit is zero, Z = R and true power = maximum power.

Finally as in the case of a mechanical oscillator, the *quality factor Q* of the circuit is [(see Eq. 4.34)].

$$Q = \frac{\omega_0 L}{R}$$

Q is a measure of the sharpness of resonance. Smaller the value of R, sharper is the resonance. The most familiar example of resonance in electrical circuits is encountered when we tune our radio to a particular broadcasting station. There are many stations which send radio waves of various frequencies which produce forced oscillations in the circuit of the receiver. A particular setting of the tuner corresponds to a particular set of values of L and C and hence to a particular frequency of the circuit. When this frequency equals the frequency of the signal from the broadcasting station, the power absorption becomes maximum and hence we hear only that station.

A study of the equations of the mechanical and electrical systems reveals a striking similarity between the various parameters. Table 4.1 shows analogous parameters of mechanical and electrical systems.

Table 4.1 Analogous Parameters of Mechanical and Electrical Systems

Mechanical System	Electrical System
Displacement ψ	Charge q
Driving force $F = F_0 \cos \omega t$	Driving voltage $V = V_0 \cos \omega t$
Mass m	Inductance L
Frictional force constant p	Resistance R
Stiffness constant K	Reciprocal capacitance $1/C$
Resonant angular frequency $\omega_0 = \sqrt{K/M}$	Resonant angular frequency $\omega_0 = \sqrt{1/LC}$
Resonance width $\gamma = p/m$	Resonance width $\gamma = R/L$
Power absorbed at resonance $F_0^2/2p$	Power absorbed at resonance $V_0^2/2R$
Quality factor $Q = \omega_0\, m/p$	Quality factor $Q = \omega_0\, L/R$
Potential energy $\frac{1}{2}K\psi^2$	Electrostatic energy $\frac{1}{2}q^2/C$
Kinetic energy of moving mass $\frac{1}{2}m\dot{\psi}^2$	Electromagnetic energy of moving charge $\frac{1}{2}L\dot{q}^2$

Parallel Resonance Circuit

A parallel resonance circuit is shown in Fig. 4.13. Here the three components R, L and C are connected in parallel rather than in series as in Fig. 4.10. An alternating voltage $V = V_0 \cos \omega t$ is applied as shown. Let us say that, at an instant of time t, the current in the circuit is I. The relation between the driving voltage V and the current I in the circuit is

$$V = Z_e\, I$$

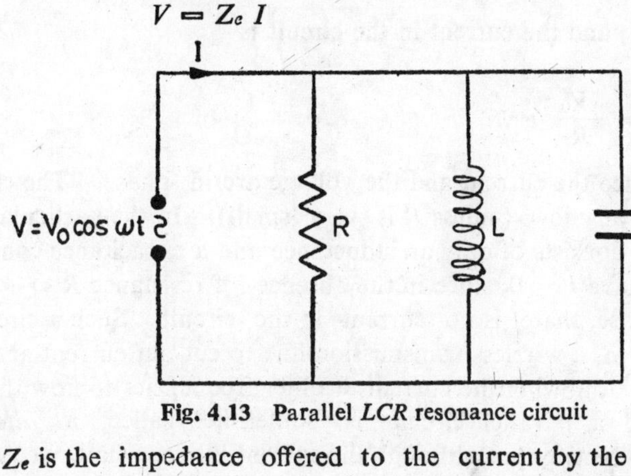

Fig. 4.13 Parallel *LCR* resonance circuit

where Z_e is the impedance offered to the current by the parallel combination of R, L and R. We have seen that $i\omega L$ and $\frac{1}{i\omega C}$ are the values of the effective resistance presented respectively by inductance L and capacitance C. Hence, impedance Z_e must be given by

$$\frac{1}{Z_e} = \frac{1}{R} + \frac{1}{i\omega L} + i\omega C$$

Therefore, the current I is given by

$$I = \frac{V}{Z_e} = V\left\{\frac{1}{R}+i\left(\omega C-\frac{1}{\omega L}\right)\right\}$$

Representing, as before, $V = V_0 \cos \omega t$ by $V = V_0\, e^{i\omega t}$, we have

$$I = V_0\, e^{i\omega t}\left\{\frac{1}{R}+i\left(\omega C-\frac{1}{\omega L}\right)\right\} \tag{4.44}$$

For extremely low frequency of the applied voltage (i.e. $\omega \to 0$), the dominant term in the square brackets in Eq. (4.44) is $-i/\omega L$ and the current in the circuit is

$$I = -\frac{iV_0}{\omega L}\, e^{i\omega t} = \frac{V_0}{\omega L}\, e^{i(\omega t-\pi/2)}$$

Thus, at extremely low frequencies of the driving voltage, the current lags the voltage by $\pi/2$. On the other hand, for extremely high frequencies ($\omega \to \infty$) the dominant term in the square brackets in Eq. (4.44) is $i\omega C$ and the current in the circuit is

$$I = i\omega C\, V_0\, e^{i\omega t} = \omega C V_0\, e^{i(\omega t+\pi/2)}$$

Thus, at extremely high frequencies of the driving voltage, the current leads the voltage by $\pi/2$. Notice that in both these limiting cases the magnitude of the current ($V_0/\omega L$ in the case $\omega \to 0$ and $\omega C V_0$ in the case $\omega \to \infty$) is quite large. But at the resonant frequency $\omega = \omega_0$, the imaginary term in the square brackets in Eq, (4.44) vanishes (because at this value of ω, $\omega C = \frac{1}{\omega L}$) and the current in the circuit is

$$I = \frac{V_0}{R}\, e^{i\omega t}$$

Thus, at resonance the current and the voltage are in phase. The current in the circuit is very low (unless R is very small). In the particular case when the circuit consists of only an inductance and a capacitance connected in parallel, we have $I = 0$, since in the absence of resistance $R = \infty$. In this particular case, thare is no current in the circuit. Such a circuit is, therefore, used in a wireless transmission line to cut out current at a particular frequency, allowing the current at other frequencies to flow through the line. Such a parallel circuit is sometimes called a *filter* or *rejector* circuit. Notice that although the current is very small, yet we use the word resonance, because the circuit absorbs maximum power at the resonant frequency, since at this frequency the current and the voltage are in phase. The power absorbed can be computed, in the case of resonance, as follows :

At resonance the current in the circuit is the real part of $I = \frac{V_0}{R}\, e^{i\omega t}$

i.e. $I = \frac{V_0}{R} \cos \omega t$. Hence the power absorbed is given by

$$P(t) = VI = V_0 \cos \omega t \times \frac{V_0}{R} \cos \omega t = \frac{V_0^2}{R} \cos^2 \omega t$$

The time-averaged power absorbed is $P = \langle P(t) \rangle = \frac{V_0^2}{R} \langle \cos^2 \omega t \rangle$

$$= \frac{1}{2} \frac{V_0^2}{R}$$

The time-averaged power absorbed is high if R is sufficiently low In the absence of R ($R = \infty$) the circuit (even when driven· at the resonant frequency $\omega_0 = 1/\sqrt{LC}$) cannot absorb any power from the applied voltage as a result of which no current flows in the circuit which, thus, acts as a filter circuit. Notice that the time-averaged power absorption is zero at very low and very high frequencies of the driving voltage because, in both these cases, the current and the voltage differ in phase by $\pi/2$. In the case $\omega = \omega_0$, the power absorption is maximum. Thus the condition $\omega = \omega_0$ is the condition of resonance in a parallel LCR circuit.

4.5 OTHER EXAMPLES OF RESONANCE

In physics we come across a variety of resonances in both mechanical and non-mechanical systems. The air-column in a resonance tube can vibrate in a number of modes (see Chap. 6). Resonance will occur in this system if frequency of any one of its modes is equal to the frequency of the driving tuning fork. When this happens, the amplitude of vibration of air becomes very large and a loud sound is heard. In the previous article we have learnt about resonance in electrical circuits. A sodium chloride crystal, which consists of positively and negatively charged ions, can absorb energy if it is subjected to an oscillating electric field. When the frequency of the relative oscillations of ions matches with the frequency of the electric field the crystal absorbs maximum energy from the field. Some resonances can cause disaster. A column of an army marching over a bridge can set in forced oscillations in the bridge If the frequency of their footsteps happens to match with one of the natural frequencies of the bridge, (which is determined by the dimensions and elastic properties of the material of the structure) resonance will take place and the bridge will start oscillating with a destructively large amplitude. This is the reason why soldiers break step when crossing a bridge.

In the examples cited above the resonant response of a system is measured by the amplitude of oscillations or power absorbed by the system which have a maximum value at or near the resonant frequency. The term resonance is, however, not restricted to mean (in a narrow sense) sudden amplification of amplitude or power. We can take a broad view and define *resonance as the phenomenon of driving a system under such suitable conditions that the interaction between the driving agency and the driven system is maximised.* The interaction has its maximum value at resonance ($\omega = \omega_0$) and its most marked changes occur over a range of about $\pm\gamma$ about the maximum. In the following examples of resonance, the response of the system is not necessarily measured in terms of amplitude or power absorbed.

Optical Resonance

If white light is passed through a gas or vapour at low pressure and the emergent light is analysed with a spectroscope, the spectrum of the emergent light consists of dark lines across a continuous background. These dark lines are due to the absorption of certain wavelengths by the gas or vapour. We say that we have got the *absorption* spectrum of the substance. The solar spectrum is an excellent example of an absorption spectrum. The dark lines in the solar spectrum are called *Fraunhofer lines.* The process of absorption is concerned with the interaction of light with matter. We know that an electron in an atom can oscillate. The frequency ν_0 of its oscillations is determined by the way it is bound to its atom. This binding provides the restoring force. When light of frequency $\nu = \nu_0$ falls on such an atom, resonance takes place. The atom extracts maximum energy from the light resulting in its absorption by the atom. The fact that the incident light must have the right frequency (or the right wavelength) to be absorbed, accounts for the narrow dark lines in the absorption spectrum.

Nuclear Resonance

In nuclear physics we come across innumerable examples of what is called *nuclear resonance.* Let us consider one simple example. When energetic protons (with energies around 875 ke V) are bombarded on a fluorine target, the fluorine nucleus captures (absorbs) a proton to form $^{20}N_e$ which is unstable and hence decays by emitting gamma rays.

The nuclear reaction can be written as

$$^{1}_{1}H + ^{19}_{9}F \rightarrow ^{20}_{10}N_e + \gamma$$

The relative yield of the emitted gamma rays is observed to change markedly with the energy of the bombarding proton. If the energy of the proton is 873.5 keV the yield of gamma rays shows a maximum. The fact that a proton with a right energy is captured accounts for an enhanced yield of gamma rays. This is a kind of a resonance in which the controlling parameter is not the frequency but the energy of the bombarding proton. Experiments of Herb, Snowden and Sala give a resonance curve (which is a graph of the relative yield of gamma rays versus the energy of the incident proton) which is almost identical in form to that of a mechanical oscillator (for high Q values) (see Fig. 4.9).

Nuclear Magnetic Resonance

This resonant phenomenon was discovered independently By F. Bloch and E.M. Purcell for which they were honoured with the Nobel Prize in physics in 1952. Nuclear magnetic resonance finds a wide variety of applications, particularly in experimental chemistry. It is essentially a quantum phenomenon. In very simple terms, the phenomenon can be described as follows : It is known that the nuclei of atoms behave like tinymagnets. A proton, for example, has only two possible orientations when it finds itself in a magnetic field. To turn this nuclear magnet from one orientation to the other, work is required to be done on it. This work is the energy difference between the two orientations. If an electromagnetic radiation of just the right energy (or frequency) is injected, the proton is observed to flip from one orientation to the other. This flipping is a consequence of a resonant process called nuclear magnetic resonance (*NMR* for short).

4.6 TRANSIENT BEHAVIOUR OF A FORCED OSCILLATOR

We have so far restricted our discussion only to the steady-state oscillations. In other words, we have analysed the behaviour of the forced oscillator after the driving force has been acting for a time which is long enough for the free oscillations to die out, so that the first term on the right-hand side of Eq. (4.2) does not contribute and the future oscillation is described by the second term. But in any real situation, the driving force is first applied at some instant which we shall call $t = 0$—and it is only some time later that the steady-state conditions prevail. During this time we have the *transient state* in which both the terms on the right-hand side of Eq. (4.2) contribute. This transient state may occupy a long time if the damping of the free oscillations is very small.

We have stated earlier that the steady-state solution (4.3) of Eq. (4.1) does not tell the complete history of the forced oscillator. This can be explained as follows : We have seen that the steady-state solution (4.3), contains constants B and ϕ which are respectively given by Eqs. (4.7) and (4.8). These constants are not arbitrary. They are determined by f_0, ω, ω_0 and γ, and hence are not adjustable. Now, according to the theory of differential equations, a complete and unique solution of a second-order differential equation must satisfy two requirements. First, it must satisfy the given second-order differential equation. Second, it must contain two arbitrary adjustable constants to match arbitrary initial conditions on displacement and velocity at time $t = 0$. Solution (4.3) with constants B and ϕ given by Eqs. (4.7) and (4.8) does satisfy Eq. (4.1), the differential equation of a harmonically driven harmonic oscillator. But the constants

B and ϕ are not adjustable. Hence solution (4.3) is incomplete and not unique. That the steady-state solution (4.3) is, not the complete solution of Eq. (4.1) can also be understood as follows : Let us suppose that the oscillator is a rest at its equilibrium position at time $t = 0$. At $t = 0$, the driving force $F = F_0 \cos \omega t$ is switched on and thereafter the motion of the oscillator is governed by Eq. (4.1) Let us assume that the solution (4.3) completely describes the behaviour of the oscillator. Let us emperically see what the consequences will be. If $\omega \leqslant \omega_0$ Eq. (4 8) gives $\phi \to 0$ so that Eq. (4.8) becomes

$$\psi_s = B \cos \omega t$$

But from Eq. (4.7) we find that, for $\omega \leqslant \omega_0$, the constant B is positive. Thus ψ_s assumes a *positive* value at $t = 0$ ($\because \cos \omega t = +1$ at $t = 0$). Thus the displacement at $t = 0$ immediately assumes a positive value. But no system with a finite mass, acted upon by a finite force can be displaced through a finite distance in zero time and if we suppose that $\omega \gg \omega_0$ the result is still more absurd. For $\omega \gg \omega_0$, $\phi \to \pi$ so that

$$\psi_s = -B \cos \omega t$$

Since B [see Eq. (4.7)] is positive for $\omega \gg \omega_0$, we find that the oscillator suddenly moves to a negative displacement under the action of a positive force. Quite clearly, therefore, Eq. (4.3) does not tell the whole story. Thus the complete solution of Eq. (4.1) is given by

$$\psi(t) = A\ e^{-\gamma t/2} \cos(\omega^* t - \delta) + B \cos(\omega t - \phi) \qquad 4.45)$$

where $\omega^* = \left(\omega_0^2 - \frac{\gamma^2}{4}\right)^{1/2}$. A and δ are the two arbitrary constants and B and ϕ are respectively given by Eqs. (4.7) and (4.8). This solution (4.45) satisfies the two requirements for it to be a unique solution. It is a straight forward procedure (but a little involved) to see that it indeed satisfies Eq. (4.1) and it contains two adjustable constants A and δ.

Transient Beats

Equation (4.45) describes the motion of a damped harmonic oscillator driven harmonically by an external force. The first part of the solution represents free damped oscillations at an effective frequency ω^*. This part of the solution involves two constants A and δ which have to be determined from the initial conditions $\psi(t = 0)$ and $\frac{d\psi}{dt}$ at $t = 0$. The second part of the solution represents the steady-state harmonic oscillation at frequency ω of the driving force. This part also involves two constants B and ϕ which do not depend upon the initial conditions but are determined by the constants ω_0 and γ of the oscillator and f_0 and ω of the driving force, as is evident from Eqs. (4.7) and (4.8). Since the general solution (4.45) is a superposition of two oscillations at frequencies ω^* and ω, we would expect beats between the two kinds of oscillations. These beats are called *transient beats* and they will occur as long as the oscillations corresponding to the first term persist, i.e. during the relaxation time $\tau = 2/\gamma$. If damping is small, time τ will be large and the transient state will occupy a long time. If $\gamma = 0$, the transient state persists for all time to come and the oscillator will never settle down to a steady-state.

We shall now discuss the beats for a special situation in which the oscillator is initially (i.e. at $t = 0$) at rest at its equilibrium position. In other words, at time $t = 0$

$$\psi\,(t = 0) = 0$$

$$\left(\frac{d\psi}{dt}\right)_{t=0} = 0$$

In order to simplify our analysis, we shall assume that (i) the damping is so small that the term $e^{-\gamma t/2}$ in Eq. (4.45) is practically constant over a couple of cycles of oscillations of frequency ω^* and (ii) the driving frequency ω is close to the effective frequency ω^*. Thus we will set $\omega/\omega^* \simeq 1$ wherever appropriate. These assumptions are justified since we are not really interested in large damping and driving force frequency too far from resonance. We shall now proceed to compute the energy of the oscillator under the above simplifying assumptions. We rewrite Eq. (4.45) as

$$\psi(t) = e^{-\gamma t/2}\,(A_1 \cos \omega^* t + A_2 \sin \omega^* t) + B_{el} \cos \omega t + B_{ab} \sin \omega t \qquad (4.46)$$

where constants A_1 and A_2 are yet undetermined and constants B_{el} and B_{ab} are given by Eqs (4.20) and (4.21). Differentiating Eq. (4.46) with respect to t, under the assumption that $e^{-\gamma t/2}$ is sensibly constant during one cycle of oscillation, we obtain the velocity of oscillator

$$\frac{d\psi}{dt} = \omega^*\, e^{-\gamma t/2}\,(A_2 \cos \omega^* t - A_1 \sin \omega^* t) + \omega(B_{ab} \cos \omega t - B_{el} \sin \omega t) \quad (4.47)$$

Setting $t = 0$ in Eqs (4.46) and (4.47) and using the initial conditions, namely, $\psi = \dfrac{d\psi}{dt} = 0$ at $t = 0$, we obtain

and
$$A_1 = -B_{el}$$

$$A_2 = -\frac{\omega}{\omega^*}\, B_{ab} \simeq -B_{ab} \qquad (\because \;\; \omega \simeq \omega^*)$$

Substituting for A_1 and A_2 in Eq. (4.47) we have

$$\psi(t) = B_{ab}\,(\sin \omega t - e^{-\gamma t/2} \sin \omega^* t) + B_{el}\,(\cos \omega t - e^{-\gamma t/2} \cos \omega^* t) \qquad (4.48)$$

Now, as in Chap. 2, let us define

$$\omega_m = \tfrac{1}{2}\,(\omega^* - \omega)$$

and $\qquad \omega_a = \frac{1}{2}(\omega^* + \omega)$

so that $\qquad \omega^* = \omega_a + \omega_m$

and $\qquad \omega = \omega_a - \omega_m$

Angular frequency ω_a is the average of ω^* and ω and ω_m is called the *modulated angular frequency.* In terms of ω_a and ω_m, Eq. (4.48), on simplification, reads

$$\psi(t) = \{B_{ab}(1 - e^{-\gamma t/2}) \cos \omega_m t + B_{el}(1 + e^{-\gamma t/2}) \sin \omega_m t\} \sin \omega_a t$$
$$+ \{B_{el}(1 - e^{-\gamma t/2}) \cos \omega_m t + B_{ab}(1 + e^{-\gamma t/2}) \sin \omega_m t\} \cos \omega_a t \qquad (4.49)$$

Now, let

$$B_{el}(1 - e^{-\gamma t/2}) \cos \omega_m t - B_{ab}(1 + e^{-\gamma t/2}) \sin \omega_m t = R \cos \theta \qquad (4.50)$$

and

$$B_{ab}(1 - e^{-\gamma t/2}) \cos \omega_m t + B_{el}(1 + e^{-\gamma t/2}) \sin \omega_m t = R \sin \theta \qquad (4.51)$$

In terms of R and θ, Eq. (4.49) becomes

$$\psi(t) = R \cos(\omega_a t - \theta) \qquad (4.52)$$

where R and θ can be determined from Eqs (4.50) and (4.51). Squaring and adding these equations and carrying out some mathematical simplifications, we obtain

$$R^2 = (B_{ab}^2 + B_{el}^2)(1 + e^{-\gamma t} - 2 e^{-\gamma t/2} \cos 2\omega_m t) \qquad (4.53)$$

The formal resemblance of Eq. (4.52) with that of *SHM* is indeed misleading. The motion described by Eq. (4.52) is really not harmonic since R and θ are not constants but they vary with time. The motion of the oscillator can, at best, be described as periodic with an angular frequency ω_a, the average of ω^* and ω. Near resonance ($\omega \simeq \omega^*$), $\omega_a \simeq \omega$ and ω_m is small. Notice that, although, we have set $\omega/\omega^* = 1$, we do not take $\omega = \omega^*$ in expressions like $\cos \omega_m t$ because even if $\omega \simeq \omega^*$, the cosine term can vary appreciably with time. Now, as in Chapter 2, we still write down the expression for energy of the oscillator as

$$E = \frac{1}{2} m\omega_a^2 R^2$$

which, near resonance, reads

$$E = \frac{1}{2} m\omega^2 R^2$$

Substituting for R^2 from Eq. (4.53) we get

$$E = E_s\{1 + e^{-\gamma t} - 2e^{-\gamma t/2} \cos(\omega^* - \omega) t\} \qquad (4.54)$$

where
$$E_s = \tfrac{1}{2} m\omega^2 (B_{ab}^2 + B_{el}^2) \qquad (4.55)$$

Notice that E_s is the value of E when the terms like $e^{-\gamma t}$ and $e^{-\gamma t/2}$ in Eq. (4.54) have died out. In other words, E_s is the steady-state energy. Expression (4.54) is indeed the same as Eq. (4.30) obtained earlier. Notice also that at $t = 0$, Eq. (4.54) gives $E = 0$, which is expected since initially, the oscillator has no energy (it is at rest at its equilibrium position at $t = 0$).

The expression (4.54) reveals that, if the oscillator starts with zero energy initially, the energy does not build up to its steady-state value E_s in a smooth fashion. In fact, the energy increases and decreases periodically, i.e. there are beats as expected. The energy E is maximum when [see Eq. (4.54)]

$$\cos(\omega^* - \omega)\, t = -1$$

or $$(\omega^* - \omega)\, t = \pi, 3\pi, 5\pi, \ldots$$

Therefore, the maxima of energy occur at times t given by

$$t = \frac{\pi}{\omega^* - \omega}, \frac{3\pi}{\omega^* - \omega}, \frac{5\pi}{\omega^* - \omega}, \ldots$$

Hence, the time period of occurrence of maxima is

$$T_b = \frac{2\pi}{\omega^* - \omega}$$

and the frequency ν_b of the occurrence of maxima is

$$\nu_b = \frac{1}{T_b} = \frac{1}{2\pi}(\omega^* - \omega) = \nu^* - \nu$$

It is easy to see that this is also the frequency of the occurrence of minima of energy. Hence $\nu_b = \nu^* - \nu$ is the *beat frequency.* These beats are called *transient beats.* They vanish when $e^{-\gamma t}$ and $e^{-\gamma t/2}$ have damped out to zero and a steady-state is established. Figure 4.14 shows the beating effect. The energy rises and falls upto time $t = \tau$, after which it assumes a constant steady-state value.

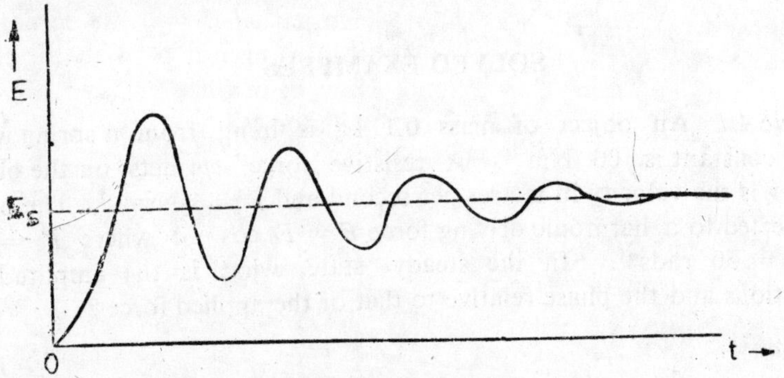

Fig. 4.14 Transient build-up of energy, transient beats

Let us examine the following special cases :

Case 1: *Driving frequency equal to the natural frequency*

Setting $\omega = \omega^*$ in Eq. (4.49) we obtain (since $\omega_m = 0$, $\omega_a = \omega$)

$$\psi(t) = (1 - e^{-\gamma t/2})\,(B_{ab} \sin \omega t + B_{el} \cos \omega t)$$

$$= (1 - e^{-\gamma t/2})\; \psi_s$$

where ψ_s is the steady-state displacement [see Eq. (4.19)]. The energy of the oscillator, in this case, is given by [see Eq. (4.53)]

$$E = E_0 \,(1 + e^{-\gamma t} - 2e^{-\gamma t/2}) = E_0 \,(1 - e^{-\gamma t/2})^2$$

Thus we conclude that, when the driving frequency is equal to the frequency of the free damped oscillations, the steady state is present right from the beginning. The amplitude and energy build up smoothly from zero to their steady-state values.

Case 2 : Absence of damping

Setting $\gamma = 0$ in Eq. (4.54) we have

$$E = 2E_s \,\{1 - \cos\,(\omega_0 - \omega)\,t\}$$

since $\omega^* = \omega_0$ when $\gamma = 0$.

Thus we conclude that, if damping were absent, the beats will continue for ever. The energy E will oscillate periodically between the maximum value $E_{max} = 4E_s$ and the minimum value which is zero. In the presence of damping, however, the oscillator keeps adjusting its phase with the driving force at initial times. After a time when the transient oscillations have died out, the oscillator settles into a steady-state and beats disappear. In the steady state, the oscillator oscillates exactly at the driving frequency with its phase having stabilised to a constant value relative to that of the driving force. The amount of energy delivered to the oscillator in each cycle of the driving force becomes exactly equal to the amount of energy lost in each cycle due to friction. Figure (4.14) shows how the transient build-up of the energy takes place.

SOLVED EXAMPLES

Example 4.1 An object of mass 0.1 kg is hung from a spring whose spring constant is 100 Nm^{-1}. A resistive force $-pv$ acts on the object, where v is the velocity in metres per second and $p = 1$ Nsm^{-1}. The object is subjected to a harmonic driving force $F = F_0 \cos \omega t$ where $F_0 = 2$ N and $\omega = 50$ rads^{-1}. In the steady state, what is the amplitude of oscillations and the phase relative to that of the applied force ?

Solution

Given

$m = 0.1$ kg.

$K = 100$ Nm^{-1}

$p = 1$ N s m^{-1}

$F_0 = 2$ N

$\omega = 50$ rad s^{-1}

Now

$$\omega_0 = \sqrt{\frac{K}{m}} = \sqrt{1000} \text{ rad s}^{-1}$$

$$f_0 = \frac{F_0}{m} = 20 \text{ N kg}^{-1}$$

$$\gamma = \frac{p}{m} = 10 \text{ s}^{-1}$$

In the steady-state, the amplitude B of oscillations is

$$B = \frac{f_0}{\{(\omega_0^2 - \omega^2)^2 + \omega^2\gamma^2\}^{1/2}}$$

$$\simeq 0.0126 \text{ m} = 1.26 \text{ cm}$$

The phase ϕ by which the oscillations lag the applied force is given by

$$\tan\phi = \frac{\omega\gamma}{\omega_0^2 - \omega^2}$$

$$= -\tfrac{1}{3}$$

or $$\phi = 161.7°$$

i.e. the oscillations lag the applied force by an angle 161.7°.

Example 4.2 A weekly damped harmonic oscillator is driven by a force $F = F_0 \cos \omega t$ whose amplitude F_0 is kept constant but its angular frequency is varied. It is experimentally observed that, the amplitude of the steady-state oscillations is 0.1 mm at very low values of ω and attains a maximum value of 10 cm when ω = 100 rads^{-1}. Calculate (a) The Q of the system (b) the time during which the energy of the oscillator falls to $\frac{1}{e}$ of its initial value and (c) half-width of the power resonance curve.

Solution

(a) We know that, if damping is weak, Q becomes very large, so that

$$\frac{B_{max}}{B_0} \simeq Q$$

where B_{max} is the maximum amplitude and B_0 is the amplitude at very low frequency of the applied force. It is given that

$$B_0 = 0.1 \text{ mm} = 0.01 \text{ cm}$$

and

$$B_{max} = 10 \text{ cm}$$

Hence

$$Q = 1000$$

(b) We know that

$$Q = \omega_0 \Gamma$$

where ω_0 is the resonant angular frequency which is given to be 100 rad s^{-1} and Γ is the energy decay time, i.e. the time during which energy falls to $\frac{1}{e}$ of its initial value. Therefore

$$\Gamma = \frac{Q}{\omega_0} = 10 \text{ s}$$

(c) The full width of the power resonance curve is

$$\Delta \omega = \gamma = \frac{1}{\Gamma}$$

The half-width of the power resonance curve is

$$\frac{\Delta\omega}{2} = \frac{1}{2\Gamma}$$

$$= 0.05 \text{ rad s}^{-1}$$

Example 4.3 An object of mass 2 kg hangs from a spring of negligible mass. The spring is extended by 2.5 cm when the object is attached. The top end of the spring is oscillated up and down in *SHM* with an amplitude of 2 mm. The Q of the system is 20. If g = 10 ms^{-2},

(a) what is the angular frequency ω_0 of the free undamped oscillations ?
(b) what is the amplitude of forced oscillations at $\omega = \omega_0$?
(c) what is the mean power input to maintain the forced oscillations at an angular frequency (ω) one per cent greater than ω_0?
(d) what is the mean power loss by friction at this frequency?

Solution

(a) Given m = 2 kg, g = 10 ms^{-2}

Since the spring extends by 2.5 cm (=2.5×10^{-2}m) when the mass m is attached, the spring constant is

$$k = \frac{2\times 10}{2.5\times 10^{-2}} = 800 \text{ Nm}^{-1}$$

Hence

$$\omega_0 = \sqrt{\frac{k}{m}} = 20 \text{ rad s}^{-1}$$

(b) From Eq. (4.7) the amplitude of the forced oscillations at $\omega = \omega_0$ is given by

$$B = \frac{F_0/m}{\omega_0 \gamma}$$

To evaluate B, we need to know F_0, the amplitude of the driving force and γ which measures the damping in the system. Since the amplitude of the top end (the driver) of the spring is 2 mm $= 2\times10^{-3}$m, the amplitude F_0 of the force applied to the spring to produce an amplitude of 2×10^{-3}m = spring constant × displacement amplitude

or $$F_0 = 800\times2\times10^{-3} = 1.6 \text{ N}$$

Now, since $Q = \dfrac{\omega_0}{\gamma}$, we have

$$\gamma = \frac{\omega_0}{Q}$$

But, $Q = 20$, therefore, $\gamma = 1.0 \text{ s}^{-1}$

Substituting for F_0, m, ω_0 and γ in the expression for B, we have

$$B = 4.0 \text{ cm}$$

Thus the amplitude of forced oscillations at $\omega = \omega_0$ is 4.0 cm.

(c) Given

$$\omega = \omega_0 + 0.01\,\omega_0 = 20.2 \text{ rad s}^{-1}$$

From Eqs (4.21) and (4.24) the mean power input is given by

$$P = \frac{1}{2}\,\frac{F_0^2}{m}\,\frac{\omega^2\gamma}{\{(\omega_0^2-\omega^2)^2+\omega^2\gamma^2\}}$$

Substituting for F_0, m, γ, ω_0 and ω we find that

$$P \simeq 0.55 \text{ W}.$$

Thus the mean power input per cycle is 0.55 W.

(d) In the steady state the mean power dissipated per cycle

= the mean power input per cycle

= 0.55 W.

Example 4.4 The graph in Fig. 4.15 shows the power resonance curve of a certain mechanical system which is driven by a force of constant

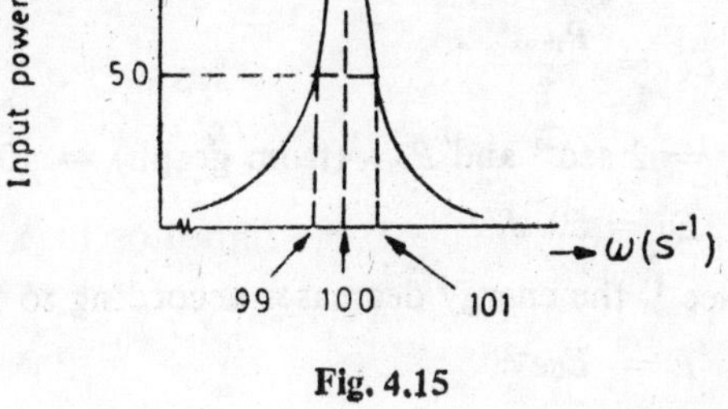

Fig. 4.15

magnitude but variable angular frequency ω. Answer the following questions on the basis of the information given in the graph.

(a) What the resonant angular frequency ω_0 of the system?
(b) What is the Q of the system?
(c) At exact resonance, what is the total mechanical energy E_0 of the oscillator ?
(d) If the driving force is removed, the energy decreases according to the equation

$$E = E_g e^{-\gamma t}$$

How many seconds does it take for the energy to decrease to a value

$$E = E_0 e^{-1} \ ?$$

Solution

(a) We know that the maximum value of the mean power input occurs when the angular frequency ω of the driving force is equal to the resonant angular frequency ω_0 of the oscillator. From the graph it is clear that this happens at

$$\omega = \omega_0 = 100 \text{ rad s}^{-1}$$

(b) It is obvious from the graph that the angular frequency width $\Delta\omega$ at half maximum power is $101-99 = 2$ rad s^{-1}.

Now, by definition,

$$Q = \frac{\omega_0}{\Delta\omega} = 50$$

(c) From Eqs (4.20), (4.21) and (4.30) the energy E_0 of the oscillator at resonance ($\omega = \omega_0$) is given by

$$E_{res} = E_0 = \tfrac{1}{2} m\omega_0^2 \frac{f_0^2}{\omega_0^{2}\gamma^2}$$

$$= \frac{F_0^2}{2m\gamma^2}$$

From Eq (4.32) the value of P at resonance is

$$P_{max} = \frac{F_0^2}{2m\gamma}$$

Hence $$E_0 = \frac{P_{max}}{\gamma}$$

Now, since $\Delta\omega = \gamma = 2$ sec^{-1} and P_{max} (from graph) $= 100$ W.

$$E_0 = 50 \ J.$$

(d) Since $\gamma = 2$ sec^{-1}, the energy decreases according to the equation

$$E = E_0 e^{-2t}$$

Hence, the time t taken for energy to fall to a value $E = E_0 e^{-1}$ is given by

$$2t = 1$$

or $$t = 0.5 \text{ s}$$

Example 4.5 A series LCR circuit is connected to an ac mains supply at 220 V and 50 Hz. If resistance $R = 10$ ohms and inductance $L = 100$ millihenry, what should be the capacitance C so as to obtain maximum current? How much is this current?

Solution :

We know that the impedance of an LCR circuit is given by

$$Ze = \left\{ R^2 + \left(\omega L - \frac{1}{\omega C} \right)^2 \right\}^{1/2}$$

For maximum current, Ze must be minimum ($= R$), which implies that

$$\omega L = \frac{1}{\omega C}$$

This is the condition of resonance. Thus

$$C = \frac{1}{\omega^2 L}$$

Now, since the frequency of the applied voltage is 50 Hz, its angular frequency $\omega = 2\pi \times 50$ rad s^{-1} and $L = 100$ mH $= 0.1$ H.

Therefore

$$C = 101.35 \ \mu\text{F}$$

Maximum current $I_0 = \dfrac{\text{voltage}}{\text{impedance}} = \dfrac{220}{10} = 22$ A

($\because Z_e = R$ at resonance)

Example 4.6 A series LCR circuit with $L = 0.05$ H, $C = 50$ μF and $R = 10\ \Omega$ is connected to an alternating supply at 200 V and 50 Hz. Find (a) the peak value of the current in the circuit, (b) the power factor of the circuit, (c) the average power delivered to the circuit in each cycle and (d) the rate of production of heat in the circuit.

Solution

Given $L = 0.05$ H, $C = 50 \times 10^{-6}$ F and $R = 10\ \Omega$. The magnitude of impendance in the circuit is

$$Z_0 = \left\{ R^2 + \left(\omega L - \frac{1}{\omega C} \right)^2 \right\}^{1/2}$$

Since $\omega = 2\pi \times 50 = 100\pi$ rad s^{-1}, we have

$$Z_0 = 48.97\ \Omega$$

(a) The peak value of the current is

$$I_0 = \frac{V_0}{Z_0} \simeq 4.08\text{A}$$

(b) The power factor cos δ is

$$\cos \delta = \frac{R}{Z_0} = 0.204$$

(c) The average power delivered in each cycle is

$$P = \tfrac{1}{2} V_0 I_0 \cos \delta$$

$$= 83.23 \text{ W}$$

(d) The heat is produced only in resistance R. Since the current varies as $I = I_0 \cos (\omega t - \delta)$, the average heat produced in resistance R in time t is given by

$$H = I^2 R t$$

The rate of heat production is, therefore, equal to

$$IR = I_0^2 R \cos^2 (\omega t - \delta)$$

The average rate of heat production is

$$\tfrac{1}{2} I_0^2 R = 83.23 \text{ W}.$$

Notice that, as expected, average power consumed=rate of dissipation of heat.

Example 4.7 An alternating current $I = I_0 \cos \omega t$ is passed through a parallel LCR circuit as shown in Fig. 4.16. If $R = 10{,}000\ \Omega$, $L = 0.2\ H$ and $C = 5\ \mu$ F and $I_0 = 2$ A, find (a) the resonant angular frequency ω_0, (b) the width (γ) of the power response curve at high maximum power, (c) the Q of the circuit and (d) the mean power absorbed at resonance.

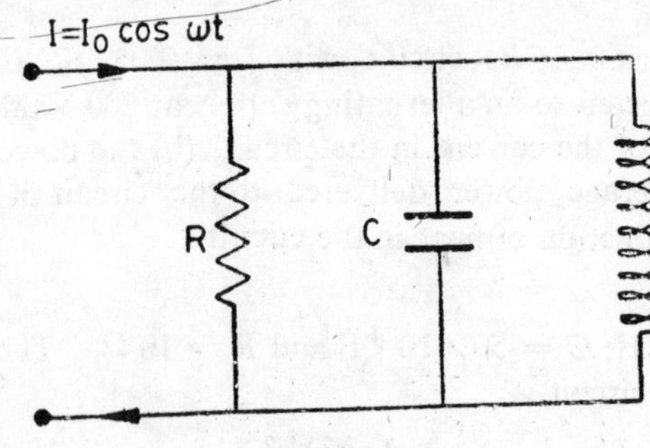

Fig. 4.16

Solution

It is clear that the potential difference across each component of the circuit is the same. Let it be denoted by V. Then the currents through

R, L and C respectively are $\frac{V}{R}$, $\frac{V}{i\omega L}$ and $I\omega CV$. Hence,

$$\frac{V}{R}+\frac{V}{i\,\omega L}+i\,\omega CV = I$$

or
$$V\left\{\frac{1}{R}+i\left(\omega C-\frac{1}{\omega L}\right)\right\} = I_0 \cos \omega t$$

Representing $I_0 \cos \omega t$ by $I_0\, e^{i\omega t}$, the potential difference is given by the real part of the expression

$$V = IZe \tag{i}$$

where the impedance Ze of the circuit is given by

$$Ze = \frac{1}{\frac{1}{R}+i\left(\omega C-\frac{1}{\omega L}\right)} = \frac{R}{1+i\,R\left(\omega C-\frac{1}{\omega L}\right)}$$

Separating the magnitude Z_0 and phase δ of Ze. we have

$$Ze = Z_0e^{-i\delta} = \frac{R}{1+iR\left(\omega C-\frac{1}{\omega L}\right)}$$

and

$$Ze^* = Z_0e^{i\delta} = \frac{R}{1-iR\left(\omega C-\frac{1}{\omega L}\right)}$$

From these equations we have

$$Z_0 = \frac{R}{\left\{1+R^2\left(\omega C-\frac{1}{\omega L}\right)^2\right\}^{1/2}}$$

and

$$\cos \delta = \frac{1}{\left\{1+R^2\left(\omega C-\frac{1}{\omega L}\right)^2\right\}^{1/2}}$$

The resonance condition corresponds to Z_0 becoming maximum which happens at a value of $\omega = \omega_0$ satisfying

$$\omega_0 C-\frac{1}{\omega_0 L} = 0 \quad \text{or} \quad \omega_0 = \frac{1}{\sqrt{LC}}$$

From Eq. (i) we have

$$V = I_0e^{i\omega t}Z_0e^{-i\delta} = I_0Z_0e^{i(\omega t-\delta)}$$

Extracting the real part of this expression we have

$$V = V_0 \cos (\omega t - \delta)$$

where $\qquad V_0 = I_0 Z_0$

Now, the instantaneous power absorbed is

$$P(t) = VI = V_0 I_0 \cos(\omega t - \delta) \cos \omega t$$

As before, the average power absorbed is given by

$$P = \langle P(t) \rangle = \tfrac{1}{2} V_0 I_0 \cos \delta$$

Substituting for $V_0 = I_0 Z_0$ and $\cos \delta$, we have

$$P = \frac{I_0^2}{2} \frac{R}{\left\{1 + R^2\left(\omega C - \dfrac{1}{\omega L}\right)^2\right\}}$$

At resonance $\left(\omega = \dfrac{1}{\sqrt{LC}}\right)$ we find that

$$P = P_{max} = \tfrac{1}{2} I_0^2 R$$

Following the arguments outlined in the text, the width $\Delta\omega$ of the power response curve at half maximum power can easily be computed. The result is

$$\Delta\omega = \gamma = \frac{1}{CR}$$

With $R = 10{,}000\ \Omega$, $L = 0.2$ H and $C = 5 \times 10^{-6}$ F, we have

(a) $\omega_0 = \dfrac{1}{\sqrt{LC}} = 1000 \text{ rad s}^{-1}$

(b) $\gamma = \Delta\omega = \dfrac{1}{CR} = 20 \text{ s}^{-1}$

(c) $Q = \dfrac{\omega_0}{\gamma} = 50$

(d) $P_{res} = P_{max} = \tfrac{1}{2} I_0^2 R = 20{,}000$ W

QUESTIONS

1. A mechanical harmonic oscillator of mass m and stiffness constant K is subjected to a viscous damping force that is proportional to its velocity; the coefficient of the damping force being p. The oscillator is driven by a force $F = F_0 \cos \omega t$. In the steady state, the displacement of the oscillator is given by

$$\psi = A \cos(\omega t - \delta)$$

   (a) Set up the differential equation of the motion of the oscillator and obtain the expressions for amplitude A and phase constant δ. Show that the values of A and δ are independent of the initial conditions.

   (b) Discuss the dependence of amplitude A on the angular frequency ω of the driving force. For an arbitrary (but small) value of p, sketch the behaviour of A as the value of ω is changed from a low ($\omega \ll \omega_0$) to a high ($\omega \gg \omega_0$) value, where $\omega_0 = \sqrt{K/m}$ is the angular frequency of the undamped free oscillations.

(c) Obtain the value of ω for amplitude resonance and show that, at resonance, the amplitude of the driven oscillator may be written as $A = F_0/\omega' p$

where $\omega' = \left(\omega_0^2 - \frac{p^2}{4m^2} \right)^{1/2}$

(d) Discuss the dependence of δ on the frequency of the driving force. For an arbitrary (but small) value of p, sketch the variation of δ as the value of ω is changed from a low to a high value. At exact displacement resonance, what will be the value of δ?

(e) Define the term impedance of the oscillator. Obtain an expression for it and discuss its variation with ω. Show that

$$A = \frac{F_0}{\omega Z_0}$$

where Z_0 is the magnitude of impedance.

(f) Show that the steady-state response of the oscillator is independent of (i) its mass if $\omega \ll \omega_0$ and (ii) its stiffness if $\omega \gg \omega_0$.

(g) Obtain the expression for the velocity amplitude of the forced oscillator. What is the value of ω for velocity resonance? Show that, at velocity resonance, the amplitude V_0 of the velocity equals F_0/p. Also show that the velocity of the oscillator is in phase with the driving force at velocity resonance.

(h) Show that the average power supplied to the oscillator during one cycle of the driving force is given by

$$P = \tfrac{1}{2} F_0 \,\omega\, A_{ab}$$

where A_{ab} is the amplitude of that part of the steady-state displacement which is 90° out of phase with the driving force.

(i) Show that, in the steady state, the time-averaged input power equals the time-averaged power dissipated through friction.

(j) What are time-averaged kinetic and potential energies of the oscillator? For what value of ω are they equal? What is the total energy stored in the oscillator under these conditions ?

(k) How does the total instantaneous energy vary with time for an arbitrary value of ω? For what value of ω is the total energy constant in time?

(l) Show that, at resonance, the average energy stored in the oscillator equals the average power dissipated through friction times the mean decay time of the free damped oscillations.

(m) What are half-power points of the power resonance curve for a driven oscillator? Show that the angular frequency width of the power resonance curve at half maximum power is equal to the reciprocal of the mean decay time of the free damped oscillations. What is the importance of this relation in experiment?

(n) Define the Q factor of a driven oscillator and show that (for low damping)

$$A_{\max} = Q\, A_0$$

where A_{mxa} is the amplitude of the oscillator at resonance and A_0 that at very low frequency of the driving force.

(o) Show that, for a forced oscillator in the steady state, the dipslacement amplitude at low frequencies ($\omega \to 0$), the velocity amplitude at velocity resonance ($\omega = \omega_0$) and the acceleration amplitude at high frequencies ($\omega \to \infty$) are independent of the frequency of the driving force.

2. Solve parts (a) to (o) of Question 1 if the driving force is of the form $F = F_0 \sin \omega t$ instead of $F = F_0 \cos \omega t$.
3. A series *LCR* circuit is driven by an alternating voltage $V = V_0 \cos \omega t$.
   (a) Obtain the expression for the instantaneous current in the circuit.
   (b) Show that the impedance offered to the current by the circuit is given by

$$Z_e = \left[R^2 + \left(\omega L - \frac{1}{\omega C} \right)^2 \right]^{1/2}$$

   (c) What is the resonant angular frequence (ω_0) of the circuit?
   (d) Obtain an expression for the phase difference between the current and the applied voltage and hence show that when $\omega = \omega_0$, the current and voltage are in phase.
   (e) What is the power factor of the circuit?
   (f) Show that, at resonance, the power delivered to the circuit is maximum and is equal to $1/2\ V_0 I_0$ where I_0 is the peak value of the current.
   (g) Define Q of the circuit and show that it is given by

$$Q = \frac{\omega_0 L}{R}$$

4. Solve parts (a) to (g) of Question 3 if the applied voltage is of the form $V = V_0 \sin \omega t$ instead of $V = V_0 \cos \omega t$.
5. A parallel *LCR* circuit is driven by an alternating current $I = I_0 \cos \omega t$. Obtain the expression for (a) the resonant frequency, (b) the frequency width of the power resonance curve at half maximum power, (c) the power absorbed at resonance and (d) the power factor and Q factor of the circuit.
6. Attempt Question 5 if the parallel *LCR* circuit is driven by a current of the form $I = I_0 \sin \omega t$ instead of $I = I_0 \cos \omega t$.
7. A mechanical oscillator of mass m and stiffness constant K is subjected to a viscous damping force $= -p \times$ velocity. The oscillator is driven by a force $F = F_0 \cos \omega t$. The motion of the oscillator is governed by the equation

$$m \frac{d^2\psi}{dt^2} + p \frac{d\psi}{d_t} + \frac{K}{m} \psi = F \cos \omega t$$

   (a) Verify, by direct substitution, that

$$\psi(t) = Ae^{-\gamma t/2} \cos(\omega^* t - \delta) + B \cos(\omega t - \phi)$$

   is the complete solution of the equation of motion Here. $\gamma = p/m$ and $\omega^* = \left(\omega_0^2 - \frac{\gamma^2}{4} \right)^{1/2}$ and $\omega_0 = \sqrt{K/m}$ Hence obtain the expression for constants B and ϕ.
   (b) Evaluate constants A and δ under the initial conditions $\psi = 0$ and $d\psi/dt = 0$ at $t = 0$.
   (c) What do you understand by the 'transient state'? What are transient beats ?
   (d) Assuming weak damping, obtain the expression for the energy of the oscillator under the above initial conditions. Sketch the behaviour of energy with time during the first few cycles to show how the energy builds up to its steady-state value. Hence show that, if damping were absent, the energy will keep on 'beating' never attaining any steady value.

PROBLEMS

1. An object of mass 0.1 kg is hung from a spring whose constant is 10 N m^{-1}. A resistive force $-pv$ acts on the object where v is the velocity in ms^{-1} and $p = 0.1$ N m^{-1} s. The object is subjected to a driving force $F = F_0 \cos \omega t$ where $F_0 = 1$N. Find the amplitude of the oscillator in the steady state if $\omega =$ (a) 1 rad s^{-1}, (b) 10 rad s^{-1} and (c) 50 rad s^{-1}.

2. A weakly damped harmonic oscillator is driven by a force $F = F_0 \sin \omega t$, whose amplitude F_0 is kept constant but its angular frequency ω is varied. It is observed that the amplitude of the steady state oscillations is 1 mm at very low values of ω and attains a maximum value of 20 cm at $\omega = 50$ rad s^{-1}. Calculate (a) The Q of the system (b) The time during which the energy of the oscillator falls to $1/e$ of its initial value and (c) The full-width of the power resonance curve.
3. An object of mass 0.1 kg hangs from a massless spring of spring constant 10 N m^{-1}. The top end of the spring is oscillated harmonically with an amplitude of 1 cm and an angular frequency ω. The Q of the system is 50. (a) What is the amplitude of the oscillator in the steady state when $\omega = \omega_0$ where ω_0 is the angular freqtiency of free undamped oscillations? (b) What is the mean power input to maintain forced oscillations at $\omega = \omega_0$?
4. The steady state displacement of a forced damped oscillator is given by $\psi = A \cos \omega t$. The resistive force is $-pv$. In terms of A, ω and p, calculate the work done against the resistive force during one cycle of oscillation.
5. The graph in Fig. 4.17 shows how the mean input power of a slightly damped system consisting of a mass attached to a spring varies as the angular frequency ω of the harmonic driving force (with constant amplitude F_0) is varied. Answer the following questions on the basis of the information given in the graph.

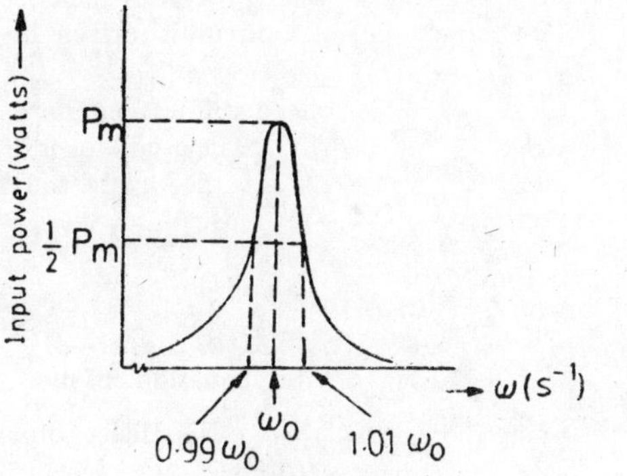

Fig. 4.17

(a) What is the Q of the system?
(b) What is the amplitude at resonance?
(c) What is the total mechanical energy at resonance?
(d) At resonance how much work per cycle is being done against the resistive force?
(e) If the driving force is removed, the energy decreases according to the equation

$$E = E_0 \, e^{-\gamma t}$$

What is the value of γ?

(f) If the driving force is removed, what fraction of energy is lost per cycle?
(g) If the driving force is removed, how many seconds does it take for the energy to fall to $1/e$ of initial value?
(h) If the driving force is removed, after how many cycles of force-free oscillations will the energy fall to $1/e^4$ of its initial value?

6. A new system is made in which the mass of the oscillator is reduced to half but the spring, the driving force and the viscous medium are the same as in Problem 5. In terms of the corresponding quantities of the original system, find the values of the following, quantities of the new system.
(a) The new resonant frequency ω_0', (b) The new Q' of the system, (c) the new value of γ', (d) The maximum mean power input of the new system P_m' and (e) the total mechanical energy of the system at resonance ?
7. A series *LCR* circuit with $C = 50\mu\ F$ and $R = 10\ \Omega$ is connected to an ac supply of 20 V at 100 Hz. What should be the value of L so as to obtain maximum current? How much is this current?
8. The power factor of a series *LCR* circuit with $R = 10\Omega$ driven by an ac supply at 100 V is 0.2. Calculate (a) the impedance of the circuit (b) the peak value of the current and (c) the rate of production of heat in the circuit.

5

Coupled Oscillations

5.1 INTRODUCTION

In Chapter 1 we have discussed the free oscillations of systems with one degree of freedom. The oscillation of such systems is characterized by a single natural frequency. In the preceding chapter we have shown in some detail how such a system responds to an externally applied periodic force. We had assumed that the driving system remains practically unaffected by the forced oscillations of the driven system; the former only serving as a source of a periodic force with a negligible feed-back of energy from the latter. In this chapter we shall analyse the oscillations of coupled systems in which we cannot neglect the feedback of energy from the driven system to the driver. Some examples of two coupled oscillating systems are illustrated in Fig. 5.1. Figure 5.1a shows two simple pendulums with their bobs attached to each other by means of a spring, in Fig. 5.1b we have two masses attached to each other by three springs (the middle spring provides the coupling) and Fig. 5.1c shows two coupled *LC* circuits.

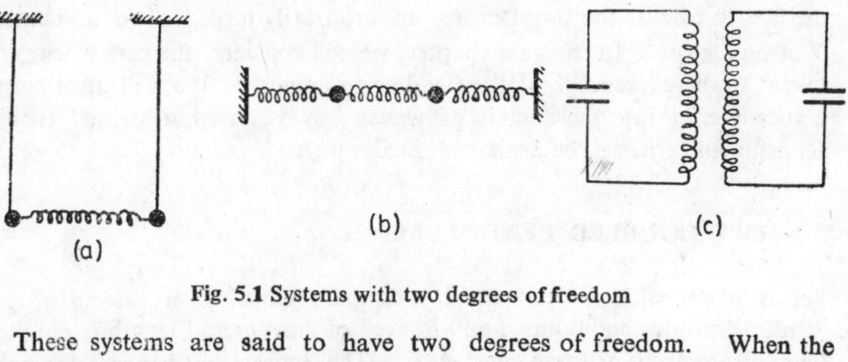

Fig. 5.1 Systems with two degrees of freedom

These systems are said to have two degrees of freedom. When the moving parts of such a system oscillate in one plane, say the plane of the paper, there are exactly two variables in terms of which the motion can be completely specified. In the case of two coupled pendulums or two coupled masses, the two variables are the displacements of the two bobs or the two masses and in the case of two coupled LC circuits the two variables are the charges on two capacitors or currents in the two circuits. In these examples the two oscillators are coupled together so tightly that we neglect the feedback of energy from the driven oscillator to the driver. In fact we can no longer distinguish the driven oscillator from the one that provides the driving force. We cannot label one as the driver and the other as the driven. Both the oscillators have now to be treated on an equal footing.

In physics we come across a variety of coupled systems that can oscillate. In fact, oscillators rarely exist in complete isolation. A string, for instance, consists of a very large number of particles coupled to one another. Atoms in a crystal lattice are capable of oscillating in many different ways just as a string is. Wave motion, which we will discuss in Chapter 7, is due to the fact that the neighbouring oscillating particles are coupled to each other and are able to transmit their energy to each other.

Let us look at the example of a string once again. We know that a string is capable of vibrating in many different ways or modes. Each mode has its own characteristic frequency. How can we account for these numerous modes and calculate the frequency of each mode? This question will be answered in the next chapter. The clue lies in the fact that a string (or any extended object) can be regarded as an extended array of large number of simple oscillators coupled together. A solid body, for example, is composed of many atoms. Each atom behaves as an oscillator, vibrating about an equilibrium position. The motion of each atom affects its neighbours. Thus all the atoms in a solid are coupled together. We have, therefore, to answer the following specific question: What is the effect of the coupling on the behaviour of individual oscillators? To answer this question, we shall begin by discussing in some detail the behaviour of a system with just two coupled oscillators. We will develop on analytical method of solving the problem. We will then use the method to discuss the problem of an arbitrarily large (but finite) number N of oscillators. In the next chapter, we will consider the case when N is taken to be exceedingly large (tending to infinity). We will then automatically enter into the domain of waves. Wave motion, which is our ultimate concern, will be dealt with in Chap. 7.

5.2 TWO COUPLED PENDULUMS

Let us begin with a very simple example. Consider a system of two identical simple pendulums A and B, each of mass m and length l, coupled by a linear spring of force constant k. The separation between the bobs is such that the spring is relaxed in the equilibrium shown in Fig. 5.2a. The system is disturbed slightly from its equilibrium position as shown in

Fig. 5.2b and released. The two pendulums begin to oscillate. Let $\psi_a(t)$ and $\psi_b(t)$ be the displacements of their bobs at an instant of time t. The spring is stretched or compressed depending on whether $\psi_b > \psi_a$ or

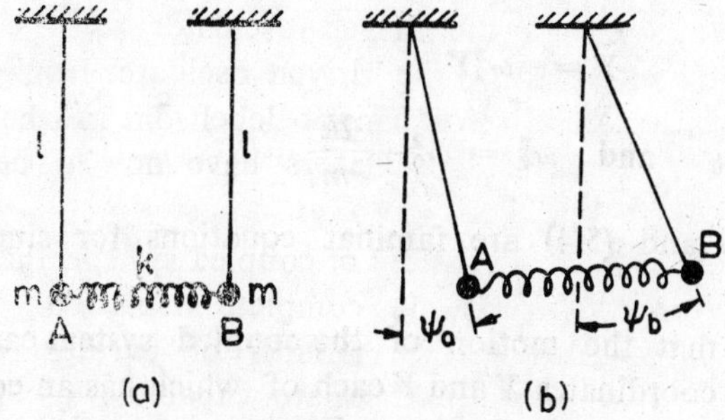

Fig. 5.2 Two identical coupled pendulums

$\psi_b < \psi_a$. The magnitude of tension in the spring is $k(\psi_b - \psi_a)$. If $\psi_b > \psi_a$ the tension will act against the acceleration $\ddot{\psi}_b$ of pendulum B but in favour of the acceleration $\ddot{\psi}_a$ of pendulum A. The equations of motion of pendulums A and B, for small oscillations in a plane, are

$$m\,\ddot{\psi}_a = -\frac{mg}{l}\,\psi_a + k(\psi_b - \psi_a)$$

$$m\,\ddot{\psi}_b = -\frac{mg}{l}\,\psi_b - k(\psi_b - \psi_a)$$

These equations do not represent simple harmonic motion since the acceleration of a pendulum is not proportional to its own displacement alone but depends also on the displacement of the other pendulum. If the spring were absent ($k = 0$), the two pendulums will execute harmonic oscillations at an angular frequency given by

$$\omega_0 = \sqrt{\frac{g}{l}}$$

In terms of ω_0, the two coupled equations are

$$\ddot{\psi}_a = -\omega_0^2\psi_a + \frac{k}{m}(\psi_b - \psi_a) \tag{5.1}$$

$$\ddot{\psi}_b = -\omega_0^2\psi_b - \frac{k}{m}(\psi_b - \psi_a) \tag{5.2}$$

In order to find out the effect of coupling on each pendulum, these equations must be solved for ψ_a and ψ_b. Instead of solving these equations directly, we shall choose two new coordinates

$$X = \psi_b + \psi_a$$

$$Y = \psi_b - \psi_a$$

The significance of this approach will become obvious a little later in the chapter. Adding and subtracting Eqs (5.1) and (5.2) we get

$$\ddot{X} = -\omega_1^2 X \tag{5.3}$$

$$\ddot{Y} = -\omega_2^2 Y \tag{5.4}$$

where $\omega_1 = \omega_0$ and $\omega_2^2 = \omega_0^2 + \dfrac{2k}{m}$

Equations (5.3) and (5.4) are familiar equations for simple harmonic motion.

Thus we find that the motion of the coupled system can be described in terms of two coordinates X and Y each of which has an equation which describes simple harmonic motion. These two motions are completely uncoupled and independent although the moving parts of the system are not. One motion can exist without the other.

If $Y = 0$, $\psi_b = \psi_a$ at all times, so that the motion is completely described by Eq. (5.3). The angular frequency of the oscillation is given by

$$\omega_1 = \omega_0 = \sqrt{\frac{g}{l}}$$

which is the same as that of either pendulum oscillating in isolation. The effect of the spring is absent. This is because both pendulums are always in phase and the spring is at its natural length throughout the motion (Fig. 5.3a).

If $X = 0$, $\psi_b = -\psi_a$ at all times, so that the motion is completely described by Eq. (5.4). The angular frequency, in this case, is

$$\omega_2 = \left(\omega_0^2 + \frac{2k}{m}\right)^{1/2}$$

which is greater than ω_1. The two pendulums are oscillating harmonically at the same frequency ω_2 and are always out of phase as shown in Fig. 5.3b. The spring is either extended or compressed during the motion and the coupling is effective.

Normal Coordinates and Normal Modes

The significance of choosing X and Y to describe the motion of a coupled system is that these parameters give a very simple illustration of what we call the normal coordinates. *The normal coordinates of a coupled system are the parameters in terms of which the equations of motion of the system can be written as a set of linear differential equations with constant coefficients in which each equation contains only one dependent variable.* In our example $X = \psi_b + \psi_a$ and $Y = \psi_b - \psi_a$ are the normal coordinates and the equations that contain only one dependent variable are Eqs (5.3) and (5.4).

The simple harmonic motion associated with each normal coordinate is called a *normal mode* of the coupled system. Each normal mode has its own characteristic frequency called the *normal mode frequency*. The ratio of the displacements (ψ_b/ψ_a) of the moving parts of the system remains constant for a normal mode and is called the *shape* or *configuration* of that mode. In our example, there are two normal modes, one associated with normal coordinate X and the other with Y. The shape of the first mode is shown in Fig. 5.3a. Both pendulums are displaced, say, to the right by the same amount so that $\psi_b/\psi_a = 1$. When they are released, each pendulum will execute simple harmonic motion at the same frequency

$$\omega_1 = \omega_0 = \sqrt{\frac{g}{l}}$$

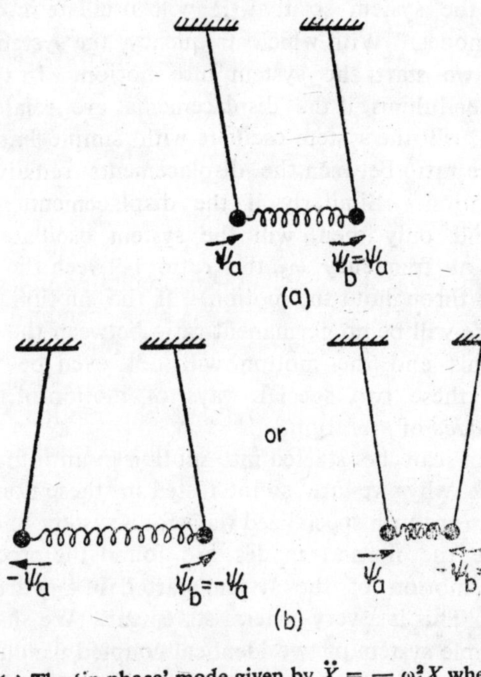

Fig. 5.3 (a) The 'in phase' mode given by $\ddot{X} = -\omega_1^2 X$ where $\omega_1^2 = g/l$ and $X = \psi_a + \psi_b$ is the normal co-ordinate.

(b) The 'out of phase' mode given by $\ddot{Y} = -\omega_2^2 Y$ where

$$\omega_2^2 = \frac{g}{l} + \frac{2k}{m}$$ and $Y = \psi_b - \psi_a$ is the normal co-ordinate.

and will always stay in phase and ψ_b/ψ_a will remain constant equal to unity throughout the motion. This is called the 'in phase' mode of the system. The shape of the second normal mode is shown in Fig. 5.3b The pendulums are displaced in opposite directions by the same amount so that $\psi_b/\psi_a = -1$. When they are released, each pendulum will execute simple harmonic motion at the same frequency

$$\omega_2 = \left(\frac{g}{l} + \frac{2k}{m}\right)^{1/2}$$

and will always be out of phase by 180° and ψ_b/ψ_a will remain constant equal to -1 throughout the motion. Both pendulums will pass through their respective equilibrium positions simultaneously, except that the amplitude of one is opposite in sign. Hence, *in a given normal mode, both the moving parts execute simple harmonic motions at the same frequency and the same phase constant.*

In conclusion one can say that although the general motion of a coupled system is not periodic, yet if the system is started in just the right way, the moving parts of the system will execute simple harmonic motion. For a system having two degrees of freedom, there are exactly two possible ways of starting the system so that it will oscillate in one of the two possible normal modes. With which frequency the system will oscillate depends on how we start the system into motion. In the case of two identical coupled pendulums, if the displacements are related as $\psi_b = \psi_a$, then and only then, will the system oscillate with simple harmonic motion at frequency ω_1, the ratio between the displacements remaining the same throughout the motion. Similarly if the displacements are related as $\psi_b = -\psi_a$, then and only then, will the system oscillate with simple harmonic motion at frequency ω_2, the ratio between the displacements remaining the same throughout the motion. If the motion is started in any other way, there will be no permanent ratio between the displacements of the two pendulums and the motion will not even be periodic. As mentioned earlier, these two special ways of motion of the system are called its *normal modes* of vibration.

Since any system can be started into motion in an infinite number of ways, one may ask why we are so interested in these normal modes of vibration when they are such specialized ways for a system to oscillate. The answer is that once the normal modes are found the problem of determining the general motion of the system started in any arbitrary way is completely solved. This is a very general statement. We shall prove it by considering the simple system of two identical coupled pendulums.

In an attempt to solve the Eqs (5.1) and (5.2) of motion for ψ_a and ψ_b, we chose to write them in terms of the normal coordinates X and Y. The Eqs (5.3) and (5.4) for X and Y can be easily solved. We know that the solutions are:

$$X = X_0 \cos(\omega_1 t + \phi_1)$$
$$Y = Y_0 \cos(\omega_2 t + \phi_2)$$

where X_0 and Y_0 are the amplitudes of the two normal modes and ϕ_1 and ϕ_2 their phase constants. The other two constants ω_1 and ω_2 are already determined. They are the angular frequencies of the modes. Now, since $X = \psi_b + \psi_a$ and $Y = \psi_b - \psi_a$ it is clear that

$$\psi_a = \tfrac{1}{2}(X - Y)$$

and $\quad \psi_b = \frac{1}{2}(X+Y)$

or $\quad \psi_a = \frac{1}{2} X_0 \cos(\omega_1 t+\phi_1) - \frac{1}{2} Y_0 \cos(\omega_2 t+\phi_2) \quad (5.5)$

and $\quad \psi_b = \frac{1}{2} X_0 \cos(\omega_1 t+\phi_1) + \frac{1}{2} Y_0 \cos(\omega_2 t+\phi_2) \quad (5.6)$

The four undetermined constants are X_0, Y_0, ϕ_1 and ϕ_2. These are determined from any arbitrary initial conditions, namely, the initial displacements of the two pendulums and their initial velocities (see section 5.7). Thus knowing X_0, Y_0, ϕ_1 and ϕ_2, the future motion of the two pendulums described by ψ_a and ψ_b is completely determined. Equations (5.5) and (5.6) show that ψ_a and ψ_b are given by the linear combination (or superposition) of the two normal modes. Thus the general motion of any coupled system can always be represented as a superposition of all its possible normal modes. It is for this reason (see also Sec. 5.7) that the determination of the normal modes of a system is of vital importance.

For a very simple system such as the one under consideration it is not even necessary to search for the normal coordinates. In fact one can easily guess the normal modes from symmetry considerations. By definition, in a given normal mode, both the pendulums execute simple harmonic motions with the same frequency. Now, for a simple harmonic motion, the square of angular frequency is given by ω^2 = restoring force per unit mass per unit displacement.

Thus, if both pendulums experience the same restoring force per unit mass per unit displacement, they will oscillate hormonically with the same frequency. It is easy to guess that this will happen if the pendulums are displaced by the same amount with the spring always at its natural length. This is the 'in phase' mode of vibration. If the displacement of either pendulum is ψ, the restoring force for small oscillations is $mg\psi/l$. Therefore, the angular frequency ω_1, of this mode is given by

$$\omega_1^2 = \frac{\text{restoring force}}{\text{mass} \times \text{displacement}} = \frac{mg\,\psi/l}{m\psi} = \frac{g}{l}$$

There exists another possible configuration for which both pendulums will execute simple harmonic motion. It is easy to guess that this will happen if the pendulums are displaced in opposite directions by the same amount, say, ψ. The tension in the spring is then equal to $k\times$change in the length $= 2k\psi$. The restoring force on either pendulum

$$= mg\frac{\psi}{l} + 2k\psi = \left(\frac{g}{l} + \frac{2k}{m}\right) m\psi$$

This is the 'out of phase' mode whose angular frequency is clearly given by

$$\omega_2^2 = \frac{\text{restoring force}}{\text{mass} \times \text{displacement}} = \frac{\left(\frac{g}{l} + \frac{2k}{m}\right) m\psi}{m\psi} = \frac{g}{l} + \frac{2k}{m}$$

One can argue that there can be no other configuration for which both moving parts move with the same frequency. We will show below that for a system with two degrees of freedom, there exists just two normal modes of vibration.

5.3 THE GENERAL METHOD OF FINDING NORMAL MODES

Sometime it is not easy to guess the normal modes from symmetry considerations nor is it easy to choose the normal coordinates to solve the differential equations of the coupled system. As an example let us consider a slightly more difficult problem of two coupled pendulums of equal string lengths l but unequal bob masses m_a and m_b. The equations governing the motion of the pendulums are:

$$\ddot{\psi}_a = -\omega_0^2 \psi_a + \frac{k}{m_a}(\psi_b - \psi_a) \tag{5.7}$$

$$\ddot{\psi}_b = -\omega_0^2 \psi_b - \frac{k}{m_b}(\psi_b - \psi_a) \tag{5.8}$$

How then can we find the normal coordinates of this system and solve the differential equations? This can be done by using a more systematic method. This method is a general one and is applicable to any coupled that system with any number of degrees of freedom. The only restriction is the differential equations governing the motion of the system must be linear. We shall illustrate the method by applying it to the above example. We make use of the property of a normal mode, namely, that in a normal mode all the moving parts of the system execute simple harmonic motions at the same frequency and the same phase constant.

Let us assume there exists a normal mode at an angular frequency ω and phase constant ϕ and find out under what condition this assumption can be realized. The existence of a normal mode at an angular frequency ω and phase constant ϕ implies that both pendulums move with a simple harmonic motion at the same angular frequency ω and the same phase constant ϕ. Hence for that mode, we have

$$\psi_a = A\cos(\omega t + \phi) \tag{5.9}$$

$$\psi_b = B\cos(\omega t + \phi) \tag{5.10}$$

where the amplitudes A and B of the two pendulums may, in general, be different. Equations (5.9) and (5.10) give

$$\ddot{\psi}_a = -\omega^2 \psi_a$$

$$\ddot{\psi}_b = -\omega^2 \psi_b$$

Substituting for $\ddot{\psi}_a$ and $\ddot{\psi}_b$ in Eqs (5.9) and (5.10) we get

$$\left(\omega_0^2+\frac{k}{m_a}-\omega^2\right)\psi_a=\frac{k}{m}\psi_b$$

and
$$\left(\omega_0^2+\frac{k}{m_b}-\omega^2\right)\psi_b=\frac{k}{m_b}\psi_a$$

From the first equation, we have

$$\frac{\psi_b}{\psi_a}=\frac{k/m_a}{\omega_0^2+\dfrac{k}{m_a}-\omega^2} \tag{5.11}$$

and from the second equation. we have

$$\frac{\psi_b}{\psi_a}=\frac{\omega_0^2+\dfrac{k}{m_b}-\omega^2}{k/m_b} \tag{5.12}$$

If ψ_a and ψ_b are not both zero, the right-hand sides of these equations must be equal. Thus

$$\frac{k/m_a}{\omega_0^2+\dfrac{k}{m_a}-\omega^2}=\frac{\omega_0^2+\dfrac{k}{m_b}-\omega^2}{k/m_b}$$

giving

$$\omega^4-\left(2\omega_0^2+\frac{k}{m_a}+\frac{k}{m_b}\right)\omega^2+\left(\omega_0^2+\frac{k}{m_a}+\frac{k}{m_b}\right)\omega_0^2=0 \tag{5.13}$$

Equation (5.13) is a quadratic equation in the variable ω^2. It has two roots which we call ω_1^2 and ω_2^2. These roots are

$$\omega_1^2=\omega_0^2$$

$$\omega_2^2=\omega_0^2+k\left(\frac{1}{m_a}+\frac{1}{m_b}\right)$$

Thus we have found that our assumption that there exists a normal mode can be realized in exactly two ways. In other words, there are just two modes. The angular frequency of mode of 1 ω_1 and that of mode 2 is ω_2. Notice that if we set $m_a=m_b=m$, we recover the results of the previous section.

The shape or configuration of mode 1 can be found by substituting $\omega^2=\omega_1^2$ back in Eq. (5.11) or Eq. (5.12) [they are equivalent because of Eq. (5.13)]. Thus

$$\left(\frac{\psi_b}{\psi_a}\right)_{\text{mode 1}}=\left(\frac{B}{A}\right)_{\text{mode 1}}=\frac{B_1}{A_1}=+1$$

Similarly, the shape of mode 2 is obtained by setting $\omega^2 = \omega_2^2$ in either Eq. (5.11) or Eq. (5.12) which gives

$$\left(\frac{\psi_b}{\psi_a}\right)_{\text{mode } 2} = \left(\frac{B}{A}\right)_{\text{mode } 2} = \frac{B_2}{A_2} = -\frac{m_a}{m_b}$$

If m_a were equal to m_b, $(\psi_b/\psi_b)_{\text{mode } 2} = -1$, the same as in the previous section.

The displacements of the oscillators in mode 1 is given by

$$(\psi_a)_1 = A_1 \cos(\omega_1 t + \phi_1)$$

$$(\psi_b)_1 = B_1 \cos(\omega_1 t + \phi_1)$$

with $B_1 = A_1$

For mode 2 the displacement is given by

$$(\psi_a)_2 = A_2 \cos(\omega_2 t + \phi_2)$$

$$(\psi_b)_2 = B_2 \cos(\omega_2 t + \phi_2)$$

with $B_2 = -\dfrac{m_a}{m_b} A_2$.

The most general solution of Eqs. (5.7) and (5.8) is then given by the superposition of the two normal modes, i.e.

$$\psi_a = (\psi_a)_1 + (\psi_a)_2 = A_1 \cos(\omega_1 t + \phi_1) + A_2 \cos(\omega_2 t + \phi_2) \tag{5.14}$$

$$\psi_b = (\psi_b)_1 + (\psi_b)_2 = B_1 \cos(\omega_1 t + \phi_1) + B_2 \cos(\omega_2 t + \phi_2) \tag{5.15}$$

Notice that whereas we have complete freedom in choosing the four constants A_1, A_2, ϕ_1 and ϕ_2 we have no freedom at all in choosing B_1 and B_2 since they are determined from A_1 and A_2. The four unknown constants A_1, A_2, ϕ_1 and ϕ_2 are to be determined from the given initial conditions i.e. the initial values of ψ_a, ψ_b, $\dot{\psi}_a$ and $\dot{\psi}_b$ as shown in the following section.

5.4 BEATS (ENERGY EXCHANGE) IN COUPLED OSCILLATIONS

Let us perform a simple experiment. Take two identical pendulums *A* and *B* and connect them with a spring whose relaxed length is exactly equal to the distance between the pendulum bobs, as shown in Fig. 5.2a. Hold pendulum *B* at its equilibrium position and give a displacement say, *a*, to pendulum *A*. Now release both bobs from rest at the same time, which we call $t = 0$. What happens? A very interesting phenomenon is observed. Pendulum *A* swings from side to side but its amplitude continuously decreases. Pendulum *B*, which was initially undisplaced, begins to oscillate and its amplitude continuously increases. After some time, the displacement of *A* is momentarily zero and that of *B* becomes equal to a, the displacement originally given to A. The starting condition is reversed. The motion of B is transferred back to A and the sequence of events continues. The energy is transferred from one pendulum to the other. The energy originally given to A does not stay with A but is gradually transferred to B and continues to shuttle back and forth between A and B. One round trip for the energy from A to B and back to A is technically called a *beat*. The beat period is the time for one round trip and is the inverse of the beat frequency.

If the mass m_a of bob A is different from mass m_b of bob B, a similar behaviour is observed. The energy keeps bouncing back and forth between A and B except that the amplitude of A never becomes zero, i.e. the energy exchange is not complete.

These observations are predicted by our Eqs (5.14) and (5.15) which give the displacements as a function of time. The instantaneous velocity of each pendulum is obtained by differentiating these equations w. r. t. time.

$$\dot{\psi}_a = -A_1\omega_1 \sin(\omega_1 t + \phi_1) - A_2\omega_2 \sin(\omega_2 t + \phi_2) \tag{5.16}$$

$$\dot{\psi}_b = -B_1\omega_1 \sin(\omega_1 t + \phi_1) - B_2\omega_2 \sin(\omega_2 t + \phi_2) \tag{5.17}$$

The initial conditions are

$$\psi_a(t = 0) = a$$

$$\psi_b(t = 0) = 0$$

$$\dot{\psi}_a(t = 0) = \dot{\psi}_b(t = 0) = 0$$

Using these conditions in Eqs (5.14) and (5.17) and setting $B_1 = A_1$ and $B_2 = -\frac{m_a}{m_b} A_2$, we have

$$a = A_1 \cos\phi_1 + A_2 \cos\phi_2$$

$$0 = +A_1 \cos\phi_1 - \frac{m_a}{m_b} A_2 \cos\phi_2$$

$$0 = -A_1\omega_1 \sin\phi_1 - A_2\omega_2 \sin\phi_2$$

and

$$0 = -A_1\omega_1 \sin\phi_1 + \frac{m_a}{m_b}\omega_2 A_2 \sin\phi_2$$

These four equations determine the four constants A_1, A_2, ϕ_1 and ϕ_2. They give

$$\sin\phi_1 = \sin\phi_2 = 0$$

$$A_1 \cos\phi_1 = \frac{am_a}{m_a + m_b}$$

$$A_2 \cos\phi_2 = \frac{am_b}{m_a + m_b}$$

Substituting these values in Eqs (5.14) and (5.15), we get

$$\psi_a = \frac{a}{(m_a+m_b)}(m_a \cos \omega_1 t + m_b \cos \omega_2 t) \tag{5.18}$$

$$\psi_b = \frac{a\, m_a}{(m_a+m_b)}(\cos \omega_1 t - \cos \omega_2 t) \tag{5.19}$$

Knowing the values of m_a, m_b, ω_1, ω_2, and a ; ψ_a and ψ_b are determined for all times t. Notice that ψ_a and ψ_b do not vary harmonically with time. In fact ψ_a and ψ_b represent superposition of two harmonic oscillations at frequencies ω_1 and ω_2 of the two modes. It is obvious that there are beats in the system.

Let us recast Eqs (5.18) and (5.19) in terms of two frequencies ω_m and ω_a defined by

$$\omega_m = \tfrac{1}{2}(\omega_2 - \omega_1)$$

$$\omega_a = \tfrac{1}{2}(\omega_2 + \omega_1)$$

So that $\omega_2 = \omega_a + \omega_m$ and $\omega_1 = \omega_a - \omega_m$. In terms of ω_a and ω_m Eq. (5.18) becomes

$$\psi_a = a \cos \omega_m t \cos \omega_a t + a\frac{(m_a-m_b)}{(m_a+m_b)} \sin \omega_m t \sin \omega_a t$$

This equation can be written in a simpler form if we set

$$a \cos \omega_m t = A_m \cos \theta$$

and

$$a\frac{(m_a-m_b)}{(m_a+m_b)} \sin \omega_m t = A_m \sin \theta$$

In terms of A_m and θ the equation for ψ_a becomes

$$\psi_a = A_m \cos(\omega_a t - \theta) \tag{5.20}$$

where A_m is given by

$$A_m^2 = a^2 \cos^2 \omega_m t + a^2\frac{(m_a-m_b)^2}{(m_a+m_b)^2} \sin^2 \omega_m t$$

or

$$A_m^2 = \frac{a^2}{(m_a+m_b)^2}(m_a^2 + m_b^2 + 2m_a m_b \cos 2\,\omega_m t) \tag{5.21}$$

Similarly Eq. (5.19) can be written as

$$\psi_b = B_m \sin \omega_a t \tag{5.22}$$

where

$$B_m = \frac{2am_a}{(m_a+m_b)} \sin \omega_m t \tag{5.23}$$

Case of Weak Coupling

It must be understood that Eqs (5.20) and (5.22) do not represent simple harmonic motions since A_m and B_m vary with time. But if the coupling were very weak (i.e. k were small) than ω_2 will be only slightly higher than ω_1 and $\omega_m = \frac{1}{2}(\omega_2 - \omega_1)$ will be very small so that A_m and B_m will take a long time to change appreciably. Hence, if k is small, A_m and B_m remain sensibly constant during many fast oscillations of angular frequency ω_a and the motion associated with Eqs (5.20) and (5.22) can be treated as almost harmonic and we can use the results of Chap. 1 and write down the expression for the energy of each pedulum. The energy E_a of pendulum A is given by [see Eq. (5.20)].

$$E_a = \frac{1}{2} m_a \omega_a^2 A_m^2$$

Substituting for A_m from Eq (5.21) we have

$$E_a = \frac{1}{2} m_a \omega_a^2 a^2 \left\{ \frac{m_a^2 + m_b^2 + 2m_a m_b \cos 2\omega_m t}{(m_a + m_b)^2} \right\}$$

Let E be the energy of pendulum A at $t = 0$. This equation gives at $(t = 0)$

$$E = \frac{1}{2} m_a \omega_a^2 a^2$$

In terms of E the expression for E_a becomes (since $2\omega_m = \omega_2 - \omega_1$)

$$E_a = E \left\{ \frac{m_a^2 + m_b^2 + 2m_a m_b \cos(\omega_2 - \omega_1)t}{(m_a + m_b)^2} \right\} \tag{5.24}$$

Similarly the energy of pendulum B is given by [using Eqs. (5.22) and (5.23)].

$$E_b = \frac{1}{2} m_b \omega_a^2 B_m^2$$

or
$$E_b = \frac{2m_a m_b E}{(m_a + m_b)^2} \{1 - \cos(\omega_2 - \omega_1)t\} \tag{5.25}$$

Equations (5.24) and (5.25) tell us that at time $t = 0$, $E_a = E$ and $E_b = 0$ which agree with our initial conditions. As time passes E_a begins to decrease and E_b begins to increase. At a value of t when $\cos(\omega_2 - \omega_1)t = -1$, E_a becomes minimum given by

$$(E_a)_{\min} = E \frac{(m_a - m_b)^2}{(m_a + m_b)^2}$$

and E_b becomes maximum given by

$$(E_b)_{\max} = E \frac{4m_a m_b}{(m_a + m_b)^2}$$

After this value of t, E_a increases until it again becomes equal to E, the energy pendulum A had at time $t = 0$ and E_b becomes zero; and the process repeats. The values of t when E_a becomes maximum equal to E are given by

$$\cos(\omega_2-\omega_1)t = +1$$

i.e. $(\omega_2-\omega_1)\,t = 0, 2\pi, 4\pi, \ldots$

or $$t = 0, \frac{2\pi}{\omega_2-\omega_1}, \frac{4\pi}{\omega_2-\omega_1}, \ldots\ldots$$

Thus the time period of beats is

$$t_b = \frac{2\pi}{\omega_2-\omega_1} = \frac{1}{\nu_2-\nu_1}$$

where ν_1 and ν_2 are the frequencies corresponding of angular frequencies ω_1 and ω_2. The frequency of beats is

$$\nu_b = \nu_2-\nu_1$$

which is equal to the difference between the two normal mode frequencies of the system. Notice from Eqs (5 24) and (5.25) that

$$E_a+E_b = E$$

i.e. the total energy of the system remains constant. This is so because we have assumed the friction to be absent. Thus we conclude that the energy keeps bouncing back and forth between the two pendulums at a frequency $\nu_b = \nu_2-\nu_1$. Since $(E_a)_{\min} \neq 0$, the energy of pendulum A is not completely transferred to pendulum B. In other words, the energy exchange is not complete.

Case of Identical Pendulums

In the case of two indentical pendulums ($m_a = m_b = m$) the Eqs (5.18), (5.19), (5.24) and (5.25) become

$$\psi_a = \frac{a}{2}(\cos\omega_1 t+\cos\omega_2 t)$$

$$\psi_b = \frac{a}{2}(\cos\omega_1 t-\cos\omega_2 t)$$

$$E_a = \frac{E}{2}[1+\cos(\omega_2-\omega_1)t]$$

$$E_b = \frac{E}{2}[1-\cos(\omega_2-\omega_1)t]$$

Notice that $(E_a)_{\min} = 0$, and $(E_b)_{\max} = E$. Thus, if the bob masses are equal, the energy exchange is complete. The energy of each pendulum oscillates between a minimum value of zero and a maximum value of E.

Figure 5.4 shows how ψ_a, ψ_b, A_m, B_m, E_a and E_b vary with time. Notice that ψ_a and ψ_b do not vary harmonically with time.

In conclusion we may remark that although we have taken the example of two coupled pendulums, the mathematical technique developed above is very general and is applicable to any coupled system with two degrees of freedom. The method is that we first find out the normal modes of the system, write down the displacement of its moving parts in each mode and obtain the general solution by constructing a superposition of the normal mode displacements. The general solution contains four adjustable constants which allows us to fit these solutions to arbitrary values of the initial displacements and velocities of both the moving parts of the system. We shall consider some other examples of coupled oscillations and obtain the configuration and frequency of the normal modes of the system under consideration.

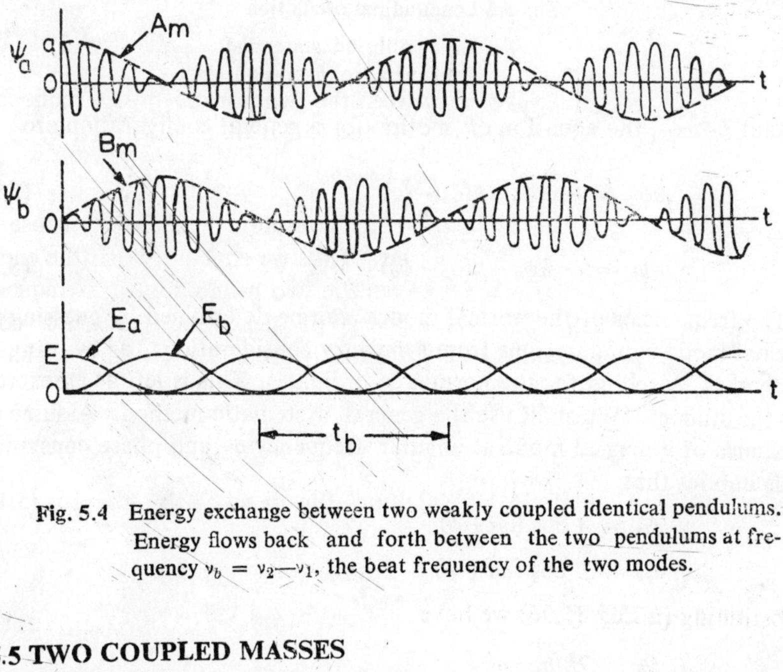

Fig. 5.4 Energy exchange between two weakly coupled identical pendulums. Energy flows back and forth between the two pendulums at frequency $\nu_b = \nu_2 - \nu_1$, the beat frequency of the two modes.

5.5 TWO COUPLED MASSES

Let us consider a system consisting of two masses and three springs as shown in Fig. 5.5. We shall first analyse the simplest case when the two masses are equal and the three springs are identical. Let each mass be m and each spring has a spring constant k. This system is capable of longitudinal as well as transverse oscillations.

Normal Modes of Longitudinal Oscillations

Figure 5.5a shows the equilibrium state of the system. Each spring (assumed massless) has a length, say, a. Figure 5.5b shows the general configuration of the system in oscillation. Let ψ_a and ψ_b be the displacements of the masses A and B at any instant of time. Assuming that at this

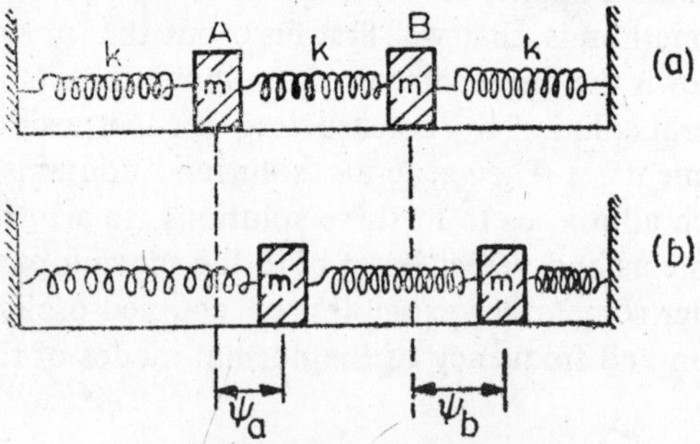

Fig. 5.5 Longitudinal oscillation

(a) equilibrium state, and
(b) general configuration

instant $\psi_b > \psi_a$, the equation of motion for a general configuration are

$$m\ddot{\psi}_a = -k\psi_a + k(\psi_b - \psi_a) \tag{5.26}$$

$$m\ddot{\psi}_b = -k\psi_b - k(\psi_b - \psi_a) \tag{5.27}$$

The frequencies of the normal modes can be determined by guessing the normal mode configurations from symmetry considerations or by using the method of searching for the normal co-ordinates. This is left as an exercise for the student. We shall use the general systematic method. Assume the existence of a normal mode at angular frequency ω and phase constant φ. This implies that

$$\psi_a = A\cos(\omega t + \phi)$$

$$\psi_b = B\cos(\omega t + \phi)$$

Substituting in Eqs. (5.26) we have

$$\frac{\psi_b}{\psi_a} = \frac{2k/m - \omega^2}{k/m} \tag{5.28}$$

Similarly Eq. (5.27) gives

$$\frac{\psi_b}{\psi_a} = \frac{k/m}{2k/m - \omega^2} \tag{5.29}$$

Equating the right-hand sides of Eqs (5.28) and (5.29). we have

$$\left(\frac{2k}{m} - \omega^2\right)^2 = \frac{k^2}{m^2}$$

or

$$-\omega^2 + \frac{2k}{m} = \pm\frac{k}{m}$$

Thus, there are two possible values of ω^2 which are

$$\omega_1^2 = \frac{k}{m}$$

and

$$\omega_2^2 = \frac{3k}{m}$$

These are the angular frequencies of the two normal modes. The shape or configuration of mode 1 with frequency ω_1 is obtained from either Eq. (5.28) or Eq. (5.29) by setting $\omega^2 = \omega_1^2 = k/m$.

$$\left(\frac{\psi_b}{\psi_a}\right)_{\text{mode }1} = +1$$

This mode is shown in Fig. 5.6 a. The shape of mode 2 which has a higher frequency ω_2 is similarly given by

$$\left(\frac{\psi_b}{\psi_a}\right)_{\text{mode }2} = -1$$

Figure 5.6 b depicts the configuration of this mode.

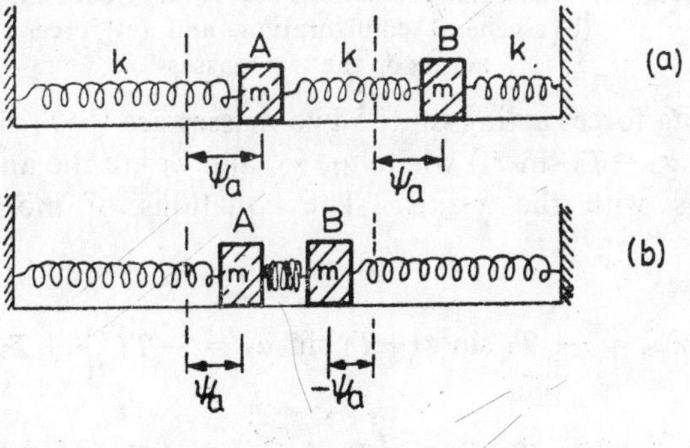

Fig. 5.6 Normal modes of longitudinal oscillations
(a) mode with lower frequency and
(b) mode with higher frequency

Normal Modes of Transverse Oscillations

A coupled system consisting of three identical massless springs and two equal masses is shown in Fig. 5.7. Each spring has relaxed length a_0 and spring constant k. A and B have the same mass m. Figure 5.7a shows the equilibrium state in which each spring is extended to a length a so that $T_0 = k\,(a - a_0)$ is the equilibrium tension in each spring. Figure 5.7b shows a general configuration of the system in transverse oscillations. The displacements of the two masses at a certain instant of time are ψ_a and ψ_b. Figure 5.7c shows the forces acting on each mass in the general configuration. The new lengths of the springs are l_1, l_2 and l_3 as shown and the tensions in the springs are

$$T_1 = k\,(l_1 - a_0)$$
$$T_2 = k\,(l_2 - a_0)$$
$$T_3 = k\,(l_3 - a_0)$$

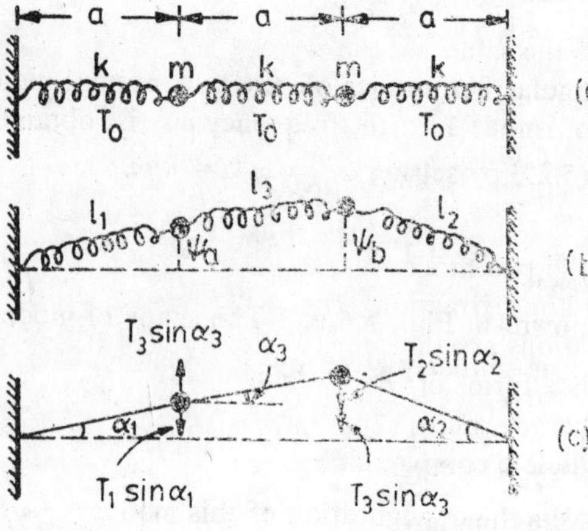

Fig. 5.7 Transverse oscillations : (a) equilibrium state, (b) a general configuration, and (c) forces acting on the two masses

The restoring forces acting on the two masses are $(-T_1 \sin\alpha_1 + T_3 \sin\alpha_3)$ and $(-T_2 \sin\alpha_2 - T_3 \sin\alpha_3)$ where α_1, α_2 and α_3 are the angles subtended by the springs with the x-axis. The equations of motion of the two masses are

$$m\ddot{\psi}_a = -T_1 \sin\alpha_1 + T_3 \sin\alpha_3 = -T_1 \frac{\psi_a}{l_1} + T_3 \frac{(\psi_b - \psi_a)}{l_3}$$

$$m\ddot{\psi}_b = -T_2 \sin\alpha_2 - T_3 \sin\alpha_3 = -T_2 \frac{\psi_b}{l_2} - T_3 \frac{(\psi_b - \psi_a)}{l_3}$$

Substituting for T_1, T_2 and T_3, we have

$$m\ddot{\psi}_a = -k\left(1 - \frac{a_0}{l_1}\right)\psi_a + k\left(1 - \frac{a_0}{l_3}\right)(\psi_b - \psi_a) \tag{5.30}$$

$$m\ddot{\psi}_b = -k\left(1 - \frac{a_0}{l_2}\right)\psi_b - k\left(1 - \frac{a_0}{l_3}\right)(\psi_b - \psi_a) \tag{5.31}$$

we shall find the normal modes under the following approximations:

Slinky Approximation

In the slinky approximation $a_0 \ll a$. Hence terms like $\dfrac{a_0}{l_1} = \dfrac{a_0}{a} \cdot \dfrac{a}{l_1}$

are of second order in smallness ($\because a < l_1$) and can therefore, be neglected compared to unity. Under this approximation, Eqs (5.30) and (5.31) reduce to

$$m\ddot{\psi}_a = -k\psi_a + k(\psi_b - \psi_a)$$

$$m\ddot{\psi}_b = -k\psi_b - k(\psi_b - \psi_a)$$

These equations are the same as the Eqs (5.26) and (5.27) of longitudinal oscillations. It follows, therefore, that the two normal mode frequencies are

$$\omega_1 = \sqrt{\frac{k}{m}} \quad \text{and} \quad \omega_2 = \sqrt{\frac{3k}{m}}$$

and the shapes of the two modes are $\psi_b = \psi_a$ and $\psi_b = -\psi_a$ respectively. Thus, in the slinky approximation, the frequencies of the normal modes of transverse oscillations are the same as those of longitudinal oscillations. We say that there is a form of degeneracy. In the following we shall see that no such degeneracy occurs for the small-oscillation approximation, where a_0 is not negligible compared to a.

Small-oscillation Approximation

In this approximation, the oscillating parts of the system stay close to the equilibrium position. Hence $l_1 \simeq l_2 \simeq l_3 \simeq a$. Under this approximation, Eqs (5.30) and (5.31) reduce to

$$m\ddot{\psi}_a = -k\left(1 - \frac{a_0}{a}\right)\psi_a + k\left(1 - \frac{a_0}{a}\right)(\psi_b - \psi_a)$$

$$m\ddot{\psi}_b = -k\left(1 - \frac{a_0}{a}\right)\psi_b - k\left(1 - \frac{a_0}{a}\right)(\psi_b - \psi_a)$$

These equations are the same as the Eqs (5.26) and (5.27) if k is replaced by its effective value $k^* = k\left(1 - \frac{a_0}{a}\right)$. It follows, therefore, that the two normal mode frequencies are

$$\omega_1 = \sqrt{\frac{k^*}{m}} = \sqrt{\frac{k}{m}\left(1 - \frac{a_0}{a}\right)}$$

and

$$\omega_2 = \sqrt{\frac{3k^*}{m}} = \sqrt{\frac{3k}{m}\left(1 - \frac{a_0}{a}\right)}$$

and the shapes of the modes are $\psi_b = \psi_a$ and $\psi_b = -\psi_a$ respectively. The mode frequencies of the transverse small oscillations are smaller than their corresponding frequencies of the longitudinal small oscillations. Figure 5.8 shows the configurations of the two modes. The student is advised to guess these configurations from symmetry considerations and also by choosing the normal co-ordinates. In Sec. 5.7 we shall solve the most general case of two masses and three springs and we shall prove a very important property of the normal modes which states that the total energy of the coupled system is just equal to the sum of the energies associated with its normal modes.

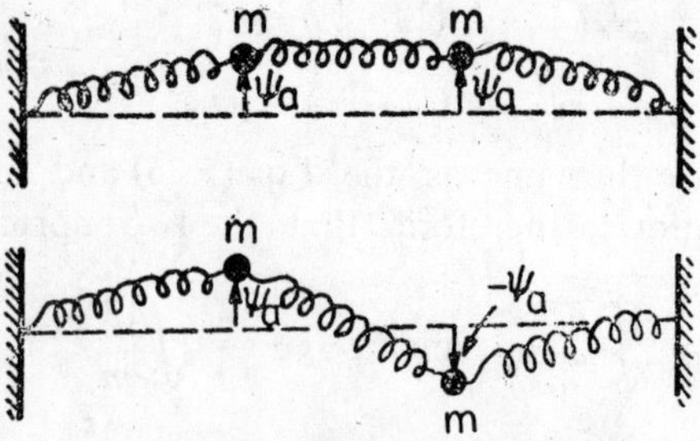

Fig. 5.8 Normal modes of transverse oscillations :
(a) mode with lower frequency, and
(b) mode with higher frequency

5.6 TWO COUPLED LC CIRCUITS

We shall consider two ways of coupling two LC circuits.

Capacitative Coupling

Consider two resistance-free *CL* circuits shown in Fig. 5.9. For simplicity we assume that the circuits have identical values of inductance L and capacitance C. The circuits are coupled to each other by the middle capacitor whose capacitance is also taken to be C. In a general configu-

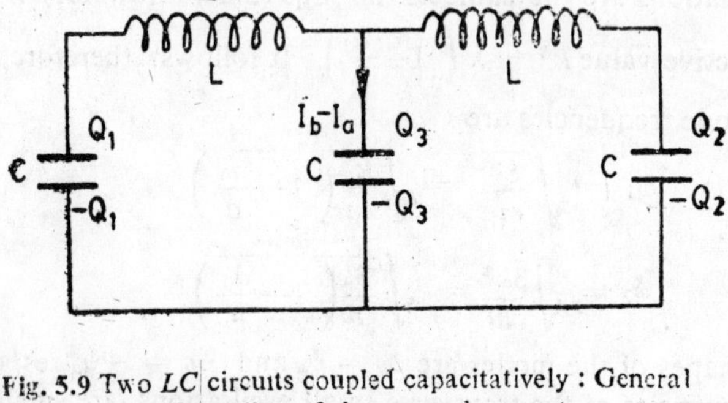

Fig. 5.9 Two *LC* circuits coupled capacitatively : General configuration of charges and currents.

ration let Q_1, Q_2 and Q_3 be the charges on the capacitor plates (as shown in the diag am) at a certain instant of time. Let us find the equation of motion of charges in each circuit. Let I_a and I_b be the instantaneous values of the currents in each circuit.

The *emf* across the left-hand inductance is $L\,\dfrac{dI_a}{dt}$. A charge Q_1 on the left-hand capacitor produces a potential difference $-\dfrac{Q_1}{C}$ which tends to decrease I_a. A charge Q_3 on the middle capacitor produces a potential difference $\dfrac{Q_3}{C}$ which tends to increase I_a. Thus

$$L\,\frac{dI_a}{dt} = -\frac{Q_1}{C} + \frac{Q_3}{C}$$

Similarly

$$L\,\frac{dI_b}{dt} = -\frac{Q_2}{C} - \frac{Q_3}{C}$$

Differentiating with respect to time gives

$$L\,\frac{d^2I_a}{dt^2} = -\frac{1}{C}\,\frac{dQ_1}{dt} + \frac{1}{C}\,\frac{dQ_3}{dt}$$

$$L\,\frac{d^2I_b}{dt^2} = -\frac{1}{C}\,\frac{dQ_2}{dt} - \frac{1}{C}\,\frac{dQ_3}{dt}$$

Since charge must be conserved, we have

$$\frac{dQ_1}{dt} = I_a$$

$$\frac{dQ_2}{dt} = I_b$$

and

$$\frac{dQ_3}{dt} = I_b - I_a$$

Substituting them in the above equations, we get

$$L\,\frac{d^2I_a}{dt^2} = -\frac{1}{C}\,I_a + \frac{1}{C}\,(I_b - I_a) \tag{5.32}$$

$$L\,\frac{d^2I_b}{dt^2} = -\frac{1}{C}\,I_b - \frac{1}{C}\,(I_b - I_a) \tag{5.33}$$

If we replace L by m, $1/C$ by k and I by ψ these equations become identical with Eqs (5.26) and (5.27). Hence the two normal modes of the system are as follows :

	Shape	*Angular frequency*
Mode 1	$I_b = I_a$	$\omega_1 = \sqrt{\frac{1}{LC}}$
Mode 2	$I_b = -I_a$	$\omega_2 = \sqrt{\frac{3}{LC}}$

Notice that in mode 1, the middle capacitor never has any charge (Q_3 is always zero) and it could be removed without affecting the currents in the circuits which are always equal. In mode 2, the current in one circuit is equal and opposite to that in the other and the charge Q_3 is twice in magnitude compared to either Q_1 or Q_2 which are both equal.

Inductive Coupling

Consider two resistance-free $L\,C$ circuits shown in Fig. 5.10. A mutual inductance exists between the two circuits when the magnetic flux due to current in one circuit threads the second circuit. When the current changes, the magnetic flux changes and induces an emf in both the circuits. What happens in one circuit affects the second circuit and the two circuits are thus coupled. This is called *inductive coupling*. Let L_1 and L_2 be the inductances of the circuits and C_1 and C_2 their capacitances. If M is the mutual inductance then the strength of the coupling is measured by the *coupling coefficient* μ which is defined as

$$\mu = \frac{M}{\sqrt{L_1 L_2}}$$

We shall now obtain the equation of motion of charges in each circuit. Let Q_1 and Q_2 be the charges on capacitors C_1 and C_2 respectively at any instant of time. Let I_a and I_b be the instantaneous values of the currents

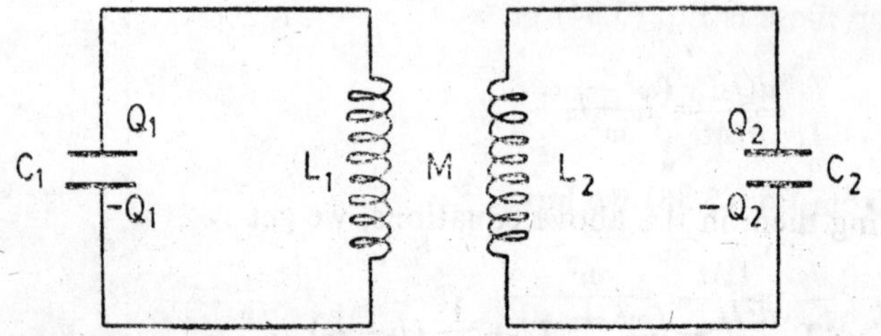

Fig. 5.10 Two inductively coupled LC circuits with mutual inductance M.

in the two circuits. The *emf* across inductance L_1 is $L_1 \frac{dI_a}{dt}$. A charge Q_1 on capacitance C_1 produces a potential difference $-\frac{Q_1}{C_1}$ which tends to decrease I_a. The *emf* produced in this circuit due to the current I_b in the other circuit is $M\dfrac{dI_b}{dt}$ which tends to increase I_a. Thus the equation governing the balance of voltages in the circuit involving C_1 and L_1 is

$$L_1\frac{dI_a}{dt} = -\frac{Q_1}{C_1} + M\frac{dI_b}{dt}$$

Differentiating with respect to time and setting $\dfrac{dQ_1}{dt} = I_a$, we have

or
$$\frac{d^2I_a}{dt^2} = -\omega'^2 I_a + \frac{M}{L_1}\frac{d^2I_b}{dt^2} \tag{5.34}$$

where $\omega' = \dfrac{1}{\sqrt{L_1C_1}}$ is the natural angular frequency of this circuit.

Similarly, for the circuit involving L_2 and C_2, we have

$$\frac{d^2I_b}{dt^2} = -\omega''^2 I_b + \frac{M}{L_2}\frac{d^2I_a}{dt^2} \tag{5.35}$$

where $\omega'' = \dfrac{1}{\sqrt{L_2C_2}}$ is the natural angular frequency of the second circuit.

Equations (5.34) and (5.35) are the two coupled equations. To obtain the normal modes of current oscillations we assume, as before, that there exists a normal mode at angular frequency ω and phase constant ϕ. Therefore,

$$I_a = A\cos(\omega t+\phi)$$

$$I_b = B\cos(\omega t+\phi)$$

giving $\dfrac{d^2I_a}{dt^2} = -\omega^2 I_a$ and $\dfrac{d^2I_b}{dt^2} = -\omega^2 I_b$ (5.36)

Substituting these is Eq. (5.34) gives

$$\frac{I_b}{I_a} = \frac{L_1}{M}\frac{(\omega^2-\omega'^2)}{\omega^2}$$

Similarly from Eq. (5.35) we have

$$\frac{I_b}{I_a} = \frac{M}{L_2}\frac{\omega^2}{(\omega^2-\omega''^2)} \tag{5.37}$$

Equating the right-hand sides of the above equations we get

$$(\omega^2-\omega'^2)(\omega^2-\omega''^2) = \frac{M^2}{L_1L_2}\omega^4 = \mu^2\omega^4$$

where μ is the coupling coefficient. This is a quadratic equation in ω^2; the roots of this equation give the normal mode frequencies. If the two circuits were identical, i.e. $L_1 = L_2 = L$ and $C_1 = C_2 = C$, then $$\omega' = \omega'' = \omega_0 = \frac{1}{\sqrt{LC}}$$

In this special case, the above equation reduces to

$$(\omega^2 - \omega_0^2)^2 = \mu^2\omega^4$$

or $$\omega^2 - \omega_0^2 = \pm\mu\omega^2$$

or $$\omega = \pm\frac{\omega_0}{\sqrt{1 \pm \mu}}$$

Since frequency cannot be negative, the two allowed frequencies are

$$\omega_1 = \frac{\omega_0}{\sqrt{1+\mu}}$$

and

$$\omega_2 = \frac{\omega_0}{\sqrt{1-\mu}}$$

These are the angular frequencies of the normal modes of two identical *LC* circuits coupled inductively. If the coupling is very weak ($\mu \to 0$), $\omega_1 = \omega_2 = \omega_0$, the natural frequency of either circuit. It is easy to see that in mode 1 with angular frequency ω_1. $I_b = -I_a$, i.e. the currents in the two circuits are always equal and in opposite directions. However, in mode 2 with angular frequency ω_2, $I_b = I_a$, i.e. the currents in the two circuits are always equal and in the same direction.

5.7 ENERGY RELATIONS IN COUPLED OSCILLATIONS

In Sec. 5.2 we established a very important property of the normal modes of a coupled system. Taking the example of two coupled pendulums we showed that the general motion of the system is determined if its normal modes are known. The general displacement of each moving part of the system, under any arbitrary initial conditions, is simply given by a linear superposition of its displacements in the normal modes of vibration. We shall now establish another equally important property of the normal modes. We will show that the total energy of a coupled system, started in any way, is just equal to the sum of the energies of its normal modes of vibration. Remember, it is very easy to find the energy of the normal modes since the motion associated with a normal mode is simple harmonic. To prove this property of the normal modes, we will consider (for the purpose of illustration) the oscillation of two masses coupled together by means of three springs. But we will treat the most general case in which the masses and the springs are not identical. We will restrict ourselves to only longitudinal oscillations.

Consider the system shown in Fig. 5.11. A and B are two masses m_a and m_b respectively and the springs (assumed massless) have constants k_1, k_2 and k_3. The spring k_3 provides the coupling. Figure 5.11a shows the equilibrium state of the system. The masses are pulled along a

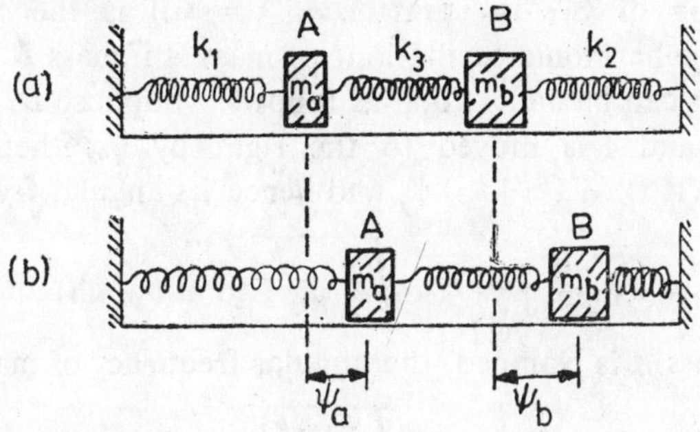

Fig. 5.11 Longitudinal oscillations : (a) equilibrium state, and (b) general configuration

frictionless horizontal surface and released. They begin to oscillate. Let ψ_a and ψ_b respectively be the displacements of A and B at a certain instant of time. Figure 5.11b shows a general configuration of the system at a time t. Let $\psi_b > \psi_a$ at this instant.

The equations of motion of A and B are

$$m_a \ddot{\psi}_a = -k_1 \psi_a + k_3 (\psi_b - \psi_a) = -(k_1 + k_3) \psi_a + k_3 \psi_b$$

$$m_b \ddot{\psi}_b = -k_2 \psi_b - k_3 (\psi_b - \psi_a) = -(k_2 + k_3) \psi_b + k_3 \psi_a$$

These equations can be written in a simpler form in terms of 'reduced' co-ordinates x and y which are related to actual co-ordinates ψ_a and ψ_b as

$$x = \sqrt{m_a}\, \psi_a \qquad (5.38)$$

$$y = \sqrt{m_b}\, \psi_b$$

In terms of these co-ordinates, the above equations reduce to the following simple form

$$\ddot{x} = -a_{11}x - a_{12}y \qquad (5.39)$$

and $$\ddot{y} = -a_{22}y - a_{21}x \qquad (5.40)$$

where $$a_{11} = \frac{k_1 + k_3}{m_a}$$

$$a_{22} = \frac{k_2 + k_3}{m_b} \qquad (5.41)$$

and $$a_{12} = a_{21} = -\frac{k_3}{\sqrt{m_a m_b}}$$

It may be mentioned that the equations of motion of the moving parts of any system with two degrees of freedom can always be reduced to the above general form with appropriate values of the coefficients a_{11}, a_{22}, a_{12} and a_{21}.

The meaning of a_{11} is that $\sqrt{a_{11}}$ ($= \omega_a$) is the natural angular frequency of the harmonic oscillations of mass A if mass B is not allowed to move. This can be understood as follows. Suppose B is clamped so that ψ_b is zero and A is moved to the right by ψ_a, then the restoring force acting on A is $-(k_1+k_3)\psi_a$ and hence its angular frequency

$$\omega_a = \sqrt{\frac{k_1+k_3}{m_a}}.$$

Similarly if mass A is clamped, the angular frequency of mass B is

$$\omega_b = \sqrt{\frac{k_2+k_3}{m_b}}$$

which is $\sqrt{a_{22}}$. A displacement of mass A produces a force on mass B. This force is $k_3\,\psi_a$. From symmetry the force on A by a displacement of B is $k_3\,\psi_b$. This is what is meant by coupling.

Frequencies of the Normal Modes

We assume that we have oscillation in a single mode. This means that both degrees of freedom, namely, x and y, oscillate with simple harmonic motions with the same frequency ω and same phase constant ϕ, i.e.

$$x = A\cos(\omega t+\phi)$$

$$y = B\cos(\omega t+\phi)$$

Then we have

$$\ddot{x} = -\omega^2 x$$

$$\ddot{y} = -\omega^2 y$$

Substituting them in Eqs (5.39) and (5.40) we obtain two homogeneous linear equations in x and y

$$(a_{11}-\omega^2)\,x+a_{12}\,y=0 \tag{5.42}$$

$$a_{21}\,x+(a_{22}-\omega^2)\,y=0 \tag{5.43}$$

Each of these equations gives the ratio $\frac{y}{x}$:

$$\frac{y}{x} = \frac{\omega^2-a_{11}}{a_{12}} \tag{5.44}$$

$$\frac{y}{x} = \frac{a_{21}}{\omega^2-a_{22}} \tag{5.45}$$

For consistency, the right-hand sides of these equations must be equal, which gives

$$(a_{11}-\omega^2)(a_{22}-\omega^2)-a_{21}a_{12}=0$$

This equation can also be obtained directly by requiring that the determinant of coefficients of the linear homogeneous Equations (5 42) and (5.43) must vanish:

$$\begin{vmatrix} a_{11}-\omega^2 & a_{12} \\ a_{21} & a_{22}-\omega^2 \end{vmatrix} \equiv (a_{11}-\omega^2)(a_{22}-\omega^2)-a_{21}a_{12}=0$$

or $$\omega^4-(a_{11}+a_{22})\,\omega^2+(a_{11}a_{22}-a_{21}a_{12})=0$$

This is a quadratic equation in ω^2. It has two solutions :

$$\omega_1^2=\frac{1}{2}(a_{11}+a_{22})-\frac{1}{2}\left\{(a_{11}-a_{22})^2+4a_{12}a_{21}\right\}^{1/2} \tag{5.46}$$

and $$\omega_2^2=\frac{1}{2}(a_{11}+a_{22})+\frac{1}{2}\left\{(a_{11}-a_{22})^2+4a_{12}a_{21}\right\}^{1/2} \tag{5.47}$$

The angular frequency of mode 1 is ω_1 and that of mode 2 is ω_2.

Shapes or Configurations of the Normal Modes

The shape of x and y in mode 1 is obtained by substituting $\omega^2=\omega_1^2$ back into either Eq. (5.44) or Eq. (5.45):

$$\left(\frac{y}{x}\right)_{\text{mode 1}}=\left(\frac{B}{A}\right)_{\text{mode 1}}$$

$$=\frac{B_1}{A_1}=\frac{\omega_1^2-a_{11}}{a_{12}}=\frac{a_{21}}{\omega_1^2-a_{22}} \tag{5.48}$$

Similarly

$$\left(\frac{y}{x}\right)_{\text{mode 2}}=\left(\frac{B}{A}\right)_{\text{mode 2}}$$

$$=\frac{B_2}{A_2}=\frac{\omega_2^2-a_{11}}{a_{12}}=\frac{a_{21}}{\omega_2^2-a_{22}} \tag{5.49}$$

Displacements in the Two Normal Modes

For mode 1, x and y are written as

$$x_1=A_1\cos(\omega_1 t+\phi_1)$$

$$y_1=B_1\cos(\omega_1 t+\phi_1)$$

where B_1 and A_1 are related through Eq. (5.48)

Similarly for mode 2 we have

$$x_2 = A_2 \cos(\omega_2 t+\phi_2)$$

$$y_2 = B_2 \cos(\omega_2 t+\phi_2)$$

where B_2 and A_2 are related through Eq. (5.49)

The Most General Solution

The most general solution of Eqs (5.39) and (5.40) is given by the superposition of the two normal modes as follows:

$$x = x_1+x_2 = A_1 \cos(\omega_1 t+\phi_1)+A_2 \cos(\omega_2 t+\phi_2) \tag{5.50}$$

$$y = y_1+y_2 = B_1 \cos(\omega_1 t+\phi_1)+B_2 \cos(\omega_2 t+\phi_2) \tag{5.51}$$

Since B_1 and B_2 are related to A_1 and A_2, there are four undermined constants A_1, A_2, ϕ_1 and ϕ_2 which are determined from the four initial conditions given by the values of x, y, $\dot{x}$ and $\dot{y}$ at time $t = 0$. The fact that Eqs (5.50) and (5.51) constitute the general solution can be verified by substitution in Eqs (5.39) and (5.40).

Normal Coordinates

The normal coordinates of the system can be immediately discovered once we are able to recognize that the amplitude ratios of the two modes are related as

$$\frac{A_1}{B_1} = -\frac{B_2}{A_2}$$

This follows from Eqs (5.46) and (5.47) which on addition give

$$\omega_1^2+\omega_2^2 = a_{11}+a_{22}$$

i.e.

$$\omega_1^2-a_{11} = -(\omega_2^2-a_{22})$$

Now it is clear from Eqs (5.48) and (5.49) that

$$\frac{A_1}{B_1} = -\frac{B_2}{A_2} = \tan\alpha \tag{5.52}$$

say, where the significance of angle α will emerge as we proceed. Thus we can write

$$A_1 = C_1 \sin\alpha \quad , \quad B_1 = C_1 \cos\alpha$$

$$B_2 = -C_2 \sin\alpha \quad , \quad A_2 = C_2 \cos\alpha$$

In terms of C_1, C_2 and α Eqs (5.50) and (5.51) become

$$x = C_1 \sin\alpha \cos(\omega_1 t+\phi_1)+C_2 \cos\alpha \cos(\omega_2 t+\phi_2)$$

$$y = C_1 \cos\alpha \cos(\omega_1 t+\phi_1)-C_2 \sin\alpha \cos(\omega_2 t+\phi_2)$$

Now let

$$X = C_1 \cos(\omega_1 t + \phi_1) \tag{5.53}$$

$$Y = C_2 \cos(\omega_2 t + \phi_2) \tag{5.54}$$

Then

$$x = X \sin \alpha + Y \cos \alpha \tag{5.55}$$

$$y = X \cos \alpha - Y \sin \alpha \tag{5.56}$$

These are the familiar equations we use in analytical geometry to represent the transformation of co-ordinates of a point in a plane with respect to axes x and y to a new set of axes X and Y inclined at an angle α to the first set, as shown in Fig. 5.12.

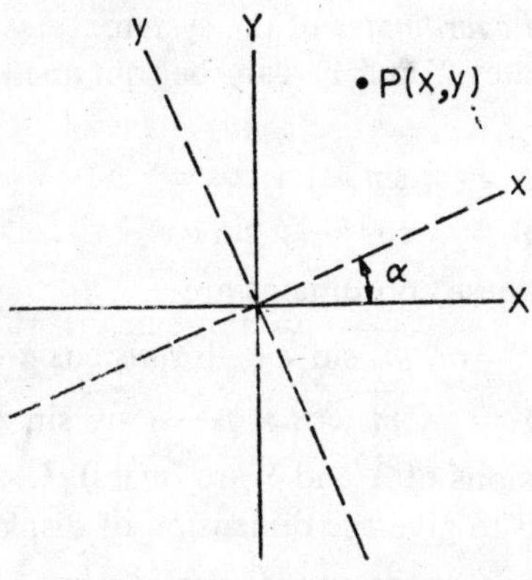

Fig. 5.12 Transformation to normal co-ordinates for two coupled oscillators

Notice that the motion associated with x or y is not even periodic but that associated with X as well as Y is harmonic. The simple harmonic motion associated with X has angular frequency ω_1 given by Eq (5.46) which is the frequency of the first normal mode. If the amplitudes of motion of the two masses are related as given by Eq. (5.48), then and only then, will the system oscillate with simple harmonic motion with frequency ω_1. The motion associated with Y has angular frequency ω_2 given by Eq. (5.47) which is the frequency of the second normal mode. If the amplitudes of the motion of the two masses are related as given by Eq. (5.49), then and only then, will the system oscillate with simple harmonic motion of frequency ω_2. If the motion is started in any other way, there will be no permanent ratio between the amplitudes of the two masses and the motion will not even be periodic.

Referring to Fig. 5.12 let us suppose that we represent the position of the system at any time t by a point P on this plane. Let the co-ordinates of point P at time t be (x, y). The abscissa x is a measure of the displacement of the first mass m_a (the actual displacement of this mass is

$\psi_a = x/\sqrt{m_a}$) and the ordinate y measures the displacement of the second mass m_b. The motion of the system corresponds to the motion of the point P on this plane. As time passes the point P moves; its motion being governed by the general oscillation of the system. The projection of P on the x-axis (which is the displacement of the first mass) moves back and forth in a complicated non-periodic fashion and so does the projection of P on the y-axis. However, the projection of P on the X-axis always moves back and forth with SHM of angular frequency ω_1 and amplitude C_1 [since $X = C_1 \cos(\omega_1 t + \phi_1)$]. The projection of P on the Y-axis also always moves back and forth with SHM with a higher angular frequency ω_2 and amplitude C_2 [since $Y = C_2 \cos(\omega_2 t + \phi_2)$]. Only when the system is started in such a way that the point P moves either along X-axis or along Y-axis is its motion harmonic. The plane in which the point P moves is called the *configuration plane* of the system and the axes X and Y are called the *normal coordinates* of the system.

The normal coordinates X and Y can be obtained from Eqs (5.55) and (5.56) which give

$$X = x \sin\alpha + y \cos\alpha$$

$$Y = x \cos\alpha - y \sin\alpha$$

In terms of ψ_a and ψ_b these coordinates are

$$X = \sqrt{m_a} \sin\alpha\, \psi_a + \sqrt{m_b} \cos\alpha\, \psi_b \tag{5.57}$$

$$Y = \sqrt{m_a} \cos\alpha\, \psi_a - \sqrt{m_b} \sin\alpha\, \psi_b \tag{5.58}$$

Notice that the dimensions of X and Y are $(\text{mass})^{1/2} \times$ displacement. They are suitably normalized to give the dimension of displacement if necessary.

The general motion of the system is given by

$$\psi_a = \frac{x}{\sqrt{m_a}} = \frac{C_1}{\sqrt{m_a}} \sin\alpha \cos(\omega_1 t + \phi_1) + \frac{C_2}{\sqrt{m_a}} \cos\alpha \cos(\omega_2 t + \phi_2) \tag{5.59}$$

$$\psi_b = \frac{y}{\sqrt{m_b}} = \frac{C_1}{\sqrt{m_b}} \cos\alpha \cos(\omega_1 t + \phi_1) - \frac{C_2}{\sqrt{m_b}} \sin\alpha \cos(\omega_2 t + \phi_2) \tag{5.60}$$

In Eqs (5.57) to (5.60) angle α is determined from the relation

$$\tan\alpha = \frac{A_1}{B_1} = \frac{\omega_1^2 - a_{22}}{a_{21}} = \frac{a_{12}}{\omega_1^2 - a_{11}} \tag{5.61}$$

where ω_1 is given by Eq (5.46) and a_{11}, a_{22}, a_{12}, etc. are given by Eq. (5.41). The general solution involves four arbitrary constants C_1, C_2, ϕ_1 and ϕ_2, whose values are fixed by specifying the initial displacements and velocities of the two masses.

Energy Relations

We shall now show that when the motion of a coupled system is described in terms of the normal coordinates X and Y (rather than in terms of x and

y), the expression for the energy of the system becomes remarkably simple.

The kinetic energy of the system in oscillation is equal to the kinetic energy of mass m_a plus that of mass m_b. If ψ_a and ψ_b are their displacements at a time t, then the kinetic energy of the system at that time is given by

$$KE = \tfrac{1}{2}\, m_a\, \dot{\psi}_a^2 + \tfrac{1}{2}\, m_b\, \dot{\psi}_b^2$$

To obtain the expression for the potential energy of the system at time t, we find the amount of work necessary to push the system from equilibrium to a position where the displacements are ψ_a and ψ_b. We first push m_a through a distance $\psi = \psi_a$ keeping m_b fixed. The work done is the potential energy of mass m_a which is given by the expression:

$$\int_0^{\psi_a} (k_1\psi + k_3\psi)\, d\psi = \tfrac{1}{2}\, (k_1 + k_3)\psi_a^2$$

Then next step is to push m_b through a distance $\psi = \psi_b$, keeping m_a fixed at its new position. The work done now is the potential energy of m_b which is given by the expression.

$$\int_0^{\psi_b} k_2\psi\, d\psi + \int_0^{\psi_b} k_3(\psi - \psi_a)\, d\psi$$

$$= \tfrac{1}{2}\, (k_2 + k_3)\, \psi_b^2 - k_3\psi_a\psi_b$$

The sum of these two energies is the potential energy of the system at time t. Thus

$$PE = \tfrac{1}{2}\, (k_1 + k_3)\, \psi_a^2 + \tfrac{1}{2}(k_2 + k_3)\, \psi_b^2 - k_3\psi_a\psi_b$$

The total instantaneous energy of the system is then given by

$$E = KE + PE$$

$$= \tfrac{1}{2}\, m_a\, \dot{\psi}_a^2 + \tfrac{1}{2}\, m_b\, \dot{\psi}_b^2 + \tfrac{1}{2}(k_1 + k_3)\psi_a^2 + \tfrac{1}{2}(k_2 + k_3)\psi_b^2 - k_3\psi_a\psi_b$$

or $$E = \tfrac{1}{2}\, (\dot{x}^2 + \dot{y}^2 + a_{11}x^2 + a_{22}y^2 + 2a_{12}xy) \tag{5.62}$$

This expression is complicated by the presence of the product term $x\,y$.

We shall now write down the expression (5.62) in terms of the normal coordinates X and Y. Using Eqs (5.55) and (5.56) the first two terms in the brackets of Eq. (5.62) can be written as

$$\dot{x}^2 + \dot{y}^2 = (\dot{X} \sin \alpha + \dot{Y} \cos \alpha)^2 + (\dot{X} \cos \alpha - \dot{Y} \sin \alpha)^2$$

$$= \dot{X}^2 + \dot{Y}^2 \tag{5.63}$$

Similarly the last three terms in the brackets of Eq. (5.62) can be written as

$$a_{11}x^2 + a_{22}y^2 + 2a_{12}xy = a_{11}\,(X \sin \alpha + Y \cos \alpha)^2 + a_{22}(X \cos \alpha - Y \sin \alpha)^2$$
$$+ 2a_{12}\,(X \sin \alpha + Y \cos \alpha)(X \cos \alpha - Y \sin \alpha)$$

$$= (a_{11} \sin^2 \alpha + a_{22} \cos^2 \alpha + 2a_{12} \sin \alpha \cos \alpha)\, X^2$$
$$+ (a_{11} \cos^2 \alpha + a_{22} \sin^2 \alpha - 2a_{12} \sin \alpha \cos \alpha)\, Y^2$$
$$+ 2(a_{11} \sin \alpha \cos \alpha - a_{22} \sin \alpha \cos \alpha - a_{12} \sin^2 \alpha$$
$$+ a_{12} \cos^2 \alpha)\, XY$$
$$= \{\tfrac{1}{2} a_{11}(1-\cos 2\alpha) + \tfrac{1}{2} a_{22}(1+\cos 2\alpha) + a_{12} \sin 2\alpha\}\, X^2$$
$$+ \{\tfrac{1}{2} a_{11}(1+\cos 2\alpha) + \tfrac{1}{2} a_{22}(1-\cos 2\alpha) \quad a_{12} \sin 2\alpha\} Y^2$$
$$+ \{(a_{11}-a_{22}) \sin 2\alpha + 2a_{12} \cos 2\alpha\}\, XY \qquad (5.64)$$

Now we shall simplify this expression by using Eq. (5.52) which reads

$$\tan \alpha = \frac{a_{12}}{\omega_1^2 - a_{11}} = -\frac{\omega_2^2 - a_{11}}{a_{12}}$$

[See Eqs (5.48) and (5.49)]

giving $$\tan^2 \alpha = -\frac{\omega_2^2 - a_{11}}{\omega_1^2 - a_{11}}$$

Expressing the trignometric function tan α in terms of tan 2α, it is easy to see that

$$\tan 2\alpha = \frac{2\tan \alpha}{1-\tan^2 \alpha} = \frac{2a_{12}}{a_{22}-a_{11}}$$

where we have used the fact that [see Eqs (5.46) and (5.47)]

$$\omega_1^2 + \omega_2^2 = a_{11} + a_{22}$$

Using the trignometric relations,

$$\sin^2 2\alpha = \frac{\tan^2 2\alpha}{1+\tan^2 2\alpha}$$

and $$\cos^2 2\alpha = \frac{1}{1+\tan^2 2\alpha}$$

and the above value of tan 2α we can show that

$$\sin 2\alpha = -\frac{2a_{12}}{\{(a_{11}-a_{22})^2 + 4a_{12}^2\}^{1/2}}$$

and $$\cos 2\alpha = \frac{a_{11}-a_{22}}{\{(a_{11}-a_{22})^2 + 4a_{12}^2\}^{1/2}}$$

Now remembering that $a_{12} = a_{21}$ and subtracting Eqs. (5.46) and (5.47) we have

$$\{(a_{11}-a_{22})^2 + 4a_{12}^2\}^{1/2} = \omega_2^2 - \omega^2$$

Therefore

$$\sin 2\alpha = -\frac{2a_{12}}{\omega_2^2 - \omega_1^2}$$

$$\cos 2\alpha = \frac{a_{11}-a_{22}}{\omega_2^2-\omega_1^2}$$

Using these values of $\sin 2\alpha$ and $\cos 2\alpha$ and Eqs (5.46) and (5.47) in Eq. (5.64) we find (after some laborious algebra) that the coefficients of X, Y and $X\,Y$ respectively simplify as follows:

$$\tfrac{1}{2}\, a_{11}(1-\cos 2\alpha)+\tfrac{1}{2}\, a_{22}(1+\cos 2\alpha)+a_{12} \sin 2\alpha = \omega_1^2$$

$$\tfrac{1}{2}\, a_{11}(1+\cos 2\alpha)+\tfrac{1}{2}\, a_{22}(1-\cos 2\alpha)-a_{12} \sin 2\alpha = \omega_1^2$$

and $\quad (a_{11}-a_{22}) \sin 2\alpha+2a_{12} \cos 2\alpha = 0$

Thus Eq. (5.64) simplifies to read

$$a_{11}x^2+a_{22}y^2+2a_{12}xy = \omega_1^2X^2+\omega_2^2Y^2 \tag{5.65}$$

Using Eqs. (5.63) and (5.65) in Eq. (5.62) we get

$$E = \tfrac{1}{2}\,(\dot{X}^2+\dot{Y}^2+\omega_1^2X^2+\omega_2^2Y^2) \tag{5.66}$$

Notice that there is no $X\,Y$ term in this expression for energy.

Now since

$$X = C_1 \cos(\omega_1 t+\phi_1)$$

we have

$$\dot{X}^2+\omega_1^2X^2 = C_1^2\omega_1^2$$

Similarly $\quad \dot{Y}^2+\omega_2^2Y^2 = C_2^2\omega_2^2$

Substituting them in Eq. (5.66) we get

$$E = \tfrac{1}{2}\, C_1\omega_1^2+\tfrac{1}{2}\, C_2\omega_2^2$$

Remember that dimensionally $C_1 = \sqrt{m_a}\times$ amplitude of the displacement of mass m_a in mode 1 and $C_2 = \sqrt{m_b}\times$ amplitude of the displacement of mass m_b in mode 2. The above equation, therefore, states that the total energy of the coupled system is simply equal to the sum of the energies of harmonic vibration along the normal coordinates. Thus the problem of finding the energy of the system simply reduces to that of finding the energy associated with each normal mode. Since the oscillation of the system in a normal mode is harmonic, this indeed is a significant simplification.

Difference Between forced and Coupled Oscillations

We have obtained a general solution for the motion of two coupled oscillators. It would be interesting to see how the results in coupled oscillation correspond to those of forced oscillations discussed in Chap. 4. We have mentioned in the beginning of this chapter that if the coupling is strong, the feedback of energy cannot be neglected and one cannot label one oscillator as the 'driver' and the other as the 'driven' oscillator. If the energy feedback is negligible then one could call the oscillator with a larger amplitude as the driver and the other whose amplitude is small may be called the driven oscillator. Experiment tells us that two conditions must be satisfied for the feedback of energy to be negligible. Firstly, the coupling must be weak. Secondly, the natural frequencies of the two oscillators must not be equal because then resonance will occur and energy transfer will be large. In Sec. 5.4 we have discussed the problem of energy exchange between two pendulums having equal natural angular frequency $\omega_0 = \sqrt{g/l}$. Only their masses were taken to be different. We found that there is considerable feedback of energy. In fact if $m_a = m_b$, all the energy of one oscillator transfers to the other and this process is repeated at a frequency called the beat frequency.

Referring to the longitudinal oscillations of two coupled masses one finds that if the coupling is small and the angular frequency ω_a $(= \sqrt{a_{11}})$ of mass m_a, is different from the angular frequency ω_b $(= \sqrt{a_{22}})$ of mass m_b, then the amplitude of the motion of one will be much larger than that of the other and the feedback of energy will be negligible. The oscillator whose amplitude is smaller can be considered as the 'driven' oscillator and its amplitude of oscillation should be given by Eq. (4.7) of Chap. 4. If the oscillator A is to be called the 'driver' its amplitude of motion should be much larger than that of B.

If the coupling is weak α will be very small, i.e. $\sin\alpha$ is very small and $\cos\alpha \to 1$. In this case the first term on the right hand of Eq. (5.59) can be neglected and we have

$$\psi_a = \frac{C_2}{\sqrt{m_a}} \cos\alpha \cos(\omega_2 t + \phi_2) \tag{5.67}$$

The angular frequency of the driver is ω_2. We have seen earlier that when mass A moves through ψ_a, it exerts a force $F(t)$ on the mass B (i.e. the driven oscillator) which is equal to $k_3\psi_a$. Thus

$$F(t) = k_3\psi_a = \frac{k_3 C_2}{\sqrt{m_a}} \cos\alpha \cos(\omega_2 t + \phi_2)$$

or

$$F(t) = F_0 \cos(\omega_2 t + \phi_2) \tag{5.68}$$

where $F_0 = \dfrac{k_3 C_2 \cos\alpha}{\sqrt{m_a}}$ is the amplitude of force $F(t)$

Now, *in the steady state*, the driven oscillator must oscillate at the frequency (ω_2) of the applied force. Hence the first term (that oscillates at frequency ω_1) on the right-hand side of Eq. (5.60) is not relevant as far as the steady state behaviour of ψ_b is concerned. Thus in the steady state ψ_b must be given by [see Eq. (5.60)]

$$\psi_b = -\frac{C_2}{\sqrt{m_b}} \sin\alpha \cos(\omega_2 t + \phi_2) \tag{5.68a}$$

Eliminating $\cos(\omega_2 t + \phi_2)$ from Eqs (5.67) and (5.68*a*) we get

$$\psi_b = -\sqrt{\frac{m_a}{m_b}} \tan\alpha\, \psi_a \tag{5.68b}$$

Since α is small $\psi_b \ll \psi_a$ unless $m_a \gg m_b$. Now substituting for $\tan\alpha$ from Eq. (5.52) in Eq. (5.68*b*) we have

$$\psi_b = \frac{k_3\psi_a}{(k_2+k_3) - m_b\omega_2^2} = \frac{\frac{k_3}{m_b}\psi_a}{\omega_b^2 - \omega_2^2}$$

where $\omega_b = \sqrt{\dfrac{k_2+k_3}{m_b}}$ is the angular frequency of free oscillations of mass *B*.

Now using Eqs (5.67) and (5.68) in the above expression, we get

$$\psi_b = \frac{F_0/m_b}{\omega_b^2 - \omega_2^2} \cos(\omega_2 t + \phi_2)$$

$$= B\cos(\omega_2 t + \phi_2)$$

where the amplitude *B* of the forced oscillator (i.e. mass *B*) is

$$B = \frac{F_0/m_b}{\omega_b^2 - \omega_2^2}$$

This is the same as Eq. (4.7) of the forced oscillator if the friction is neglected ($\gamma = 0$). Thus forced oscillations are a special case of coupled oscillations; the case when feedback of energy can be neglected.

Energy Transfer in the Case of Resonance

While discussing the problem of exchange of energy between two coupled pendulums we found that, even when the coupling between them is weak, there is a large transfer of energy, if the natural frequencies of the two pendulums are equal. This is the case of resonance In such cases, where the feedback of energy from one oscillator to another is appreciable, the formulae of Chap. 4 do not hold. For instance, the Eq (4.7) in this case predicts that amplitude of the driven oscillator becomes infinite in the absence of friction. Since the driven oscillator cannot extract more energy than the driver has, it is clear that Eq. (4.7) cannot be used in this case. To solve this problem, we must use the formulae of this chapter. We have already seen how two coupled pendulums behave if their natural frequencies are equal and coupling is weak. There are beats in the system.

Let us consider the case when $a_{11} = a_{22}$, i.e. $\omega_a^2 = \omega_b^2 = \omega_0^2$. Equations (5.46) and (5.47) reduce to

$$\omega_1^2 = a_{11} - a_{12} = \omega_0^2 - a_{12}$$
$$\omega_2^2 = a_{11} + a_{12} = \omega_0^2 + a_{12}$$

Equation (5.52) then gives

$$\tan\alpha = \frac{a_{12}}{\omega_1^2 - a_{11}} = \frac{a_{12}}{-a_{12}} = -1$$

i.e. $$\alpha = -\frac{\pi}{4}$$

If in addition, the coupling is weak, i.e. $a_{12} \ll a_{11}$, the above expressions for ω_1 and ω_2 give

$$\omega_1 = \omega_0\left(1 - \frac{a_{21}}{\omega_0^2}\right)^{1/2} \simeq \omega_0 - \frac{a_{12}}{2\omega_0}$$

$$\omega_2 \simeq \omega_0 + \frac{a_{12}}{2\omega_0}$$

where only the first two terms in the Binomial expansion are retained. With these two approximations the displacements and energies of the two masses can be obtained (using the procedure followed in the problem of two pendulums) under any initial conditions. If the initial conditions are

$$\psi_a(t=0) = a$$
$$\psi_b(t=0) = 0$$
$$\dot{\psi}_a(t=0) = \dot{\psi}_b(t=0) = 0$$

then the constants C_1, C_2, ϕ_1, and ϕ_2 in Eqs. (5.59) and (5.60) can be evaluated and these equations reduce to[1]

$$\psi_a = \frac{a}{2}(\cos\omega_2 t + \cos\omega_1 t) \tag{5.69}$$

$$\psi_b = \frac{a}{2}\sqrt{\frac{m_a}{m_b}}(\cos\omega_2 t - \cos\omega_1 t) \tag{5.70}$$

Define, as before,

$$\omega_m = \tfrac{1}{2}(\omega_2 - \omega_1)$$
$$\omega_a = \tfrac{1}{2}(\omega_2 + \omega_1)$$

In terms of ω_m and ω_a, Eqs (5.69) and (5.70) become

$$\psi_a = a\cos\omega_m t\cos\omega_a t = A_m\cos\omega_a t \tag{5.71}$$

1. The student is advised to deduce Eqs (5.69) and (5.70) following the procedure outlined in Sec. 5.4.

$$\psi_b = a\sqrt{\frac{m_a}{m_b}}\sin\omega_m t \sin\omega_a t = B_m \sin\omega_a t \quad (5.72)$$

Here $$\omega_m = \tfrac{1}{2}(\omega_2 - \omega_1) = \frac{a_{12}}{2\omega_0}$$

If the coupling is weak A_m and B_m vary slowly with time and we can obtain the expression for energy by using the relations

$$E_a = \tfrac{1}{2}\, m_a \omega_a^2 A_m^2 = \tfrac{1}{2} m_a a^2 \omega_a^2 \cos^2 \omega_m t \quad (5.73)$$

$$E_b = \tfrac{1}{2}\, m_b \omega_a^2 B_m^2 = \tfrac{1}{2} m_a a^2 \omega_a^2 \sin^2 \omega_m t \quad (5.74)$$

Notice that the total energy $E = E_a + E_b = \frac{1}{2} m_a a^2 \omega_a^2 =$ constant as expected.

Also notice that at $t = 0$, $E_a = \frac{1}{2} m_a a^2 \omega_a^2$ and $E_b = 0$, as it should be, since at $t = 0$, the mass A only has the energy. It is clear from Eqs (5.73) and (5.74) that the energy bounces back and forth from one oscillator to the other at a beat frequency $\nu_b = \nu_2 - \nu_1$, the frequency difference.

Thus we find that, when there is an appreciable transfer of energy from one oscillator to the other, the motion of the two coupled oscillators can be very different from the usual forced oscillations. This happens in either of the following two situations: (*i*) when the coupling is strong (*ii*) when the coupling is weak but the natural frequencies of the two oscillators are equal.

5.8 FORCED OSCILLATIONS OF TWO COUPLED OSCILLATORS

It will be interesting to study the response of a system of two coupled oscillators when a periodic force is applied to one of them. For simplicity we shall consider the simplest system of two identical coupled pendulums with a force $F(t) = F_0 \cos \omega t$, applied to either one of the pendulums (see Fig. 5.13). This simple example reveals the important features of the response. If the force is applied to bob A, the equations of motion of the two pendulums are given by

$$m\ddot{\psi}_a = -\frac{mg}{l}\psi_a + k(\psi_b - \psi_a) + F_0 \cos \omega t \quad (5.75)$$

$$m\ddot{\psi}_b = -\frac{mg}{l}\psi_b - k(\psi_b - \psi_a) \quad (5.76)$$

Let us transform to the normal coordinates X and Y, where

$$X = \psi_a + \psi_b$$

$$Y = \psi_a - \psi_b$$

From Eqs (5.75) and (5.76) we get

$$\ddot{X} = -\omega_1^2 X + f_0 \cos \omega t \tag{5.77}$$

$$\ddot{Y} = -\omega_2^2 Y + f_0 \cos \omega t \tag{5.78}$$

where $\omega_1^2 = g/l,\ \omega_2^2 = \dfrac{g}{l} + \dfrac{2k}{m}$ and $f_0 = F_0/m$.

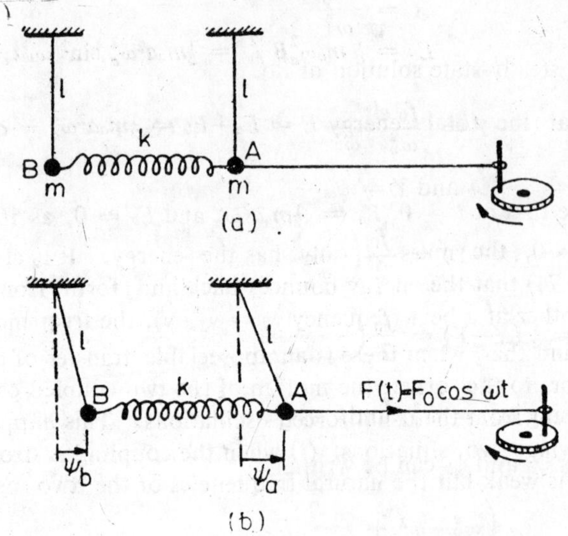

Fig. 5.13 Forced oscillations of two coupled pendulums: (a) equilibrium state (b) general configuration

We recall that ω_1 and ω_2 are angular frequencies of the two normal modes of the system. Notice that Eqs (5.77) and (5.78) are uncoupled, Comparing them with Eq. (4.1) of Chap. 4 we see that Eqs (5.77) and (5.78) are each of the form appropriate to a driven harmonic oscillator if damping is neglected ($\gamma = 0$). For simplicity, we have neglected damping although some amount of friction is always present in any system.

Thus the normal coordinate X behaves like a harmonic oscillator of mass m angular frequency ω_1 driven by a force $F_0 \cos \omega t$. Similarly the normal coordinate Y behaves like a harmonic oscillator of mass m, angular frequency ω_2 driven by a force $F_0 \cos \omega t$. The oscillations associated with X and Y are independent and we can write down the steady-state solutions for X and Y as follows:

As in Chap. 4, we rewrite Eq. (5.77) in the form

$$\ddot{X} + \omega_1^2 X = f_0\, e^{i\omega t} \tag{5.79}$$

and its steady-state solution is,

$$X = X_0 \, e^{i(\omega t + \phi)} \tag{5.80}$$

where ϕ is the phase of X relative to that of $F(t)$. Substituting Eq. (5.80) in Eq. (5.79) we get

$$X_0 = \frac{f_0 e^{-i\phi}}{\omega_1^2 - \omega^2}$$

so that

$$X = \frac{f_0 e^{i\omega t}}{\omega_1^2 - \omega^2}$$

Similarly the steady-state solution of Eq. (5.78) is

$$Y = \frac{f_0 \, e^{i\omega t}}{\omega_2^2 - \omega^2}$$

Since $X = \psi_a + \psi_b$ and $Y = \psi_a - \psi_b$, we have

$$\psi_a = \tfrac{1}{2}(X+Y) = \frac{f_0}{2}\left(\frac{1}{\omega_1^2 - \omega^2} + \frac{1}{\omega_2^2 - \omega^2}\right) e^{i\omega t} \tag{5.81}$$

$$\psi_b = \tfrac{1}{2}(X-Y) = \frac{f_0}{2}\left(\frac{1}{\omega_1^2 - \omega^2} - \frac{1}{\omega_2^2 - \omega^2}\right) e^{i\omega t} \tag{5.82}$$

Now using $\omega_1^2 = \omega_0^2$ and $\omega_2^2 = \omega_0^2 + \dfrac{2k}{m}$, the real parts of the displacements ψ_a and ψ_b can be written as

$$\psi_a = \frac{\left(\omega_0^2 - \omega^2 + \dfrac{k}{m}\right) f_0 \cos \omega t}{(\omega_1^2 - \omega^2)(\omega_2^2 - \omega^2)} = A \cos \omega t \tag{5.83}$$

$$\psi_b = \frac{\dfrac{k}{m} f_0 \cos \omega t}{(\omega_1^2 - \omega^2)(\omega_2^2 - \omega^2)} = B \cos \omega t \tag{5.84}$$

where

$$A = \frac{f_0 \left(\omega_0^2 - \omega^2 + \dfrac{k}{m}\right)}{(\omega_1^2 - \omega^2)(\omega_2^2 - \omega^2)}$$

and

$$B = \frac{f_0 \dfrac{k}{m}}{(\omega_1^2 - \omega^2)(\omega_2^2 - \omega^2)}$$

are the amplitudes of ψ_a and ψ_b respectively.

Resonance and Normal Modes

It is evident from Eqs (5.83) and (5.84) that each of the amplitudes A and B become infinite when the driving frequency ω equals the normal mode frequency ω_1 or ω_2. Thus there are two resonance frequencies ω_1 and ω_2. The infinity is due to our neglect of frictional forces. If we had included a small amount of friction, the amplitudes would have been large but not infinite. We shall now make a very important observation. Equations (5.83) and (5.84) give the ratio

$$\frac{\psi_b}{\psi_a} = \frac{B}{A} = \frac{k/m}{\omega_0^2 - \omega^2 + k/m} \tag{5.85}$$

If $\quad \omega = \omega_1 = \omega_0 \, ; \quad \dfrac{\psi_b}{\psi_a} = +1$

and if $\quad \omega = \omega_2 = \left(\omega_0^2 + \dfrac{2k}{m}\right)^{1/2} ; \quad \dfrac{\psi_b}{\psi_a} = -1.$

These are just the configurations of the two normal modes of the system. Thus we conclude that when the system is driven by a harmonic force of frequency equal to one of the normal mode frequencies of the system, its response is large and the relationship between the displacements of the moving parts (i.e. the configuration of the system) is the same as if the system were in free oscillation at that frequency. This important observation can be made use of in finding the possible frequencies of free oscillations of a coupled system and the configurations of the corresponding normal modes. We simply drive the system by a harmonic driving force and vary its frequency until the amplitudes becomes very large i.e. until the resonance takes place. The resonance frequencies are just the frequencies of the normal modes and the configurations at resonance are those of the normal modes. This is a general result and is applicable to any system with any number of degrees of freedom.

Filters

If we drive the pendulum A with a force whose frequency is considerably less than the frequency of the lower mode (i.e. if $\omega \ll \omega_1$ where $\omega_1 = \omega_0 = \sqrt{g/l}$) then Eq. (5.85) tells us that the ratio ψ_b/ψ_a is positive indicating that the pendulums oscillate in phase with the force but $\psi_b < \psi_a$ since in this limit

$$\frac{\psi_b}{\psi_a} = \frac{B}{A} \propto \frac{k/m}{\omega_0^2 + k/m}$$

This we find that the amplitude A of the pendulum to which the force of a low frequency is applied, is larger than that of the other pendulum. The pendulum to which the force is directly applied may be called the *input-pendulum*. The other pendulum is then the *output-pendulum*. Thus when $\omega \ll \omega_1$, the amplitude of the output pendulum is always less than that of the input pendulum. If, instead of two, we have a linear arrangement of a large number of pendulums, an analogous situation is obtained. If we drive this system at a frequency less than that of the lowest mode, each bob moves in the same phase as its neighbour but the amplitude of successive bobs decreases as one progresses farther from the driven end. The amplitude of the last pendulum (the output pendulum) will be negligibly small. The amplitude is *attenuated* with increasing distance from the input end of the system. The system is then called a *filter*. The frequency of the lowest mode is the *cut-off* frequency. This arrangement is a low frequency mechanical filter and it filters out the undesirable frequencies.

On the other hand, if we drive the pendulum A with a force whose frequency is considerably higher than the higher mode frequency (i.e. if $\omega \gg \omega_2$ where $\omega_2 = \left(\omega_0^2 + \frac{2k}{m}\right)^{1/2}$, the Eq. (5.85) tells us that the ratio ψ_b/ψ_a becomes negative indicating that the two pendulums are out of phase by 180°. It is clear that in this high frequency limit, the amplitude of ψ_b is again much less than that of ψ_a. Thus when $\omega \gg \omega_2$, the amplitude of the output pendulum is much less than that of the input pendulum. If we have a linear array of pendulums and if we drive this system at a frequency higher than that of its highest mode, each pendulum will be in opposite phase as its neighbour but the amplitude of successive pendulums decreases as one progresses farther from the driven end. The amplitude of the last (output) pendulum will be negligibly small. The amplitude is attenuated with increasing distance from the input end of the system. The frequency of the highest mode is the cut-off frequency. This arrangement thus also behaves as a high frequency mechanical filter.

The band of frequencies between the low and the high cut-off frequencies is called the *pass band* of the filter. For driving frequencies lying in the pass band, the amplitude at the output end is comparable with that at the input end. For driving frequencies outside the pass band, the amplitude at the output end is much smaller than that at the input end. The system is, therefore, called a ***bandpass filter***.

Example of an electrical filter. We can find an electrical analogue for the mechanical example of two coupled pendulums of Fig 5.13. For each mass m we substitute an inductance L and for the spring we substitute a capacitance C where spring constant $k = 1/C$. Figure 5.14 shows two LC circuits coupled through a capacitor C. The circuit is driven by voltage $V(t) = V_0 \cos \omega t$. We neglect the resistance of inductances. The equations for the currents I_a and I_b in the two circuits can be written. The normal modes are then found. This is left as an exercise for the student. Here we simply guess the results by analogy with coupled pendulums.

$$\text{Mode 1} \quad I_a = I_b; \qquad \omega_1^2 = \frac{1}{C_0 L}$$

$$\text{Mode 2} \quad I_b = -I_a; \qquad \omega_2^2 = \frac{1}{C_0 L} + \frac{2}{CL}$$

The capacitor C provides the coupling. The analogue of Eq. (5.85) is

$$\frac{I_b}{I_a} = \frac{\frac{1}{LC}}{\frac{1}{LC_0} + \frac{1}{LC} - \omega^2}$$

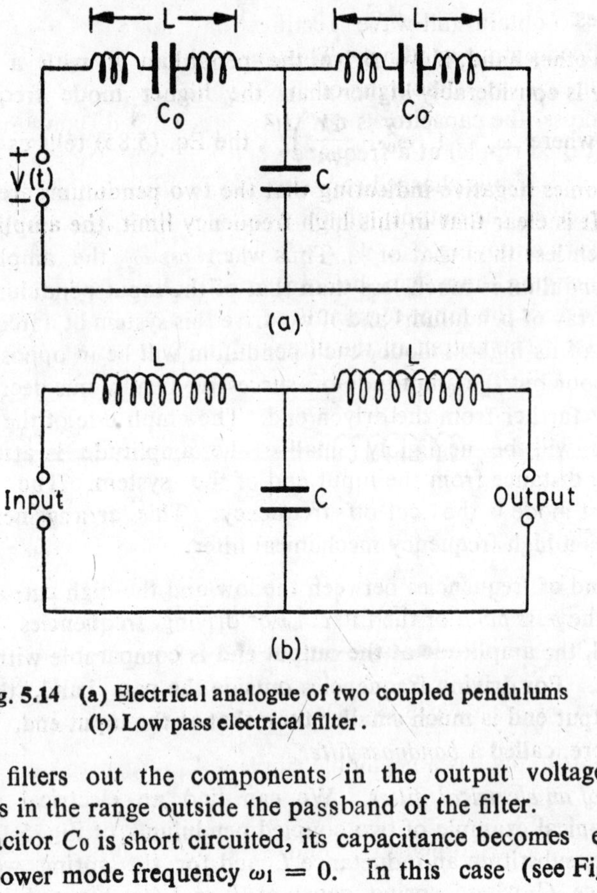

Fig. 5.14 (a) Electrical analogue of two coupled pendulums
(b) Low pass electrical filter.

This circuit filters out the components in the output voltage whose frequency lies in the range outside the passband of the filter.

If the capacitor C_0 is short circuited, its capacitance becomes effectively infinity and lower mode frequency $\omega_1 = 0$. In this case (see Fig. 5.14 *b*)

$$\frac{I_b}{I_a} = \frac{\frac{1}{LC}}{\frac{1}{LC} - \omega^2} \tag{5.86}$$

The low-cut-off frequency of this circuit it zero which corresponds to *DC* (direct current). The electrical circuit whose low frequency cut-off is zero is called a *low-pass* filter.

A very important practical application of Eq. (5.86) is a low-pass filter for a DC power supply. Electronic devices such as transistor radios and amplifiers require a direct-current (DC) source for their operation. In a typical DC power supply, one starts with alternating current (AC) from the mains which supplies power at a frequency of 50 Hz and at an rms (root-mean square) voltage of about 220 volts. This voltage is applied to the input winding (or primary) of a transformer which may be a step-up transformer or a step-down transformer depending on what DC voltage we require for the given device. The output winding (or secondary) of the transformer is connected across a diode which would pass current in only one direction. Thus we have what is called 'half-wave rectified' current. To obtain 'full-wave rectification' we use two diodes with a centre-tapped output winding. This current is used to charge a capacitor which then acts as a source of DC voltage. In actual practice, the voltage across the capacitor is not truly constant; it contains an oscillating part (called ripple) at a frequency of 100 Hz in the case of full-wave rectification. If this charged capacitor is used to provide power to operate a radio receiver, the output of the radio will include a 100 Hz 'hum' which is very annoying.

This annoying hum can be substantially reduced by connecting the capacitor across the input terminals of a low-pass filter circuit as shown in Fig. 5.15. The output of the filter is then used as a source of DC

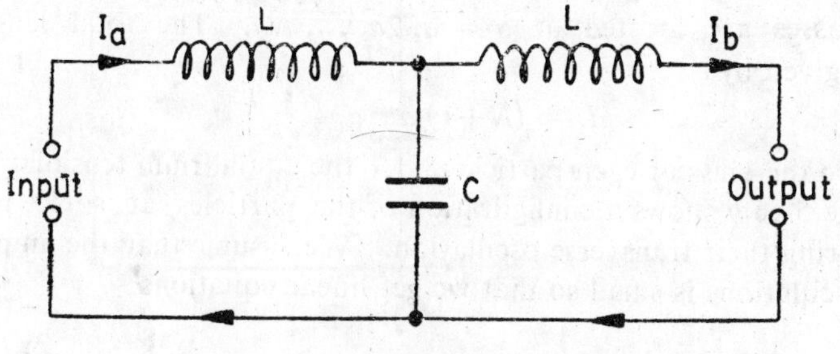

Fig. 5.15 A low-pass electrical filter

voltage which now has a negligible ripple. In a typical filter circuit, the values of L and C are

$$L \simeq 10 \text{ henry} \quad \text{and} \quad C \simeq 5 \text{ microfarad.}$$

The high-frequency cut-off is given by

$$\nu_2 = \frac{\omega_2}{2\pi} = \frac{1}{2\pi}\sqrt{\frac{2}{LC}} = \frac{1}{2\times 3.14}\left(\frac{2}{10\times 5\times 1^{-6}}\right)^{1/2}$$

$$\simeq 32 \text{ Hz}$$

The attenuation factor (I_b/I_a) for the 100 Hz component of the current is given by [see Eq. (5.86)]

$$\frac{I_b}{I_a} = \frac{1/LC}{1/LC - \omega^2} = \frac{\nu_2^2}{\nu_2^2 - 2\nu^2}$$

$$= \frac{(32)^2}{(32)^2 - 2(100)^2} = -0.054$$

Thus the ripple component is reduced by a factor of about 5.4%. The DC component is not affected by the filter.

5.9 MANY COUPLED OSCILLATORS

Any real physical system, such as a piece of string or a volume of a fluid, contains many particles bound (or coupled) to each other by forces of cohesion. We are therefore naturally motivated to tackle the problem of an arbitrary number (not just two) of similar oscillators coupled together. We shall investigate this problem by using the ideas developed in the preceding sections relating to two coupled oscillators.

Normal Modes of Transverse Oscillations of *N* Coupled Oscillators

Consider a flexible elastic string of negligible mass to which are attached N identical particles spaced at equal distances a along its length. The string is fixed at two points at $x = 0$ and $x = L$ as shown in Fig. 5.16 *a*. The masses are located at $x = a, 2a, \ldots, Na$. The total length L is clearly given by

$$L = (N+1)\,a$$

Let m be the mass of each particle and T the equilibrium tension.

Figure 5.16 *b* shows a configuration of the particles at some instant of time during their transverse oscillation. We assume that the amplitude of these oscillations is small so that we get linear equations.

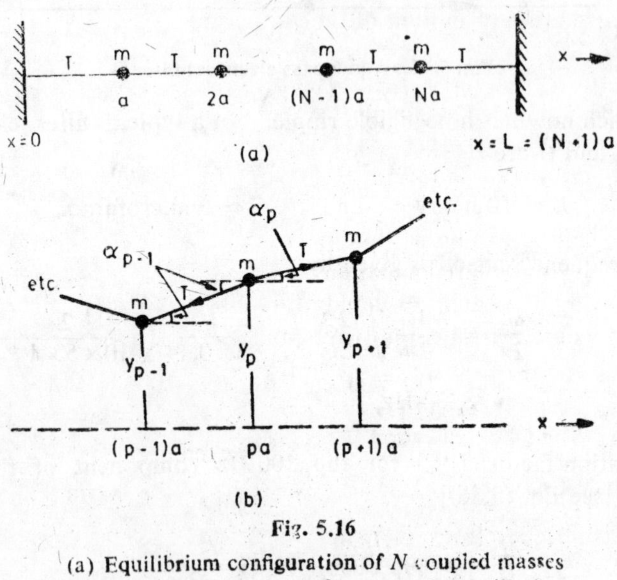

Fig. 5.16

(a) Equilibrium configuration of N coupled masses

(b) General configuration for transverse oscillations

Equation of Motion. Let us focus our attention on the pth particle together with two of its immediate neighbours $(p-1)$th and $(p+1)$th particles. Let their respective displacements from the equilibrium state be y_p, y_{p-1} and y_{p+1}. The possible vaules p can take are

$$p = 1, 2, 3, \ldots, (N-1), N$$

Referring to Fig. 5.16*b*, we have

$$\tan \alpha_p = \frac{y_{p+1} - y_p}{a}$$

and $$\tan \alpha_{p-1} = \frac{y_p - y_{p-1}}{a}$$

The resultant y component of force on the pth particle is

$$F_p = -T \sin \alpha_{p-1} + T \sin \alpha_p$$

Restricting ourselves to small oscillation approximation so that y displacements are small compared to a, we have

$$\tan \alpha_p \simeq \sin \alpha_p \text{ and } \tan \alpha_{p-1} \simeq \sin \alpha_{p-1} \text{ ; giving}$$

$$F_p = -T \tan \alpha_{p-1} + T \tan \alpha_p$$

$$= -\frac{T}{a} (y_p - y_{p-1}) + \frac{T}{a} (y_{p+1} - y_p)$$

or $$F_p = \frac{T}{a} (y_{p+1} + y_{p-1} - 2y_p)$$

Hence the equation of motion of the pth particle is

$$m \ddot{y}_p = \frac{T}{a} (y_{p+1} + y_{p-1} - 2y_p)$$

or $$\ddot{y}_p = \frac{T}{ma} (y_{p+1} + y_{p-1} - 2y_p) \tag{5.87}$$

A similar equation can be written for each of the N particles. Thus we have a set of N differential equations, one for each value of p from 1 to N. For a string fixed at $x = 0$ and $x = (N+1)\,a$, the corresponding y displacements are zero, i.e. the two boundary conditions are:

$$y_0 = 0$$

$$y_{N+1} = 0$$

Normal Modes

To find the normal modes we assume, as before, that there exists a mode with angular frequency ω and phase constant ϕ. In a normal mode all particles execute harmonic oscillations with the same frequency and phase constant. Thus for the pth particle, we have

$$y_p = A_p \cos(\omega t + \phi) \tag{5.88}$$

where A_p is the amplitude of the harmonic oscillations of the pth particle. Similarly

$$y_{p-1} = A_{p-1} \cos(\omega t + \phi)$$

$$y_{p+1} = A_{p+1} \cos(\omega t + \phi)$$

Using these values in Eq. (5.87) we have

$$-\omega^2 A_p \cos(\omega t + \phi) = \frac{T}{ma}(A_{p+1} + A_{p-1} - 2A_p) \cos(\omega t + \phi)$$

For this equation to hold for all values of t, we must have

$$-\omega^2 A_p = \frac{T}{ma}(A_{p+1} + A_{p-1} - 2A_p)$$

or
$$A_{p+1} + A_{p-1} = \left(2 - \frac{ma\omega^2}{T}\right) A_p \tag{5.89}$$

where $p = 1, 2, \ldots, N$ and our boundary conditions that the ends of the string are fixed become

$$A_0 = 0$$

$$A_{N+1} = 0$$

Equation (5.89) is the fundamental equation. It represents a set of N equations which have to be simultaneously solved to give the possible mode frequencies. We have seen above that with two coupled oscillators there are just two normal modes. With N coupled oscillators there will be N normal modes. The set of N equations, when solved, will give N different values of ω^2; each value of ω being the frequency of a normal mode.

The solution of N simultaneous equations (so that they all can be satisfied with the same value of ω^2) is clearly a formidable task. We shall, therefore, seek a solution for N coupled oscillators by making a bold guess and see if it works. In the next chapter we shall justify the correctness of the guess that we are going to make.

Before we venture to guess the solution of Eq. (5.89) let us check it for one oscillator ($N = 1, p = 1$) and two coupled oscillators ($N = 2$, $p = 1$ and 2). When $p = 1$ we have one mass m in the middle of a string of length $L = 2a$. Setting $p = 1$ in Eq. (5.89) we have

$$A_2 + A_0 = \left(2 - \frac{ma\omega^2}{T}\right) A_1$$

Now for a string fixed at both ends, $A_0 = A_2 = 0$. Hence we have

$$\left(2 - \frac{ma\omega^2}{T}\right) A_1 = 0$$

giving $$\omega^2 = \frac{2T}{ma}$$

Thus a single oscillator has only one allowed frequency of vibration (Fig. 5.17*a*).

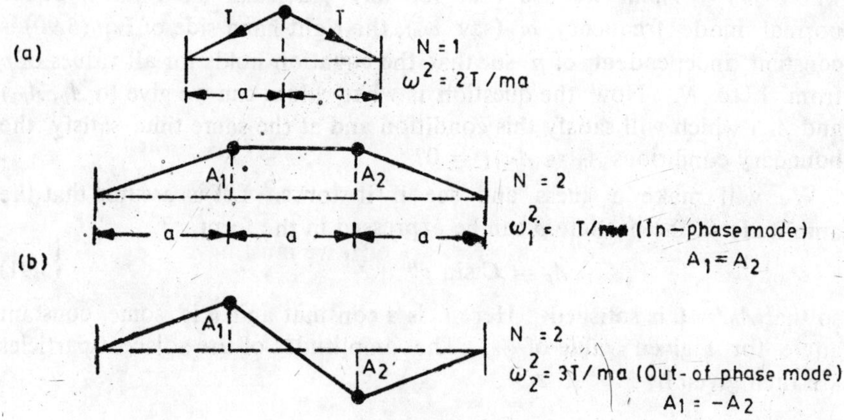

Fig. 5.17

(a) Transverse oscillations of a single mass ($N = 1$)
(b) Normal modes of transverse oscillations of two coupled masses ($N = 2$)

When $N =$,2 the string length $L = 3a$ (Fig. 5.17*b*) we need the equations for both $p = 1$ and $p = 2$. Since the string is fixed at its ends ($A_0 = A_3 = 0$), these two equations are

$$\left(2 - \frac{ma\omega^2}{T}\right) A_1 - A_2 = 0$$

and $$-A_1 + \left(2 - \frac{ma\omega^2}{T}\right) A_2 = 0$$

giving $$\frac{A_2}{A_1} = \left(2 - \frac{ma\omega^2}{T}\right) = \frac{1}{\left(2 - \frac{ma\omega^2}{T}\right)}$$

Thus we have

$$\left(2 - \frac{ma\omega^2}{T}\right)^2 = 1$$

$$\therefore \left(2 - \frac{ma\omega^2}{T} - 1\right)\left(2 - \frac{ma\omega^2}{T} + 1\right) = 0$$

Thus there are two normal modes of frequencies given by

$$\omega_1^2 = \frac{T}{ma} \quad \text{and} \quad \omega_2^2 = \frac{3T}{ma}$$

For the first mode ($\omega = \omega_1$) we have $A_2/A_1 = 1$ (the in-phase mode) and for the second ($\omega = \omega_2$) we have $A_2/A_1 = -1$ (the out-of-phase mode) as shown in Fig. 5.17*b*.

We will now find a general solution of Eq. (5.89) for any value of N. If we rewrite Eq. (5.89) as

$$\frac{A_{p-1}+A_{p+1}}{A_p} = \frac{2\omega_0^2-\omega^2}{\omega_0^2} \tag{5.90}$$

where $\omega_0^2 = T/ma$, we see that for any particular fixed value of the normal mode frequency ω (say ω_n), the right-hand side of Eq. (5.90) is constant independent of p so that the equation holds for all values of p from 1 to N. Now the question is what values can we give to A_p, A_{p+1} and A_{p-1} which will satisfy this condition and at the same time satisfy the boundary conditions $A_0 = A_{N+1} = 0$?

We will make a guess and see if it works. Let us assume that the amplitude of the particle p can be expressed in the form

$$A_p = C \sin p\theta \tag{5.91}$$

so that $A_0 = 0$ is satisfied. Here C is a constant and θ is some constant angle for a given value of ω_n. The amplitudes of the adjacent particles are then given by

$$A_{p-1} = C \sin (p-1)\,\theta$$

and

$$A_{p+1} = C \sin (p+1)\,\theta$$

The left-hand side of Eq. (5.90) then becomes

$$\frac{A_{p-1}+A_{p+1}}{A_p} = \frac{C\{\sin (p-1)\,\theta + \sin (p+1)\,\theta\}}{C \sin p\theta}$$

$$= \frac{2C \sin p\theta \cos \theta}{C \sin p\theta} = 2 \cos \theta \tag{5.92}$$

which is constant and independent of p as demanded by Eq. (5.90). This means that our guess is successful. Thus our solution (5.91) will satisfy all the N equations (5.89).

All that remains now is to find the value of θ. This can be done by using the boundary conditions namely $A_p = 0$ for $p = 0$ and $p = N+1$. As stated above the first boundary condition is automatically satisfied; the second will be satisfied if $(N+1)\,\theta$ is an integral multiple of π, i.e.

$$(N+1)\,\theta = n\pi$$

or

$$\theta = \frac{n\pi}{N+1} \tag{5.93}$$

where $n = 1, 2, 3, \ldots$

Substituting for θ in Eq. (5.91) we get

$$A_p = C \sin\left(\frac{pn\pi}{N+1}\right) \tag{5.94}$$

The allowed frequencies of the normal modes can now be obtained from Eqs (5.90) to (5.93) which give

$$\frac{2\omega_0^2 - \omega^2}{\omega_0^2} = 2\cos\left(\frac{n\pi}{N+1}\right)$$

Therefore

$$\omega^2 = 2\omega_0^2\left\{1-\cos\left(\frac{n\pi}{N+1}\right)\right\}$$

$$= 4\omega_0^2 \sin^2\left\{\frac{n\pi}{2(N+1)}\right\}$$

or

$$\omega = 2\omega_0 \sin\left\{\frac{n\pi}{2(N+1)}\right\}$$

where $\omega_0 = \sqrt{T/ma}$ and n may take the values 1, 2, 3, ..., N.

There are N possible frequencies, each corresponding to a single mode n with $n = 1, 2, 3, \ldots, N$. In other words, different values of the integer n define different normal mode frequencies. We will, therefore, label a mode and its characteristic frequency by the value of n. Thus we will write the angular frequency of the nth mode as

$$\omega_n = 2\omega_0 \sin\left\{\frac{n\pi}{2(N+1)}\right\} \tag{5.95}$$

At each frequency ω_n, the pth particle has the amplitude

$$A_p = C \sin\left(\frac{pn\pi}{N+1}\right) \tag{5.96}$$

It is instructive to see how the results for $N = 1$ and $N = 2$ can be recovered from Eqs (9.95) and (5.96).

Single Oscillator ($N = 1, p = 1, n = 1$): In this case Eqs (5.95) and (5.96) reduce to

$$\omega_1 = 2\omega_0 \sin\frac{\pi}{4} = \sqrt{2}\,\omega_0 = \sqrt{\frac{2T}{ma}}$$

and

$$A_1 = C \sin\frac{\pi}{2} = C$$

There is a single frequency of transverse oscillations (see Fig. 5.17*a*).

Two Coupled Oscillators ($N = 2$. $p = 1$ and 2, $n = 1$ and 2): In this case Eqs (5.95) and (5.96) become

$$\omega_n = 2\omega_0 \sin\left(\frac{n\pi}{6}\right)$$

and

$$A_p = C \sin\left(\frac{pn\pi}{3}\right)$$

For the first mode ($n = 1$) the frequency is

$$\omega_1 = 2\omega_0 \sin\left(\frac{\pi}{6}\right) = \omega_0 = \sqrt{\frac{T}{ma}}$$

In this mode, the amplitudes of the two masses are

$$A_1 = C \sin\left(\frac{\pi}{3}\right)$$

$$A_2 = C \sin\left(\frac{2\pi}{3}\right) = C \sin\left(\frac{\pi}{3}\right)$$

i.e. $A_1 = A_2$. This is the in-phase mode (see Fig. 5.17*b*).

The frequency of the second mode ($n = 2$) is given by

$$\omega_2 = 2\omega_0 \sin\left(\frac{\pi}{3}\right) = \sqrt{3}\,\omega_0 = \sqrt{\frac{3T}{ma}}$$

In this mode, the amplitudes of the two masses are

$$A_1 = C \sin\left(\frac{2\pi}{3}\right) = C \sin\left(\frac{\pi}{3}\right)$$

$$A_2 = C \sin\left(\frac{4\pi}{3}\right) = -C \sin\left(\frac{\pi}{3}\right)$$

i.e. $A_1 = -A_2$. This is the out-of-phase mode (see Fig. 5.17*b*).

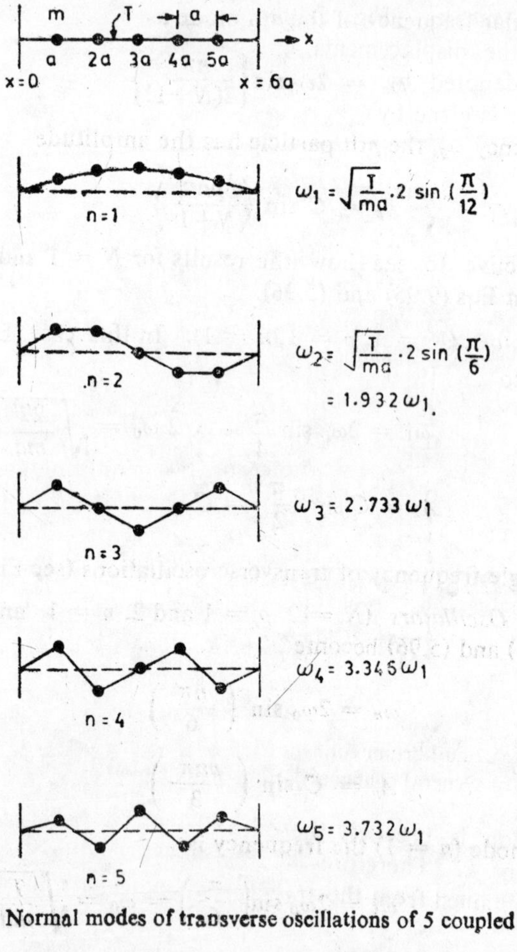

Fig. 5.18 Normal modes of transverse oscillations of 5 coupled masses

As an illustration, let us find the normal modes for transverse oscillations of $N = 5$ coupled masses. In this case the five modes are obtained by giving values to n equal to 1, 2, 3, 4 and 5. For each mode (i.e. each value of n) the amplitudes of the first through the fifth particle are obtained by setting $p = 1, 2, 3, 4$ and 5 respectively. Figure 5.18 shows the configurations for the five modes.

Normal Modes of Longitudinal Oscillations of N Coupled Oscillators

We shall conclude our discussion on normal modes of discrete systems by considering longitudinal oscillations of N particles each of mass m spaced equally a distance a apart and coupled by means of $N+1$ springs (Fig. 5.19). The importance of this problem lies in the fact that this very simple model helps us understand sound waves which consist of longitudinal oscillations.

In chapter 1 we considedre the case $N = 1$ and in this chapter (Sec. 5.5) we have considered the case $N = 2$. We now consider the general case of N coupled masses. Fig. 5.19b shows a general configuration of the system. Let the displacements of the particles from their equilibrium positions be denoted by $x_1, x_2, ..., x_N$. Then the equation of motion of the pth particle is gizne by

$$m\ddot{x}_p = k(x_{p+1}-x_p)-k(x_p-x_{p-1})$$

or
$$\ddot{x}_p = \frac{k}{m}\left(x_{p+1}+x_{p-1}-2x_p\right) \tag{5.97}$$

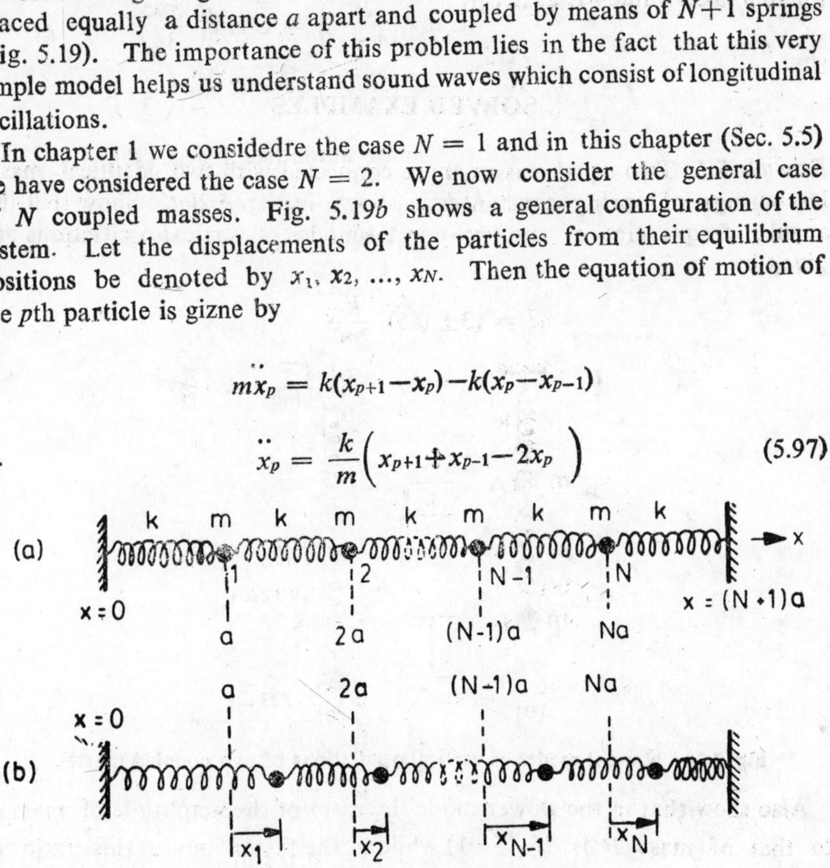

Fig. 5.19

(a) Equilibrium configuration of N coupled masses
(b) General configuration for longitudinal oscillations

This equation is precisely the same as (Eq. 5.87) if we replace T/a by spring constant k. Therefore all the results in the longitudinal case are immediately obtained from the transverse case by making the replacement T/a by k.

Thus the angular frequency of the nth normal mode is

$$\omega_n = 2\omega_0 \sin\left\{\frac{n\pi}{2(N+1)}\right\}$$

where $\omega_0 = \sqrt{k/m}$.

The amplitudes of particles in a given mode (i.e. given value of n) are given by

$$A_p = C \sin\left(\frac{pn\pi}{N+1}\right)$$

where p takes values 1, 2, ..., N.

SOLVED EXAMPLES

Example 5.1 Two equal masses m are connected with two identical massless springs of spring constant k as shown in Fig. 5.20a. Show that the angular frequencies of the two normal modes of vertical oscillations are given by

$$\omega^2 = (3 \pm \sqrt{5})\,\frac{k}{2m}$$

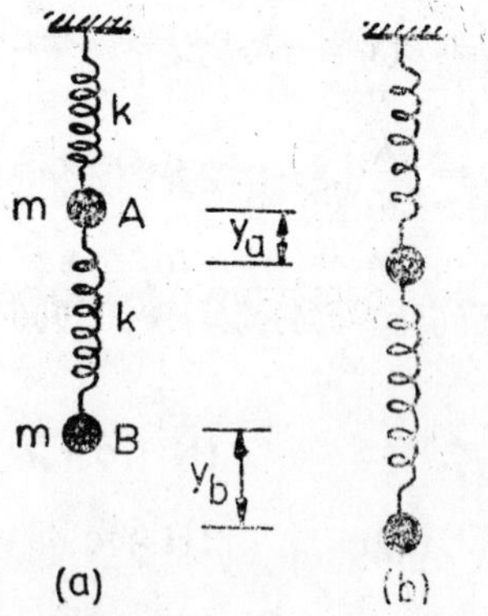

Fig. 5.20 Normal modes of vertical oscillations of two coupled masses

Also show that in the slower mode the ratio of the amplitude of mass A to that of mass B is $\frac{1}{2}(\sqrt{5}-1)$ while in the faster mode this ratio is $\frac{1}{2}(\sqrt{5}+1)$.

Solution

To solve this problem we need not take into account the gravitational forces acting on the two masses because they are independent of the displacements and hence do not contribute to the restoring forces responsible for the oscillations.

Fig. 5.20b shows a general configuration of the system. It is clear that the restoring force acting on mass A is $-ky_a+k(y_b-y_a)$ whereas that acting on mass B is $-k(y_b-y_a)$. Here y_a and y_b are the displacements of the masses at a certain instant of time. Hence, the equations of motion are

$$m\ddot{y}_a = -ky_a+k(y_b-y_a)$$

and

$$m\ddot{y}_b = -k(y_b-y_a)$$

or

$$\ddot{y}_a = -\frac{2k}{m}y_a+\frac{k}{m}y_b \qquad (i)$$

and

$$\ddot{y}_b = -\frac{k}{m}y_b+\frac{k}{m}y_a \qquad (ii)$$

In a normal mode with frequency ω and phase constant ϕ, we have

$$y_a = A\cos(\omega t+\phi)$$
$$y_b = B\cos(\omega t+\phi)$$

where A and B are the amplitudes of the two masses. Using these values in Eqs (i) and (ii) we have

$$\frac{y_a}{y_b} = \frac{A}{B} = \frac{k/m}{2k/m-\omega^2} = \frac{k/m-\omega^2}{k/m} \qquad (iii)$$

whence

$$\left(\frac{2k}{m}-\omega^2\right)\left(\frac{k}{m}-\omega^2\right) = \frac{k^2}{m^2}$$

or

$$\omega^4-\frac{3k}{m}\omega^2+\frac{k^2}{m^2} = 0$$

The angular frequencies of the normal modes are the two roots of this equation which are given by

$$\omega^2 = (3\pm\sqrt{5})\frac{k}{2m}$$

Now to find the ratio of the amplitudes in the two modes we substitute $\omega^2 = (3-\sqrt{5})\frac{k}{2m}$ (for the slower mode) and $\omega^2 = (3+\sqrt{5})\frac{k}{2m}$ (for the faster mode) in Eq. (iii). Thus, for the slower mode, we have

$$\frac{A}{B} = \frac{\frac{k}{m}-(3-\sqrt{5})\frac{k}{2m}}{\frac{k}{m}} = \tfrac{1}{2}(\sqrt{5}-1)$$

and for the faster mode the ratio is

$$\frac{A}{B} = \frac{\frac{k}{m}(-3+\sqrt{5})\frac{k}{2m}}{\frac{k}{m}} = -\tfrac{1}{2}(\sqrt{5}+1)$$

The minus sign indicates that this is the out-of-phase mode. The displacements y_a and y_b are oppositely directed. The amplitude ratio of these displacement is just $\frac{1}{2}(\sqrt{5}+1)$.

Example 5.2 Two masses m_1 and m_2 are connected by a massless spring of spring constant k as shown in Fig. 5.21*a*. The system is free to oscillate along the length of the spring. Show that the system oscillates with a frequency

$$\nu = \frac{1}{2\pi}\sqrt{\frac{k}{\mu}}$$

where
$$\mu = \frac{m_1 m_2}{m_1+m_2}$$

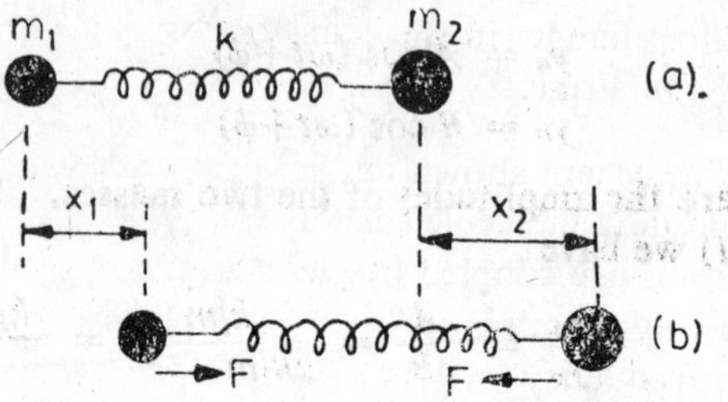

Fig. 5.21 (a) Equilibrium state (b) General configuration

Solution

Fig. 5.21*b* shows a general configuration where x_1 and x_2 are the displacements of the masses from their equilibrium positions shown in Fig. 5.21*a* Let us assume that $x_2 > x_1$ so that the spring is stretched and the increase in its length $= x_2 - x_1$. The tension in the spring is $F = k(x_2 - x_1)$. The forces (F) exerted on the two masses are obviously equal in magnitude and opposite in direction as shown in the figure. Hence the equations of motion are

$$m_1\ddot{x}_1 = k(x_2 - x_1) \tag{i}$$

$$m_2\ddot{x}_2 = -k(x_2 - x_1) \tag{ii}$$

Multiplying Eq. (*i*) by m_2 and Eq. (*ii*) by m_1 and subtracting we have

$$m_1 m_2(\ddot{x}_2 - \ddot{x}_1) = -k(m_1+m_2)(x_2 - x_1)$$

or
$$\ddot{x} = -\frac{k}{\mu}x \tag{iii}$$

where $x = x_2 - x_1$ is the relative displacement of the two masses and $\mu = \dfrac{m_1 m_2}{m_1+m_2}$ is called the reduced mass of the system (being smaller than either of the two masses m_1 and m_2).

Equation (*iii*) is identical in form with the equation of a single harmonic oscillator (see Fig. 1.9 Chap. 1) with the difference that μ here is the reduced mass of the system instead of mass m of the single oscillator and x is the relative displacement of the two masses from their equilibrium positions instead of the displacement of mass m alone from its equilibrium position.

Thus the system oscillates harmonically with a frequency

$$\nu = \frac{1}{2\pi}\sqrt{\frac{k}{\mu}}$$

It may be remarked that the system shown in Fig. 1.9 is in reality a two-body system where one of the bodies (i.e. the wall) is rigidly connected to the earth and thus has effectively an infinite mass and is, therefore, immovable. Since the end of the spring is connected to the wall that cannot move, the change in the length of the spring is given by the displacement of mass m alone.

Example 5.3 The system shown in Fig. 5.21 can serve as a simple model to represent a diatomic molecule such as that of H_2, CO, HCl, NaCl, etc. where the two atoms are coupled by electromagnetic forces which may be represented by a tiny massless spring. A sedium chloride molecule has a natural vibration frequency $= 1.14 \times 10^{13}$ Hz. Calculate the interatomic force constant. Mass of Na atom = 23 a.m.u. Mass of Cl atom = 35 a.m.u. (1 a.m.u. $= 1.67 \times 10^{-27}$ kg).

Solution

Here $m_1 = 23 \times 1.67 \times 10^{-27}$ kg

$$m_2 = 35 \times 1.67 \times 10^{-27} \text{ kg}$$

The reduced mass of the system is

$$\mu = \frac{m_1 m_2}{m_1 + m_2} = 23.18 \times 10^{-27} \text{ kg}$$

The frequency of vibration is

$$\nu = 1.14 \times 10^{13} \text{ Hz}$$

The interatomic force constant k is obtained from the equation

$$\nu = \frac{1}{2\pi}\sqrt{\frac{k}{\mu}}$$

which gives

$$k = 4\pi^2\nu^2\mu = 118.9 \text{ Nm}^{-1}$$

This very simple model gives a higher value of k than more refined methods which take into account the other interactions between the two atoms of the molecule.

Example 5.4 The CO_2 molecule may be represented by a system consisting of a central mass m_2 connected by identical springs of spring constant k to two masses m_1 and m_3 (with $m_1 = m_3$) as shown in Fig. 5.22*a*. Write down the equations of motion of each mass and solve them for the two normal modes in which the masses oscillate along the line joining their centres. Putting $m_1 = m_3 = 16$ units and $m_2 = 12$ units and assuming that this classical model is applicable, find the ratio of the frequencies of the two modes.

Solution

Figure 5.22*b* shows a general configuration of the system where x_1, x_2 and x_3 are the displacements of the three masses from their respective equilibrium positions shown in Fig. 5.22*a*. Assuming that $x_3 > x_2 > x_1$, the equations of motion are

$$m_1\ddot{x}_1 = k(x_2 - x_1) \tag{i}$$

$$m_2\ddot{x}_2 = k(x_3 - x_2) - k(x_2 - x_1) \tag{ii}$$

$$m_3\ddot{x}_3 = -k(x_3 - x_2) \tag{iii}$$

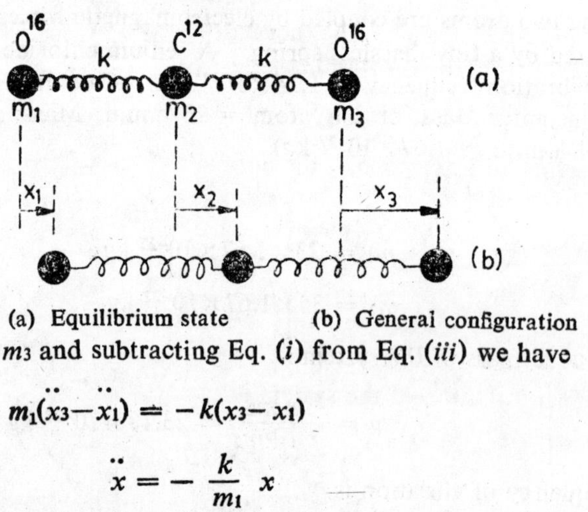

Fig. 5.22 (a) Equilibrium state (b) General configuration

Setting $m_1 = m_3$ and subtracting Eq. (*i*) from Eq. (*iii*) we have

$$m_1(\ddot{x}_3 - \ddot{x}_1) = -k(x_3 - x_1)$$

or

$$\ddot{x} = -\frac{k}{m_1}x$$

where $x = x_3 - x_1$ is the relative displacement of the masses m_1 and m_3. This equation gives one of the normal modes of the system with angular frequency ω_1 given by

$$\omega_1^2 = \frac{k}{m_1} \tag{iv}$$

To find the other mode (or modes) let us assume that

$$x_1 = A\cos(\omega t + \phi)$$
$$x_2 = B\cos(\omega t + \phi)$$
$$x_3 = C\cos(\omega t + \phi)$$

Substituting these values in Eqs. (*i*) to (*iii*) we have

$$m_1\omega^2 A = k(A-B)$$

$$m_2\omega^2 B = k(2B-A-C)$$

$$m_3\omega^2 C = k(B-C)$$

Eliminating A, B and C from these equations we obtain the angular frequency of the second mode which is given by

$$\omega_2^2 = k\left(\frac{m_2+2m_1}{m_1 m_2}\right) \qquad (v)$$

From Eqs (*iv*) and (*v*) the ratio of the frequencies of the two modes is

$$\frac{\omega_2}{\omega_1} = \left(\frac{m_2+2m_1}{m_2}\right)^{1/2}$$

$$= \left(\frac{12+2\times 16}{12}\right)^{1/2}$$

$$= 1.91$$

Example 5.5 Determine the normal coordinates of the system shown in Fig. 5.20 if $k = 10\ \text{Nm}^{-1}$ and $m = 1$ kg.

Solution

Referring to Example 5.1, the normal mode frequencies are

$$\omega_1 = \left\{(3-\sqrt{5})\frac{k}{2m}\right\}^{1/2} = 1.95\ \text{rad s}^{-1}$$

$$\omega_2 = \left\{\left(3+\sqrt{5}\right)\frac{k}{2m}\right\}^{1/2} = 5.12\ \text{rad s}^{-1}$$

The general motion of the system is given by

$$y_a = A_1 \cos(\omega_1 t+\phi_1)+A_2 \cos(\omega_2 t+\phi_2)$$

$$y_b = B_1 \cos(\omega_1 t+\phi_1)+B_2 \cos(\omega_2 t+\phi_2)$$

where

$$\frac{B_1}{A_1} = \frac{k/m}{k/m-\omega_1^2} = \frac{10}{10-(1.95)^2} = 1.62$$

$$\frac{B_2}{A_2} = \frac{k/m}{k/m-\omega_2^2} = \frac{10}{10-(5.12)^2} = -0.62$$

Hence we have

$$y_a = A_1 \cos(1.95t+\phi_1)+A_2 \cos(5.12t+\phi_2)$$

$$y_b = 1.62\, A_1 \cos(1.95t+\phi_1)-0.62\, A_2 \cos(5.12t-\phi_2)$$

Define a new set of coordinates X and Y so that

$$X = A_1 \cos(1.95t + \phi_1)$$

$$Y = A_2 \cos(5.12t + \phi_2)$$

The motions associated with coordinates X and Y are simple harmonic at frequencies $\omega_1 = 1.95$ rad s^{-1} and $\omega_2 = 5.12$ rad s^{-1} respectively. Hence X and Y are the two normal coordinates of the system. Now

$$y_a = X + Y$$

$$y_b = 1.62\,X - 0.62\,Y$$

which give

$$X = 0.28\,y_a + 0.45 y_b$$

$$Y = 0.72 y_a - 0.45 y_b$$

Example 5.6 The mass B of the system shown in Fig. 5.20 is held at rest at a position 10 cm from its equilibrium position and the mass A is at its equilibrium position. The system is then released. Determine the subsequent motion of each mass if k = 10 Nm^{-1} and m = 1 kg.

Solution

Referring to Examples 5.1 and 5.6, the general displacements and velocities of the two masses are given by

$$y_a = A_1 \cos(\omega_1 t + \phi_1) + A_2 \cos(\omega_2 t + \phi_2) \qquad (i)$$

$$y_b = B_1 \cos(\omega_1 t + \phi_1) + B_2 \cos(\omega_2 t + \phi_2) \qquad (ii)$$

$$\dot{y}_a = -A_1\omega_1 \sin(\omega_1 t + \phi_1) - A_2\omega_2 \sin(\omega_2 t + \phi_2) \qquad (iii)$$

$$\dot{y}_b = -B_1\omega_1 \sin(\omega_1 t + \phi_1) - B_2\omega_2 \sin(\omega_2 t + \phi_2) \qquad (iv)$$

where

$$B_1 = 1.62\,A_1 \qquad B_2 = -0.62\,A_2$$

$$\omega_1 = 1.95 \text{ rad s}^{-1} \qquad \omega_2 = 5.12 \text{ rad s}^{-1}$$

The initial conditions are:

$$y_a(t=0) = 0, \qquad y_b(t=0) = 10 \text{ cm} = 0.1 \text{ m}$$

$$\dot{y}_a(t=0) = \dot{y}_b(t=0) = 0$$

Using these conditions in Eqs (i) to (iv) we have

$$0 = A_1 \cos\phi_1 + A_2 \cos\phi_2$$

$$0.1 = B_1 \cos\phi_1 + B_2 \cos\phi_2$$

$$= 1.62\,A_1 \cos\phi_1 - 0.62\,A_2 \cos\phi_2$$

$$0 = A_1\omega_1 \sin\phi_1 + A_2\omega_2 \sin\phi_2$$

$$0 = 1.62\,A_1\omega_1 \sin\phi_1 - 0.62\,A_2\omega_2 \sin\phi_2$$

Hence

$$\sin \phi_1 = \sin \phi_2 = 0$$

$$A_1 \cos \phi_1 = 0.0446 \simeq 0.045$$

$$A_2 \cos \phi_2 = -0.0446 \simeq -0.045$$

Substituting them in Eqs (i) and (ii) we get

$$y_a = 0.045 (\cos 1.95\ t - \cos 5.12\ t) \quad (v)$$

$$y_b = 0.072 \cos 1.95\ t + 0.028 \cos 5.12\ t \quad (vi)$$

Equations (*v*) and (*vi*) determine the subsequent motion of the system. In these equations, the displacement is in metres and time in seconds.

Example 5.7 Two pendulums are suspended one below the other to form a double pendulum as shown in Fig. 5.23(*a*). If $m_1 = m_2 = m$ and $l_1 = l_2 = l$ show that the frequencies of the two normal modes for small oscillations are given by

$$\omega^2 = \left(2 \pm \sqrt{2}\right)\frac{g}{l}$$

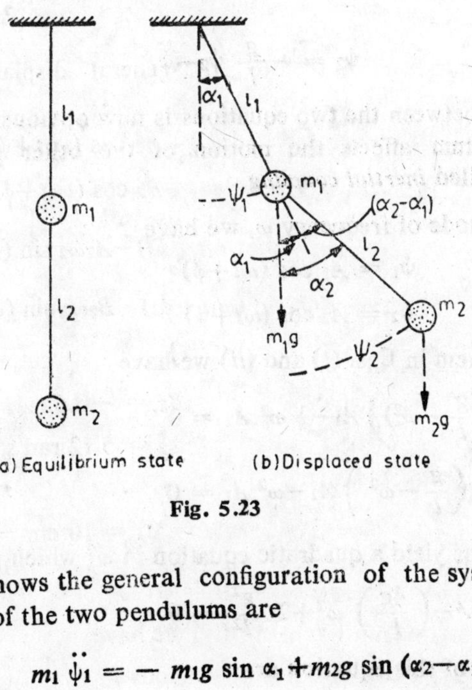

Fig. 5.23

Solution

Figure 5.23 (*b*) shows the general configuration of the system. The equations of motion of the two pendulums are

$$m_1 \ddot{\psi}_1 = -m_1 g \sin \alpha_1 + m_2 g \sin (\alpha_2 - \alpha_1)$$

$$m_2 \ddot{\psi}_2 = -m_2 g \sin \alpha_2 - m_1 \ddot{\psi}_1 \cos (\alpha_2 - \alpha_1)$$

For small displacements

$$\sin \alpha_1 \simeq \alpha_1 = \frac{\psi_1}{l_1}$$

$$\sin \alpha_2 \simeq \alpha_2 = \frac{\psi_2}{l_2}$$

$$\sin(\alpha_2-\alpha_1) \simeq \alpha_2-\alpha_1 = \frac{\psi_2}{l_2}-\frac{\psi_1}{l_1}$$

$$\cos(\alpha_2-\alpha_1) \simeq 1$$

For small displacements and for $m_1 = m_2 = m$ and $l_1 = l_2 = l$, the equations of motion simplify as follows.

$$\ddot{\psi}_1 = -\frac{g}{l}\psi_1 + \frac{g}{l}(\psi_2-\psi_1)$$

$$\ddot{\psi}_2 = -\frac{g}{l}\psi_2 - \ddot{\psi}_1$$

The coupling between these equations is not apparent as the equations are not symmetric. Symmetry can be achieved if we replace the first equation by the sum of the two equations. In the symetricm form, the equations can be written as

$$\ddot{\psi}_1 = -\frac{g}{l}\psi_1 - \tfrac{1}{2}\ddot{\psi}_2 \qquad (i)$$

$$\ddot{\psi}_2 = -\frac{g}{l}\psi_2 - \ddot{\psi}_1 \qquad (ii)$$

The coupling between the two equations is now obvious. The acceleration of one pendulum affects the motion of the other pendulum. Such a coupling is called *inertial coupling*.

In a normal mode of frequency ω, we have

$$\psi_1 = A_1 \cos(\omega t+\phi)$$

$$\psi_2 = A_2 \cos(\omega t+\phi)$$

Substituting them in Eqs. (*i*) and (*ii*) we have

$$\left(\frac{g}{l}-\omega^2\right)A_1 - \tfrac{1}{2}\omega^2 A_2 = 0$$

$$\left(\frac{g}{l}-\omega^2\right)A_2 - \omega^2 A_1 = 0$$

These equations yield a quadratic equation in ω^2 which reads

$$\omega^4 - \left(\frac{4g}{l}\right)\omega^2 + 2\frac{g^2}{l^2} = 0$$

The two roots of this equation are

$$\omega^2 = (2 \pm \sqrt{2})\frac{g}{l}$$

which determine the frequencies of the two normal modes.

Example 5.8 Two identical pendulums of mass m and length l are connected by a spring of spring constant k as shown in Fig. 5.24. Determine the frequencies of the normal modes of the system

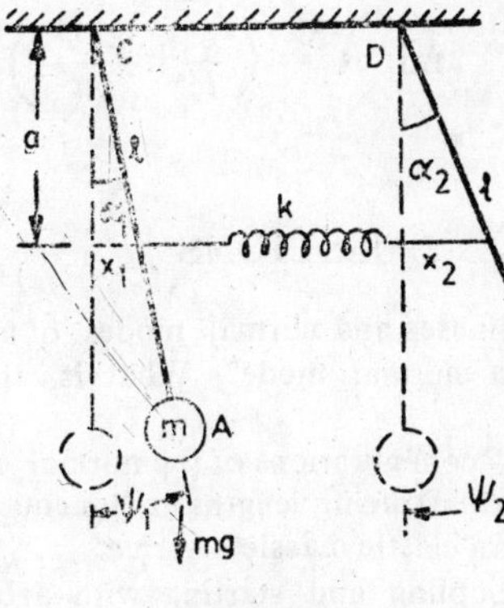

Fig 5.24

Solution

Let ψ_1 and ψ_2 be the displacements of the two masses at a certain instant of time. Referring to Fig. 5.24, the change in the length of the spring at this time $= x_2 - x_1$. But $x_1 = \dfrac{a}{l}\psi_1$ and $x_2 = \dfrac{a}{l}\psi_2$.

Hence

$$x_2 - x_1 = \frac{a}{l}(\psi_2 - \psi_1)$$

The force in the spring is $F = k(x_2 - x_1) = k\dfrac{a}{l}(\psi_2 - \psi_1)$.

We shall solve this problem by considering the moments of the forces about the pivoted points C and D.

Clockwise moment of mass A about $C = (m\ddot{\psi}_1)\times l + (mg \sin \alpha_1)\times l$

Anticlockwise moment of mass A about $C = F\times a = \dfrac{ka^2}{l}(\psi_2 - \psi_1)$. Equating the moments and setting $\sin \alpha_1 \simeq \alpha_1 = \dfrac{\psi_1}{l}$, we have

$$ml\,\ddot{\psi}_1 + mgl\frac{\psi_1}{l} = \frac{ka^2}{l}(\psi_2 - \psi_1)$$

or

$$\ddot{\psi}_1 = -\frac{g}{l}\psi_1 + \frac{ka^2}{ml^2}(\psi_2 - \psi_1) \qquad (i)$$

Similarly for mass B we have

$$\ddot{\psi}_2 = -\frac{g}{l}\psi_2 - \frac{ka^2}{ml^2}(\psi_2 - \psi_1) \qquad (ii)$$

The frequencies of the normal modes are determined from these equations of motion; they are

$$\omega_1 = \left(\frac{g}{l}\right)^{1/2} \text{ and } \omega_2 = \left(\frac{g}{l} + \frac{2ka^2}{ml^2}\right)^{1/2}$$

QUESTIONS

1. What are normal coordinates and normal modes of a coupled system? What are the properties of a normal mode? What is the significance of normal modes?
2. Obtain frequencies and configurations of the normal modes of the oscillations of two pendulums of equal string lengths and unequal masses with their bobs connected by means of an elastic massless spring.
3. (a) Assuming weak coupling and starting with arbitrary initial conditions, obtain the expressions for the energy of each pendulum of Question 2.
   (b) Show that the energy bounces back and forth from one pendulum to the other with a frequency equal to the difference between the two mode frequencies.
   (c) Show that energy exchange is complete if the masses of the bobs are equal.
4. Two equal masses m are connected by three identical massless springs of spring constant k. The free ends of the springs are rigidly fixed. Find the frequencies and configurations of the two normal modes if the masses oscillate (a) along the line joining the centres of the masses and (b) perpendicular to the line joining the centres of the masses.
5. Obtain frequencies of the normal modes of oscillations of two coupled LC circuits if the coupling is (a) capacitative and (b) inductive.
6. Two masses A and B are connected by three massless springs as shown in Fig. 5.11 in the text. If $m_a = m_b = m$ and $k_3 = \sqrt{k_1 k_2}$
   (a) find the frequencies of the normal modes for longitudinal oscillations,
   (b) find the normal coordinates of the system,
   (c) show that the total energy of the system is equal to the sum of the energies associated with the two modes.
7. How will you distinguish coupled oscillations form forced oscillations ?
8. Analyse the problem of energy exchange between two coupled oscillators whose natural frequencies are equal.
9. Set up the equations of motion of two coupled oscillators when a periodic force is applied to one of them. Discuss the response of the system when the driving frequency is (a) equal to one of the mode frequencies, (b) considerably less than the frequency of the lower mode, and (c) considerably higher than the higher mode frequency.
10. What is a low-pass electrical filter? Draw the circuit diagram of a low-pass filter for a DC power supply and explain how it acts as a filter.
11. A uniform string fixed at both ends is beaded with N identical particles of mass m with a constant spacing a between two neighbouring particles. Show that the frequencies of the normal modes of transverse oscillations are given by

$$\omega_n = 2\sqrt{\frac{T}{ma}} \sin\left\{\frac{n\pi}{2(N+)1}\right\}$$

where T is the equilibrium tension and $n = 1, 2, 3, \ldots, N$.

12. Draw the configurations of the normal modes of the system of Question 11 for $N = 3, 4$ and 5.

13. N identical particles of mass m are connected together by $(N+1)$ identical massless springs of spring constant k. The free ends of the extreme springs are rigidly fixed. Show that the frequencies of the normal modes of longitudinal oscillations are given by

$$\omega_n = 2\sqrt{\frac{k}{m}} \sin\left\{\frac{\pi n}{2(N+1)}\right\}$$

where $n = 1, 2, 3, \ldots, N$.

14. Draw the configurations of the normal modes of the system of Question 13 for $N = 3$.

15. A harmonic force $F = F_0 \cos \omega t$ acts on a mass M which is connected to a mass m by a massless spring of spring constant k. Both masses are free to oscillate along the length of the spring. Show that the condition for M to remain stationary is

$$\omega = \sqrt{\frac{k}{m}}$$

PROBLEMS

1. Two identical pendulums each of length 0.5 m have their bobs connected to each other by a weak spring. With one pendulum clamped, the time period of the other is found to be 1.40 s. With neither pendulum clamped, what are the periods of the two normal modes? Take $g = 9.8$ m s^{-2}.

2. One of the pendulums of Problem 1 is drawn aside while the other is held at its equilibrium position. Both pendulums are then released. What is the time interval between successive maximum possible amplitudes of either pendulum?

3. Find the normal mode frequencies ω_1 and ω_2 of the system shown in Fig. 5.11 in the text if $m_a = m_b = m$ and $k_3 = \sqrt{k_1 k_2}$

4. Two equal masses on a frictionless horizontal surface are held between rigid supports by three identical springs as shown in Fig. 5.5 in the text. If one mass is clamped, the frequency of longitudinal oscillations of the other is found to be 0.5 Hz. What are the frequencies of the two normal modes of the system if both are free to oscillate?

5. Mass A in Problem 4 is held at its equilibrium position and mass B is pulled aside a distance of 0.04 m from its equilibrium position. The masses are released from rest at time $t = 0$. After how long will the energy of mass A be transferred to mass B?

6. Two identical oscillators A and B, each of mass m and stiffness constant K are coupled in such a way that the coupling force exerted on A is $\mu m \ddot{\psi}_b$ and that exerted on B is $\mu m \ddot{\psi}_a$ where ψ_a and ψ_b are the instantaneous displacements of the two masses from their equilibrium positions and μ is a coupling constant. Find the frequencies of the normal modes of the coupled system.

7. Two identical LC circuits each having a natural frequency of 500 Hz are coupled inductively in such a way that their coupling coefficient is 0.5. Find the frequencies of the two normal modes of the system.

8 Two masses 0.01 kg and 0.03 kg are connected by a massless spring of constant 10 Nm^{-1}. Find the frequency of oscillation if the masses are free to move along the line joining their centres.

9. What is the ratio of the kinetic energies of the two masses of the system of Problem 8 above? (Hint: Use the law of conservation of linear momentum.)

10. The interatomic force constant of an HCl molecule is 5.4×10^2 Nm^{-1}. Assuming that the simple classical model of a diatomic molecule is applicable, calculate the fundamental frequency of the vibration of molecule. Mass of H atom = 1.67×10^{-27} kg and mass of Cl atom = 5.845×10^{-26}kg.

11. Two identical masses are connected with two identical massless springs and placed on a horizontal frictionless surface as shown in Fig. 5.25 Considering only motion along the line joining their centres, find (a) the ratio of the normal mode frequencies of the two masses in each mode.

Fig. 5.25

12. Three equal masses m are equally spaced along a string of length $4a$. The tension in the string is T. Show that the three normal mode frequencies and the corresponding normal mode displacements are

$$\omega_1 = \omega_0 \sqrt{1 - \frac{1}{\sqrt{2}}} \; ; \; y_1 = y_3 = \frac{y_2}{\sqrt{2}}$$

$$\omega_2 = \omega_0 \; ; \; y_1 = -y_3 \text{ and } y_2 = 0$$

$$\omega_3 = \omega_0 \sqrt{1 + \frac{1}{\sqrt{2}}} \; ; \; y_1 = y_3 = -\frac{y_2}{\sqrt{2}}$$

$$\text{where } \omega_0 = \sqrt{\frac{2T}{ma}}$$

6

Normal Modes of Continuous Systems: Fourier Method

6.1. INTRODUCTION

In Chap. 5 we studied oscillations of systems having two degrees of freedom. Later in that chapter we studied oscillations of systems having N degrees of freedom where N is a finite number We have seen that a system with N degrees of freedom has exactly N normal modes. Each mode has its own characteristic frequency and its own characteristic 'shape' or configuration. In each mode, all moving parts of the system pass through the equilibrium positions simultaneously, i.e. each moving part has the same constant phase which is determined by the initial conditions.

The results of Chap. 5 are applicable to 'discrete' systems, i.e. systems whose mass is concentrated at two or a finite number of points. The systems we come across in nature are not so simple. Consider, for example, the vibrating string of a violin. The mass of the string is not concentrated at a finite number of points along the string. In other words, the string is not a discrete system, it is 'continuous', i.e. the mass of the string is spread *uniformly* along its length. Similarly, the mass of a loud-speaker diaphragm is distributed uniformly over its extent. In these cases, each particle of the system will vibrate with somewhat different motion from that of any other particle. It will not be sufficient to specify the motion of a few particles, the motion of every particle of the system must be specified.

In principle, the problem of a continuous system can be tackled by considering the motion of N equally spaced coupled masses (see Chap. 5) and letting N go to infinity and inter-particle separation go to zero. We would then have an infinite number of equations of motion (one equation for each particle). These equations could, in principle, be solved to give the motion of every one of the infinite particles. We would expect the system to have an infinite number of normal modes; each mode having its own characteristic frequency. This will be a hopeless task and a very awkward way of solving a problem that is essentially simple. What is needed is a new point of view; a new method of attack.

The new point of view can be called the *continuous approximation.* If a system consists of a very large number (which we shall loosely call infinity) of particles and if these particles are packed within a limited region of space, the average distance between neighbouring particles becomes very small (tending to zero). As an approximation, we may imagine the number of particles as becoming infinity and inter-particle separation tending to zero. We assume that the system, then, behaves as if it were continuous. This assumption implies that the motion of near neighbours is very nearly the same. We can then divide the system into infinitesimally small cubes or elements and describe the motion of the elements; ignoring the details of motion of the individual particles in a given element. This is called the *continuous approximation.* When this description is used to replace the description using the displacements of individual particles we say that we are dealing with waves. In other words, waves will automatically emerge from our analysis of a continuous system. However, waves will be separately dealt with in the next chapter. We shall now use this description to study normal modes and general motion of a few continuous systems.

6.2 TRANSVERSE VIBRATION OF A STRING

Consider a uniform string stretched with a tension T and lying along the x-axis in the equilibrium position. The string is divided into a large number of infinitesimally small elements each of length Δx. Each element is characterized by its x-coordinate. Figure 6.1 shows one such element of the string. We shall now obtain the equation of motion of such an element. In the equilibrium state, the forces acting on element AB are

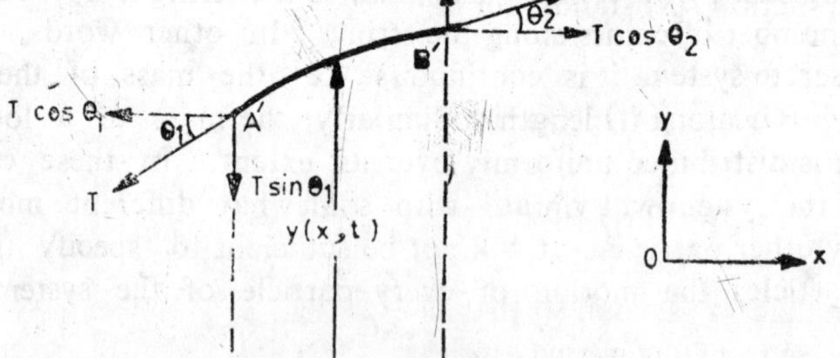

Fig. 6.1 *AB* is an infinitesimal element of the string in equilibrium position along x-axis. *A′B′* is the displaced position of the element.

equal and opposite. The element is displaced along the y-axis to a new position $A'B'$. The element is no longer exactly straight; it has a slight curvature and the tension is tangential to the curve at A' and B'. If the displacement is small, the tension remains unchanged when the string is deformed.

Let $y(x, t)$ be the transverse displacement of the element located at x at time t. When the string is released, the displacement of the element will change with time. Thus the displacement is a function of two variables x and t. For an extremely small element ($\Delta x \to 0$) the net forces on the element along x and y directions are

$$F_x = T(\cos\theta_2 - \cos\theta_1)$$

$$F_y = T(\sin\theta_2 - \sin\theta_1)$$

where θ_1 and θ_2 are the directions of tangents to the string at the ends of the element, i.e. at $x = x_1$ and $x = x_2$ with $x_2 - x_1 = \Delta x$.

We are assuming that the transverse displacement is very small, so that θ_1 and θ_2 are small angles. We may, therefore, make the following approximations

$$\cos\theta_1 \simeq 1 \qquad \cos\theta_2 \simeq 1$$

$$\sin\theta_1 \simeq \tan\theta_1 \qquad \sin\theta_2 \simeq \tan\theta_2$$

The transverse component of tension on the element is, therefore, given by

$$\begin{aligned} F_y &= T(\tan\theta_2 - \tan\theta_1) \\ &= T\left\{ \left.\frac{\partial y}{\partial x}\right|_{x = x_2} - \left.\frac{\partial y}{\partial x}\right|_{x = x_1} \right\} \\ &= T\{f(x_2) - f(x_1)\} \end{aligned} \tag{6.1}$$

where the function $f(x)$ stands for $\dfrac{\partial y}{\partial x}$. Using Taylor's series, we have

$$\begin{aligned} f(x_2) &= f(x_1) + (x_2 - x_1)\left.\frac{\partial f}{\partial x}\right|_{x = x_1} + \tfrac{1}{2}(x_2 - x_1)^2 \left.\frac{\partial^2 f}{\partial x^2}\right|_{x = x_1} + \ldots \\ &= f(x_1) + \Delta x \left.\frac{\partial f}{\partial x}\right|_{x = x_1} + \tfrac{1}{2}(\Delta x)^2 \left.\frac{\partial^2 f}{\partial x^2}\right|_{x = x_1} + \ldots \end{aligned} \tag{6.2}$$

We assume that Δx is small so that the terms of order $(\Delta x)^2$ and higher may be ignored. Thus we may write

$$\begin{aligned} f(x_2) - f(x_1) &= \Delta x \left.\frac{\partial f}{\partial x}\right|_{x = x_1} \\ &= \Delta x \frac{\partial}{\partial x}\left(\frac{\partial y}{\partial x}\right) = \Delta x \frac{\partial^2 y}{\partial x^2} \end{aligned} \tag{6.3}$$

where we have dropped the subscript x_1 because, having ignored the higher order derivatives in Eq. (6.2), it does not matter now where in the interval Δx we evaluate the x-derivative. Notice that the space derivative is written as a partial derivative because now the displacement $y(x, t)$ is a function of two variables x and t. In a complete form we would write Eq. (6.3) as

$$f(x_2)-f(x_1) = \Delta x \frac{\partial^2 y\,(x, t)}{\partial x^2}$$

and Eq. (6.1) as

$$F_y = T\,\Delta x \frac{\partial^2 y\,(x, t)}{\partial x^2} \tag{6.4}$$

The equation of motion of the element is obtained from Newton's second law, namely, force = mass × acceleration. Let μ be the linear density (mass per unit length) of the string. Therefore, the mass of the element of length $\Delta x = \mu \Delta x$.

The acceleration of the element is $\frac{\partial^2 y(x, t)}{\partial t^2}$. Hence Newton's force $= \mu \Delta x \frac{\partial^2 y\,(x, t)}{\partial t^2}$. For dynamic equilibrium this force must balance the force F_y given by Eq. (6.4). This gives the equation of motion of an element of the string, namely,

$$\mu \Delta x \frac{\partial^2 y\,(x, t)}{\partial t^2} = T \Delta x \frac{\partial^2 y\,(x, t)}{\partial x^2}$$

or

$$\frac{\partial^2 y\,(x, t)}{\partial t^2} = \frac{T}{\mu}\frac{\partial^2 y\,(x, t)}{\partial x^2} \tag{6.5}$$

or

$$\frac{\partial^2 y\,(x, t)}{\partial t^2} = v^2 \frac{\partial^2 y\,(x, t)}{\partial x^2} \tag{6.6}$$

where $v = \sqrt{\frac{T}{\mu}}$ has the dimensions of velocity.

Equation (6.6) is a very famous second-order linear partial differential equation. It is called the *classical wave equation*. This equation relates the acceleration of a simple harmonic oscillator of a medium to the second derivative of its displacement with respect to position x (called the second space derivative). We shall encounter this equation in many physical situations. The positive constant T/μ (having dimensions of square of a velocity) is special to the string. In other physical situations, some other positive constant appears in its place in the wave equation. So far we have not explicitly stated which velocity the constant $v = \sqrt{T/\mu}$ represents.

In the next chapter (which deals with wave-motion) we shall identify v to be the velocity of transverse waves on a string. For the time being, however, all that we can say is that v is a constant having dimensions of a velocity.

As stated at the beginning of this chapter, we can obtain the equation of a continuous medium by taking the continuous limit $N \to \infty$ and $a \to 0$ of a discrete system dealt with in Sec. 5.9. We show below how this can be done. The equation of motion of the pth mass is given by (see Eq. 5.87)

$$\frac{d^2 y_p}{dt^2} = \frac{T}{ma}(y_{p+1} + y_{p-1} - 2y_p)$$

Now the separation $a = \Delta x$ and we take the limit $\Delta x \to 0$ as the masses merge into a continuous string. We can then write

$$\begin{aligned}\frac{d^2 y_p}{dt^2} &= \frac{T}{m}\left(\frac{y_{p+1} - 2y_p + y_{p-1}}{\Delta x}\right)\\ &= \frac{T}{m}\left\{\frac{(y_{p+1} - y_p)}{\Delta x} - \frac{(y_p - y_{p-1})}{\Delta x}\right\}\\ &= \frac{T}{m}\left\{\left.\frac{\partial y}{\partial x}\right|_{p+1} - \left.\frac{\partial y}{\partial x}\right|_{p}\right\}\\ &= \frac{T}{m}\left\{\left.\frac{\partial y}{\partial x}\right|_{x+\Delta x} - \left.\frac{\partial y}{\partial x}\right|_{x}\right\}\\ &= \frac{T}{m}\Delta x \frac{\partial^2 y}{\partial x^2}\end{aligned}$$

In the limit $\Delta x \to 0$, we may drop the subscript p and write

$$\frac{\partial^2 y}{\partial t^2} = \frac{T}{m}\Delta x \frac{\partial^2 y}{\partial x^2} \tag{6.7}$$

Now the total length of the string is

$$L = (N+1)\,a \simeq N\Delta x \qquad (\because N \gg 1 \text{ and } a = \Delta x)$$

The total mass of the string is

$$M = mN$$

where m is the mass of each bead. Hence, mass per unit length of the string is given by

$$\mu = \frac{M}{L} = \frac{mN}{N\Delta x} = \frac{m}{\Delta x}$$

Substituting this in Eq. (6.7) gives Eq. (6.5). But this would be a very awkward and unnecessarily difficult way of obtaining an equation which can be obtained in a very simple manner if we adopt a new view-point expressed by the so-called *continuous approximation*.

Normal Modes

We shall now obtain the normal modes of transverse vibrations of a uniform string fixed at a both ends and stretched with a tension.

Consider a uniform string of length L and linear density μ stretched along the x-axis with a tension T and fixed at its ends $x = 0$ and $x = L$. Equation (6.6) describes the motion of any part of the string lying between $x = 0$ and $x = L$. In order to find the normal modes, we assume the existence of a normal mode at angular frequency ω and phase constant ϕ. This means that every particle of the string executes SHM of angular frequency ω and phase constant ϕ. Thus, for a normal mode we have

$$y(x, t) = A(x) \cos (\omega t+\phi) \tag{6.8}$$

These are infinite number of equations one for each particle characterized by its x value in the range 0 and L. The variable $y(x, t)$ is the displacement at time t of a particle located at x and $A(x)$ is the amplitude of its motion. The amplitudes of all particles of the string will determine the shape or configuration of the mode. Differentiating twice with respect to x and t we have

$$\frac{\partial^2 y\,(x, t)}{\partial x^2} = \frac{d^2A(x)}{dx^2} \cos (\omega t+\phi)$$

and

$$\frac{\partial^2 y\,(x, t)}{\partial t^2} = -\omega^2 A(x) \cos (\omega t+\phi)$$

Notice that, since $A(x)$ is by definition a function x only, we can write the total derivative $\frac{d^2A(x)}{dx^2}$ instead of a partial derivative. Substituting these derivatives in Eq. (6.6) gives

$$-\omega^2 A(x) \cos (\omega t+\phi) = v^2 \frac{d^2A(x)}{dx^2} \cos (\omega t+\phi)$$

or

$$\frac{d^2A(x)}{dx^2} = -\frac{\omega^2}{v^2} A(x) = -k^2 A(x) \tag{6.9}$$

where $k = \omega/v$. The parameter k will be identified to be the *wave number* of the wave (see Chap. 7). At the moment k just stands for ω/v where

$$v = \sqrt{\frac{T}{\mu}}.$$

Equation (6.9) governs the shape of the mode. This is the familiar differential equation of SHM except that it represents oscillation in space (x) rather than in time (t). The general form of the harmonic oscillation in space can be written as

$$A(x) = A \sin kx + B \cos kx \tag{6.10}$$

where A and B are undetermined constants.

The general solution for the displacement $y(x, t)$ of the string, in a given mode, is obtained by using Eq. (6.10) in Eq. (6.8).

$$y(x, t) = (A \sin kx + B \cos kx) \cos (\omega t + \phi) \tag{6.11}$$

Boundary Conditions. Equation (6.11) is a bit too general because the boundary conditions have not been used so far. Our string is fixed at both ends. Suppose the string has total length L and the ends of the string are at $x = 0$ and $x = L$. Since these ends are rigidly fixed, there can be no displacement at these ends. In other words, the boundary conditions are

$$y(0, t) = y(L, t) = 0 \text{ for all values of } t$$

Using the first boundary condition, namely $y(0, t) = 0$ for all t in Eq. (6.11) we have

$$B = 0$$

Thus for a string fixed at $x = 0$, Eq. (6.11) reduces to

$$y(x, t) = A \sin kx \cos (\omega t + \phi) \tag{6.12}$$

Normal Mode Frequencies. The frequencies of the normal modes of transverse vibrations of the string can be obtained by using the second boundary condition, namely $y(L, t) = 0$ for all t, in Eq. (6.12) This requires

$$A \sin kL = 0$$

This equation is satisfied by choosing $A = 0$. But this corresponds to a trivial situation of a string permanently at rest. Hence the only way we can satisfy the boundary condition at $x = L$ is to have

$$\sin kL = 0$$

or

$$kL = n\pi$$

where n is an integer having values 1, 2, 3, ..., ∞. Thus

$$k = \frac{n\pi}{L} \tag{6.13}$$

We have excluded the case $n = 0$, i.e. $k = 0$, because this case also corresponds to an uninteresting situation of a string permanently at rest as is obvious from Eq. (6.12). Notice that the condition that the string is fixed at $x = L$ permits only some values of k, namely, those given by Eq. (6.13). But $k = \omega/v$, where $v = \sqrt{\dfrac{T}{\mu}}$ and ω is the angular frequency of the normal mode. The fact that only definite values of k [dictated by Eq. (6.13)] are permitted implies that only definite values of ω are allowed. These values are given by

$$\omega = kv$$

or

$$\omega_n = \frac{n\pi}{L}\sqrt{\frac{T}{\mu}} \tag{6.14}$$

with $n = 1, 2, 3, ..., \infty$. Here we have used subscript n to indicate the value of ω for a particular integral value of n. Equation (6.14) gives the angular frequencies of the normal modes for transverse vibrations of a string fixed at both ends. The corresponding frequencies (in hertz) of the modes are given by

$$\nu_n = \frac{n}{2L}\sqrt{\frac{T}{\mu}} \tag{6.15}$$

The mode with $n = 1$ is called the *fundamental* mode; its frequency is

$$\nu_1 = \frac{1}{2L}\sqrt{\frac{T}{\mu}}$$

The modes with $n = 2, 3, 4, ...$ are *harmonics* of the fundamental frequencies ν_1 which follows from the fact that

$$\nu_2 = 2\nu_1,\ \nu_3 = 3\nu_1,\ \nu_4 = 4\nu_1, \text{ etc.}$$

Thus the string has an infinite number of possible frequencies of vibration which are harmonics of the fundamental frequency ν_1. The fact that the mode frequencies ν_2, ν_3, etc. are harmonics of the fundamental mode frequency ν_1 is a result of our assumption that the string is perfectly uniform and flexible. Real strings do not strictly obey this simple sequence of frequencies The reason is that the strings in real physical systems (such as the strings of a violin or a piano) are not perfectly uniform and flexible.

Normal Mode Shapes. Equation (6.12) gives the displacement of the particles of the string in a normal mode. For the nth mode this equation reads

$$y_n(x, t) = A_n \sin k_n x \cos(\omega_n t + \phi_n) \tag{6.16}$$

where $k_n = \dfrac{n\pi}{L}$ and $\omega_n = k_n v = k_n\sqrt{\dfrac{T}{\mu}}$. The constants A_n and ϕ_n are to be determined from the initial conditions. We shall subsequently consider a set of initial conditions. Let us now obtain the shape of the first few modes.

In the fundamental mode ($n = 1$), also called the first harmonic, the particle displacements are given by

$$y_1(x, t) = A_1 \sin k_1 x \cos(\omega_1 t + \phi_1)$$

or

$$y_1(x, t) = A_1 \sin\left(\frac{\pi x}{L}\right) \cos(\omega_1 t + \phi_1)$$

This equation gives the displacements (in the fundamental mode) of all particles of the string (i e. all x values) as a function of time t. Notice that $y = 0$ at $x = 0, L$. There is no other value of x in the range 0 to L where y_1 can vanish. The displacement y_1 is maximum at $x = L/2$ for a given value of t. Figure 6.2 (a) shows a plot of $y_1(x, t)$ against x for different values of t. This is the shape of the fundamental mode. When the string is vibrating in one segment, as shown in Fig. 6.2 (*a*) its frequency of vibration is ν_1 given by

$$\nu_1 = \frac{1}{2L}\sqrt{\frac{T}{\mu}}$$

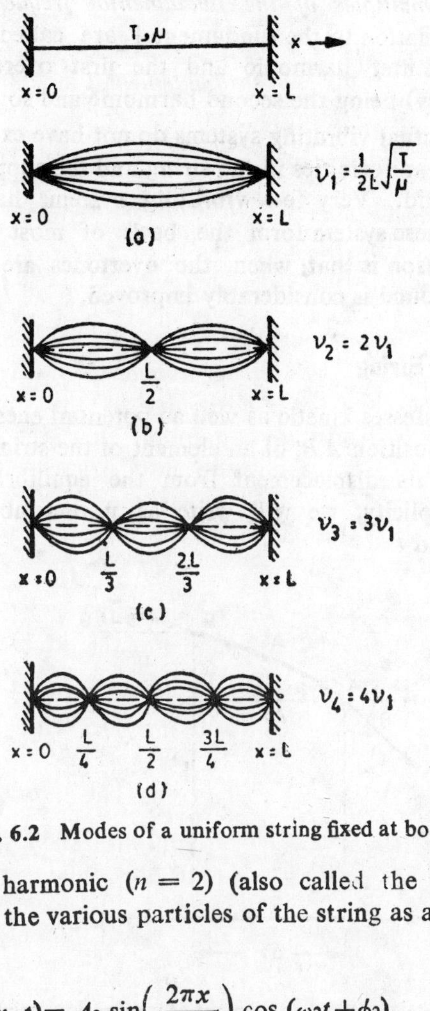

Fig. 6.2 Modes of a uniform string fixed at both ends

In the second harmonic ($n = 2$) (also called the first overtone), the displacements of the various particles of the string as a function of time are given by

$$y_2(x, t) = A_2 \sin\left(\frac{2\pi x}{L}\right) \cos(\omega_2 t + \phi_2)$$

Notice that y_2 is zero at $x = 0$, $L/2$ and L. Figure 6.2 (*b*) gives the shape of this mode; the string now vibrates in two segments at a frequency ν_2 which is twice the frequency ν_1 of the fundamental mode. Figures 6.2 (*c*) and (*d*) show the next two harmonics. These figures reveal that there are certain points on the string which are permanently at rest; the number of these points depends upon the number of the mode under study. These points are called *nodes*. The points where the displacement is maximum (at a given time) are called *antinodes*. In the next chapter we shall show that the *standing waves* on a string are nothing but its normal modes.

It may be remarked that Eq. (6.15) for allowed frequencies expresses a very important property of a uniform flexible string stretched between rigid supports. It states that the frequencies of all the overtones of such a string are *integral multiples of the fundamental frequency.* Overtones bearing this simple relation to the fundamental are called *harmonics*; the fundamental being the first harmonic and the first overtone (twice the fundamental frequency) being the second harmonic and so on.

As stated earlier, actual vibrating systems do not have exactly harmonic overtones due to non-uniformities in the string and the supports at its ends being not perfectly rigid. Very few vibrating systems have nearly harmonic overtones. These system form the basis of most of the musical instruments. The reason is that, when the overtones are harmonic, the tonal quality of the sound is considerably improved.

Energy of a Vibrating String

A vibrating string possesses kinetic as well as potential energy. Figure 6.3 shows the displaced position $A'B'$ of an element of the string at a certain instant of time when its displacement from the equilibrium position is $y(x, t)$ which, for simplicity, we will write as y remembering that y is a function of both x and t.

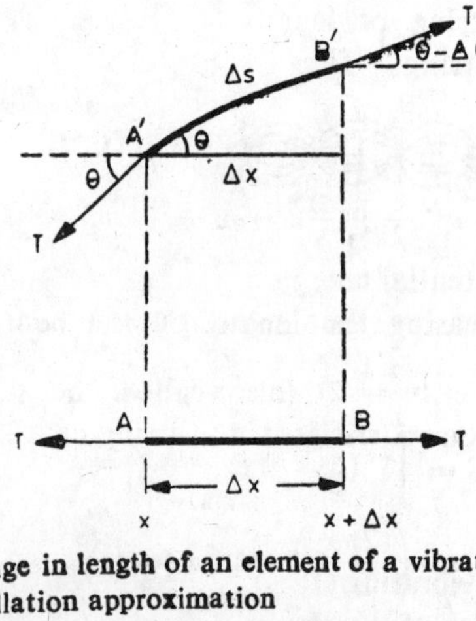

Fig. 6.3 Change in length of an element of a vibrating string under small oscillation approximation

The kinetic energy is due to the velocity the element has at that instant of time and the potential energy is the work done by the tension T in extending the element AB of length Δx to a new length Δs when the string is vibrating. The length of the curved element $A'B'$ is

$$\Delta s = \left[1+\left(\frac{\partial y}{\partial x} \right)^2 \right]^{1/2} \Delta x$$

This can be proved as follows. It is clear from Fig. 6.3 that

$$\frac{\Delta x}{\Delta s} = \cos\theta$$

or
$$\frac{(\Delta s)^2}{(\Delta x)^2} = \sec^2\theta = 1+\tan^2\theta = 1+\left(\frac{\partial y}{\partial x}\right)^2$$

giving
$$\Delta s = \left[1+\left(\frac{\partial y}{\partial x}\right)^2\right]^{1/2}\Delta x$$

Since $\frac{\partial y}{\partial x}$ is small we can binomially expand $\left[1+\left(\frac{\partial y}{\partial x}\right)^2\right]^{1/2}$ and neglect terms of order higher than $\left(\frac{\partial y}{\partial x}\right)^2$. We can then write

$$\Delta s = \left[1+\tfrac{1}{2}\left(\frac{\partial y}{\partial x}\right)^2\right]\Delta s$$

Therefore, the change in the length of the element is

$$\Delta s-\Delta x = \tfrac{1}{2}\left(\frac{\partial y}{\partial x}\right)^2\Delta x$$

The instantaneous kinetic energy of an element of length Δs and linear density μ is given by $\frac{1}{2}\mu\Delta s\left(\frac{\partial y}{\partial t}\right)^2$. Since y and its derivatives are assumed small we can write $\Delta s = \Delta x$ in this expression. The total kinetic energy of the whole string of length L is the integral of this expression over the length of the string. Thus

$$KE = \tfrac{1}{2}\mu\int_0^L\left(\frac{\partial y}{\partial t}\right)^2 dx$$

The instantaneous potential energy of the element is the work done by the tension T in increasing the length of the element from Δx to Δs. Thus

$$PE = \int_0^L T(\Delta s-\Delta x) = \frac{T}{2}\int_0^L\left(\frac{\partial y}{\partial x}\right)^2 dx$$

The total energy of a vibrating string is the sum of the energies associated with all the normal modes of the string. The energy in each mode can be obtained by substituting the value of y for that mode in the above expressions. Thus the kinetic energy of the nth mode is

$$(KE)_n = \tfrac{1}{2}\mu\int_0^L\left(\frac{\partial y_n}{\partial t}\right)^2 dx$$

and the potential energy is

$$(PE)_n = \frac{1}{2} T \int_0^L \left(\frac{\partial y_n}{\partial x} \right)^2 dx$$

The displacement in the nth mode is given by [see Eq. (6.16)]

$$y_n = A_n \sin k_n x \cos (\omega_n t + \phi_n)$$

which may be recast in the form

$$y_n = (B_n \cos \omega_n t + C_n \sin \omega_n t) \sin k_n x$$

with $B_n = A_n \cos \phi_n$ and $C_n = -A_n \sin \phi_n$. Then

$$\frac{\partial y_n}{\partial t} = \omega_n (C_n \cos \omega_n t - B_n \sin \omega_n t) \sin k_n x$$

and

$$\frac{\partial y_n}{\partial x} = k_n (B_n \cos \omega_n t + C_n \sin \omega_n t) \cos k_n x$$

$\therefore$

$$(KE)_n = \frac{1}{2} \mu \omega_n^2 (C_n \cos \omega_n t - B_n \sin \omega_n t)^2 \int_0^L \sin^2 (k_n x)\, dx$$

and

$$(PE)_n = \frac{1}{2} T k_n^2 (B_n \cos \omega_n t + C_n \sin \omega_n t)^2 \int_0^L \cos^2 (k_n x)\, dx$$

where

$$k_n = \frac{\omega_n}{v} = \omega_n \sqrt{\frac{\mu}{T}}$$

or

$$T k_n^2 = \mu \omega_n^2$$

Total energy in nth mode (E_n) is given by

$$E_n = (KE)_n + (PE)_n$$

$$= \frac{1}{4} \mu L\, \omega_n^2 (B_n^2 + C_n^2)$$

or

$$E_n = \frac{1}{4} m \omega_n^2 A_n^2 \tag{6.17}$$

where $m = \mu L$ is the mass of the string and $A_n^2 = B_n^2 + C_n^2$. A_n is the amplitude of the nth mode.

To find the exact value of E_n we would need to know the exact value of constants B_n and C_n which, as mentioned earlier, are to be evaluated from the specified initial conditions. This will be done in Sec 6 6 by using Fourier method.

The total energy of the vibrating string is given by the sum of the energies associated with each of its normal modes. Thus

$$E_{\text{total}} = \frac{1}{4}m \sum_{n=1}^{\infty} \omega_n^2 (B_n^2 + C_n^2) = \frac{1}{4} m \sum_{n=1}^{\infty} \omega_n^2 A_n^2$$

Forced Vibration of a Stretched String

So far we have studied the particularly simple case of the free vibrations of an idealized string. The string was assumed flexible so that the restoring force was due only to the transverse component of tension. The string was assumed uniform so that the mass per unit length was constant throughout the length of the string. We have seen above that the free vibrations of such a string, with both ends rigidly fixed, are strictly limited to the fundamental frequency and its integral multiples.

We shall now study the response of a string to a periodic driving force. In Chap. 4 we have studied the response of a harmonic oscillator to a periodic driving force. We found that, as along as the frequency of the driving force is not very near the natural frequency of the oscillator, the effect of friction on the motion of the oscillator can be neglected. We shall nevertheless assume that there is some friction whose effect can be neglected if we avoid describing the behaviour of the string when the frequency of the driving force is exactly equal to one of the normal mode frequencies of the string. We know, from our results in Chap. 4, that if the friction is actually absent, the forced oscillator will continue forever in the transient state and will never reach the steady state.

Consider a uniform flexible string of mass per unit length equal to μ and stretched with a tension T. The equation of motion of the string is given by Eq. (6.6). We shall imagine that the end $x = L$ is rigidly fixed and the periodic force is applied at $x = 0$ (by attaching this end to a tuning fork at shown in Fig. 6.4 (*a*) which imparts a simple harmonic motion to the end at $x = 0$ of amplitude A_0 and angular frequency ω. In other words, our boundary conditions are

$$y(0, t) = A_0 \cos \omega t$$
$$y(L, t) = 0$$

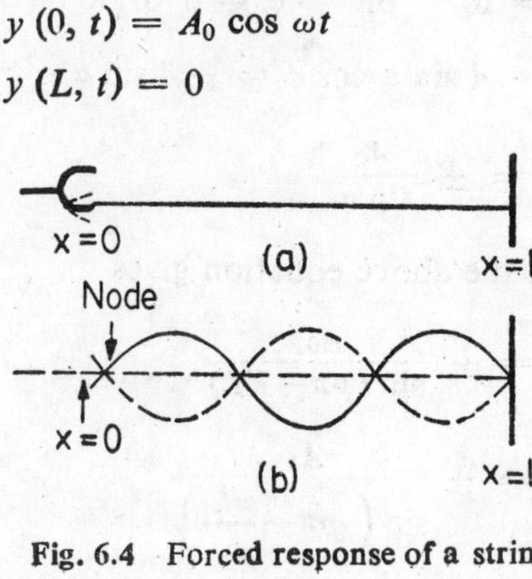

Fig. 6.4 Forced response of a string

Let us suppose that, in the steady state, the displacement of the string is given by

$$y(x, t) = A(x)\cos(\omega t + \phi)$$

where $A(x)$ is the amplitude of the various particles of the string [the amplitude at $x = 0$ is $A(0) = A_0$] and ϕ is the phase constant. Using this equation in Eq. (6.6) gives

$$\frac{d^2A(x)}{dx^2} = -\frac{\omega^2}{v^2}A(x) = -K^2A(x)$$

where $K = \frac{\omega}{v}$. Here ω is the angular frequency of the applied force and $v = \sqrt{T/\mu}$. The solution of the above equation may be written as

$$A(x) = A\sin(Kx + \alpha)$$

where A and α are constants to be determined. Thus

$$y(x, t) = A\sin(Kx + \alpha)\cos(\omega t + \phi) \tag{6.18}$$

Using the boundary condition at $x = L$ we have

$$\sin(KL + \alpha) = 0$$

or

$$KL + \alpha = p\pi \tag{6.19}$$

where p is an integer. The boundary condition at $x = 0$ gives

$$A_0\cos\omega t = A\sin\alpha\cos(\omega t + \phi)$$

or

$$(A_0 - A\sin\alpha\cos\phi)\cos\omega t + (A\sin\alpha\sin\phi)\sin\omega t = 0$$

which can be satisfied for all values of t if

$$A_0 - A\sin\alpha\cos\phi = 0$$

and

$$A\sin\alpha\sin\phi = 0$$

giving

$$\sin\phi = 0 \quad \text{or} \quad \phi = 0 \text{ or } \pi$$

and

$$A_0 = A\sin\alpha\cos\phi = \pm A\sin\alpha \tag{6.20}$$

Therefore

$$A = \pm\frac{A_0}{\sin\alpha}$$

The use of Eq. (6.19) in the above equation gives

$$A = \pm\frac{A_0}{\sin(p\pi - KL)}$$

or

$$A = \pm\frac{A_0}{\sin\left(p\pi - \frac{\omega L}{v}\right)} \tag{6.21}$$

We know that the frequencies of the normal modes of the string are given by

$$\omega_n = \frac{n\pi v}{L}$$

where n is an integer with values $= 1, 2, 3, \ldots, \infty$. Thus we have

$$A = \pm \frac{A_0}{\sin\left(p\pi - \frac{n\pi\omega}{\omega_n}\right)} \tag{6.22}$$

Notice that if $\omega \cong \omega_n$, $A \gg A_0$, This means that, for a given amplitude A_0 of the forced displacement at $x = 0$, the response of the string as a whole will be very large (since A is much larger than A_0) whenever the driving frequency ω is close to one of the normal mode frequencies, ω_n. Remember that Eq. (6.22) is not valid when ω is exactly equal to ω_n. An important feature of this result is that we can build up a large forced response with a small driving amplitude at $x = 0$. This point is not exactly a node but quite close to it as shown in Fig. 6.4*b* which shows a large amplitude response for a real string.

Another important feature is the fact that the phase ϕ of the string relative to that of the driving force is either 0 or π. In other words, the displacement (in the steady state) of any particle of the string is either in phase or 180° out of phase with the driving force. This is expected since we have assumed negligible damping. In the absence of damping, there is no energy dissipation. Therefore, the driving force is not doing any work on any moving part of the string. This implies that the displacement of every moving part of the string is either in phase or 180° out of phase with the driving force. If we let amplitude A to be either positive or negative, we need not talk about 180° phase constant at all, in which case, all moving parts vibrate in phase (with either a positive amplitude or a negative amplitude) at the same frequency. Notice that this is exactly the conditions for the existence of a normal mode of a freely oscillating undamped system.

6.3 LONGITUDINAL VIBRATIONS OF A ROD

Consider a cylindrical metal rod of uniform cross-sectional area. When the rod is struck lengthwise (or stroked) it begins to vibrate. The vibrations are in the audible range of frequencies. To find the frequencies of the normal modes of vibration of the rod, we must first obtain the equation of motion of the particles of the rod. This can be done as follows.

Suppose the rod is lying along the x-axis. We divide the rod into a large number of small slices each of length, say, Δx. Consider one such slice or element AB as shown in Fig. 6.5*a*. The particles at the end A are at a distance of $x_1 = x$ from some origin and those at end B are at a distance $x_2 = x+\Delta x$. When the rod is struck lengthwise the particles of the rod are displaced along the x-axis, i.e. along the length of the rod. In other words, the particle displacements are longitudinal rather than

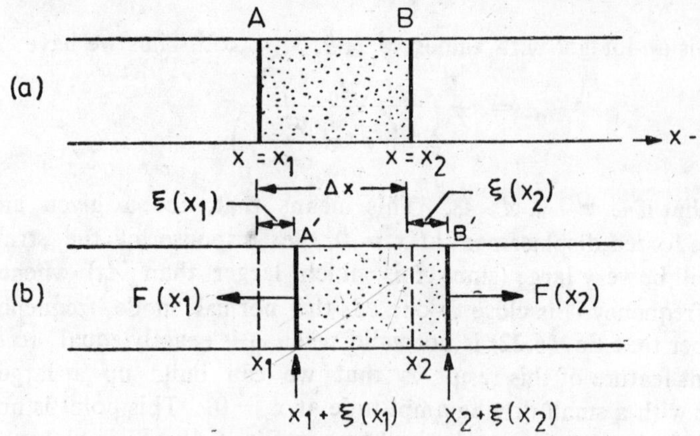

Fig. 6.5 Longitudinal vibration of a slice of the rod
(*a*) Equilibrium state (*b*) Disturbed state
The shaded portion contains the same number of particles in (*a*) and (*b*).

transverse as in the case of a string. We shall use the symbol ξ to denote the displacement, along the x-axis, of the particles in a plane at $x = x_1$ perpendicular to the axis of the rod. We shall now obtain the equation of motion of a thin slice AB of the rod which, in the undisturbed state, is contained between x and $x+\Delta x$ as shown in Fig. 6.5*a*.

Figure 6.5*b* shows the displaced position of the slice. Let the x-coordinate of the end A of the slice in the displaced position be $x_1+\xi(x_1)$ so that $\xi(x_1)$ represents the displacement of the particles at the plane x_1. Similarly let $x_2+\xi(x_2)$ be the x-coordinate the particles originally located at a plane $x = x_2$ so that $\xi(x_2)$ is the displacement of the particles at the plane x_2. Therefore,

$$\text{Change in length of the slice} = \xi(x_2)-\xi(x_1)$$

$$= \left(\frac{\partial \xi}{\partial x}\right)_{x_1} \Delta x$$

where we have used Taylor's expansion

$$\xi(x_2) = \xi(x_1) + \left(\frac{\partial \xi}{\partial x}\right)_{x_1} \Delta x + \ldots$$ and retained terms up to order (Δx), as in the case of a string. Since the original length of the slice is Δx. The longitudinal strain ε at x_1 is given by

$$\varepsilon(x_1) = \frac{\text{change in length}}{\text{original length}} = \frac{\left(\frac{\partial \xi}{\partial x}\right)_{x_1} \Delta x}{\Delta x} = \left(\frac{\partial \xi}{\partial x}\right)_{x_1}$$

If Y is the Young's modulus of the material of the rod, the stress S at x_1 is given by

$$S(x_1) = Y\left(\frac{\partial \xi}{\partial x}\right)_{x_1}$$

Similarly the stress at x_2 is given by

$$S(x_2) = Y\left(\frac{\partial \xi}{\partial x}\right)_{x_2}$$

The net stress on the element AB is

$$S(x_2)-S(x_1) = Y\,[\,f(x_2)-f(x_1)\,]$$

where $f(x) \equiv \frac{\partial \xi}{\partial x}$. Since $f(x_2)-f(x_1) = \frac{\partial f}{\partial x}\,.\,\Delta x$, we have

$$S(x_2)-S(x_1) = Y\frac{\partial f}{\partial x}\,.\,\Delta x$$

$$= Y\frac{\partial}{\partial x}\left(\frac{\partial \xi}{\partial x}\right)\Delta x$$

$$= Y\,\frac{\partial^2 \xi}{\partial x^2}\,.\,\Delta x$$

If α is the cross-sectional area of the rod, the net longitudinal force on the element is given by [see Fig. 6.5 (*b*)]

$$F = F(x_2)-F(x_1)$$

$$= \text{stress} \times \text{area} = Y\,\alpha\,\Delta x\,\frac{\partial^2 \xi}{\partial x^2}$$

To obtain the equation of motion, we apply Newton's law to the slice. If ρ is the density of the rod, the mass of the slice is $\rho\alpha\,\Delta x$. Therefore, Newton's force F' is mass $\times$ acceleration, i.e.

$$F' = \rho\alpha\,\Delta x\,\frac{\partial^2 \xi}{\partial t^2}$$

In dynamic equilibrium, the two force F and F' must balance, giving

$$\rho\alpha\,\Delta x\,\frac{\partial^2 \xi}{\partial t^2} = Y\,\alpha\Delta x\,\frac{\partial^2 \xi}{\partial x^2}$$

or
$$\frac{\partial^2 \xi}{\partial t^2} = \frac{Y}{\rho}\,\frac{\partial^2 \xi}{\partial x^2}$$

or
$$\frac{\partial^2 \xi}{\partial t^2} = v^2\,\frac{\partial^2 \xi}{\partial x^2} \tag{6.23}$$

with
$$v = \sqrt{\frac{Y}{\rho}} \tag{6.24}$$

Notice that v has the dimensions of a velocity which will be identified as the velocity of longitudinal (i.e. sound) waves travelling in the rod (see chapter 7). Also notice that Eq (6.23) is exactly similar to Eq. (6.6) except that now the displacements are longitudinal rather than transverse.

In the next chapter we shall learn that, when a rod is struck lengthwise a sound wave travels in the rod with a velocity v given by Eq. (6.24). As the wave travels along the rod, different particles acquire a displacement $\xi(x, t)$ which is a function of both x and t. It may be remarked that the picture presented above is a crude one because the particles of the rod are bouncing back and forth due to thermal agitation, even when there is no sound. The quantity $\xi(x, t)$ actually measures the *average* displacement, due to sound wave, of those particles whose *average* position was originally at x.

Normal Modes

Having obtained the equation of motion of the particles of the rod we can now proceed to find the possible normal modes of longitudinal vibrations of the rod under the given boundary conditions which we will impose a little later. Let us assume that there exists a normal mode at angular frequency ω and phase constant ϕ. This means that, in that mode, every particle of the rod executes SHM of angular frequency ω and phase constant ϕ. Thus we have [see also Eq. (6.8)]

$$\xi(x, t) = A(x) \cos (\omega t+\phi)$$

Differentiating this equation twice w.r.t. x and t and substituting these derivatives in Eq. (6.23) gives, as before,

$$\frac{d^2A(x)}{dx^2} = -k^2A(x)$$

where $k = \frac{\omega}{v}$, with v given by Eq. (6.24). The general solution of the equation is

$$A(x) = A \sin kx+B \cos kx$$

where A and B are constants to be determined from the given boundary conditions.

The general displacement $\xi(x, t)$ of the rod, in a given mode, is then given by

$$\xi(x, t) = (A \sin kx+B \cos kx) \cos (\omega t+\phi) \tag{6.25}$$

We shall now determine the frequencies and shapes of the normal modes of longitudinal vibrations of the rod under the following boundary conditions:

Rod Clamped at one end. Let the uniform rod of length L be rigidly clamped at $x = 0$ and let $x = L$ be the free end. Since the end $x = 0$ is rigidly fixed there can be no particle displacement at this end, i.e.

$$\xi(0, t) = 0$$

The condition at $x = L$ must express the fact that this end is free. This means that the particles at $x = L$ are free to vibrate and are under no stress. No neighbouring particles are pulling on the end of the rod at $x = L$. In other words,

$$\text{Stress at } x(= L) = Y\left(\frac{\partial \xi}{\partial x}\right)_{x=L} = 0$$

i.e.
$$\left(\frac{\partial \xi}{\partial x}\right)_{x=L} = 0$$

Using the first boundary condition, nomely, $\xi(0, t) = 0$ for all t in Eq. (6.25) gives

$$B = 0$$

Thus for a rod fixed at $x = 0$, Eq. (6.25) reduces to

$$\xi(x, t) = A \sin kx \cos(\omega t + \phi)$$

Differentiating this equation wrt. x, we have

$$\frac{\partial \xi}{\partial x} = Ak \cos kx \cos(\omega t + \phi) \tag{6.26}$$

The frequencies of the normal modes are obtained by using the second boundary condition, namely, $\left(\frac{\partial \xi}{\partial x}\right)_{x=L} = 0$ for all t in Eq. (6.26). This requires

$$Ak \cos kL = 0$$

or
$$A\frac{\omega}{v} \cos\left(\frac{\omega L}{v}\right) = 0$$

where $v = \sqrt{\frac{Y}{\rho}}$; Y being the Young's modulus and ρ the density of the rod. The above condition can be satisfied by choosing $A = 0$ or $\omega = 0$ which corresponds to trivial situations. Hence the only way we can satisfy the boundary condition at $x = L$ is to have

$$\cos\left(\frac{\omega L}{v}\right) = 0$$

$$\frac{\omega L}{v} = \frac{\pi}{2}, \frac{3\pi}{2}, \frac{5\pi}{2}, \ldots$$
$$= (n - \tfrac{1}{2})\pi$$

where n is an integer having values 1, 2, 3, ..., ∞. Thus

$$\omega_n = (n - \tfrac{1}{2})\frac{\pi v}{L}$$

or
$$\omega_n = (n-\tfrac{1}{2})\frac{\pi}{L}\sqrt{\frac{Y}{\rho}} \tag{6.27}$$

Equation (6.27) gives the angular frequencies of the normal modes for longitudinal vibrations of a rod fixed at one end and free at the other. The corresponding frequencies (in Hz) are given by

$$\nu_n = \frac{(n-\tfrac{1}{2})}{2L}\sqrt{\frac{Y}{\rho}} \tag{6.28}$$

In the nth mode, the displacements of the particles of the rod are given by

$$\xi_n(x, t) = A_n \sin\left\{(n-\tfrac{1}{2})\frac{\pi x}{L}\right\}\cos(\omega_n t+\phi_n) \tag{6.29}$$

since
$$\omega_n = 2\pi\nu_n \text{ and } k_n = \frac{\omega_n}{v} = (n-\tfrac{1}{2})\frac{\pi}{L}$$

The constants A_n and ϕ_n are to be determined from the prescribed initial conditions.

The lowest mode ($n = 1$) of the rod has a frequency given by

$$\nu_1 = \frac{1}{4L}\sqrt{\frac{Y}{\rho}} \tag{6.30}$$

The particle displacements in this mode are given by

$$\xi_1(x, t) = A_1 \sin\left(\frac{\pi x}{2L}\right)\cos(\omega_1 t+\phi_1)$$

Notice that ξ_1 is zero at $x = 0$ and maximum at $x = L$. There is no other value of x in the range 0 to L where ξ_1 can be zero. The shape of this mode is shown schematically in Fig. 6.6*a* which represents a graph of ξ_1 against x. Remember the displacements ξ_1 are longitudinal and not transverse.

The next mode ($n = 2$) has a frequency given by

$$\nu_2 = \frac{3}{4L}\sqrt{\frac{Y}{\rho}} = 3\nu_1$$

which is three times the frequency of the fundamental mode. The particle displacements in this mode are given by

$$\xi_2(x, t) = A_2 \sin\left(\frac{3\pi x}{2L}\right)\cos(\omega_2 t+\phi_2)$$

Notice that ξ_2 is zero at $x = 0$ as well as at $x = L/3$. At $x = L$, ξ_2 is maximum (but negative). The shape of this mode is depicted in Fig. 6.6*b*. Figure 6.6*c* shows the mode with frequency ν_3 which is $5\nu_1$. There are two modes at $x = 2L/5$ and $x = 4L/5$ in addition to one at $x = 0$. Thus, a rod fixed at one end and free at the other, has only odd harmonics of frequencies 3, 5, 7,... times the fundamental frequency. All the even harmonics of frequencies $2\nu_1$, $4\nu_1$, $6\nu_1$...are absent.

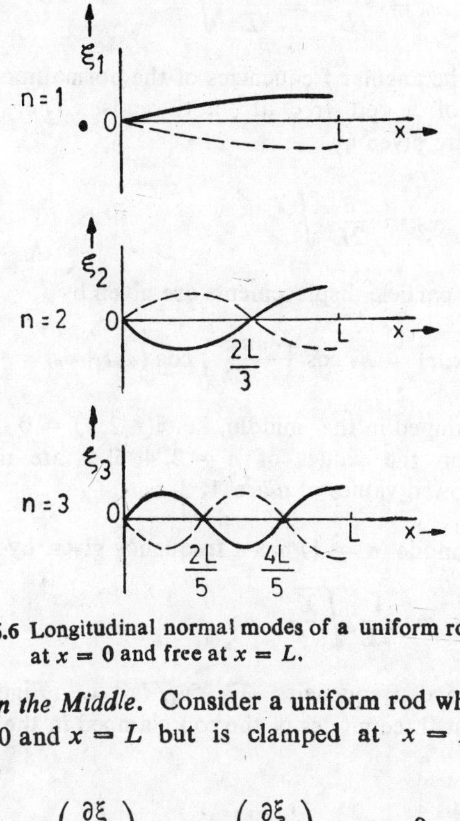

Fig. 6.6 Longitudinal normal modes of a uniform rod fixed at $x = 0$ and free at $x = L$.

Rod Clamped in the Middle. Consider a uniform rod which is free at its ends at $x = 0$ and $x = L$ but is clamped at $x = L/2$. The boundary conditions are

$$\left(\frac{\partial \xi}{\partial x}\right)_{x=0} = \left(\frac{\partial \xi}{\partial x}\right)_{x=L} = 0$$

Equation (6.25) gives the displacement of the particles of the rod in a mode of angular frequency ω and phase constant ϕ. Differentiating with respect to x we have

$$\frac{\partial \xi}{\partial x} = k(A \cos kx - B \sin kx) \cos(\omega t + \phi)$$

Using the boundary condition at $x = 0$ gives $A = 0$. Therefore

$$\frac{\partial \xi}{\partial x} = -Bk \sin kx \cdot \cos(\omega t + \phi)$$

The boundary condition at $x = L$ can be satisfied if

$$\sin kL = 0$$

or

$$kL = n\pi$$

or

$$\frac{\omega L}{v} = n\pi$$

where n is an integer having values 1, 2, 3,...., ∞. Thus

$$\omega_n = \frac{n\pi v}{L} = \frac{n\pi}{L}\sqrt{\frac{Y}{\rho}} \tag{6.31}$$

This equation gives the angular frequencies of the normal modes for longitudinal vibrations of a rod free at both ends. The corresponding frequencies (in Hz) are given by

$$\nu_n = \frac{n}{2L}\sqrt{\frac{Y}{\rho}}$$

In the nth mode the particle displacements are given by

$$\xi(x, t) = B_n \cos\left(\frac{n\pi x}{L}\right) \cos(\omega_n t + \phi_n) \tag{6.32}$$

Now the rod is clamped in the middle i.e. $\xi(L/2, t) = 0$ for all t. To satisfy this condition the values of $n = 2, 4, 6,...$ are not allowed in Eq. (6.32). The allowed values of n are 1, 3, 5, . .

The fundamental mode ($n = 1$) has a frequency given by

$$\nu_1 = \frac{1}{2L}\sqrt{\frac{Y}{\rho}}$$

The higher modes have frequencies $3\nu_1$, $5\nu_1$, $7\nu_1$, ... Figure 6.7 shows the shapes of the first three modes of the rod clamped in the middle.

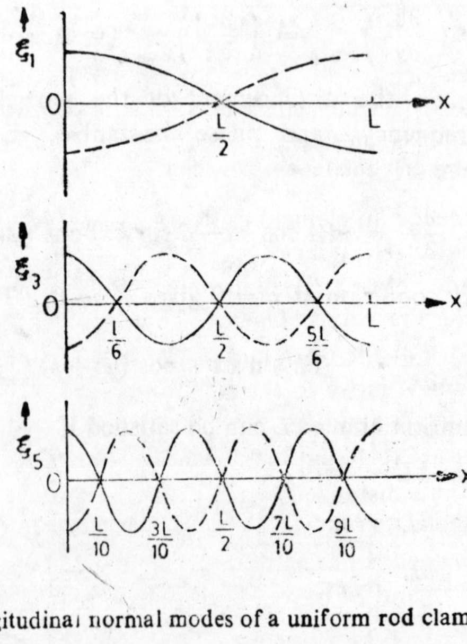

Fig. 6.7 Longitudinal normal modes of a uniform rod clamped in the middle

6.4 VIBRATIONS OF AIR COLUMNS

Air or any other gas can be regarded as a continuous medium at ordinary pressures. One cubic centimetre or air, under normal pressure, contains as many as 2.7×10^{19} molecules. The average distance between the molecules is very small and, as an approximation, we imagine the number of molecules becoming infinite and the distance between them tending to zero. A column of gas then behaves as if it were continuous.

It is clear that a column of gas represents a system almost equivalent to a solid rod. Both are continuous systems and have internal elasticity. In the case of a gas, however, the relevant modulus of elasticity is the bulk modulus or volume elasticity of the gas.

Consider a gas contained in a cylindrical tube of cross-sectional area α. Let ρ be the density of the gas. As before, we divide the column of gas into small elements or slices, each of a small length Δx. Figure 6.8*a* shows one such slice *AB* in the undisturbed state. Let P_0 be the equilibrium (atmospheric) pressure and V the volume of the slice. Thus

$$V = \alpha \Delta x$$

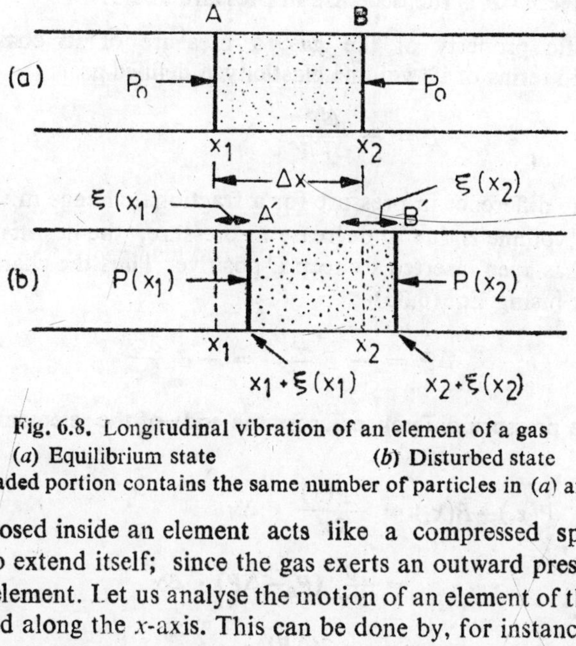

Fig. 6.8. Longitudinal vibration of an element of a gas
(*a*) Equilibrium state (*b*) Disturbed state
The shaded portion contains the same number of particles in (*a*) and (*b*).

The gas enclosed inside an element acts like a compressed spring which would like to extend itself; since the gas exerts an outward pressure on the walls of the element. Let us analyse the motion of an element of the gas when it is disturbed along the x-axis. This can be done by, for instance, placing a vibrating tuning fork at the open end of the tube containing the gas. The particles at the plane *A* move to the right and oscillate about *A*. Particles at plane *B* also oscillate about *B*.

Let the particles at *A* be at a distance $x = x_1$ from some origin and those at end *B* be at a distance $x = x_2$. The length of the element *AB* is $\Delta x = x_2 - x_1$. Figure 6.8(*b*) shows the disturbed state of the element at an instant of time when *A* has shifted to *A′* and *B* has shifted to *B′*. Let the coordinates of *A′* and *B′* be $x_1 + \xi(x_1)$ and $x_2 + \xi(x_2)$ respectively where

$\xi(x_1)$ and $\xi(x_2)$ are the displacements of the particles originally at x_1 and x_1 respectively. If at this instant of time, $\xi(x_2) > \xi(x_1)$, there is an increase in the length of the element and hence an increase in its volume. Let the increase in volume be ΔV. Then

$$\Delta V = \text{area of cross-section} \times \text{increase in length}$$

$$= \alpha \{\xi(x_2) - \xi(x_1)\}$$

$$= \alpha \frac{\partial \xi}{\partial x} \cdot \Delta x$$

Therefore the volume strain is given by

$$\text{Volume strain} = \frac{\Delta V}{V} = \frac{\alpha \frac{\partial \xi}{\partial x} \Delta x}{\alpha \Delta x} = \frac{\partial \xi}{\partial x}$$

The increase in the volume of the element is due to a decrease in pressure. Let the new pressure at A' be $P(x_1) = P_0 - \Delta P$ while that at B' is $P(x_2)$. Here ΔP is the decrease in pressure at A.

The elastic property of the gas, a measure of its compressibility, is expressed in terms of its volume elasticity E defined as

$$E = -\frac{\Delta P}{\Delta V / V}$$

which is the difference in pressure for a fractional change in volume. Since increase in volume is due to decrease in pressure, the negative sign in this definition has been inserted to keep E positive. Thus the change in pressure is given by [using Eq. (6.33)]

$$\Delta P = -E \frac{\Delta V}{V} = -E \frac{\partial \xi}{\partial x} \tag{6.34}$$

Now the pressure difference across the ends of the element $A'B'$ is given by

$$P(x_2) - P(x_1) = \frac{\partial P(x)}{\partial x} \cdot \Delta x$$

$$= \frac{\partial}{\partial x} (P_0 - \Delta P) \cdot \Delta x$$

$$= -\frac{\partial (\Delta P)}{\partial x} \cdot \Delta x$$

Substituting for ΔP from Eq. (6.34) we get

$$\text{Excess pressure} = -\frac{\partial}{\partial x}\left(-E \frac{\partial \xi}{\partial x}\right) \cdot \Delta x$$

$$= E \Delta x \frac{\partial^2 \xi}{\partial x^2}$$

Therefore the net force acting on the element is

$$F = \text{cross-sectional area} \times \text{excess pressure}$$

$$= \alpha E\,\Delta x\,\frac{\partial^2 \xi}{\partial x^2}$$

If ρ is the density of the gas, the mass of the element of length Δx is equal to $\alpha\Delta x\rho$. The Newton's force on the element is given by $\alpha\Delta x\rho\,\dfrac{\partial^2 \xi}{\partial t^2}$.

Equating the two forces gives the equation of motion of the element of the gas. Thus

$$\alpha\Delta x\rho\,\frac{\partial^2 \xi}{\partial t^2} = \alpha\Delta x E\,\frac{\partial^2 \xi}{\partial x^2}$$

or

$$\frac{\partial^2 \xi}{\partial t^2} = v^2\,\frac{\partial^2 \xi}{\partial x^2} \tag{6.35}$$

where

$$v = \sqrt{\frac{E}{\rho}} \tag{6.36}$$

The dimensions of v are

$$v = \left\{\frac{\text{dimensions of } E}{\text{dimensions of } \rho}\right\}^{1/2}$$

$$= \left\{\frac{ML^{-1}\,T^{-2}}{ML^{-3}}\right\}^{1/2} = \{LT^{-1}\}$$

Thus the constant v has the dimensions of a velocity. In the next chapter we will identify this v as the velocity of longitudinal (sound) waves in a gas.

Normal Modes

Having deduced the equation of motion of the particles of the gas we can now obtain the possible normal modes of longitudinal vibrations of the column of a gas under specified boundary conditions. As before, the particle displacements in a normal mode of angular frequency ω and phase constant ϕ are given by

$$\xi(x, t) = (A \sin kx + B \cos kx) \cos (\omega t + \phi) \tag{6.37}$$

where $k = \omega/v$ with v given by Eq. (6.36). Hence, in that mode, the variation in pressure along the tube is given by [see Eq. (6.34)]

$$\Delta P = -E\,\frac{\partial \xi}{\partial x}$$

Using Eq. (6.37) we have

$$\Delta P = kE\,(B \sin kx - A \cos kx) \cos (\omega t + \phi) \tag{6.38}$$

We shall now determine the frequencies and shapes of the normal modes of longitudinal vibrations of a column of a gas enclosed in a tube under the following boundary conditions.

Tube Closed at one end and open at the other. Consider a gas enclosed in a uniform cylindrical tube of length L lying along the x-axis. The end $x = 0$ is open while the end $x = L$ is closed. An open end represents a condition of zero pressure variation during the oscillation and maximum movement of gas particles. This implies that at the open end

$$\Delta P = 0$$

or

$$\frac{\partial \xi}{\partial x} = 0$$

In other words, the particles at the open end are under no strain and, therefore, free to move. The closed end, on the other hand, is the place of zero movement and maximum pressure variation. Thus at the closed end; $\xi = 0$. So the boundary conditions are (for all t values)

$$(\Delta P)_{x=0} = 0$$

or

$$\left(\frac{\partial \xi}{\partial x}\right)_{x=0} = 0$$

and

$$\xi(L, t) = 0$$

Using the first boundary condition in Eq. (6.38) gives

$$A = 0$$

Therefore, Eq. (6.37) reduces to

$$\xi(x, t) = B \cos kx \cos(\omega t + \phi) \tag{6.39}$$

The frequencies of the normal modes are obtained by using the second boundary condition in Eq. (6.39). This requires

$$B \cos kL = 0$$

which gives

$$\omega_n = (n - \tfrac{1}{2}) \frac{\pi}{L} \sqrt{\frac{E}{\rho}} \tag{6.40}$$

where $n = 1, 2, 3, \ldots, \infty$. The corresponding frequencies (in Hz) are given by

$$\nu_n = \frac{(n - \tfrac{1}{2})}{2L} \sqrt{\frac{E}{\rho}}$$

Figure 6 9 (*a*) shows the first three modes of the tube. Remember the particle displacements are longitudinal. Notice that, a tube closed at one end, has only odd harmonics of frequencies 3, 5, 7, ... times the fundamental frequency ν_1. All the even harmonics of frequencies $2\nu_1$, $4\nu_1$, $6\nu_1$, etc. are absent. The frequency of the fundamental mode is given by

$$\nu_1 = \frac{1}{4L}\sqrt{\frac{E}{\rho}}$$

Tube open at both ends. For a tube open at both the ends, the boundary conditions are

$$\left(\frac{\partial \xi}{\partial x}\right)_{x=0} = 0$$

$$\left(\frac{\partial \xi}{\partial x}\right)_{x=L} = 0$$

Using Eq. (6.38) we find that these conditions can be satisfied if $A = 0$ and

$$\sin kL = 0$$

or

$$kL = n\pi$$

which gives

$$\omega_n = \frac{n\pi}{L}\sqrt{\frac{E}{\rho}}$$

and

$$\nu_n = \frac{n}{2L}\sqrt{\frac{E}{\rho}}$$

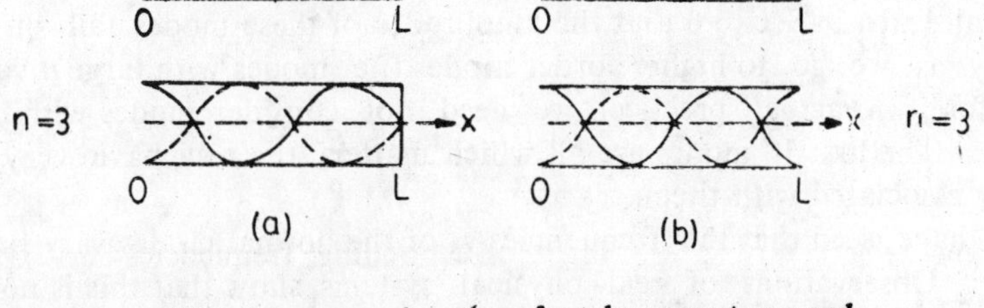

Fig. 6.9 (*a*) First three normal modes of a tube open at one end
(*b*) First three normal modes of a tube open at both ends.

where $n = 1, 2, 3, \ \ , \infty$. Figure 6.9 (*b*) shows the first three modes of the tube open at both ends. Notice that all even and odd harmonics are present. The positions where the particle displacements are maximum are called *antinodes*. The positions of particles at permanent rest are *nodes*. It may be noted that a tube with both ends closed has the same set of natural frequencies as the one with both ends open; but the positions of nodes and antinodes are interchanged.

6.5 SOME COMMENTS ON NORMAL MODES

In the preceding sections we have discussed normal modes of a few physical systems. The systems were assumed to have the following features:

1. Each system was of a limited extent and subject to boundary conditions at its ends. A system of an infinite size cannot oscillate in a normal mode. The reason is that the normal modes of a system are nothing but standing waves in the system (as we shall see in the next chapter). Standing waves are a consequence of reflections at the ends of a system. If the system is of an infinite extent, the wave travelling in it loses most of its energy before it reaches the ends of the system. Consequently there are no reflected waves to give rise to standing waves.
2. Each system was taken to be one-dimensional, i.e. the displacements of the moving parts of the system were constrained to change in one direction only. We have chosen this direction to be along the x-axis, although any other direction is equally good.
3. Each system was assumed to be continuous, which is an idealization.
4. Thermal agitations of the particles of the system were ignored.
5. The effect of gravity has been neglected. It can be shown that gravity does not modify normal mode frequencies.
6. The particle displacements were assumed small so that the restoring forces were proportional to the displacement in order that the equations governing the system were linear.

We have seen that, under these conditions, each system possesses an infinite set of distinct modes, each having its own characteristic frequency. Under suitable initial conditions each mode of the system can be excited. We shall learn in Sec. 6.6 that the amplitudes of these modes fall off very rapidly as we go to higher order modes (i.e. modes with large n values). Therefore, in actual practice, we need not consider modes with high n values. The first 10 modes are all which matter; the rest have very little energy associated with them.

We have seen that the frequencies ν_n of the normal modes vary linearly with n. Observations of real physical systems show that this is not true for modes with high n values. The reason is that a physical system is neither perfectly uniform nor strictly one-dimensional and continuous, in the true sense of the term.

We conclude our discussion of the normal modes of a continuous system by recalling the following two features of great importance.

(1) The frequencies and shapes of the normal modes of the system are determined by the boundary conditions.

(2) Any or all the normal modes of vibration can co-exist with amplitudes and phases which are determined by the prescribed initial conditions.

6.6 FOURIER METHOD: GENERAL MOTION OF A CONTINUOUS SYSTEM

If the normal modes of a system are known, the general motion of the system (under any arbitrary initial conditions) can be determined quite easily by using the technique of Fourier analysis. This technique (which is one of the most powerful techniques in mathematical physics), was developed in 1807 by a French mathematician JB Fourier. We shall use this technique to analyse the general motion of some continuous systems. The so-called Fourier method can be used to analyse the motion of any system under any arbitrary initial conditions.

Transverse Motion of a String Fixed at Both Ends

We have seen that a uniform flexible string has an infinite number of possible frequencies of vibration; and if the string is rigidly fixed at its ends, these frequencies bear a simple relationship with the fundamental frequency.

The frequencies of the higher harmonics are multiples of the fundamental frequency. If such a string is set into vibration in just the right manner, it will vibrate with just one of these frequencies. But, under arbitrary initial conditions, the general motion of the string is given by the superposition of all the normal modes. Thus the general displacement of the string is a superposition of all the normal modes displacements

$$\begin{aligned} y(x, t) &= y_1(x, t)+y_2(x, t)+\ldots+y_n(x, t) \\ &= A_1 \sin k_1x \cos(\omega_1 t+\phi_1)+A_2 \sin k_2x \cos(\omega_2 t+\phi_2)+ \\ &\quad \ldots+A_n \sin k_nx \cos(\omega_n t+\phi_n) \end{aligned}$$

or
$$y(x, t) = \sum_{n=1}^{\infty} A_n \sin k_nx \cos(\omega_n t+\phi_n)$$

where
$$k_n = \frac{n\pi}{L} \quad \text{and} \quad \omega_n = \frac{n\pi}{L}\sqrt{\frac{T}{\mu}}$$

The above equation may be recast in the following form:

$$y(x, t) = \sum_{n=1}^{\infty} \sin k_nx \,(B_n \cos \omega_n t+C_n \sin \omega_n t) \tag{6.41}$$

where the constants B_n and C_n are related to constants A_n and ϕ_n as

$$B_n = A_n \cos \phi_n$$

$$C_n = -A_n \sin \phi_n$$

Particle velocities are given by

$$V(x, t) = \frac{\partial y(x, t)}{\partial t} = \sum_{n=1}^{\infty} \omega_n \sin k_n x(-B_n \sin \omega_n t + C_n \cos \omega_n t) \tag{6.42}$$

The constants B_n and C_n in Eqs (6.41) and (6.42) are infinite arbitrary constants which are determined from the initial conditions corresponding to the infinite number of points along the string whose displacements and velocities must all be specified at time $t = 0$. We shall show below how these constants can be evaluated by the Fourier technique under any arbitrary initial displacements and velocities. Once these constants are determined, the general motion, i.e. $y(x, t)$ is determined, since the other constants, namely, k_n and ω_n in Eq. (6.41) are already known.

Before we proceed to evaluate these constants, let us make an important observation. Equation (6.41) shows that the vibrations of the string are periodic in character. This is because all overtones are harmonic, i.e. $\omega_n = n\omega_1$. So, by the time fundamental mode (of frequency ω_1) has completed one cycle, the second harmonic (of frequency $\omega_2 = 2\omega_1$) has completed two cycles and the third harmonic (of frequency $\omega_3 = 3\omega_1$) has completed three cycles and so on. Thus, during the second cycle of the fundamental, the motion is an exact repetition of the first cycle. In other words, the motion is periodic. But the motion is not harmonic because the resultant of a superposition of harmonic oscillations of different frequencies is not a harmonic oscillation.

Evaluation of Constants B_n and C_n by Fourier Analysis. Let $y(x, 0) = y_0(x)$ and $V(x, 0) = \left(\frac{\partial y}{\partial t}\right)_{t=0} = V_0(x)$ respectively be the displacements and velocities of all particles of the string (with x in the range 0 to L) at time $t = 0$. Then from Eqs (6.41) and (6.42) we have $\left(\because k_n = \frac{n\pi x}{L}\right)$

$$y_0(x) = \sum_{n=1}^{\infty} B_n \sin k_n x = \sum_{n=1}^{\infty} B_n \sin\left(\frac{n\pi x}{L}\right) \tag{6.43}$$

and

$$V_0(x) = \sum_{n=1}^{\infty} C_n \omega_n \sin k_n x = \sum_{n=1}^{\infty} C_n \omega_n \sin\left(\frac{n\pi x}{L}\right) \tag{6.44}$$

The initial conditions must satisfy these two equations. The right-hand sides of these equations are called *Fourier series*. An important feature of the series is that they have a spatial periodicity, i.e. the function $y_0(x)$ or $V_0(x)$ repeats itself in space. The spatial periodicity is $\lambda = 2L$ since, as shown below, $y_0(x)$ is recovered if x changes from a value x to a value $x+\lambda$ where $\lambda = 2L$. This is because

$$\sin \frac{n\pi}{L}(x+\lambda) = \sin \frac{n\pi}{L}(x+2L) = \sin\left(\frac{n\pi x}{L}+2n\pi\right) = \sin\left(\frac{n\pi x}{L}\right)$$

It may be remarked that the initial displacement shape $y_0(x)$ and the velocity profile $V_0(x)$ can be any sort of functions of x; the only restriction is that they go to zero at $x = 0$ and $x = L$, where the string is rigidly fixed.

We shall now evaluate constants B_n and C_n. Suppose we wish to find out these constants at a particular value of n, say, m where, like n, m is an integer having values 1, 2, 3, ..., ∞. All that we have to do is to multiply both sides of Eqs (6.43) and (6.44) by $\sin\left(\frac{m\pi x}{L}\right)$ and integrate over x in the range zero to L. That is,

$$\int_0^L y_0(x) \sin\left(\frac{m\pi x}{L}\right) dx = \sum_{n=1}^{\infty} B_n \int_0^L \sin\left(\frac{n\pi x}{L}\right)\sin\left(\frac{m\pi x}{L}\right)dx$$

Using the trignometric identity

$$\sin\alpha \sin\beta = \tfrac{1}{2}\left[\cos(\alpha-\beta)-\cos(\alpha+\beta)\right]$$

The integral on the RHS of the above equation may be written as

$$\int_0^L \sin\left(\frac{n\pi x}{L}\right)\sin\left(\frac{m\pi x}{L}\right)dx = \frac{1}{2}\int_0^L \cos\left\{(n-m)\frac{\pi x}{L}\right\}dx$$

$$-\tfrac{1}{2}\int_0^L \cos\left\{(n+m)\frac{\pi x}{L}\right\}dx$$

$$= \frac{L}{2(n-m)\pi}\sin\left\{(n-m)\frac{\pi x}{L}\right\}\Big|_0^L - \frac{L}{2(n+m)\pi}\sin\left\{(n+m)\frac{\pi x}{L}\right\}\Big|_0^L$$

If we insert the limits $x = 0$ and $x = L$, the values of

$$\frac{L}{2(n+m)\pi}\sin\left\{(n+m)\frac{\pi x}{L}\right\}\Big|_0^L$$

are all zero for all integral values of n and m. But the value of

$$\frac{L}{2(n-m)\pi}\sin\left\{(n-m)\frac{\pi x}{L}\right\}\Big|_0^L$$

is zero for $n \neq m$ but is indeterminate (of the form $\frac{0}{0}$) for $n = m$, in which case the value is $L/2$. This can be seen as follows: For $n = m$, the integral to be evaluated is

$$\int_0^L \sin^2\left(\frac{m\pi x}{L}\right) dx = \frac{1}{2}\int_0^L \left\{1-\cos\left(\frac{2\pi mx}{L}\right)\right\} dx = \frac{L}{2}$$

Since the cosine term contributes nothing at the two limits. Thus we arrive at the following identity

$$\int_0^L y_0(x) \sin\left(\frac{m\pi x}{L}\right)dx = B_m \frac{L}{2}$$

i.e.
$$B_m = \frac{2}{L}\int_0^L y_0(x) \sin\left(\frac{m\pi x}{L}\right) dx$$

which is the same as

$$B_n = \frac{2}{L}\int_0^L y_0(x) \sin\left(\frac{n\pi x}{L}\right) dx \tag{6.45}$$

Since n (like m) is an integer having values 1, 2, 3, ..., ∞.

Similarly C_n is given by

$$C_e = \frac{2}{L\omega_n}\int_0^L V_0(x) \sin\left(\frac{n\pi x}{L}\right) dx \tag{6.46}$$

Before proceeding any further, let us collect our results at one place for convenience and future reference [see Eqs (6.41), (6.45) and (6.46)]

$$\left.\begin{aligned} y(x, t) &= \sum_{n=1}^{\infty} \sin\left(\frac{n\pi x}{L}\right)(B_n \cos \omega_n t + C_n \sin \omega_n t) \\ B_n &= \frac{2}{L}\int_0^L y_0(x) \sin\left(\frac{n\pi x}{L}\right) dx \\ C_n &= \frac{2}{L\omega_n}\int_0^L V_0(x) \sin\left(\frac{n\pi x}{L}\right) dx \end{aligned}\right\} \tag{6.47}$$

where $\omega_n = \dfrac{n\pi v}{L}$ with $v = \sqrt{\dfrac{T}{\mu}}$; T is the tension and μ is the linear density of the string and n = 1, 2, 3, ..., ∞.

We shall now take some specific forms of $y_0(x)$ and $V_0(x)$ and determine the general motion of a string fixed at both ends.

Plucked String

A string is said to be 'plucked' if it is given a finite initial displacement but zero initial velocity. Let us consider a simple example which will indicate how the Fourier method works. Suppose the string is plucked (or pulled) at the centre through a distance h and released at time $t = 0$ (see Fig. 6.10). Thus, at $t = 0$, the velocity of the string is zero i.e. $V_0(x) = 0$ which gives

$$C_n = 0$$

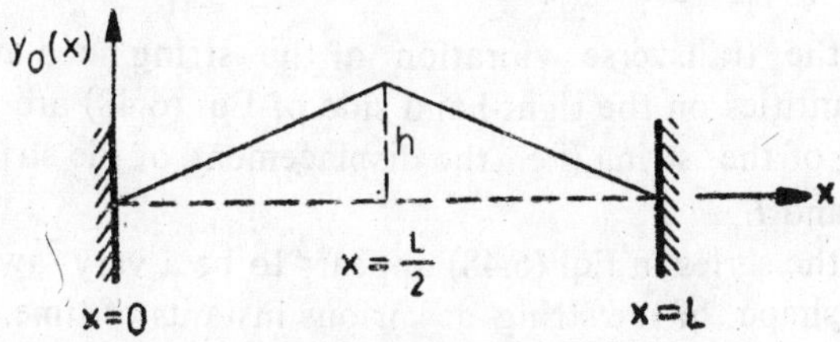

Fig. 6.10 String plucked at the centre

The initial displacement shape of the string is given by (see Fig. 6.10)

$$y_0(x) = \begin{cases} \dfrac{2hx}{L} \ ; 0 \leqslant x \leqslant \dfrac{L}{2} \\ \dfrac{2h}{L}(L-x) \ ; \dfrac{L}{2} \leqslant x \leqslant L \end{cases}$$

Substituting this in Eq. (6.45) we have

$$B_n = \frac{2}{L}\left[\int_0^{L/2} \frac{2hx}{L}\sin\left(\frac{n\pi x}{L}\right) dx + \int_{L/2}^{L} \frac{2h}{L}(L-x)\sin\left(\frac{n\pi x}{L}\right) dx\right]$$

Integration by parts gives (after some simplifications)

$$B_n = \frac{8h}{n^2\pi^2}\sin\left(\frac{n\pi}{2}\right)$$

$$\begin{cases} \text{zreo; if } n \text{ is an even integer} \\ =(-1)^{n-1/2} \dfrac{8h}{\pi^2 n^2} \ ; \text{if } n \text{ is an odd integer} \end{cases}$$

Therefore, the general motion of the string is given by [see Eq. (6.47)]

$$y(x, t) = \sum_{n=1}^{\infty} B_n \sin\left(\frac{n\pi x}{L}\right) \cos \omega n t$$

$$= \frac{8h}{\pi^2} \sum_{n=1}^{\infty} \frac{1}{n^2} \sin\left(\frac{n\pi}{2}\right) \sin\left(\frac{n\pi x}{L}\right) \cos \omega_n t$$

or

$$y(x, t) = \frac{8h}{\pi^2}\left[\sin\left(\frac{\pi x}{L}\right)\cos \omega_1 t - \frac{1}{9}\sin\left(\frac{3\pi x}{L}\right)\cos \omega_3 t + \frac{1}{25}\sin\left(\frac{5\pi x}{L}\right)\cos \omega_5 t - \ldots\right] \quad (6.48)$$

where $\omega_1 = \frac{\pi}{L}\sqrt{\frac{T}{\mu}}$, $\omega_3 = 3\omega_1$, $\omega_5 = 5\omega_1$, etc. are the normal mode frequencies of the transverse vibration of the string fixed at both ends. Since all the quantities on the right-hand side of Eq. (6.48) are known, the general motion of the string (i.e. the displacement of the string) is determined for all x and t.

At first sight, the series in Eq. (6.48) appears to be a very awkward way of finding the shape of the string at various instants of time. However, the series solution gives us a very valuable information about the motion of the string. Firstly, it tells us that all the even harmonics (second, fourth, sixth, etc) will be absent from the sound emitted by the string, for they are not present in the motion. Secondly, it tells us that the amplitude of the third harmonic is 1/9 that of the fundamental, the amplitude of the fifth harmonic is 1/25 that of the fundamental, etc. In other words, the'

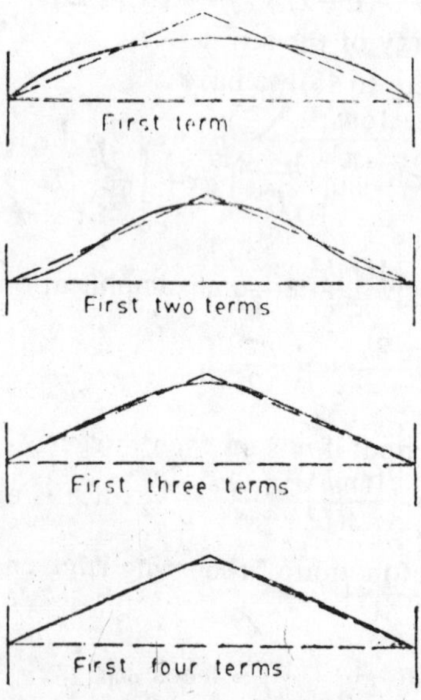

Fig 6.11 Fourier series representation of the initial shape of the string given by Eq. (6.48) at time $t = 0$

intensity of the third harmonic is 1/81 of that of the fundamental, the intensity of the fifth harmonic is 1/625 of that of the fundamental, and so on. Thus, only the first few terms in the series contribute to displacement or energy of the string. From Eq. (6.48) the value of $y(x, t)$ can be computed for all x and t. Fig. 6.11 shows how the correct form of the string is approached at a certain instant as more terms in the series are included. The first four terms are enough to describe the shape of the string at that instant.

Which harmonics are absent depend on where the string is plucked. If the string is plucked at $x = L/4$, one can show that the fourth, eighth, etc. harmonics will be absent. We shall learn in the next chapter that the absent harmonics correspond to the standing waves that have a node at the point where the string is plucked. The general rule (which can be proved by computing the value of B_n) is that, in the motion of any plucked string, all those harmonics are absent which have a node at the point where the string is plucked.

Total energy of a string plucked at the centre. We have seen that the total energy of a vibrating string is given by (see page 253)

$$E_{\text{total}} = \frac{1}{4} m \sum_{n=1}^{\infty} \omega_n^2 \, (B_n^2 + C_n^2)$$

Now, for a string plucked at the centre, $C_n = 0$ and

$$B_n = \frac{8h}{n^2\pi^2} \sin\left(\frac{n\pi}{2}\right)$$

Hence, the total energy of the string is given by

$$E_{\text{total}} = \frac{16mh^2}{\pi^2} \sum_{n=1}^{\infty} \frac{\omega_n^2}{n^4} \sin^2\left(\frac{n\pi}{2}\right)$$

$$= \frac{16mh^2}{\pi^4} \frac{\pi^2 v^2}{L^2} \sum_{n=1}^{\infty} \frac{1}{n^2} \sin^2\left(\frac{n\pi}{2}\right)$$

$$\left[\because \omega_n = \frac{n\pi v}{L} \right]$$

$$= \frac{16mh^2v^2}{\pi L^2} \left(\frac{1}{1^2} + \frac{1}{3^2} + \frac{1}{5^2} + \ldots - \right)$$

But $\dfrac{1}{1^2} + \dfrac{1}{3^2} + \dfrac{1}{5^2} + \ldots = \dfrac{\pi^2}{8}$

Hence

$$E_{\text{total}} = \frac{2mv^2h^2}{L^2} = \frac{2Th^2}{L} \qquad \left[\because v = \sqrt{\frac{T}{\mu}},\ m = \mu L \right]$$

Thus the vibrational energy of the string is directly proportional to the tension and the square of the distance through which the centre of the string is plucked, and inversely proportional to the length of the string.

Struck String

A string is said to be 'struck' if it is given a finite initial velocity but zero initial displacement. Consider a uniform string stretched taut between its ends $x = 0$ and $x = L$. Suppose the string is struck (with a hammer) at $x = L/4$ keeping the portion between $x = L/2$ to $x = L$ at rest. Let us suppose that at time $t = 0$ the hammer imparts a velocity to the string which increases linearly from zero at $x = 0$ to value V_0 at $x = L/4$ and decreases linearly to zero at $x = L/2$. The velocity profile of such a string is shown in Fig. 6.12. Thus, the initial displacement $y_0(x) = 0$, giving all $B_n = 0$ and the initial velocity $V_0(x)$ is given by

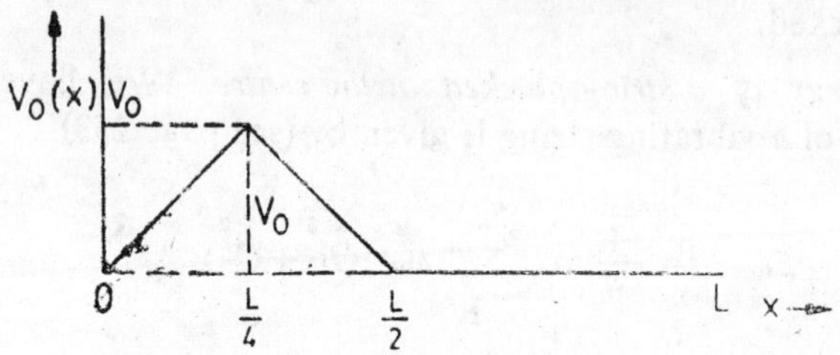

Fig. 6.12 String struck at $x = L/4$

$$V_0(x) = \begin{cases} \dfrac{4V_0 x}{L} & ; 0 \leqslant x \leqslant L/4 \\[2ex] \dfrac{4V_0}{L}\left(\dfrac{L}{2} - x\right) & ; \dfrac{L}{4} \leqslant x \leqslant L/2 \\[2ex] 0 & ; \dfrac{L}{2} \leqslant x \leqslant L \end{cases}$$

Substituting this $V_0(x)$ in Eq. (6.46) and integrating by parts, we get

$$\left(\because \quad \omega_n = \frac{n\pi v}{L}\right)$$

$$C_n = \frac{2}{n\pi v}\left[\int_0^{L/4} \frac{4V_0 x}{L} \sin\left(\frac{n\pi x}{L}\right) dx + \int_{L/4}^{L/2} \frac{4V_0}{L}\left(\frac{L}{2} - x\right) \sin\left(\frac{n\pi x}{L}\right) dx\right]$$

$$= \frac{16 V_0 L}{v(n\pi)^3} \sin\left(\frac{n\pi}{4}\right)\left\{1 - \cos\left(\frac{n\pi}{4}\right)\right\}$$

Notice that $C_n = 0$ for $n = 4, 8, 12, \ldots$ etc. Hence the general motion of the string is given by [see Eq. (6.47)]

$$y(x, t) = \frac{16V_0L}{v\pi^3} \sum_{n=1}^{\infty} \frac{1}{n^3} \sin\left(\frac{n\pi}{4}\right)\left\{1 - \cos\left(\frac{n\pi}{4}\right)\right\} \sin\left(\frac{n\pi x}{L}\right) \times \sin\left(\frac{n\pi vt}{L}\right)$$

$$= \frac{8V_0L}{v\pi^3}\left[(\sqrt{2}-1) \sin\frac{\pi x}{L} \cdot \sin\frac{\pi vt}{L} + \frac{1}{4} \sin\frac{2\pi x}{L} \sin\frac{2\pi vt}{L} + \frac{1}{27}(\sqrt{2}+1) \sin\frac{3\pi x}{L} \sin\frac{3\pi vt}{L} - \frac{1}{125}(\sqrt{2}+1) \sin\frac{5\pi x}{L} \sin\frac{5\pi vt}{L} + \ldots\right]$$

Notice that the fourth, eighth, etc. harmonics (those having a node at $x = L/4$) are absent in this case.

The more general case in which both the displacement and velocity are finite at time $t = 0$ can also be discussed; but it is much more complicated and is not of much interest.

We have discussed above how Fourier analysis can be used to analyse the transverse vibration of strings. The method can quite generally be used to deal with any bounded uniform continuous system. For example, the method will apply for longitudinal vibrations of rods and gas columns equally well (see Example 6.11).

SOLVED EXAMPLES

Example 6.1 A steel wire is stretched between two clamps 100 cm apart with a tension which produces an extension of 0.608 cm in it. Calculate the fundamental frequency of its transverse vibration. Density of steel = 7600 kgm^{-3} and its Young's modules = 2×10^{11} Nm^{-2}. Which overtone of the string will correspond to a frequency of 800 Hz?

Solution

Let L be the length of the wire and ΔL the extension produced when it is stretched with a tension T. Then Y, the Young's modulus of the wire is given by

$$Y = \frac{\text{stress}}{\text{strain}} = \frac{T/\alpha}{\Delta L/L} = \frac{TL}{\alpha \Delta L}$$

where α is the cross-sectional area of the wire. Thus

$$T = \alpha Y \frac{\Delta L}{L} \qquad (i)$$

Now the fundamental frequency of the wire is given by

$$\nu_1 = \frac{1}{2L}\sqrt{\frac{T}{\mu}} \tag{ii}$$

But μ (= mass per unit length) = $\alpha\rho$, where ρ is the density of the wire. Using this μ and Eq. (*i*) in Eq. (*ii*), we get

$$\nu_1 = \frac{1}{2L}\sqrt{\frac{Y}{\rho}\frac{\Delta L}{L}}$$

Putting the numerical values (in SI units), we have

$$\nu_1 = 200 \text{ Hz}$$

Evidently the fourth harmonic (or the third overtone) has a frequency of 800 Hz.

Example 6.2 Two strings A and B of the same material, cross-sectional area and length are fixed at their ends and subjected to tensions in the ratio of 2.89 : 1 respectively. When the strings are vibrated, 8 beats per second are heard between the third harmonic of string A and the fifth harmonic of string B. Calculate the fundamental frequency of each string.

Solution

Let T be the tension in string B. Then the tension in string A is 2.89 T. The frequency of the third harmonic of string A is given by

$$\nu_3 = \frac{3}{2L}\sqrt{\frac{2.89\,T}{\mu}} = \frac{5.1}{2L}\sqrt{\frac{T}{\mu}}$$

The frequency of the fifth harmonic of string B is given by

$$\nu_5' = \frac{5}{2L}\sqrt{\frac{T}{\mu}}$$

But $$\nu_3 - \nu_5' = 8$$

or $$\frac{5.1}{2L}\sqrt{\frac{T}{\mu}} - \frac{5}{2L}\sqrt{\frac{T}{\mu}} = 8$$

whence $$\frac{1}{L}\sqrt{\frac{T}{\mu}} = 160$$

∴ The fundamental frequency of string A is

$$\nu_1 = \frac{1}{2L}\sqrt{\frac{2.89\,T}{\mu}} = \frac{1.7}{2L}\sqrt{\frac{T}{\mu}} = 136 \text{ Hz}$$

The fundamental frequency of string B is

$$\nu_1' = \frac{1}{2L}\sqrt{\frac{T}{\mu}} = 80 \text{ Hz}$$

Example 6.3 A wire of length 100 cm and of mass 1.25 g is stretched with a tension of 100 N. If the wire is set into vibration and touched lightly with a feather at a point one-third of its length from one end, calculate the frequency of the note emitted.

Solution

It is evident that the wire is vibrating in the third harmonic whose frequency is given by

$$\nu_3 = \frac{3}{2L}\sqrt{\frac{T}{\mu}}$$

where $T = 100$ N, $L = 1$ m and $\mu = 1.25 \times 10^{-3}$ kg m^{-1}. Substituting these values, we get

$$\nu_3 \simeq 424 \text{ Hz}$$

Example 6.4 Show that the total energy of a string vibrating in the nth mode is equal to $\frac{1}{4}\, mV^2_{max}$ where V_{max} is the maximum velocity at an antinode.

Solution

From Eq. (6.16) the particle displacements in the nth mode are given by

$$y_n = A_n \sin k_n x \cos (\omega_n t + \phi_n)$$

At an antinode $\sin k_n x = 1$, therefore at these positions

$$y_n = A_n \cos (\omega_n t + \phi_n)$$

and particle velocity is

$$V_n = \frac{\partial y_n}{\partial t} = -A_n\omega_n \sin (\omega_n t + \phi_n)$$

Now, from Eq. (6.17) the total energy of the string in the nth mode is given by

$$E_n = \frac{1}{4} m A_n^2 \omega_n^2$$

But $A_n\omega_n = V_{max}$, the maximum velocity at the antinode. Hence

$$E_n = \frac{1}{4}\, mV^2_{max}$$

Example 6.5 A uniform brass rod of length 3 m is clamped at the centre. It emits a note of frequency 600 Hz when it is vibrated longitudinally. If the density of brass is 8000 kgm^{-3}, calculate its Young's modulus.

Solution

The fundamental frequency of a rod clamped at the centre is given by

$$\nu_1 = \frac{1}{2L}\sqrt{\frac{Y}{\rho}}$$

$$\therefore \quad Y = 4L^2\rho\nu_1^2$$

$$= 4\times(3)^2\times 8000\times(600)^2$$

$$= 10.37\times 10^{10}\ \text{Nm}^{-2}.$$

Example 6.6 Two open pipes of lengths 100 cm and 105 cm produce 5 beats in 6 s when each is sounding its fundamental note. Calculate the frequencies of the two notes.

Solution

The frequencies of the fundamental notes in the two pipes are given by

$$\nu_1 = \frac{1}{2L}\sqrt{\frac{E}{\rho}} = \frac{v}{2L} = \frac{v}{2} \qquad (\because L = 1\text{m})$$

and

$$\nu_1' = \frac{v}{2L'} = \frac{v}{2\times 1.05} \qquad (\because L' = 1.05\ \text{m})$$

where

$$v = \sqrt{\frac{E}{\rho}}$$

$\therefore$ Beat frequency $\nu_b = \nu_1 - \nu_1' = \frac{v}{2}\left(1 - \frac{1}{1.05}\right) = \frac{0.05v}{2\times 1.05}$

But $\nu_b = 5/6$. Therefore

$$\frac{0.05v}{2\times 1.05} = \frac{5}{6}$$

whence

$$v = 350\ \text{ms}^{-1}$$

The frequencies of the two notes are

$$\nu_1 = \frac{v}{2L} = 175\ \text{Hz}$$

$$\nu_1' = \frac{v}{2L'} = 166.7\ \text{Hz}.$$

Example 6.7 Two cylindrical pipes of the same length, but one closed and the other open are sounded together. The frequency of the second overtone in the closed pipe is 300 Hz higher than that of the first overtone of the open pipe. Calculate the fundamental frequency of the closed pipe.

Solution

The frequencies of the normal modes of a closed pipe are $\left(v = \sqrt{\frac{E}{\rho}}\right)$

$$\nu_n = \frac{\left(n-\frac{1}{2}\right)v}{2L}$$

while those of an open pipe are

$$\nu'_n = \frac{nv}{2L}$$

The frequency of the second overtone ($n = 3$) of the closed pipe is

$$\nu_3 = \frac{7v}{4L}$$

The frequency of the first overtone ($n = 2$) of the open pipe is

$$\nu_2' = \frac{v}{L}$$

Since the two overtones differ by 300 Hz, we have

$$\frac{7v}{4L} - \frac{v}{L} = 300$$

$$\therefore \qquad \frac{v}{L} = 400$$

The frequency of the fundamental mode ($n = 1$) of the closed pipe is

$$\nu_1 = \frac{v}{4L} = 100 \text{ Hz.}$$

Example 6.8 A uniform rod of length L is fixed at one end ($x = 0$) and subjected to a sinusoidal force $F = F_0 \sin \sigma t$ at the free end ($x = L$) as shown in Fig. 6.13. Show that the steady-state displacement for forced oscillations of the rod is given by

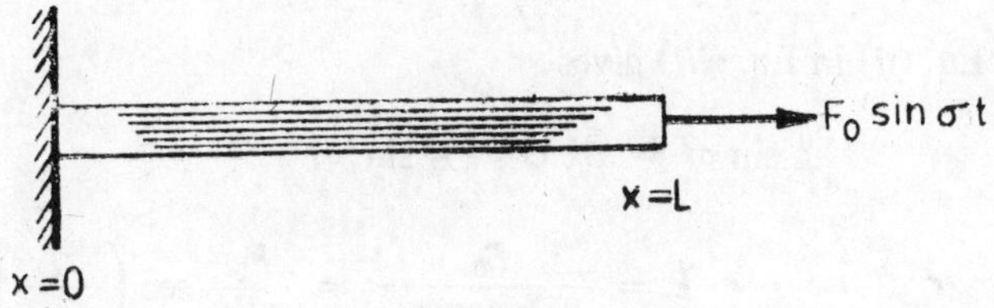

Fig. 6.13

$$\xi(x, t) = \frac{F_0 v}{\alpha Y \sigma} \sec\left(\frac{\sigma L}{v}\right) \sin\left(\frac{\sigma x}{v}\right) \sin \sigma t$$

where α is the cross-sectional area and Y the Young's modulus of the rod and $v = \sqrt{\frac{Y}{\rho}}$; ρ being the density of the rod.

Solution

The equation of motion of the rod in longitudinal vibrations is

$$\frac{\partial^2 \xi}{\partial t^2} = v^2 \frac{\partial^2 \xi}{\partial x^2}.$$

Let the steady-state displacement of the forced vibrations be

$$\xi(x, t) = A(x) \sin \sigma t$$

The solution of $A(x)$ may be written as

$$A(x) = A \cos kx + B \sin kx$$

where

$$k = \sigma/v$$

Hence

$$\xi(x, t) = (A \cos kx + B \sin kx) \sin \sigma t$$

Since the rod is fixed at $x = 0$; $\xi(0, t) = 0$ which gives $A = 0$. Therefore

$$\xi(x, t) = B \sin kx \sin \sigma t \qquad (i)$$

At the free end $(x = L)$, the strain due to longitudinal vibrations is $\dfrac{\delta\xi}{\delta x}$.

Hence the internal force at this end is $\alpha Y \dfrac{\partial \xi}{\partial x}$ which must equal the externally applied force $F_0 \sin \sigma t$. Thus the boundary condition at $x = L$ is

$$\alpha Y \left(\frac{\partial \xi}{\partial x}\right)_{x=L} = F_0 \sin \sigma t \qquad (ii)$$

Now from Eq. (*i*) we have

$$\frac{\partial \xi}{\partial x} = Bk \cos kx \sin \sigma t \qquad (iii)$$

Using Eq. (*ii*) in Eq. (*iii*) gives

$$\frac{F_0}{\alpha Y} \sin \sigma t = Bk \cos kL \sin \sigma t$$

or

$$B = \frac{F_0}{\alpha Y k \cos kL} = \frac{F_0 v}{\alpha Y \sigma} \sec\left(\frac{\sigma L}{v}\right)$$

$$\left(\because k = \frac{\sigma}{v}\right)$$

Substituting for B in Eq. (*i*) we get

$$\xi(x, t) = \frac{F_0 v}{\alpha Y \sigma} \sec\left(\frac{\sigma L}{v}\right) \sin\left(\frac{\sigma x}{v}\right) \sin \sigma t$$

Resonance will occur (i.e. $\xi \to \infty$) if $\sec\left(\dfrac{\sigma L}{v}\right) \to \infty$ or

$$\cos\left(\frac{\sigma L}{v}\right) \to 0$$

or
$$\frac{\sigma L}{v} = \left(n-\frac{1}{2}\right)\pi$$

or
$$\sigma = \left(n-\frac{1}{2}\right)\frac{\pi v}{L}$$

But $(n-\frac{1}{2})\frac{\pi v}{L} = \omega_n$ [see Eq. (6.27)] where ω_n is the angular frequency of the nth normal mode of the rod. Therefore, the resonance condition is

$$\sigma = \omega_n.$$

Hence, if the frequency of the applied force equals the frequency of any of the normal modes of the rod, the amplitude of the forced vibrations becomes theoretically infinity. If the effect of friction were included, the amplitude would become large (not infinite) at resonance.

Example 6.9 A bar of Length L and mass M is fixed at one end and carries a mass m at the other end as shown in Fig. 6.14. Derive the equation which determines the normal mode frequencies of the system. Show that; if $m \ll M$ the frequency equation reduces to the equation of normal mode frequencies of a rod fixed at one end and free at the other.

Solution

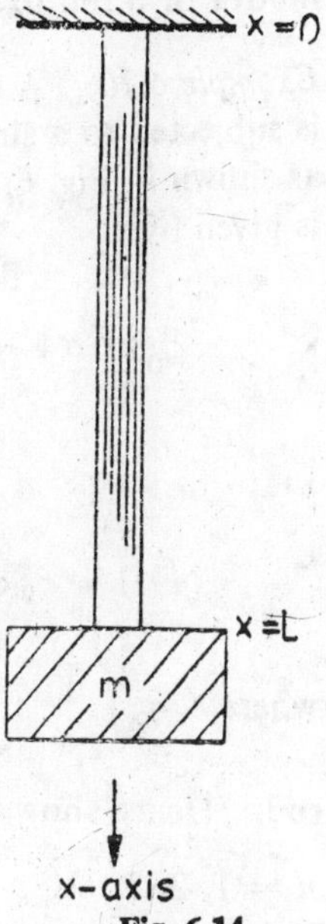

Fig. 6.14

The general displacement of the rod in the nth normal mode is given by

$$\xi(x, t) = (A_n \sin k_n x + B_n \cos k_n x)\cos(\omega_n t + \phi_n)$$

where $\qquad k_n = \omega_n/v$ with $v = \sqrt{\dfrac{Y}{\rho}}$

Since the rod is fixed at $x = 0$; $\xi(0, t) = 0$ which gives $B_n = 0$. Therefore

$$\xi(x, t) = A_n \sin k_n x \cos(\omega_n t + \phi_n) \qquad (i)$$

At the free end $x = L$, the force $\alpha Y \dfrac{\partial \xi}{\partial x}$ due to vibrations must be equal and opposite to Newton's force $=$ mass $\times$ acceleration of the mass m, i.e. $m\dfrac{\partial^2 \xi}{\partial t^2}$. Thus the boundary condition at the free end is

$$\alpha Y\left(\frac{\partial \xi}{\partial x}\right)_{x=L} = -m\left(\frac{\partial^2 \xi}{\partial t^2}\right)_{x=L} \qquad \text{(ii)}$$

From Eq. (*i*) we have

$$\left(\frac{\partial \xi}{\partial x}\right)_{x=L} = A_n k_n \cos k_n L \cos(\omega_n t + \phi_n)$$

and $$\left(\frac{\partial^2 \xi}{\partial t^2}\right)_{x=L} = -A_n\omega_n^2 \sin k_nL \cos(\omega_n t+\phi_n)$$

Substitution of these in Eq. (*ii*) gives

$$\alpha Y k_n \cos(k_nL) = m\omega_n^2 \sin(k_nL)$$

or $$\frac{L\alpha Y k_n}{v m\omega_n} = \omega_n \frac{L}{v} \tan(k_nL)$$

Now since $k_n = \frac{\omega_n}{v}$, $v = \sqrt{\frac{Y}{\rho}}$ and $M = \rho\alpha L$ we have

$$\frac{M}{m} = \omega_n \frac{L}{v} \tan\left(\frac{\omega_n L}{v}\right)$$

This equation determines the angular frequencies ω_n of the normal modes of the system.

If $m \ll M$; $\tan\left(\frac{\omega_n L}{v}\right) \to \infty$

i.e., $$\frac{\omega_n L}{v} = \frac{\pi}{2}, \frac{3\pi}{2}, \frac{5\pi}{2}, \ldots = \left(n-\frac{1}{2}\right)\pi$$

or $$\omega_n = \left(n-\frac{1}{2}\right)\frac{\pi v}{L} = \left(n-\frac{1}{2}\right)\frac{\pi}{L}\sqrt{\frac{Y}{\rho}}$$

which agrees with Eq. (6.27) for the angular frequencies of the normal modes of a rod fixed at one end and free at the other.

Example 6.10 A uniform rod of length L is free at one end ($x = L$) and is subjected to a sinusoidal displacement $\xi_0 \sin \sigma t$ at the other end ($x = 0$) as shown in Fig. 6.15. Show that the steady-state displacement of the rod is given by

Fig. 6.15

$$\xi(x, t) = \xi_0 \cos\left(\frac{\sigma x}{v}\right)\left[1+\tan\left(\frac{\sigma L}{v}\right)\tan\left(\frac{\sigma x}{v}\right)\right] \sin \sigma t$$

where $v = \sqrt{\frac{Y}{\rho}}$; Y being the Young's modulus and ρ the density of the rod. Hence show that resonance will occur if $\sigma = (n-\frac{1}{2})\frac{\pi v}{L}$; $n = 1, 2, 3, \ldots$.

Solution

The equation of motion for longitudinal vibration of the rod is given by

$$\frac{\partial^2 \xi}{\partial t^2} = v^2 \frac{\partial^2 \xi}{\partial x^2}$$

where ξ is the displacement at any cross-section of the bar.

In the steady state, the general form of the solution will be

$$\xi(x, t) = A(x) \sin \sigma t$$

Using this in the equation of motion we have

$$\frac{d^2 A(x)}{dx^2} = -k^2 A(x), \text{ with } k = \frac{\sigma}{v}$$

$$\therefore \qquad A(x) = A \sin kx + B \cos kx$$

Hence $\qquad \xi(x, t) = (A \sin kx + B \cos kx) \sin \sigma t$

Now the boundary conditions for this problem are

$$\xi(0, t) = \xi_0 \sin \sigma t$$

and $$\left(\frac{\partial \xi}{\partial x}\right)_{x=L} = 0$$

The first boundary condition is satisfied if $B = \xi_0$, The second boundary is satisfied if

$$k(A \cos kL - B \sin kL) = 0$$

or $\qquad A = B \tan kL = \xi_0 \tan kL$

Therefore, the steady-state displacement is given by

$$\xi(x, t) = (\xi_0 \tan kL \sin kx + \xi_0 \cos kx) \sin \sigma t$$

$$= \xi_0 \cos\left(\frac{\sigma x}{v}\right)\left[1 + \tan\left(\frac{\sigma L}{v}\right) \tan\left(\frac{\sigma x}{v}\right)\right] \sin \sigma t$$

Resonance occurs (i.e. ξ becomes theoretically infinity) if

$$\tan\left(\frac{\sigma L}{v}\right) = \infty$$

or $$\frac{\sigma L}{v} = \left(n - \frac{1}{2}\right)\pi \; ; \; n = 1, 2, 3, \ldots$$

or $$\sigma = \left(n - \frac{1}{2}\right)\frac{\pi v}{L}$$

Example 6.11 A uniform rod of length L is fixed at both ends. The rod is set into longitudinal vibrations by giving it a constant velocity V_0 at time $t = 0$ at all points along the rod in the x-direction. Find the subsequent longitudinal vibration of the rod.

Solution

Since the rod is fixed at both ends. Eq. (6.47) apply to the rod. They are

$$\xi(x, t) = \sum_{n=1}^{\infty} \sin\left(\frac{n\pi x}{L}\right)\left[B_n \cos\left(\frac{n\pi vt}{L}\right)+C_n \sin\left(\frac{n\pi vt}{L}\right)\right]$$

$$B_n = \frac{2}{L}\int_0^L \xi_0(x) \sin\left(\frac{n\pi x}{L}\right) dx$$

$$C_n = \frac{2}{n\pi v}\int_0^L V_0(x) \sin\left(\frac{n\pi x}{L}\right) dx$$

where $v = \sqrt{\dfrac{Y}{\rho}}$; Y is the Young's modulus and ρ is the density of the rod.

The initial conditions are

$$\xi(x, 0) = \xi_0(x) = 0 \text{ for all } x\text{; giving } B_n = 0$$

and $$V_0(x) = V_0 \text{ for all } x$$

Therefore,

$$C_n = \frac{2}{n\pi v}\int_0^L V_0 \sin\left(\frac{n\pi x}{L}\right) dx = \frac{2V_0L}{(n\pi)^2 v}(1-\cos n\pi)$$

$$= \frac{4V_0L}{(n\pi)^2 v} \qquad \text{when } n = 1, 3, 5, \ldots$$

$$= 0 \qquad \text{when } n = 2, 4, 6, \ldots$$

Thus the resulting motion of the bar is given by (for n = odd)

$$\xi(x, t) = \frac{4V_0L}{\pi^2 v}\sum_{n=1}^{\infty}\frac{1}{n^2}\sin\left(\frac{n\pi x}{L}\right)\sin\left(\frac{n\pi vt}{L}\right)$$

$$= \frac{4V_0L}{\pi^2 v}\left[\sin\left(\frac{\pi x}{L}\right)\sin\left(\frac{\pi vt}{L}\right)+\frac{1}{9}\sin\left(\frac{3\pi x}{L}\right)\sin\left(\frac{3\pi vt}{L}\right) + \ldots\right]$$

Notice that all the even harmonics are absent.

Example 6.12 A string of length L is tightly stretched with a tension T between its two fixed ends $x = 0$ and $x = L$. A hammer blow is given to a small part of length b of the string at a distance a from the end $x = 0$; as a result of which the part b of the string has an initial velocity V_0, the rest of the string is initially undisturbed. Show that the subsequent vibration of the string is given by

$$y(x, t) = \frac{4V_0L}{\pi^2 v}\sum_{n=1}^{\infty}\frac{1}{n^2}\sin\left(\frac{n\pi a}{L}\right)\sin\left(\frac{n\pi b}{2L}\right)\sin\left(\frac{nx\pi}{L}\right)\sin\left(\frac{n\pi vt}{L}\right)$$

where $v = \sqrt{\dfrac{T}{\mu}}$; μ being the linear density of the string. Assume that the tension T remains constant.

Solution

The vibration of the string is given by Eq. (6.47). The initial conditions for this problem are (see Fig. 6.16)

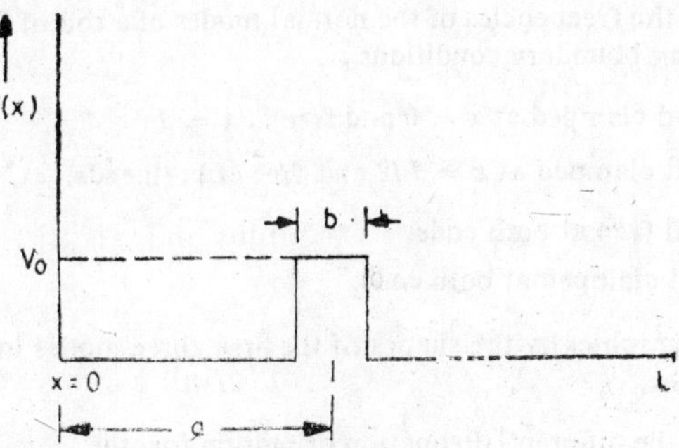

Fig. 6.16

$$y_0(x) = 0 \text{ for all } x \text{ ; giving } B_n = 0$$

$$V_0(x) = V_0 \text{ from } x = a - \frac{b}{2} \text{ to } x = a + \frac{b}{2}$$

Using this value of $V_0(x)$ we have

$$C_n = \frac{2}{n\pi v}\int_{a-b/2}^{a+b/2} V_0 \sin\left(\frac{n\pi x}{L}\right) dx$$

$$= \frac{2V_0L}{(n\pi)^2 v}\left| -\cos\left(\frac{n\pi x}{L}\right)\right|_{a-b/2}^{a+b/2}$$

$$= \frac{4V_0L}{(n\pi)^2 v}\sin\left(\frac{n\pi a}{L}\right)\sin\left(\frac{n\pi b}{2L}\right)$$

Hence,

$$y(x, t) = \frac{4V_0L}{\pi^2 v}\sum_{n=1}^{\infty}\frac{1}{n^2}\sin\left(\frac{n\pi a}{L}\right)\sin\left(\frac{n\pi b}{2L}\right)\sin\left(\frac{n\pi x}{L}\right)\sin\left(\frac{n\pi vt}{L}\right)$$

QUESTIONS

1. (a) Derive the differential equation of motion for the transverse vibrations of a uniform flexible stretched string.

   (b) Obtain an expression for the frequencies of the normal modes of the string which is rigidly fixed at its ends.

   (c) Sketch graphically the shapes of first three modes of the string.

(d) Show that the total energy of the string vibrating in the nth mode is equal to $\frac{1}{4}mV_n^2$ where m is the mass of the string and V_n the maximum velocity at an antinode.

(e) Discuss the resonant response of the string driven at $x = 0$ by a force which imparts harmonic displacement at $x = 0$. Neglect damping.

2. (a) Derive the differential equation of motion for the longitudinal vibrations of a uniform rod.

(b) Obtain the frequencies of the normal modes of a rod of length L under the following boundary conditions :

(i) Rod clamped at $x = 0$ and free at $x = L$.

(ii) Rod clamped at $x = L/2$ and free at both ends.

(iii) Rod free at both ends.

(iv) Rod clamped at both ends.

(c) Sketch graphically the shapes of the first three-modes in each of the above four cases.

3. (a) Derive the differential equation of motion for the longitudinal vibrations of air in a pipe.

(b) Obtain the frequencies of the normal modes of a pipe of length L under the following boundary conditions:

(i) Pipe closed at one end and open at the other.

(ii) Pipe open at both ends.

(c) Sketch graphically the shapes of the first three modes in the two cases.

4. A uniform string of length L and linear density μ is stretched with a tension T between the fixed ends $x = 0$ and $x = L$. Obtain the equation for transverse displacement $y(x,t)$ of the string which is set into vibration under the following initial conditions:

(i) The string is plucked at $x = a$ through a transverse height b at time $t = 0$ and released so that the initial conditions are: $V_0(x) = 0$, and

$$y_0(x) = \begin{cases} \dfrac{bx}{a} & ; 0 \leqslant x \leqslant a \\ \dfrac{b(L-x)}{(L-a)} & ; a \leqslant x \leqslant L \end{cases}$$

(ii) The string is struck at the centre ($x = L/2$) so as to give a velocity which varies linearly from zero at $x = 0$ to V_0 at $x = L/2$, while the rest of the string is initially at rest.

5. (a) Deduce an expression for the total energy of a vibrating string which is fixed at its ends and initially displaced a distance h at the centre.

(b) Show that, in the absence of damping, this energy equals the potentia energy of the string in the displaced state at time $t = 0$.

6. Using energy considerations, derive the equation $\dfrac{\partial^2 y}{\partial t^2} = v^2 \dfrac{\partial^2 y}{\partial x^2}$ for a flexible string stretched with a constant tension.

PROBLEMS

1. A uniform steel wire of length L metre and mass m kg m^{-1} is stretched with a load of mass mL^2 kg. If the ends of the wire are fixed, calculate the frequency of the fundamental.

2. A steel wire is stretched between two clamps 100 cm apart, with a tension which produces an extension of 0.038 cm. The fundamental frequency of transverse vibrations of the wire is 50 Hz. If the density of steel is 7600 kg m^{-3}, calculate its Young's modulus.

3. Two strings A and B of the same material, cross-sectional area and length are subjected to tensions in the ratio of 2.56 : 1. When the strings are vibrated together 8 beats per second are heard between the third harmonic of string A and the fifth harmonic of string B. What is the beat frequency if each string were vibrating in the fundamental mode?

4. A string of length 100 cm is stretched with a force of 2 N. It vibrates in the fundamental mode with a maximum amplitude of 1 cm. Calculate the energy of the vibrating string.

5. A stone hangs freely from one end of a sonometer wire whose vibrating length, when tuned to a tuning fork, is 100 cm. When the stone hangs wholly immersed in water, the vibrating length of the wire has to be reduced to 90 cm in order to tune it to the same fork. Calculate the relative density of the stone.

6. An iron bar of density 7700 kg m^{-3} and length 100 cm, when clamped in the middle, and stroked, emits a note which is in resonance with 16.2 cm length of a sonometer wire. A fork of frequency 540 Hz resonates with 72.6 cm of the same wire under the same tension. Find the frequency of the note and Young's modulus of iron.

7. (a) A uniform rod of length L is fixed at one end and connected at the other end by a spring of constant K as shown in Fig. 6.17. Show that the frequencies (ω_n) of the longitudinal vibrations of the rod are given by the equation

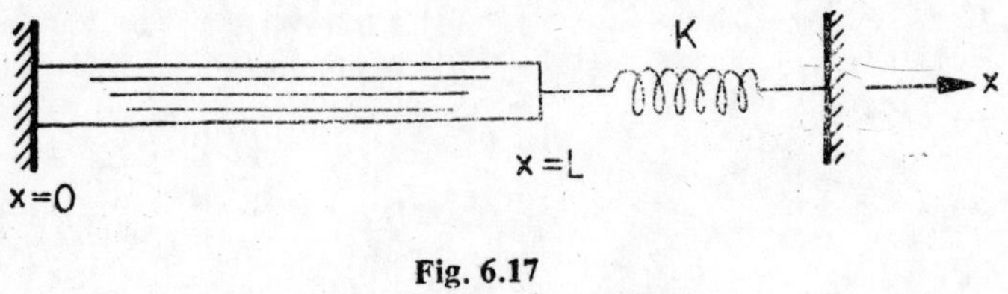

Fig. 6.17

$$\tan\left(\frac{\omega_n L}{v}\right) = -\frac{\omega_n M}{LK}$$

where M = mass of the rod and $v = \sqrt{\dfrac{Y}{\rho}}$; Y being the Young's modulus and ρ the density of the rod.

(Hint: The boundary condition at the end $x = L$ is $\alpha Y\left(\dfrac{\partial \xi}{\partial x}\right)_{x=L} = -K(\xi)_{x=L}$

where α is the cross-sectional area of the rod)

(b) Show that, if K is negligibly small, the frequency equation reduces to the one for longitudinal vibrations of a uniform rod fixed at one end and free at the other.

8. A uniform flexible string of length L and linear density μ is stretched between its two fixed ends ($x = 0$ and $x = L$) with a tension T. The string is plucked at $x = L/3$ with an amplitude h and released. Obtain the equation for the subsequent motion of the string.

9. A uniform flexible string of length L and linear density μ is stretched between its two fixed ends $x = 0$ and $x = L$ with a tension T. if the string is given an initial displacement

$$y_0(x) = y_0 \sin\left(\frac{\pi\lambda}{L}\right)$$

find the equation for the subsequent motion of the string.

10. A uniform flexible string of length L and linear density μ is stretched with a tension T between its two fixed ends at $x = 0$ and $x = L$. The string is struck at the centre so as to obtain an initial velocity which varies linearly from zero at the ends to V_0 at the centre. Show that the resulting motion of the string is given by

$$y(x,t) = \frac{8V_0L}{\pi^3 v}\sum_{n=1}^{\infty}\frac{1}{n^3}\sin\left(\frac{n\pi}{2}\right)\sin\left(\frac{n\pi x}{L}\right)\sin\left(\frac{n\pi vt}{L}\right)$$

where $v = \sqrt{\dfrac{T}{\mu}}$

7

Wave Motion

7.1 INTRODUCTION

There are essentially two ways of transporting energy from the place where it is produced to the place where it is desired to be utilized. The first involves the actual transport of matter. For example, a bullet fired from a gun carries its kinetic energy with it which can be used at another location. The second method by which energy can be transported is much more useful and important; it involves what we call a wave process. The wave carries energy but there is no transport of matter. When a drummer beats a drum its sound is heard at distant places. The sound carries energy as it can move the diaphragm of the ear. When a stone is dropped in the still water in a pond, water waves move steadily along the water until they reach the shore. If there is a small floating object, such a piece of cork, it will move up and down near its own location, which indicates that the molecules of water do not move along with the wave. When a bulb is switched on the room is filled with light. Light waves also carry energy. It is possible to transmit an electric signal (or a message) from one place to another. Although these various processes of transport of energy are different yet they have a common feature which we shall call wave motion.

The key word in wave motion is disturbance or perturbation. The reason why a variety of wave motions are possible is that there are various ways of disturbing the physical state of a body. In the case of water waves, the disturbance is the change in the position of water molecules relative to the equilibrium state (flat surface) brought about by a stone. For sound waves, the changes of pressure can be regarded as a disturbance. For this disturbance to travel from one point of a medium to another, the particles of the medium must be coupled to one another by some force so that the disturbance created at one point can be handed down to its neighbours.

It is essential, at this stage to clarify the meaning of the word 'disturbance'. The scope of its meaning need not be limited, in a narrow sense, to mean physical displacement of particles of a medium. In fact, a material medium is not even necessary if some physical property of space can exist in vacuum. We know that electric and magnetic fields can exist in vacuum. Disturbance, in this case, could be a perturbation (or change) in these fields. The disturbance travels in vacuum, needing no material medium for its propagation. Electromagnetic waves are an example of such a wave.

7.2 WAVES IN A CONTINUOUS MEDIUM

The systems we discussed in Chap. 6 were *closed* or *bounded* systems, i.e. systems enclosed by definite boundaries so that all the energy remains within the system. We found that the general motion of such systems was given by a superposition of the normal modes. The frequencies and shapes of the modes are determined by the boundary conditions. In this chapter we shall consider oscillations of *open* or *unbounded* systems, i.e. systems having no outer boundaries. We shall show that, if such a system is disturbed, waves travel in the system with a speed which is determined by the properties of the system. Strictly speaking, our results will hold for systems of infinite extent so that the waves are not reflected back into the system. In practice, however, an open system need not be of infinite extent. A bounded system will behave as an open system if the waves travelling in the system are, somehow, absorbed at the boundaries so that no reflected waves are present in the system. In the next chapter we will show that, if reflected waves are present in the system, standing or stationary waves are formed. It will be shown that the standing waves in a bounded system are nothing but the normal modes of the system discussed in Chap. 6.

We shall now consider forced oscillations of open systems. The waves generated by a driving force (that is coupled to the open system) are called *travelling waves*; these waves travel away from the point where the driving force produces the disturbance. If the driving force produces a harmonic disturbance in the system, the travelling waves it produces are called *harmonic* travelling waves. In the steady state, all the moving parts of the system oscillate with simple harmonic motion at the driving frequency. We shall be dealing with only harmonic waves.

Travelling Waves in a Long Stretched String

Hold one end of a long string whose other end is fixed to a peg in the wall (Fig. 7.1). Move your hand suddenly in the upward direction and bring it back to the original position. A kink (or pulse) is created in the

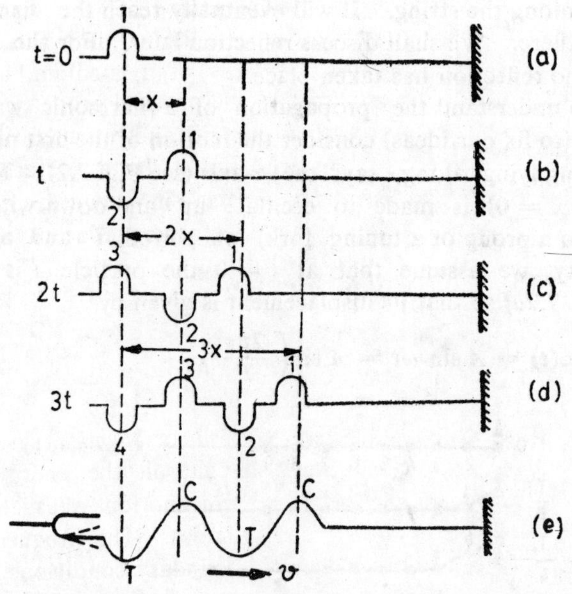

Fig. 7.1 Travelling waves on an infinitely long string.

string near your hand and you will see it move along the length of the string. The particles of the string simply move up and down as the pulse passes by. The pulse will be seen to move with a finite velocity. Let the pulse labelled *1* be created at time $t = 0$ (see Fig. 7.1*a*). It travels a distance x in time t when a second pulse labelled 2 is created by moving the hand in the downward direction and bringing it back to its original position (Fig. 7.1b). Whatever may be the shape or size, the second pulse will travel the same distance x in time t. The velocity of each pulse is $v = x/t$. At time $2t$ another pulse labelled 3, of the same shape as pulse *1*, is produced and so on (Fig. 7.1*c*). The sequence till time $3t$ is shown in Fig. 7.1*d*.

On the basis of these observations, the following two conclusions can be drawn: (*i*) the pulse travels on the string carrying energy with it but the string does not bodily move with it and no matter is transported; and (*ii*) the velocity of the pulse is independent of the shape or size of the pulse.

We shall now discuss how harmonic waves are produced. You will notice that a succession of pulses were created in the string at regular interval of time and this was easily accomplished by jerking our hand upward or downward at regular intervals. Now suppose we create pulses on the string so that there is no time interval between them. We do not wait between two pulses. This can easily be done by tying the free end of the string to a prong of a tuning fork and setting it into harmonic vibration. The result is depicted in Fig. 7.1*e*. Crests (marked *C*) and troughs (marked *T*) will be seen to move along the string. A harmonic wave travels along the string. It will eventually reach the fixed end and get reflected there. We shall discuss reflection later. For the moment we assume that no reflection has taken place.

In order to understand the propagation of a harmonic wave on the string, let us (to fix our ideas) consider the motion of the first nine particles of a long string lying along, say, the x-axis (see Fig. 7.2). The particle labelled 1 (at $x = 0$) is made to oscillate up and down with SHM (by attaching it to a prong of a tuning fork) of period T and amplitude A. For simplicity, we assume that at $t = 0$ the particle *1* is at its mean position (Fig. 7.2*a*) so that its displacement is given by

$$\psi(t) = A \sin \omega t = A \sin\left(\frac{2\pi t}{T}\right)$$

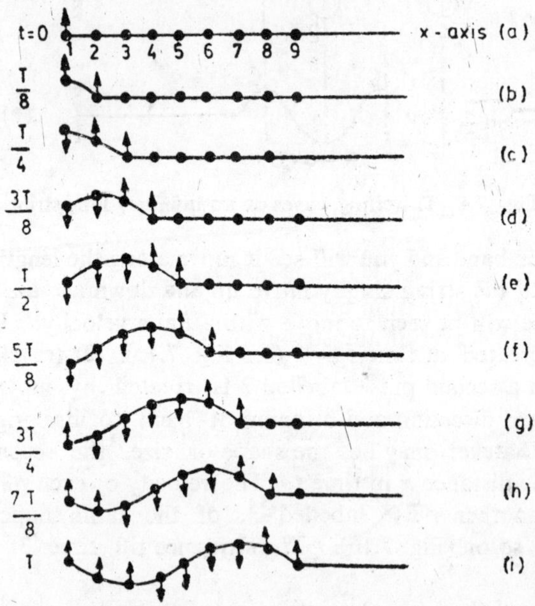

Fig. 7.2 State of particle oscillations and wave propagation.

This disturbance created at particle 1 will be transmitted to particles 2, 3, 4,..., 9. The particles 1 to 9 are identical and equispaced. Let the interparticle spacing be such that the disturbance takes time $T/8$ to travel from one particle to the next. The state of motion of the particles at time,

$$t = 0, \quad \frac{T}{8}, \frac{T}{4}, \frac{3T}{8}, \frac{T}{2}, \frac{5T}{8}, \frac{3T}{4}, \frac{7T}{8} \text{ and } T$$

is depicted in Fig. 7.2. The spacing between particles has been exaggerated for convenience. The arrows indicate the direction in which the particle will move the next moment. At $t = 0$, particle 1 is at its mean position and will move up the next moment. At $t = T/8$, the particle *1* is displaced by

$$\psi\left(t=\frac{T}{8}\right)=A\sin\frac{2\pi}{T}\cdot\frac{T}{8}=A\sin\frac{\pi}{4}=\frac{A}{\sqrt{2}}$$

and particle 2 has just felt the disturbance and will move up the next instant. At $t = T/4$, particle 1 is displaced by

$$\psi\left(t=\frac{T}{4}\right)=A\sin\frac{2\pi}{T}\cdot\frac{T}{4}=A$$

particle 2 is displaced by $A/\sqrt{2}$ and particle 3 is about to go up.

At $t = T/2$, particle 1 having completed half its cycle is again at its mean position but now having a tendency to move down ; the disturbance created at particle 1 having reached as far as the fifth particle. At $t = 5T/8$, the displacement of 1 is

$$\psi\left(t=\frac{5T}{8}\right)=A\sin\frac{5\pi}{4}=-\frac{A}{\sqrt{2}}$$

At time $t = T$ when particle 1 has completed one full vibration and is passing through its mean position upward, the disturbance has reached particle 9 which will have a tendency to move up.

In an actual string the particles are very close to one another. Figure 7.3 shows how the string will look like at $t = T$, $2T$ and $3T$ as the particle 1 completes one, two and three vibrations respectively. We notice that as time passes new crests and troughs (labelled C and T respectively) are created giving the impression that something is travelling along the string. A wave is said to travel on the string.

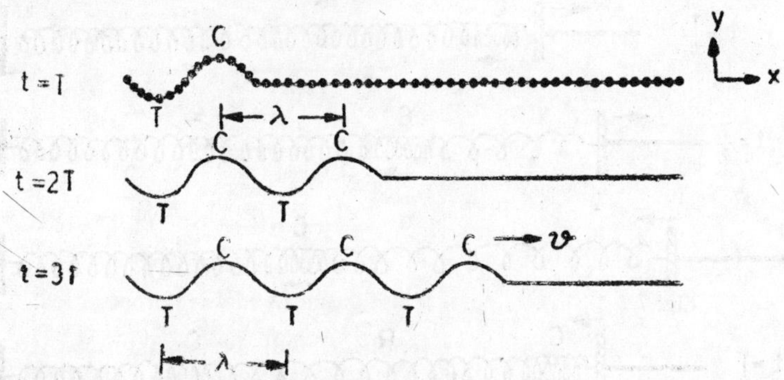

Fig. 7.3 Travelling harmonic wave

Notice that the particles execute SHM (along the y-axis) at their own location (i.e. the same x value) with the same period and amplitude while the wave travels along the x-direction. But the state of motion (i.e phase) of particle oscillation is different for different particles. This is the crucial point. Such a wave where the displacements or oscillations are transverse (i.e. perpendicular) to the direction of wave propagation is called a *transverse wave*.

We notice in Fig. 7.2*i* that the two nearest particles in the same state of motion (i.e. having the same phase) are particles 1 and 9. They are both at their mean positions and are moving up. The phase of particle 5 is opposite to that of *1* or *9* as it is moving down. The *wavelength* (denoted by λ) of a wave is defined as the distance, measured along the direction of wave propagation, between two nearest points which are in the same state of vibration (see Fig. 7.3). Referring again to Fig. 7.2*i* we find that the wavelength (λ) is just the distance travelled by the wave during one time period (T) of particle oscillation. Hence the wave velocity v is given by

$$v = \frac{\lambda}{T} = \nu\lambda \tag{7.1}$$

where $\nu = 1/T$ is the frequency of particle oscillations This relation between wave velocity, frequency and wavelength holds also for *longitudinal waves* in which the displacements or oscillations in the medium are parallel to the direction of wave propagation. The waves in a spring (discussed below) are longitudinal. The most familiar example of a longitudinal wave is a sound wave.

Waves on a Spring

Take a long spring one end of which is attached to a piston which can

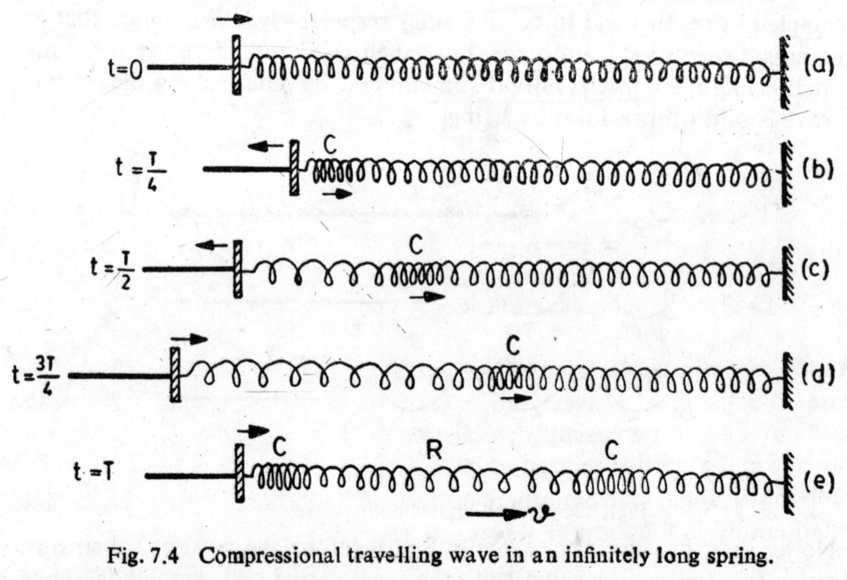

Fig. 7.4 Compressional travelling wave in an infinitely long spring.

move back and forth; the other end being fixed to the wall (see Fig. 7.4). The piston is moving back and forth with SHM of period T. At time $t = 0$ the piston is about to move to the right. As it moves, it compresses the coils of the spring immediately next to it. As time passes this compression labelled C travels along the spring but the coils only vibrate back and forth. When the piston moves to the left, it extends the coils producing a rarefaction R which also moves along the coil following the compression. We say that a compressional (longitudinal) harmonic wave travels on the spring. The wavelength of the wave is the distance between two successive compressions or rarefactions. Sound waves are also compressional. When we speak, the particles of air next to our mouth get displaced producing a compression. Thus a compressional wave (analogous to the wave on a spring) travels in air.

7.3 MATHEMATICAL DESCRIPTION OF HARMONIC WAVE: THE WAVE EQUATION

We have learnt that wave motion (transverse or longitudinal) is a result of oscillations of the moving parts of a continuous system. Figure 7.5 shows a graph of the displacements $\psi(x, t)$ of different particles against x at a given time. The length of the arrows represents the displacements of particles and the curve is the locus of displacements in a continuous medium. This is what we observe as waves. The crests and troughs

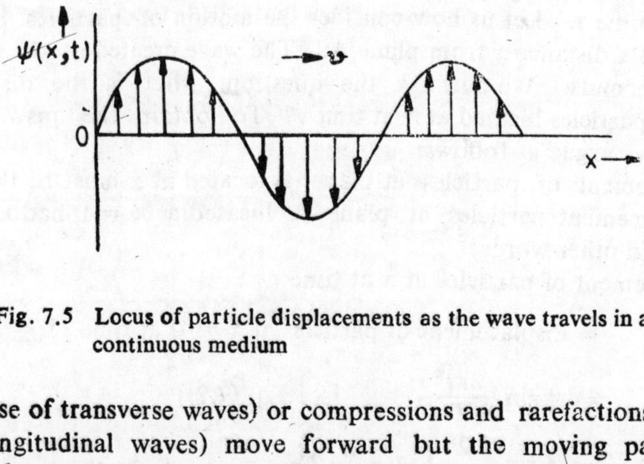

Fig. 7.5 Locus of particle displacements as the wave travels in a continuous medium

(in the case of transverse waves) or compressions and rarefactions (in the case of longitudinal waves) move forward but the moving parts of the medium do not; they simply oscillate at their own location. What we observe is the state of motion of the moving parts which have their own motion which differs from others only in phase. We observe their phase relationships in the form of waves.

We shall deal with *plane waves* only which travel in one direction. Take a plane perpendicular to the direction of wave propagation. In that plane all the particles have the same phase. In the next plane the phase of particle oscillations is different. In formulating such a wave motion in mathematical terms we shall have to relate the phase difference between particle oscillations at any two planes to their physical separation in space. This relation is called the *wave equation*.

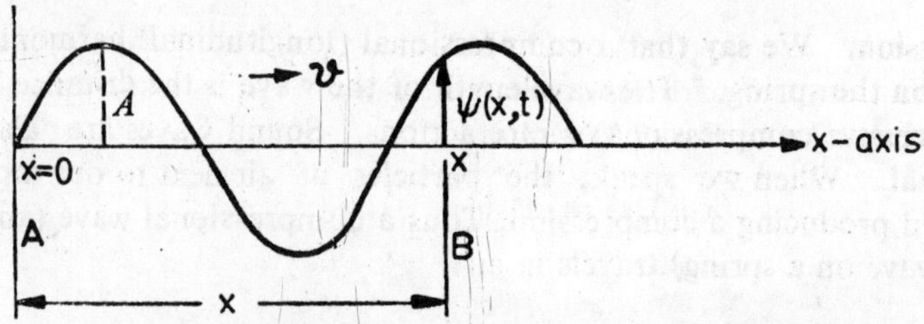

Fig. 7.6 Phase relationship between the motion of particles at planes A and B separated by a distance x as a harmonic waves travels in a medium in the $+x$ direction with a velocity v.

We will now obtain the wave equation for one-dimensional plane waves. Consider a harmonic wave travelling in a medium with a velocity v. Consider two planes A and B separated by a distance x (Fig. 7.6). Let us assume that the wave is created at $x = 0$ (plane A) by giving harmonic oscillations of amplitude A and period T to particles at this plane. Let the displacement of particles at plane A be given by

$$\psi(0, t) = A \sin \frac{2\pi t}{T} \tag{7.2}$$

where $\psi(0, t)$ stands for the displacement of particles located at plane A at $x = 0$ at time t. Let us now consider the motion of particles located at plane B at a distance x from plane A. The wave created at A will reach B in x/v seconds. We now ask the question: what is the displacement $\psi(x, t)$ of particles located at x at time t? To obtain the answer to this question we argue as follows:

Displacement of particles at plane B located at x must be the same as the displacement particles at plane A located at $x = 0$ had x/v seconds earlier. In other words;

Displacement of particles at x at time t

$$= \text{displacement of particles at } x = 0 \text{ at time } t^* = t - \frac{x}{v}$$

$$= A \sin \frac{2\pi t^*}{T} \qquad \text{[see Eq. (7.2)]}$$

$$= A \sin \left\{ \frac{2\pi}{T} \left(t - \frac{x}{v} \right) \right\}$$

or $$\psi(x, t) = A \sin \left\{ \frac{2\pi}{T} \left(t - \frac{x}{v} \right) \right\} \tag{7.3}$$

This equation gives the displacements of the particles of the continuous medium as a function of x and t as a harmonic wave travels in the $+x$ direction with a velocity v. In obtaining this equation we have assumed that (*i*) the amplitude A of particle oscillations does not change in the course of propagation of the wave and (*ii*) the medium is isotropic and homogeneous so that the wave velocity v does not change from place to place.

For a wave travelling in the negative x-direction the corresponding equation is

$$\psi(x, t) = A \sin \left\{ \frac{2\pi}{T} \left(t + \frac{x}{v} \right) \right\} \tag{7.4}$$

We may use Eq. (7.3) or (7.4) to check our definition of wavelength λ. By definition, two particles separated by a distance λ are in the same state of motion. From Eq. (7.3) we have

$$\begin{aligned}\psi(x+\lambda, t) &= A \sin\left\{\frac{2\pi}{T}\left(t-\frac{x+\lambda}{v}\right)\right\}\\ &= A \sin\left\{\frac{2\pi}{T}\left(t-\frac{x}{v}\right)-2\pi\right\} \quad \left(\because \lambda = \frac{v}{T}\right)\\ &= A \sin\left\{\frac{2\pi}{T}\left(t-\frac{x}{v}\right)\right\}\\ &= \psi(x, t)\end{aligned}$$

In other words, the function $\psi(x, t)$ repeats itself in a distance λ. Wavelength λ is also called the spatial periodicity of the wave. The wave is thus doubly periodic. It has temporal (in relation to time) periodicity T and spatial (in relation to space) periodicity λ. Eq. (7.1) gives the relation between these two very important characteristics of wave motion.

Equation (7.3) may be rewritten in an alternative form as

$$\psi(x, t) = A \sin\left\{\frac{2\pi v}{\lambda}\left(t-\frac{x}{v}\right)\right\}$$

or

$$\psi(x, t) = A \sin\left\{\frac{2\pi}{\lambda}\left(vt-x\right)\right\} \tag{7.5}$$

We may write Eq. (7.3) is yet another more compact form by defining two quantities; $k = \dfrac{2\pi}{\lambda}$ and $\omega = 2\pi/T$. In terms of k and ω we can recast Eq. (7.3) as

$$\psi(x, t) = A \sin(\omega t-kx) \tag{7.6}$$

The quantity k is called the *wave number* of the wave and ω is the angular frequency of particle oscillations in a wave.

To summarize, the particle displacements when a harmonic wave travel in the positive x-direction are given by

$$\begin{aligned}\psi(x, t) &= A \sin\left\{\frac{2\pi}{T}\left(t-\frac{x}{v}\right)\right\}\\ &= A \sin\left\{\frac{2\pi}{\lambda}(vt-x)\right\} \qquad (7.7)\\ &= A \sin(\omega t-kx)\end{aligned}$$

The corresponding equations for a wave travelling in the negative x-direction are

$$\begin{aligned}\psi(x, t) &= A \sin\left\{\frac{2\pi}{T}\left(t+\frac{x}{v}\right)\right\}\\ &= A \sin\left\{\frac{2\pi}{\lambda}(vt+x)\right\} \qquad (7.8)\\ &= A \sin(\omega t+kx)\end{aligned}$$

It may be remarked that the waves described by the above equations are equally well represented by a cosie function instead of a sine function.

The Classical Wave Equation

Let us now set up a differential equation which governs the propagation of a one-dimensional wave described by Eqs. (7.7) and (7.8).

Differentiating Eq. (7.7) twice with respect to t, keeping x constant, we get

$$\frac{\partial^2\psi}{\partial t^2} = -A\left(\frac{2\pi v}{\lambda}\right)^2 \sin\left\{\frac{2\pi}{\lambda}(vt-x)\right\} = -\left(\frac{2\pi v}{\lambda}\right)^2 \psi \tag{7.9}$$

Differentiating Eq. (7.7) twice with respect to x, keeping t constant, we get

$$\frac{\partial^2\psi}{\partial x^2} = -A\left(\frac{2\pi}{\lambda}\right)^2 \sin\left\{\frac{2\pi}{\lambda}(vt-x)\right\} = -\left(\frac{2\pi}{\lambda}\right)^2 \psi \tag{7.10}$$

Equations (7.9) and (7.10) give,

$$\frac{\partial^2\psi}{\partial t^2} = v^2\frac{\partial^2\psi}{\partial x^2} \tag{7.11}$$

This is the *classical wave equation*. We can check that the same equation is obtained also for a wave travelling along the negative x-direction. We have already come across this equation in Chap. 6 while discussing the normal modes of continuous systems.

We may draw two important inferences from the wave Eq. (7.11):

(1) Whenever the second-order time derivative $\frac{\partial^2\psi}{\partial t^2}$ of any physical quantity is related to the second-order space derivative $\frac{\partial^2\psi}{\partial x^2}$, a wave of some sort must travel in the medium.

(2) The velocity v of that wave is given by the square-root of the coefficient of the second-order space derivative.

It must be clearly understood that the individual particles which make up the medium do not progress through the medium with the wave; they merely oscillate (transversely or longitudinally) about their equilibrium positions. It is their phase relationships which we observe as waves. Therefore, the *wave velocity* is also called the *phase velocity*; it is the velocity with which planes of equal phase (crests or troughs in the case of transverse waves and compressions or rarefactions in the case of longitudinal waves) travel through the medium. The phase velocity of a wave is given by

$$v = \nu\lambda$$

$$= 2\pi\nu\,\frac{\lambda}{2\pi}$$

or

$$v = \frac{\omega}{k} \qquad (\because\ \omega = 2\pi\nu,\ k = 2\pi/\lambda) \tag{7.12}$$

The relation (7.12) may also be obtained as follows : The total phase $\phi(x, t)$ of a harmonic wave is defined as the argument of the sine (or cosine) function*. The phase of a wave travelling in the $+x$-direction is

$$\phi(x, t) = \omega t - kx$$

At a given space point x, the phase increases linearly with t (as is indicated by the term ωt). At a given time t, the phase decreases linearly with x (as is indicated by the term $-kx$). The decrease in phase is due to the fact that at greater x the phase corresponds to waves emitted at earlier times.

Let us now calculate the velocity with which, for instance, a wave crest, i.e. maximum value of sin $(\omega t - kx)$ or sin ϕ moves. This is the wave velocity. Suppose we have a crest at x at time t and it appears at $x+dx$ at a later time $t+dt$. To keep the phase constant, the total derivative $d\phi$ of the phase must be zero. Now $d\phi$ is given by

$$d\phi = \left(\frac{\partial \phi}{\partial t}\right)dt + \left(\frac{\partial \phi}{\partial x}\right) dx$$

$$= \omega dt - k dx$$

which will be zero if dx and dt are related as

$$v = \frac{dx}{dt} = \frac{\omega}{k}$$

which is Eq. (7.12).

In complete form the one-dimensional wave equation is written as

$$\frac{\partial^2 \psi(x, t)}{\partial t^2} = v^2 \frac{\partial^2 \psi(x, t)}{\partial x^2} \qquad (7.13)$$

Solution of the Wave Equation. We will now show that the possible solutions of the wave Eq. (7.13) are indeed the ones given by Eqs. (7.7) and (7.8). This can be shown as follows:

The solution of Eq. (7.13) will be a function of the variables x and t, We will show that any function of the form $\psi(x, t) = f(vt - x)$ or of the form $\psi(x, t) = F(vt + x)$ is a possible solution. Let us consider the function

$$\psi(x, t) = f(vt - x) = f(z)$$

where $z = vt - x$. Differentiating with respect to t keeping x fixed we have $\left(\text{since } \frac{\partial}{\partial t} = \frac{\partial z}{\partial t} \cdot \frac{\partial}{\partial z}\right)$

*This definition of the phase of a wave is a matter of choice or convenience. Some authors define phase as $(kx - \omega t)$.

$$\frac{\partial \psi}{\partial t} = v \frac{\partial f}{\partial z}$$

and
$$\frac{\partial^2 \psi}{\partial t^2} = v^2 \frac{\partial^2 f}{\partial z^2}$$

Now differentiating with respect to x keeping t fixed we have (since $\frac{\partial}{\partial x} = \frac{\partial z}{\partial x} \cdot \frac{\partial}{\partial z}$)

$$\frac{\partial \psi}{\partial x} = -\frac{\partial f}{\partial z}$$

and
$$\frac{\partial^2 \psi}{\partial x^2} = +\frac{\partial^2 f}{\partial z^2}$$

so that
$$\frac{\partial^2 \psi}{\partial t^2} = v^2 \frac{\partial^2 \psi}{\partial x^2}$$

which is the wave equation. Hence $\psi(x, t) = f(vt-x)$ is a solution of the wave equation. We can similarly show that $\psi(x, t) = F(vt+x)$ is also a solution of the same equation.

If $\psi(x, t)$ is a simple harmonic displacement of an oscillator at position x and time t, we would expect this function to be a sine or cosine function. The bracket $(vt-x)$ has the dimensions of a length. Since the argument of a sine or cosine function must have the dimensions of radians, the bracket $(vt-x)$ must be multiplied by a factor $2\pi/\lambda$ where λ is a length. Thus, we may write

$$\psi(x, t) = A \sin \left\{ \frac{2\pi}{\lambda} (vt-x) \right\}$$

where constant A has the dimensions of ψ. We recognise that the length λ is the wavelength and this solution represents a wave travelling in the positive x-direction. Similar arguments hold for the function $\psi(x, t) = F(vt+x)$ which represents a wave travelling in the negative x-direction.

7.4 WAVE VELOCITIES IN CONTINUOUS SYSTEMS

Recall that in Chap. 1 we recognized a simple harmonic motion by the equation $\frac{\partial^2 \psi}{\partial t^2} = -\omega^2 \psi$ and used it to identify various types of harmonic oscillations and to obtain the expression for the angular frequency ω for various oscillators. Similarly Eq. (7.13) is used to obtain the expression for the velocity of a wave travelling in a given medium. We shall illustrate this in the case of a few continuous systems.

Any system whose particle motions are governed by Eq. (7.13) is a system in which harmonic waves of any wavelength can travel with the speed v. The value of v depends on the elastic and inertial properties of the system under consideration. Let us obtain the expression for the wave velocity in a few specific media. All that we have to do is to deduce the equation of motion of the moving parts of the medium.

Transverse Waves on a Stretched String

In Sec. 6.2 we have already deduced the equation of transverse vibrations of a stretched string. We have seen that the displacement of the string is governed by the equation [see Eq. (6.5)]

$$\frac{\partial^2 y(x, t)}{\partial t^2} = \frac{T}{\mu} \frac{\partial^2 y(x, t)}{\partial x^2}$$

where T is the tension and μ is the linear density (or mass per unit length) of the string. Comparing this equation with the wave Eq. (7.13) we find that the velocity of the transverse waves on a string is given by

$$v = \sqrt{\frac{T}{\mu}} \tag{7.14}$$

We have now identified the quantity v introduced in Chap. 6. It is the velocity of the wave. Let us calculate the value of v for a particular case. Suppose a string of mass 1.0 g m^{-1} is stretched with a force of 10 N, i.e. $\mu = 1.0$ $gm^{-1} = 1.0 \times 10^{-3}$ kg m^{-1} and $T = 10$ N. The velocity of transverse waves on such a string is

$$v = \sqrt{\frac{T}{\mu}} = \sqrt{\frac{10}{10^{-3}}} = 100 \text{ ms}^{-1}$$

Characteristic Impedance of a String to Transverse Waves. Any medium through which waves travel will offer impedance (or opposition) to those waves. If the medium is loss-less, i.e. it does not have any resistive or dissipative component, the impedance is determined solely by its inertia and elasticity. These two properties are also responsible for storing and propagating energy in the medium in the form of waves.

Consider travelling waves on a string which are generated at one end (say $x = 0$) by applying a transverse harmonic force $F = F_0 \cos \omega t$ at this end (see Fig. 7.7). As a result of the application of the force, transverse waves travel in, say the $+x$-direction, in which the particle displacements are given by

$$y(x, t) = A \sin \left\{ \frac{2\pi}{\lambda} (vt - x) \right\}$$

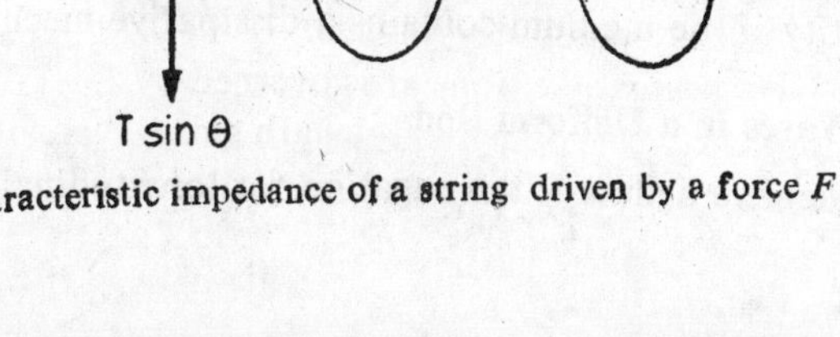

Fig. 7.7 Characteristic impedance of a string driven by a force $F = F_0 \cos \omega t$.

where $\lambda = \frac{v}{\nu} = \frac{2\pi v}{\omega}$; since, in the steady state, the frequency of particle oscillations equals that of the driving force.

Differentiating we have

$$\frac{\partial y}{\partial x} = -A\left(\frac{2\pi}{\lambda}\right)\cos\left\{\frac{2\pi}{\lambda}(vt-x)\right\}$$

$$\frac{\partial y}{\partial t} = A\left(\frac{2\pi v}{\lambda}\right)\cos\left\{\frac{2\pi}{\lambda}(vt-x)\right\}$$

Comparing these equations, we have

$$\frac{\partial y}{\partial x} = -\frac{1}{v}\frac{\partial y}{\partial t}$$

Now, referring to Fig. 7.7, the transverse force exerted at the end $x = 0$ must be equal and opposite to $T \sin\theta$, the transverse component of tension i.e.

$$F = F_0 \cos\omega t = -T\sin\theta$$

$$\simeq -T\tan\theta \qquad (\because \theta \text{ is small})$$

$$= -T\left(\frac{\partial y}{\partial x}\right)_{x=0}$$

$$= \frac{T}{v}\left(\frac{\partial y}{\partial t}\right)_{x=0} \qquad \left(\because \frac{\partial y}{\partial x} = -\frac{1}{v}\frac{\partial y}{\partial t}\right)$$

But

$$\left(\frac{\partial y}{\partial t}\right)_{x=0} = A\frac{2\pi v}{\lambda}\cos\frac{2\pi vt}{\lambda} = V_0\cos\omega t$$

where $V_0 = \frac{A2\pi v}{\lambda} = A\omega$ is the velocity amplitude. Thus we have

$$F_0\cos\omega t = V_0\cos\omega t\,\frac{T}{v}$$

or

$$F_0 = \frac{TV_0}{v}$$

The characteristic impedance of the string is defined as

$$Z = \frac{\text{transverse force amplitude}}{\text{transverse velocity amplitude}} = \frac{F_0}{V_0}$$

Hence

$$Z = \frac{T}{v} = \sqrt{\mu T} = \mu v \qquad \left(\because v = \sqrt{\frac{T}{\mu}}\right)$$

Since the velocity v is determined by the inertia and elasticity, it is clear that the impedance is also governed by these two properties of the medium. For a loss-less medium, the impedance is a real quantity. It becomes a complex quantity if the medium contains a dissipative mechanism.

Longitudinal Waves in a Uniform Rod

In Chap. 6 we have deduced the equation for longitudinal vibrations of a uniform rod. We have seen that the displacement $\xi(x, t)$ of the rod is given by [see Eq. (6.23)]

$$\frac{\partial^2 \xi}{\partial t^2} = \sqrt{\frac{Y}{\rho}} \frac{\partial^2 \xi}{\partial x^2}$$

where Y is the Young's modulus of the rod and ρ its density. Comparing this equation with the wave Eq. (7.13) we find that the velocity of longitudinal (i.e. sound) waves in the rod is given by

$$v = \sqrt{\frac{Y}{\rho}} \tag{7.15}$$

The velocity of sound in a rod made of nickel ($Y = 21.4 \times 10^{10}$ N m^{-2}) and $\rho = 8.9 \times 10^3$ kg m^{-3}) is

$$v = \sqrt{\frac{Y}{\rho}} = \sqrt{\frac{21.4 \times 10^{10}}{8.9 \times 10^3}} = 4900 \text{ m s}^{-1}$$

Thus the speed of sound in solids is fairly high.

Longitudinal Waves in a Gas

In Chap. 6, we have deduced the equation for longitudinal vibrations of a column of a gas. We have seen that the displacement $\xi(x, t)$ of the molecules of the gas is given by [see Eq. (6.35)]

$$\frac{\partial^2 \xi}{\partial t^2} = \frac{E}{\rho} \frac{\partial^2 \xi}{\partial x^2}$$

where E is the volume elasticity or bulk modulus of the gas and ρ its density. Comparing this equation with the wave Eq. (7.13) we find that the velocity of longitudinal (i.e. sound) waves in a gas is given by

$$v = \sqrt{\frac{E}{\rho}} \tag{7.16}$$

Newton's Formula. Newton gave the first theoretical expression of the velocity of sound wave in a gas. He assumed that when sound wave travels through a gaseous medium, the temperature variations in the regions of compression and rarefaction are negligible. The conditions are, therefore, isothermal and Boyle's law can be applied.

Consider a volume V of a gas at pressure P. As a result of sound having travelled in it, let the change in pressure be ΔP and the corresponding change in volume be ΔV. From Boyle's law we have

$$PV = \text{constant} \qquad \text{(isothermal change)}$$

Differentiating partially we have

$$P\Delta V + \Delta P V = 0$$

or

$$P = -\frac{\Delta P}{\Delta V/V} = \frac{\text{excess pressure}}{\text{volume strain}}$$

$$= E_i \qquad \text{(by definition)}$$

where E_i is the volume elasticity of the gas when isothermal conditions hold.

Therefore, under isothermal conditions,

$$v = \sqrt{\frac{P}{\rho}}$$

This is Newton's formula for the velocity of sound in a gas. For air at standard temperature and pressure (STP), density $\rho = 1.29$ kg m^{-3} and pressure $P = 0.76$ m of Hg $= 0.76 \times (13.6 \times 10^3) \times 9.8$ Nm^{-2}. Substituting these values in Newton's formula gives

$$v \simeq 280 \text{ m s}^{-1}$$

The experimental value, from various experiments, for the velocity of sound at STP is $\simeq 332$ ms^{-1} which is considerably higher (by about 16%) than the above theoretical value obtained from Newton's formula. Newton was unable to give a satisfactory explanation for this discrepancy. The first satisfactory explanation was given by a French scientist Laplace, in 1816.

Laplace's Correction. As sound wave travels in a gas, the region of compression is heated and the region of rarefaction is cooled. Since the thermal conductivity of a gas is small and these thermal changes occur so rapidly that the heat developed in compression and cooling produced in rarefaction is not transferred out or in to achieve thermalization during the short time-scale. The time-scale is the time required by sound to travel from compression to rarefaction. The wave 'sees' regions of unequal temperature. Laplace pointed out that the thermal changes are adiabatic which obey

$$PV^\gamma = \text{constant}$$

where $\gamma = C_P/C_V$, C_P and C_V being specific heats of gas at constant pressure and at constant volume respectively. Differentiating partially we get

$$\gamma PV^{\gamma-1}\Delta V + \Delta PV^\gamma = 0$$

or

$$\gamma P = -\frac{\Delta PV^\gamma}{V^{\gamma-1}\Delta V} = -\frac{\Delta P}{\Delta V/V} = E_a$$

where E_a is the volume elasticity, the subscript a refers to the adiabatic change.

Thus

$$v = \sqrt{\frac{\gamma P}{\rho}} \tag{7.17}$$

Taking $\gamma = 1.4$ for air, velocity of sound in air at STP is

$$v \simeq 331.6 \text{ m s}^{-1}$$

which is in close agreement with the experimental value, thereby establishing the correctness of Laplace's explanation.

Equation. (7.17) may be written in a more useful form. The relation between pressure P and volume V of an ideal gas of mass m and molecular weight M is

$$PV = \frac{m}{M} RT$$

where R is the universal gas constant and T is the absolute temperature of the gas

Since $\rho = m/V$, Eq. (7.17) may be written as

$$v = \sqrt{\frac{\gamma RT}{M}}$$

From this relation it follows that the velocity of sound in a gas (*a*) is independent of pressure or density, (*b*) is proportional to the square-root of absolute temperature, (*c*) is inversely proportional to the square root of the molecular weight and (*d*) depends on the value of γ of the gas, i.e. whether the gas is monoatomic, diatomic, etc.

Acoustic Impedance. The acoustic impedance Z offered by an elastic medium to the propagation of sound waves through it is defined as

$$Z = \frac{\text{pressure due to wave}}{\text{particle velocity}}$$

The particle displacements for a wave travelling in the $+x$-direction are given by

$$\xi(x, t) = A \sin\left\{\frac{2\pi}{\lambda}(vt-x)\right\}$$

where $v = \sqrt{E/\rho}$; A is the displacement amplitude and λ is the wave-length.

Differentiating we have

$$\text{Volume strain} = \frac{\partial \xi}{\partial x} = -A\left(\frac{2\pi}{\lambda}\right)\cos\left\{\frac{2\pi}{\lambda}(vt-x)\right\}$$

$$\text{Particle velocity} = \frac{\partial \xi}{\partial t} = A\left(\frac{2\pi v}{\lambda}\right)\cos\left\{\frac{2\pi}{\lambda}(vt-x)\right\}$$

Now referring to Eq. (6.34) of Chap. 6, we have

Pressure due to wave

$$= \Delta P = -E\frac{\partial \xi}{\partial x} = EA\left(\frac{2\pi}{\lambda}\right)\cos\left\{\frac{2\pi}{\lambda}(vt-x)\right\}$$

Hence, acoustic impedance Z is given by

$$Z = \frac{\Delta P}{\frac{\partial \xi}{\partial t}} = \frac{E}{v} = \rho v \qquad \left(\because\ v = \sqrt{\frac{E}{\rho}}\right)$$

Since the wave velocity v depends upon the inertial (mass density ρ) and elastic (bulk modulus E) properties of the medium, the characteristic impedance is also governed by these two properties of the medium. We have assumed that the medium is lossless and possesses no dissipative mechanism, in which case Z is real.

Current and Voltage Waves on Electrical Transmission Lines

The waves described so far are *mechanical* waves; they are produced by oscillations of the particles of a material medium. As stated earlier, the scope of the meaning of the function $\psi(x, t)$ in wave Eq. (7.13) need not be restricted to mean only the physical displacement of particles of a medium. In fact, waves can travel even in a particle-free space, i.e. vacuum. The current (or voltage) waves on a transmission line are an example of such waves. Electromagnetic waves (described later in this chapter) are also non-mechanical waves. These waves are produced due to a change in electric and magnetic fields which can exist even in free-space.

An electrical transmission line essentially consists of two long parallel conductors (or wires). At one end of the line, power is fed by an AC generator which provides the driving force for the waves to travel in the transmission line. Let the capacitance between the conductors be C per unit length and the inductance of the line be L per unit length. We assume that the line is loss-less, i.e. we neglect the resistance of the conductors and any possible electrical leakage between them. When a.c. currents flow in the wires, they generate magnetic flux lines which thread the region between the wires; thus giving rise to a self-inductance L per unit length. The wires themselves constitute a capacitor whose capacitance is C per unit length.

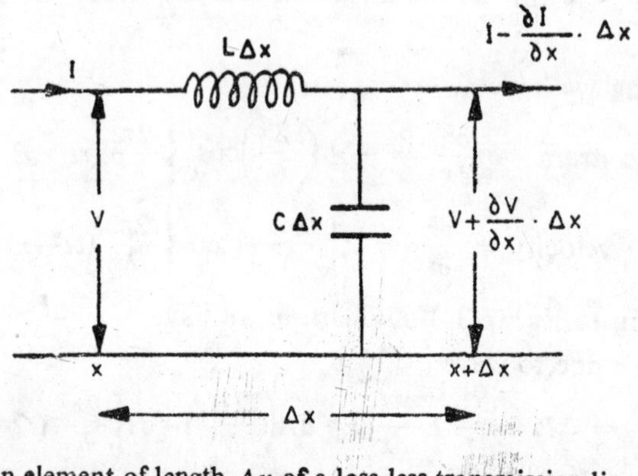

Fig. 7.8 An element of length Δx of a loss-less transmission line of inductance L per unit length and capacitance C per unit length.

Figure 7.8 shows an infinitesimally small element of length Δx of an ideal transmission line. The self-inductance of the element is $L\,\Delta x$ and its capacitance is $C\Delta x$. Suppose I is the instantaneous value of the current in either wire at a point x and V is the instantaneous voltage between the wires at this point. Let $\frac{\partial V}{\partial x}$ be the rate of change (with respect to x) of voltage at that instant of time. Then the voltage at the end $(x+\Delta x)$ of the element will be $V+\frac{\partial V}{\partial x}.\Delta x$. The voltage difference between the ends of the element is $\frac{\partial V}{\partial x}.\Delta x$ which must be equal to the voltage drop $-(L\,\Delta x)\frac{\partial I}{\partial t}$ from the self-inductance. Thus

$$\frac{\partial V}{\partial x}.\Delta x = -L\,\Delta x\,\frac{\partial I}{\partial t}$$

or

$$\frac{\partial V}{\partial x} = -L\frac{\partial I}{\partial t} \qquad (7.18)$$

Let $\frac{\partial I}{\partial x}$ be the rate of change (with respect to x) of current at that instant. Since some current is used up in charging the capacitor to a voltage V, the current at the end $(x+\Delta x)$ of the element is $I-\frac{\partial I}{\partial x}\Delta x$. The change in current is therefore equal to $-\frac{\partial I}{\partial x}\Delta x$, which has been used up in charging the capacitor of capacitance $C.\Delta x$ to a voltage V. Hence

$$-\frac{\partial I}{\partial x}\Delta x = \frac{\partial q}{\partial t} = \frac{\partial}{\partial t}(C\Delta xV)$$

where $q = (C\Delta x)V$ is the instantaneous charge on the capacitor. Hence

$$\frac{\partial I}{\partial x} = -C\frac{\partial V}{\partial t} \qquad (7.19)$$

Taking $\frac{\partial}{\partial x}$ of Eq. (7.18) and $\frac{\partial}{\partial t}$ of Eq. (7.19) and remembering that

$$\frac{\partial^2}{\partial x\partial t} = \frac{\partial^2}{\partial t\partial x}$$

we have

$$\frac{\partial^2 V}{\partial t^2} = \frac{1}{LC}\frac{\partial^2 V}{\partial x^2} \qquad (7.20)$$

Similarly taking $\frac{\partial}{\partial t}$ of Eq. (7.18) and $\frac{\partial}{\partial x}$ of Eq. (7.19) we have

$$\frac{\partial^2 I}{\partial t^2} = \frac{1}{LC}\frac{\partial^2 I}{\partial x^2} \qquad (7.21)$$

Comparing Eqs. (7.20) and (7.21) with the wave Eq. (7.13) we find the current and voltage waves propagate along the transmission line with a velocity v given by

$$v = \frac{1}{\sqrt{LC}}$$

Remember that L and C have been defined per unit length; their dimensions being henry per metre and farad per metre respectively. Thus we find that current and voltage waves travel on a transmission line into which a.c. power is fed just as mechanical waves travel on a stretched string driven by a harmonic driving force.

Characteristic Impedance of an Ideal Transmission Line

The voltage and current in a transmission line are related to each other. The relationship between them is characterized by an impedance of the line. The solutions of Eqs (7.20) and (7.21) for waves in the $+x$ direction are

$$V = V_0 \sin\left\{\frac{2\pi}{\lambda}(vt-x)\right\}$$

$$I = I_0 \sin\left\{\frac{2\pi}{\lambda}(vt-x)\right\}$$

where V_0 and I_0 are the amplitudes (i.e. maximum values) of voltage V and current I respectively. Therefore,

$$\frac{\partial V}{\partial x} = -\frac{2\pi}{\lambda}V_0 \cos\left\{\frac{2\pi}{\lambda}(vt-x)\right\}$$

Substituting this value of $\frac{\partial V}{\partial x}$ in Eq. (7.18) we get

$$\frac{\partial I}{\partial t} = \frac{2\pi}{\lambda}\frac{V_0}{L}\cos\left\{\frac{2\pi}{\lambda}(vt-x)\right\}$$

Integrating we have (apart from a constant of integration which need not be considered because we are considering only the oscillatory values of I and the constant will merely change the d.c. value)

$$I = \int \frac{\partial I}{\partial t}\,dt = \frac{2\pi}{\lambda}\frac{V_0}{L}\int \cos\left\{\frac{2\pi}{\lambda}(vt-x)\right\}dt$$

$$= \frac{V_0}{vL}\sin\left\{\frac{2\pi}{\lambda}(vt-x)\right\}$$

$$= \frac{V}{vL}$$

The ratio

$$Z = \frac{V}{I} = vL = \sqrt{\frac{L}{C}} \qquad \left(\because\ v = \frac{1}{\sqrt{LC}}\right)$$

has the dimensions of ohms and is a constant for a given transmission line. The constant Z is called the *characteristic impedance* of the line. It is a pure resistance as the ratio $\sqrt{L/C}$ does not involve dimensions of length.

The current and voltage waves are always in phase in an infinitely long transmission line (where reflected waves are not present) and hence propagate energy along the line. In a finite (or closed) transmission line energy is not propagated due to the formation of standing waves resulting from the waves reflected from the boundaries An important property of the characteristic impedance of a line is that even a finite line, which is connected at its ends to a resistance equal to its characteristic impedance, behaves as an infinite line and can carry travelling waves which dissipate their energy in the resistance, thus avoiding reflections (see Chap. 8).

Electromagnetic Waves in Space

We have seen that the velocity of waves in a material medium is determined by the inertial and elastic properties of the medium. The linear density (in the case of strings) and mass density (in the case of rods and gases) are a measure of the inertia of the system. The tension and elastic modulus are a measure of the elasticity of the system. In the case of a transmission line the inductance of the line is a measure of magnetic inertia and elasticity is provided by the capacitance of the line. These two properties are responsible for storing energy in the medium. This energy travels in the medium in the form of waves.

The wave-like behaviour of current and voltage on a transmission line helps us to understand the propagation of electromagnetic disturbances in space. The existence of voltage across the transmission line implies the existence of an electric field in the space between the cables. Since the voltage varies in space and time, the associated electric field must vary in space and time. The existence of space and time dependent current in the line implies the existence of space and time dependent magnetic field in the space around the cables. Since the current and voltage propagate in the line in the form of waves, the electric and magnetic fields must exhibit a similar wave-like behaviour.

When electric and magnetic fields vary in space and time, they produce what we call the electromagnetic waves. It is clear that an oscillating charge will produce electromagnetic waves. An oscillating charge has an oscillating electric and magnetic fields around it and hence it produces electromagnetic waves. Electrons falling from a higher to a lower energy orbit in an atom radiate electromagnetic waves of a particular wavelength and frequency. The motion of electrons in an antenna radiate electromagnetic waves by a process called 'Bremsstrahlung'.

The propagation of electromagnetic waves in a medium is also due to the inertial and elastic properties of the medium. Every medium (including vacuum) has inductive property described by what we call the *magnetic permeability* μ of the medium. In the SI system, μ is expressed in henries per metre. This property provides the magnetic inertia of the medium. The elasticity of the medium is provided by the capacitive property called the *electrical permittivity* ε of the medium. Permeability

μ stores the magnetic energy and permittivity ε stores the electric field energy. This electromagnetic energy propagates in the medium in the form of electromagnetic waves.

The electric and magnetic fields are connected by Maxwell's equations which are (for a dielectric medium)*

$$\nabla \times \vec{H} = \varepsilon \frac{\partial \vec{E}}{\partial t} \tag{7.22}$$

$$\nabla \times \vec{E} = -\mu \frac{\partial \vec{H}}{\partial t} \tag{7.23}$$

$$\varepsilon (\nabla \cdot \vec{E}) = \rho \tag{7.24}$$

$$\nabla \cdot \vec{H} = 0 \tag{7.25}$$

In these equations the electric field $\vec{E}$ is expressed in volts per metre, magnetic field $\vec{H}$ in amperes per metre and charge density ρ in coloumbs per cubic metre.

Before we proceed to obtain the expression for the velocity of an electromagnetic wave in a dielectric medium, let us look at Eqs (7.22) and (7.23). Equation (7.22) is a generalization of Ampere's law in electricity. This can be seen from dimensional consideration. The dimensions of the two sides of Eq. (7.22) are given by

$$\frac{H}{\text{length}} = \frac{\text{dimensions of } \varepsilon \times \text{dimensions of } E}{\text{time}}$$

$$= \frac{\text{farad}}{\text{length}} \times \frac{\text{volt}}{\text{length}} \times \frac{1}{\text{time}}$$

or
$$H = \frac{\text{charge}}{\text{time}} \cdot \frac{1}{\text{length}} = \frac{\text{current}}{\text{length}}$$

which is Ampere's law giving the magnetic field associated with a current carrying conductor.

Equation (7.23) is a generalization of Faraday's law in electromagnetism. The dimensions of the two sides of this equation are:

$$\frac{\mu H}{\text{time}} = -\frac{E}{\text{length}} = -\frac{\text{volts}}{(\text{length})^2}$$

or
$$\frac{\mu H \times \text{area}}{\text{time}} = -\text{volts}$$

or
$$\frac{\text{magnetic flux}}{\text{time}} = -\text{volts}$$

---

* For a detailed development of Maxwell's equations, refer to any textbook on electromagnetism.

This is dimensionally of the form of Faraday's law which states that an induced e.m.f. is produced whenever the magnetic flux changes with time. The negative sign gives the Lenz's law.

Velocity of Plane Waves in Free Space. We shall consider the simplest case of one-dimensional plane waves travelling in the x-direction in free space (or vacuum) for which $\rho = 0$. A plane wave travelling in the x-direction has the property that $\vec{E}$ and $\vec{B}$ do not vary in the $y\ z$ plane and therefore all derivatives $\frac{\partial}{\partial y}$ and $\frac{\partial}{\partial z}$ will be zero. The x, y and z components of Equations (7.22) and (7.23) then give (setting $\frac{\partial}{\partial y} = \frac{\partial}{\partial z} = 0$)

$$x\text{-component} \qquad \varepsilon_0 \frac{\partial E_x}{\partial t} = 0 \tag{7.26}$$

$$y\text{-component} \qquad \varepsilon_0 \frac{\partial E_y}{\partial t} = -\frac{\partial H_z}{\partial x} \tag{7.27}$$

$$z\text{-component} \qquad \varepsilon_0 \frac{\partial E_z}{\partial t} = \frac{\partial H_y}{\partial x} \tag{7.28}$$

and

$$x\text{-component} \qquad \mu_0 \frac{\partial H_x}{\partial t} = 0 \tag{7.29}$$

$$y\text{-component} \qquad \mu_0 \frac{\partial H_y}{\partial t} = \frac{\partial E_z}{\partial x} \tag{7.30}$$

$$z\text{-component} \qquad \mu_0 \frac{\partial H_z}{\partial t} = -\frac{\partial E_y}{\partial x} \tag{7.31}$$

From Eqs (7.24) and (7.25) we have (since $\rho = 0$ for vacuum)

$$\frac{\partial E_x}{\partial x} = 0 \tag{7.32}$$

$$\frac{\partial H_x}{\partial x} = 0 \tag{7.33}$$

where ε_0 and μ_0 are the values of ε and μ for vacuum. From Eqs (7.27) and (7.31) it is evident that E_y and H_z are not independent; they are coupled by these two equations. This means that if E_y is known as a function of x and t, then H_z is also known as a function of x and t. Taking $\frac{\partial}{\partial t}$ of Eq. (7.27) and $\frac{\partial}{\partial x}$ of Eq. (7.31) we have

$$\left(\text{since } \frac{\partial^2}{\partial x \partial t} = \frac{\partial^2}{\partial t \partial x}\right)$$

$$\frac{\partial^2 E_y}{\partial t^2} = \frac{1}{\mu_0 \varepsilon_0} \frac{\partial^2 E_y}{\partial x^2} \tag{7.34 a}$$

(which is the wave equation for E_y)

Similarly taking $\frac{\partial}{\partial x}$ of Eq. (7.27) and $\frac{\partial}{\partial t}$ of Eq. (7.31) we have

$$\frac{\partial^2 H_z}{\partial t^2} = \frac{1}{\mu_0 \varepsilon_0} \frac{\partial^2 H_z}{\partial x^2} \tag{7.34 b}$$

(which is the wave equation for H_z)

These two equations together represent an electromagnetic wave travelling in the x-direction with a velocity

$$v_0 = \frac{1}{\sqrt{\mu_0 \varepsilon_0}} \tag{7.35}$$

It is evident from Eqs. (7.28) and (7.30) that E_z and H_y are also not independent. These equations give

$$\frac{\partial^2 E_z}{\partial t^2} = \frac{1}{\mu_0 \varepsilon_0} \frac{\partial^2 E_z}{\partial x^2} \tag{7.36 a}$$

and

$$\frac{\partial^2 H_y}{\partial t^2} = \frac{1}{\mu_0 \varepsilon_0} \frac{\partial^2 H_y}{\partial x^2} \tag{7.36 b}$$

representing an electromagnetic wave associated with E_z and H_y. Notice that the set (E_y, H_z) is independent of the set (E_z, H_y); one can exist without the other. An unpolarised wave has both sets of waves; but if one set is absent (if $E_z = 0$ which implies that $H_y = 0$) the wave is said to be plane-polarised (see Fig. 7.9). The solutions for E_y and H_z are

$$E_y = E_{0y} \sin \left\{ \frac{2\pi}{\lambda} (vt - x) \right\} \tag{7.37 a}$$

$$H_z = H_{0z} \sin \left\{ \frac{2\pi}{\lambda} (vt - x) \right\} \tag{7.37 b}$$

where E_{0y} and H_{0z} are the maximum values of E_y and H_z.

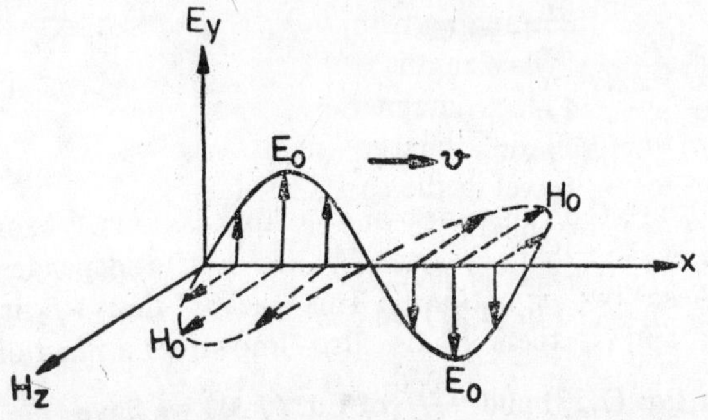

Fig. 7.9 Plane polarized electromagnetic wave travelling in the $+x$ direction

It is clear from Eqs (7.26), (7.29), (7.32) and (7.33) that E_x and H_x are constant in space and time. These constant fields can have no effect on wave motion because we consider only the oscillatory nature of the fields. We can, therefore, put $E_x = 0$ and $H_x = 0$ without altering our conclusions about the propagation of waves.

Thus we find that electric and magnetic fields cannot vary in the x-direction; variation in the field can occur in directions perpendicular to the direction in which the wave propagates. This leads us to the conclusion that electromagnetic waves are *transverse waves*.

The electromagnetic wave described by Eqs (7.34*a*, *b*) if $E_z = H_y = 0$ or by Eqs. (7.36*a*, *b*) if $E_y = H_z = 0$ is polarized. The direction of propagation being fixed (i.e. the x-direction), the choice of y and z directions is arbitrary so that the electric vector can lie in any direction perpendicular to the direction of propagation and the magnetic vector will then be perpendicular both to the electric vector and to the direction of propagation. The wave is said to be plane polarized and conventionally the direction of the magnetic vector is taken to be the direction of polarization. Thus the plane wave described by Eqs. (7.34*a*, *b*) is polarized in the z-direction and that described by Eqs. (7.36*a*, *b*) is polarized in the y direction. If both sets of waves are present with no definite phase relationship, the waves are unpolarized.

Let us calculate the value of velocity given by Eq. (7.34). For vacuum the values of ε_0 and μ_0 are

$$\varepsilon_0 = \frac{1}{36\pi \times 10^9} \text{ farad per metre}$$

$$\mu_0 = 4\pi \times 10^{-7} \text{ henry per metre}$$

$$\therefore \quad v_0 = \sqrt{\frac{1}{\mu_0 \varepsilon_0}} = \sqrt{\frac{36\pi \times 10^9}{4\pi \times 10^{-7}}} = 3 \times 10^8 \text{ m s}^{-1}$$

In other words, electromagnetic waves travel with the speed of light. Historically, the quantitative identification of the velocity v_o with the experimentally observed velocity of light established that light is indeed electromagnetic in character. Visible light is only a small part of the entire electromagnetic spectrum ranging from very short wavelengths $\sim 10^{-13}$ m (γ-rays) to very long wavelengths ~ 1 m (radio waves). Later Hertz discovered and detected electromagnetic waves produced from oscillating *LC* circuits. The oscillating charge in these circuits produces electromagnetic waves which travel in the space around the circuits. These waves can be detected at a distance.

Characteristic Impedance of Free Space to Electromagnetic Waves. If we use Eqs. (7.37*a*, *b*) and Eq. (7.27) we get

$$\varepsilon_0 \frac{2\pi}{\lambda} v_0 E_0 \cos\left\{\frac{2\pi}{\lambda}(vt - x)\right\} = \frac{2\pi}{\lambda} H_0 \cos\left\{\frac{2\pi}{\lambda}(vt - x)\right\}$$

or
$$\varepsilon_0 v_0 E_y = H_z$$

Therefore

$$Z = \frac{E_y}{H_z} = \frac{E_0}{H_0} = \sqrt{\frac{\mu_0}{\varepsilon_0}}$$

Notice that dimensions of Z are

$$\frac{\text{dimensions of } E}{\text{dimensions of } H}$$

$$= \frac{\text{volts/length}}{\text{amperes/length}} = \text{ohms}$$

The quantity $Z = \sqrt{\mu_0/\varepsilon_0}$, therefore, represents the *characteristic impedance* of the free space to the propagation of electromagnetic waves. Compare this with characteristic impedance $Z = \sqrt{L/C}$ of a transmission line. For free space the value of Z is

$$Z = \sqrt{\frac{4\pi \times 10^{-7}}{(36\pi \times 10^{9})^{-1}}} \simeq 376.7 \text{ ohms}$$

Thus, free space offers an impedance of 376.7 ohms to electromagnetic waves travelling through it.

7.5 ENERGY TRANSPORT IN TRAVELLING WAVES

We shall now discuss the transport of energy by a wave travelling in a medium. We shall discuss this problem for three different kinds of waves.

Transverse Waves on a String

Consider a wave travelling on a string, say, in the +ve x-direction. The various particles of the string along the direction of propagation are thrown into vibration in succession. Consequently, there is a transfer of energy from one part of the string to another. The particle displacements are given by

$$y(x, t) = A \sin \left\{ \frac{2\pi}{\lambda} (vt - x) \right\}$$

where $v = \sqrt{T/\mu}$; T being the tension in the string and μ its linear density.

Differentiating we have

$$\frac{\partial y}{\partial x} = -A\left(\frac{2\pi}{\lambda}\right) \cos\left\{ \frac{2\pi}{\lambda} (vt - x) \right\} \tag{7.38}$$

$$\frac{\partial y}{\partial t} = A\left(\frac{2\pi v}{\lambda}\right)^2 \cos\left\{ \frac{2\pi}{\lambda} (vt - x) \right\} \tag{7.39}$$

Before discussing the transfer of energy along the string, let us first calculate the total energy density (i.e. total energy per unit length) of the string. Consider a small element of length dx of the string. We have shown in Chapter 6 that the potential energy (dU) of this element in vibration is given by

$$dU = \tfrac{1}{2} T \left(\frac{\partial y}{\partial x}\right)^2 dx$$

and the kinetic energy (dK) of the element is given by

$$dK = \tfrac{1}{2}\mu \left(\frac{\partial y}{\partial t} \right)^2 dx$$

Therefore, potential energy density and kinetic energy density are

$$\frac{dU}{dx} = \tfrac{1}{2}T \left(\frac{\partial y}{\partial x} \right)^2$$

$$\frac{dK}{dx} = \tfrac{1}{2}\mu \left(\frac{\partial y}{\partial t} \right)^2$$

Let us calculate energy densities at time $t = 0$. At this instant we have [from Eqs. (7.38) and (7.39)]

$$\frac{dU}{dx} = \tfrac{1}{2}TA^2 \left(\frac{2\pi}{\lambda} \right)^2 \cos^2 \left(\frac{2\pi x}{\lambda} \right)$$

$$\frac{dK}{dx} = \tfrac{1}{2}\mu A^2 \left(\frac{2\pi v}{\lambda} \right)^2 \cos^2 \left(\frac{2\pi x}{\lambda} \right)$$

We shall now calculate the total energy associated with one complete wavelength of the sinusoidal wave on the string. This is done by integrating the above expressions over one wavelength and adding them. Thus

$$U = \int_0^\lambda \frac{dU}{dx}\, dx = \tfrac{1}{2}TA^2 \left(\frac{2\pi}{\lambda} \right)^2 \int_0^\lambda \cos^2 \left(\frac{2\pi x}{\lambda} \right) dx$$

$$= \frac{1}{4} TA^2 \left(\frac{2\pi}{\lambda} \right)^2 \int_0^\lambda \left\{ 1 + \cos \left(\frac{4\pi x}{\lambda} \right) \right\} dx$$

$$= \frac{1}{4} TA^2\lambda \left(\frac{2\pi}{\lambda} \right)^2$$

$$= \frac{\pi^2 TA^2}{\lambda}$$

Since $v = \nu\lambda$ and $v = \sqrt{T/\mu}$ we have

$$U = \pi^2 A^2 \mu \nu^2 \lambda \tag{7.40}$$

Similarly

$$K = \int_0^\lambda \frac{dK}{dx}\, dx = \tfrac{1}{2}\mu A^2 \left(\frac{2\pi v}{\lambda} \right)^2 \int_0^\lambda \cos^2 \left(\frac{2\pi x}{\lambda} \right) dx$$

$$= \pi^2 A^2 \mu \nu^2 \lambda \tag{7.41}$$

Notice that potential and kinetic energies over a length of the string equal to a wavelength are equal in magnitude. The total energy per wavelength is given by

$$E = U + K = 2\pi^2 A^2 \mu \nu^2 \lambda \tag{7.42}$$

The above result may also be obtained in a simple way as follows: We assume that the dimensions of the element of the string are so small compared with the wavelength that, within this element, the phase and amplitude of all particles may be considered the same. Let there be n particles in a length dx of the element and let m be the mass of each particle. Each particle executes SHM of amplitude A and angular frequency ω of the driving force. Hence the energy of oscillation of each particle is given by (see Chap. 1) $\frac{1}{2}mA^2\omega^2$.

Energy of oscillation of n particles $= \frac{1}{2}nmA^2\omega^2$.

This is the energy dE of particles in an element of length dx. Therefore

$$\text{Energy density} = \frac{dE}{dx} = \frac{\frac{1}{2}nmA^2\omega^2}{dx}$$

$$= \frac{1}{2}\mu A^2\omega^2 \qquad (\because nm = \text{mass of the element and } \mu = \text{mass/length})$$

The total energy density in a string of length equal to one wavelength is obtained by integrating over one wavelength. Hence

$$E = \int_0^\lambda \frac{dE}{dx}dx = \frac{1}{2}\mu A^2\omega^2 \int_0^\lambda dx$$

$$= \frac{1}{2}\mu A^2\omega^2\lambda$$

$$= 2\pi^2\mu A^2\nu^2\lambda$$

which is Eq. (7.42) deduced above.

We will now calculate the rate of transfer of energy in a string. Suppose a harmonic driving force F is applied at the end $x = 0$ of the string. We have seen earlier that this force must be equal and opposite to the transverse component of tension (see Fig. 7.7), i.e.

$$F = -T\sin\theta = -T\tan\theta = -T\left(\frac{\partial y}{\partial x}\right)_{x=0}$$

From Eq. (7.38) we have

$$\left(\frac{\partial y}{\partial x}\right)_{x=0} = -A\left(\frac{2\pi}{\lambda}\right)^2 \cos\left(\frac{2\pi vt}{\lambda}\right)$$

Therefore

$$F = AT\left(\frac{2\pi}{\lambda}\right)\cos\left(\frac{2\pi vt}{\lambda}\right)$$

The rate at which energy is supplied to the string at $x = 0$ or the input power P at a time t, by definition, is

$$P(t) = \text{force} \times \text{velocity}$$

$$= F\left(\frac{\partial y}{\partial t}\right)_{x=0}$$

Setting $x = 0$ in Eq. (7.39) and using it in the above equation, we have

$$P(t) = A^2Tv\left(\frac{2\pi}{\lambda}\right)^2 \cos^2\left(\frac{2\pi vt}{\lambda}\right)$$

Since $\quad \dfrac{v}{\lambda} = \dfrac{1}{\tau}$, we have

$$P(t) = A^2Tv\left(\frac{2\pi}{\lambda}\right)^2 \cos^2\left(\frac{2\pi t}{\tau}\right)$$

The mean power input $<P(t)>$ is obtained if we integrate this expression over one time period τ. This gives

$$<P(t)> = A^2Tv\left(\frac{2\pi}{\lambda}\right)^2 \frac{\int_0^\tau \cos^2\left(\frac{2\pi t}{\tau}\right) dt}{\int_0^\tau dt}$$

$$= \tfrac{1}{2}A^2 \frac{Tv}{\tau}\left(\frac{2\pi}{\lambda}\right)^2 \int_0^\tau \left\{1+\cos\left(\frac{4\pi t}{\tau}\right)\right\} dt$$

$$= \tfrac{1}{2}A^2Tv\left(\frac{2\pi}{\lambda}\right)^2$$

Since $\tau = \dfrac{1}{\nu}$, $v = \sqrt{T/\mu}$ and $v = \nu\lambda$, we have

$$<P(t)> = 2\pi^2A^2\mu\nu^2\, v = \tfrac{1}{2}\mu v\,\omega^2A^2 \tag{7.43}$$

Equation (7.42) tells us that $2\pi A^2\mu\nu^2$ is the total energy per unit length of the string. Thus we find that the time-averaged input power is equal to the total energy per unit length times the wave velocity. The energy does not stay at the driving agent; it flows along the string which thus acts as a medium for transport of energy from one point to another; the speed of transport being equal to the wave velocity.

Sound Waves in a Gas

Suppose a sound wave is travelling in a gas. Consider an infinitesimally small volume ΔV of the medium whose dimensions are so small compared with the wavelength of the wave that the phase and amplitude of oscillation of all particles within this volume element, may be considered the same. Let there be n particles in the volume element ΔV and let m be the mass of each particle. The energy of oscillation of each particle is equal to $\frac{1}{2}mA^2\omega^2$, where ω is the angular frequency of oscillation and A its amplitude. Therefore

$$\text{Energy of oscillation of } n \text{ particles} = \tfrac{1}{2}nmA^2\omega^2$$

$$= \tfrac{1}{2}(\rho\Delta V)\, A^2\omega^2$$

where $\rho = \dfrac{mn}{\Delta V}$ is the density of the medium. Hence, energy per unit volume or energy density U is given by

$$U = \tfrac{1}{2}\rho A^2\omega^2$$

Since $\omega = 2\pi\nu$, we have

$$U = 2\pi^2\rho A^2\nu^2 \tag{7.44}$$

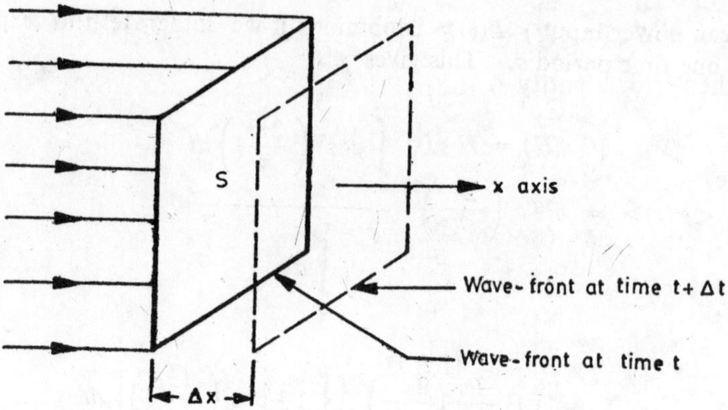

Fig. 7.10 Energy flux in sound wave

We shall now compute the energy flux or intensity of a wave. Intensity is defined as the energy transferred per unit area per second. Suppose a plane wave is travelling, say, along the positive x-direction. Consider a plane wavefront of area S at time t. After a small interval of time Δt the wavefront is displaced through a distance $\Delta x = v\Delta t$ in the direction in which the wave is travelling; v being the velocity of the wave (Fig. 7.10). Because of this the particles of the medium in volume $S\Delta x$ are set into oscillations. If U is the energy density of the medium, then in time Δt the medium of volume $S\Delta x$ receives energy equal to $U\times(S\Delta x)$. Thus the energy transmitted through area S in time $\Delta t = U\times(S\Delta x)$.

Hence the *energy flux or intensity* (which is the energy transmitted per unit area per unit time) is given by

$$I = \frac{U\times(S\Delta x)}{S\times\Delta t} = U\frac{\Delta x}{\Delta t} = Uv$$

where $v = \dfrac{\Delta x}{\Delta t}$ is the wave velocity. Thus we find that the energy flux is equal to energy density times the wave velocity. Using Eq. (7.44) we get

$$I = 2\pi^2\rho A^2\nu^2 v \tag{7.45}$$

Compare this expression with that in Eq. (7.43) for transverse waves on a string.

Electromagnetic Waves in Space

Since an electromagnetic wave has electric and magnetic fields associated with it, we would expect these fields to have stored energy. Suppose there is a region of space (of magnetic permeability μ and electrical permittivity ϵ) in which the electric and magnetic fields are space and time dependent. In other words, electromagnetic waves travel in this region. The fields are related to each other by Maxwell's equations (7.22) and (7.23). Taking a dot product of Eq. (7.22) with $\vec{E}$ we have

$$\vec{E}\,.\,(\nabla\times\vec{H}) = \epsilon\,\vec{E}\,.\left(\frac{\partial\vec{E}}{\partial t}\right) \tag{7.46}$$

Using the vector identity

$$\nabla\,.\,(\vec{E}\times\vec{H}) = \vec{H}\,.\,(\nabla\times\vec{E}) - \vec{E}\,.\,(\nabla\times\vec{H})$$

we have

$$\vec{E}\,.\,(\nabla\times\vec{H}) = \vec{H}\,.\,(\nabla\times\vec{E}) - \nabla\,.\,(\vec{E}\times\vec{H})$$

so that Eq. (7.46) becomes

$$\vec{H}\,.\,(\nabla\times\vec{E}) - \nabla\,.\,(\vec{E}\times\vec{H}) = \varepsilon\vec{E}\,.\left(\frac{\partial\vec{E}}{\partial t}\right)$$

Substituting for $\nabla\times\vec{E}$ from Eq. (7.23) we get

$$\varepsilon\vec{E}.\left(\frac{\partial\vec{E}}{\partial t}\right) + \mu\vec{H}\,.\left(\frac{\partial\vec{H}}{\partial t}\right) + \nabla\,.\,(\vec{E}\times\vec{H}) = 0$$

or

$$\frac{1}{2}\frac{\partial}{\partial t}(\varepsilon\vec{E}\,.\,\vec{E} + \mu\vec{H}\,.\,\vec{H}) + \nabla\,.\,(\vec{E}\times\vec{H}) = 0$$

or

$$\frac{1}{2}\frac{\partial}{\partial t}(\varepsilon E^2 + \mu H^2) + \nabla\,.\,(\vec{E}\times\vec{H}) = 0 \tag{7.47}$$

Now we know that the energy density U_e of the electric field and the energy density U_m of the magnetic field are given by

$$U_e = \tfrac{1}{2}\varepsilon E^2$$

and

$$U_m = \tfrac{1}{2}\mu H^2$$

so that

$$U = U_e + U_m = \tfrac{1}{2}(\varepsilon E^2 + \mu H^2) \tag{7.48}$$

is the energy density of the electromagnetic field. Equation (7.48) can be obtained by calculating the work done per unit volume in establishing an electromagnetic field. Therefore, the energy dW in a small volume dV of the medium is given by

$$dW = UdV$$

Hence the electromagnetic energy associated with a volume V of the medium is

$$W = \int_V UdV \tag{7.49}$$

Integrating Eq. (7.47) over the volume V and using Eqs (7.48) and (7.49) we have

$$\frac{\partial W}{\partial t}+\int_V \nabla \cdot (\vec{E}\times\vec{H})\, dV = 0$$

Using Gauss' (or divergence) theorem we have

$$\frac{\partial W}{\partial t}+\oint_S (\vec{E}\times\vec{H}) \cdot \hat{n}\, dS = 0$$

where S is the surface bounding volume V and $\hat{n}$ is unit outward drawn normal to the surface. We may rewrite the above equation as

$$\frac{\partial W}{\partial t}+\oint_S \vec{P} \cdot \hat{n}\, dS = 0 \tag{7.50}$$

where $\vec{P}$ stands for

$$\vec{P} = \vec{E}\times\vec{H} \tag{7.51}$$

Poynting Vector: Equation (7.49) is a statement of conservation of energy. This equation tells us that the total energy W stored in the electromagnetic field decreases with time as a result of the surface integral which must necessarily represent the rate at which the energy flows out of the bounding surface. The vector $\vec{P} = (\vec{E}\times\vec{H})$ is called the *Poynting vector*. It is evident from Eq. (7.49) that the magnitude of $\vec{P}$ gives the amount of energy which crosses per unit time per unit area of the surface in a direction normal to both $\vec{E}$ and $\vec{H}$. This is the direction in which the electromagnetic wave propagates. Thus the magnitude of Poynting vector represents the *intensity* (i.e. rate at which energy propagates per unit area held normal to it) of the wave and its direction is the direction of propagation of the wave. This may also be understood as follows:

Let us calculate the rate at which energy crosses a unit area held normal to a plane-polarized electromagnetic wave travelling in, say, the $+x$ direction. Imagine a cylinder of length dx and cross-sectional area dA with its axis along the direction of propagation of the wave. If U is the energy density of the electromagnetic field, the energy of the field in the volume $dA\, dx$ of the cylinder

$$= U dA\, dx = U dA \frac{dx}{dt} dt,$$

where dt is the time taken by the wave to traverse the cylinder. Hence,

Rate at which energy crosses a unit area $= U\dfrac{dx}{dt} = Uv$ where v is the velocity of the wave.

Now let us calculate the component of vector $\vec{P}$ in the direction of wave propagation (x-axis)

$$P_x = E_yH_z - E_zH_y \tag{7.52}$$

Now E's and H's satisfy the wave equation. Hence

$$E_y = E_{0y} \sin\left\{\frac{2\pi}{\lambda}(vt-x)\right\}$$

$$E_z = E_{0z} \sin\left\{\frac{2\pi}{\lambda}(vt-x)\right\}$$

$$H_y = H_{0y} \sin\left\{\frac{2\pi}{\lambda}(vt-x)\right\}$$

$$H_z = H_{0z} \sin\left\{\frac{2\pi}{\lambda}(vt-x)\right\}$$

Using these in Eqs. (7.22) and (7.23) we find that

$$\varepsilon v E_y = H_z \quad \text{or} \quad \sqrt{\varepsilon}\, E_y = \sqrt{\mu}\, H_z$$

$$\varepsilon v E_z = -H_y \quad \text{or} \quad \sqrt{\varepsilon}\, E_z = -\sqrt{\mu}\, H_y$$

$$\mu v H_y = -E_z \quad \text{or} \quad \sqrt{\mu}\, H_y = -\sqrt{\varepsilon}\, E_z$$

$$\mu v H_z = E_y \quad \text{or} \quad \sqrt{\mu}\, H_z = \sqrt{\varepsilon}\, E_y$$

These equations give

$$\varepsilon(E_y^2 + E_z^2) = \mu(H_y^2 + H_z^2)$$

or

$$\varepsilon E^2 = \mu H^2$$

Thus the electrostatic energy $\frac{1}{2}\epsilon E^2$ per unit volume in an electromagnetic wave equals the magnetic energy $\frac{1}{2}\mu H^2$ per unit volume and the total energy of the field per unit volume is $U = \frac{1}{2}(\varepsilon E^2 + \mu H^2)$. Using the above results in Eq. (7.52) we have

$$P_x = v\varepsilon E^2 = v\mu H^2 = \tfrac{1}{2}v(\varepsilon E^2 + \mu H^2) = vU$$

a result we have obtained above. Thus we conclude that the magnitude of the Poynting vector indeed gives the rate at which electromagnetic energy crosses per unit area held normal to its direction of propagation.

7.6 WAVES IN AN ABSORBING MEDIUM

So far we have considered one-dimensional waves travelling in a non-absorbing medium, i.e. a medium which does not absorb the energy of the wave as it progresses through the medium. But when a wave travels in an absorbing medium (e.g. transverse waves on a string immersed in a viscous fluid, sound waves in a viscous fluid and light waves in a dielectric medium or conducting medium) it loses energy as it propagates through the medium. It has been found experimentally that the amplitude of the wave decays by a constant fraction of its value when the wave progresses through a certain distance. This means that the amplitude falls exponentially with distance and we can write the amplitude at a distance x with respect to the origin at $x = 0$ as

$$A(x) = A_0 e^{-\alpha x} \tag{7.53}$$

where A_0 is the amplitude at $x = 0$. The constant α is called the *attenuation constant.* From Eq. (7.53) we find that

$$\alpha = -\frac{1}{A(x)} \frac{dA(x)}{dx}$$

Thus α is the decrease in amplitude per unit amplitude per unit length. Therefore, α is the fractional amplitude attenuation of amplitude $A(x)$ per unit length.

The inverse of α is a length χ which is the distance over which the amplitude $e^{-\alpha x} = e^{-x/\chi}$ is attenuated by a factor of $e = 2.718$. It is called the amplitude attenuation length or simply *attenuation length.*

Combining Eq. (7.53) with the exponential representation of a wave travelling in the positive x-direction we write

$$\psi(x, t) = A_0 \exp(-\alpha x) \exp\left\{\frac{2\pi i}{T}\left(t - \frac{x}{v}\right)\right\}$$

$$= A_0 \exp \frac{2\pi i}{T}\left\{t - x\left(\frac{1}{v} + \frac{i\alpha T}{2\pi}\right)\right\}$$

$$= A_0 \exp 2\pi i\nu \left\{t - x\left(\frac{1}{v} + \frac{i\alpha}{2\pi\nu}\right)\right\}$$

or

$$\psi(x, t) = A_0 \exp 2\pi i\nu\left(t - \frac{x}{v^*}\right) \tag{7.54}$$

so that we can write, instead of velocity v, a complete quantity v^* given by

$$\frac{1}{v^*} = \frac{1}{v} + \frac{i\alpha}{2\pi\nu} \tag{7.55}$$

In practice this method of introducing the complex wave velocity finds applications chiefly in optics in which we define the *refractive index* n as

$$n = \frac{c}{v}$$

where c is the velocity of light in vacuum and v is the velocity of light in the medium. For an absorbing medium Eq. (7.55) enables us to define a complex refractive index n^* as

$$n^* = \frac{c}{v^*} = \frac{c}{v} + \frac{i\alpha c}{2\pi\nu} = n + \frac{i\alpha c}{2\pi\nu}$$

This shows that the refractive index (and hence the wave velocity) in an absorbing medium depends upon the frequency ν (or wavelength λ) of the wave. Such a medium is called *dispersive* and the relation between wave velocity and frequency (or wavelength) is called the *dispersion relation.* Complex refractive index finds applications in anamolous dispersion theory.

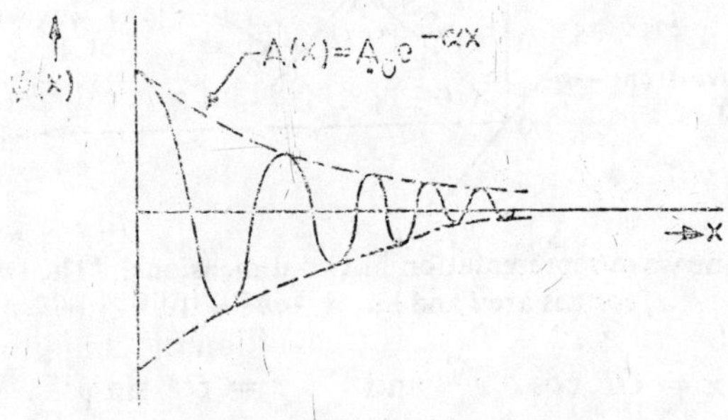

Fig. 7.11 Decay of amplitude in an absorbing medium

Figure 7.11 illustrates the decay of amplitude of the wave with distance in an absorbing medium.

The intensity (or energy flux) I in a wave is proportional to the square of the amplitude. Hence the decay of I is given by

$$I(x) = (\text{constant})\ A^2(x) = (\text{constant})\ A_0^2 \exp(-2\alpha x) = I_0 \exp(-2\alpha x).$$

Thus the intensity falls with an exponent twice as great as that for the amplitude. In other words, the attenuation length for intensity is half that for amplitude. The decrease in intensity implies a loss of energy as the wave progresses through the medium. This loss of energy mainly appears as heat in the system.

7.7 WAVES IN TWO DIMENSIONS

In order to describe a plane wave in two dimensions, the exponential notation is the most convenient. We know that a one-dimensional wave travelling in the positive x-direction can be represented in the exponential notation as

$$\psi(x, t) = A \exp\left\{ \frac{2\pi}{\lambda} (vt - x) \right\}$$

$$= A \exp \{ik(vt - x)\}$$

where $k = 2\pi/\lambda$ is the wave number.

We shall now write down the expression for a two-dimensional plane wave travelling in a direction OP making an angle θ with the x-axis in the xy plane (see Fig. 7.12). Let (x, y) be the coordinates of the point P, the origin being at O. Now

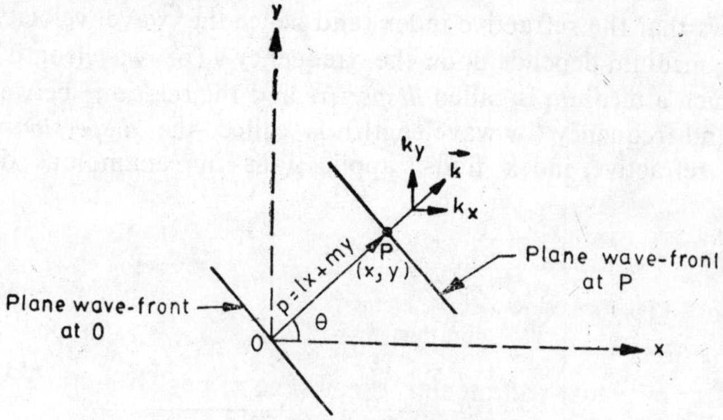

Fig. 7.12 Plane wave representation in two dimensions. The two direction cosines are l and m. $l^2+m^2 = 1$.

$$x = OP \cos\theta \quad \text{and} \quad y = OP \sin\theta$$

and
$$(OP)^2 = x^2+y^2$$
$$= (OP\cos\theta)x+(OP\sin\theta)\,y$$

or
$$OP = p = lx+my$$

where $l = \cos\theta$ and $m = \sin\theta$ are the direction cosines. Therefore the wave travelling in the direction OP can be represented as

$$\psi(x, y, t) = A\exp\{ik\,(vt-p)\}$$
$$= A\exp\{ik\,(vt-lx-my)\} \tag{7.56}$$
$$= A\exp\{i\,(kvt-klx-kmy)\}$$

or
$$\psi(x, y, t) = A\exp\{i\,(kvt-k_x x-k_y y)\}$$

where $k_x = k\cos\theta$ and $k_y = k\sin\theta$, so that $k^2 = k_x^2+k_y^2$. Thus we can write (since $kv = \omega$)

$$\psi(x, y, t) = A\exp\{i\,(\omega t-\vec{k}.\vec{r})\} \tag{7.57}$$

where $\vec{k}\,.\,\vec{r} = k_x x+k_y y$. Here $\vec{r}$ is the position vector of the point (x, y) and $\vec{k}$ is a vector in the direction l, m. The vector $\vec{k}$ is called the propagation vector; the direction of $\vec{k}$ is the direction in which the wave travels.

Wave Equation in Two Dimensions: The Stretched Membrane

The two-dimensional analogue of the stretched string is the stretched membrane. We shall obtain the two-dimensional wave equation for a stretched membrane by using arguments similar to those used for a stretched string described in Sec. 6.2.

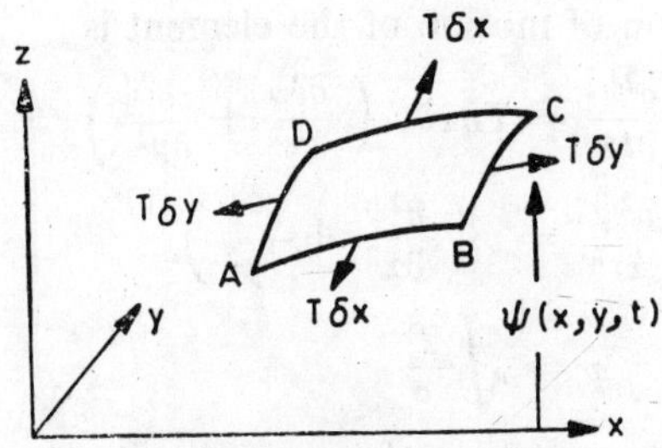

Fig. 7.13 Forces on a small element *ABCD* of a stretched membrane vibrating in the *z*-direction.

Consider a plane rectangular membrane of negligible thickness in the *xy* plane of a rectangular coordinate system (see Fig. 7.13). We shall consider transverse vibrations of the membrane, i.e. vibrations in the *z*-direction perpendicular to both *x* and *y* axes. The displacement $\psi(x, y, t)$ is a function of *x*, *y* and also of time *t*.

Suppose σ is the mass per unit area of the membrane and *T* is the uniform tension per unit length. This means that if a line of unit length is drawn in the surface of the membrane, then the material on one side of this line exerts a force *T* on the material on the other side in a direction normal to that of the line. Thus *T* is the surface tension of the membrane.

Figure 7.13 shows a small rectangular element *ABCD* of sides δx and δy of the membrane in the *xy* plane. Let $\psi(x, y, t)$ be the displacement of the element at time *t* in the *z*-direction. The forces $T\delta x$ and $T\delta y$ are acting on the sides of the element during vibration. The components of these forces in direction normal to the *xy* plane constitute the restoring force tending to bring the element back to its equilibrium position.

As discussed in Chapter 6, the resultant of forces on *AD* and *BC*, resolved in direction normal to *xy* plane

$$= T\delta y \frac{\partial^2\psi}{\partial x^2} \cdot \delta x$$

Similarly the resultant of forces on *AB* and *DC*, resolved in direction normal to *xy* plane

$$= T\delta x \frac{\partial^2\psi}{\partial y^2} \cdot \delta y$$

Thus the net transverse force on the element is

$$F_{tr} = T\delta x\delta y\left(\frac{\partial^2\psi}{\partial x^2} + \frac{\partial^2\psi}{\partial y^2}\right)$$

This constitutes the restoring force which must be equal to Newton's force for dynamical equilibrium. Since $\sigma\delta x\delta y$ is the mass of the element, Newtons force

$$= \sigma\delta x\delta y \frac{\partial^2\psi}{\partial t^2}$$

Therefore, the equation of motion of the element is

$$\sigma \delta x \delta y \frac{\partial^2 \psi}{\partial t^2} = T \delta x \delta y \left(\frac{\partial^2 \psi}{\partial x^2} + \frac{\partial^2 \psi}{\partial y^2} \right)$$

or
$$\frac{\partial^2 \psi}{\partial t^2} = v^2 \left(\frac{\partial^2 \psi}{\partial x^2} + \frac{\partial^2 \psi}{\partial y^2} \right) \tag{7.58}$$

where
$$v = \sqrt{\frac{T}{\sigma}}$$

Equation (7.58) is the equation of two-dimensional wave motion. If

$$\frac{\partial^2 \psi}{\partial y^2} = 0,$$

this equation reduces to the one-dimensional wave equation (7.11).

We will solve Eq. (7.58) by assuming an exponential solution

$$\psi(x, y, t) = A\, e^{(ax+by+ct)}$$

Using this solution in Eq. (7.58) gives

$$a^2+b^2 = c^2/v^2 \tag{7.59}$$

We are looking for a solution $\psi(x, y, t)$ which must vary periodically with time. If τ is the period of particle vibrations, then the constant c in the solution must be of the form $\frac{2\pi i}{\tau} = 2\pi i\nu$, where ν is the frequency. Equation (7.59) requires that

$$a^2+b^2 = \frac{c^2}{v^2} = -\frac{4\pi^2\nu^2}{v^2} = -k^2 \left(\because \frac{2\pi\nu}{v} = \frac{2\pi}{\lambda} = k \right)$$

Now the equation of a straight line whose perpendicular distance from the origin is p is (see Fig. 7.12)

$$p = lx+my$$

where l and m are the direction cosines of the perpendicular and

$$l^2+m^2 = 1.$$

Hence the solution can be written as

$$\psi(x, y, t) = A \exp \frac{2\pi i\nu}{v}(vt - lx - my)$$

$$= A \exp \{ik\,(vt-lx-my)\}$$

which is Eq. (7.56) as expected. The constants a and b are

$$a = -ikl$$

$$b = -ikm$$

so that $\quad a^2+b^2 = -k^2(l^2+m^2) = -k^2$ is satisfied.

The solution may also be writen as

$$y(x, y, t) = A \exp \{ik\,(vt-p)\}$$

where $p = lx+my$. This equation represents a wave motion in which the lines of constant phase are normal to the line whose direction cosines are l, m; the wave advancing in the direction l, m with a velocity v.

7.8 WAVES IN THREE DIMENSIONS

We have seen that the one-and two-dimensional wave equations respectively are

$$\frac{\partial^2\psi}{\partial x^2} = \frac{1}{v^2}\frac{\partial^2\psi}{\partial t^2}$$

and
$$\frac{\partial^2\psi}{\partial x^2} + \frac{\partial^2\psi}{\partial y^2} = \frac{1}{v^2}\frac{\partial^2\psi}{\partial t^2}$$

Similarly the three-dimensional wave equation is written as

$$\frac{\partial^2\psi}{\partial x^2} + \frac{\partial^2\psi}{\partial y^2} + \frac{\partial^2\psi}{\partial z^2} = \frac{1}{v^2}\frac{\partial^2\psi}{\partial t^2} \tag{7.60}$$

or
$$\nabla^2\psi = \frac{1}{v^2}\frac{\partial^2\psi}{\partial t^2}$$

This equation represents the wave-like behaviour of $\psi(x, y, z, t)$ in a three-dimensional medium. For an isotropic medium (i.e. a medium in which the wave velocity is the same in every direction) the quantity ψ is a scalar. For non-isotropic media ψ becomes a vector.

Sound and electromagnetic waves are three-dimensional. We shall consider the example of electromagnetic waves in vacuum. Maxwell's equations for vacuum are

$$\nabla \times \vec{H} = \varepsilon_0 \frac{\partial \vec{E}}{\partial t} \tag{7.61}$$

$$\nabla \times \vec{E} = -\mu_0 \frac{\partial \vec{H}}{\partial t} \tag{7.62}$$

$$\nabla \cdot \vec{E} = 0 \tag{7.63}$$

$$\nabla \cdot \vec{H} = 0 \tag{7.64}$$

Taking the curl of Eq. (7.62) and using Eq. (7.61) we have

$$\nabla \times (\nabla \times \vec{E}) = -\mu_0 \frac{\partial}{\partial t}(\nabla \times \vec{H}) = -\varepsilon_0\mu_0 \frac{\partial^2 \vec{E}}{\partial t^2} \tag{7.65}$$

Using the vector identity

$$\nabla \times (\nabla \times \vec{E}) = \nabla(\nabla \cdot \vec{E}) - \nabla^2 \vec{E}$$

and Eq. (7.63) we have

$$\nabla \times (\nabla \times \vec{E}) = -\nabla^2 \vec{E} \tag{7.66}$$

Using Eq. (7.66) in Eq. (7.65) we get

$$\nabla^2 \vec{E} = \frac{1}{v^2}\frac{\partial^2 \vec{E}}{\partial t^2} \tag{7.67}$$

where $v = \sqrt{\frac{1}{\mu_0 \varepsilon_0}}$ is the velocity of electromagnetic waves in vacuum.

Similarly Maxwell's equations give

$$\nabla^2 \vec{H} = \frac{1}{v^2} \frac{\partial^2 \vec{H}}{\partial t^2}$$

These equations show that the electric and magnetic field vectors satisfy the three-dimensional wave equation (7.60). Each component of the vectors satisfies an equation of the type

$$\nabla^2 \psi = \frac{1}{v^2} \frac{\partial^2 \psi}{\partial t^2}$$

The physical meaning of this result is that electromagnetic waves exist and spread in space with a velocity v which is determined by permeability μ and permittivity ϵ of the medium.

Solution of Three-Dimensional Wave Equation

The solution (in the Cartesian form) of the three-dimensional wave equation

$$\nabla^2 \psi \equiv \frac{\partial^2 \psi}{\partial x^2} + \frac{\partial^2 \psi}{\partial y^2} + \frac{\partial^2 \psi}{\partial z^2} = \frac{1}{v^2} \frac{\partial^2 \psi}{\partial t^2} \tag{7.68}$$

can be obtained if we recall the solutions of one and two-dimensional wave equations discussed earlier. In the one-dimensional case of a plane wave travelling in the x direction, ψ does not vary with y or z and

$$\frac{\partial^2 \psi}{\partial y^2} = \frac{\partial^2 \psi}{\partial z^2} = 0$$

Equation (7.68) then reduces to

$$\frac{\partial^2 \psi}{\partial x^2} = \frac{1}{v^2} \frac{\partial^2 \psi}{\partial t^2}$$

whose solution in the exponential notation is

$$\psi(x, t) = A \exp \{ik(vt - x)\}$$

where $k = \frac{2\pi}{\lambda}$ is the wave number. In the two-dimensional case, in which ψ does not vary with z, Eq. (7.68) reduces to

$$\frac{\partial^2 \psi}{\partial x^2} + \frac{\partial^2 \psi}{\partial y^2} = \frac{1}{v^2} \frac{\partial^2 \psi}{\partial t^2}$$

whose solution is [see Eq. (7.56)]

$$\psi(x, y, t) = A \exp \{ik(vt - lx - my)\}$$

where l and m are the direction cosines. Similarly a three-dimensional plane harmonic wave travelling in the direction whose cosines are l, m, n is described by the equation

$$\psi(x, y, z, t) = A \exp \{ik(vt - lx - my - nz)\} \tag{7.69}$$

with $l^2+m^2+n^2 = 1$. A straight-forward substitution shows that Eq. (7.69) is indeed a solution of Eq. (7.68).

Now l, m, n are the direction cosines of the normal on the plane $lx+my+nz = p$ from the origin and p is the length of the normal. We then have

$$\psi(x, y, z, t) = A \exp \{ik(vt-p)\}$$

which represents a plane wave progressing in the direction of its normal. In vector notation this equation can be written more compactly as

$$\psi(\vec{r}, t) = A \exp \{i(\omega t-\vec{k} \cdot \vec{r})\} \tag{7.70}$$

where $\vec{r}$ is the position vector of the point (x, y, z) and $\vec{k}$ is a vector in the direction (l, m, n) whose magnitude is $2\pi/\lambda$, where λ is the wavelength. This vector which specifies the wavelength and the direction of propagation of the wave is called the *propagation vector* or more appropriately the *wave vector*.

Transformation into Other Co-ordinate Systems

It is often necessary to transform Eq. (7.68) into other co-ordinate systems while dealing with waves of a particular symmetry. For example, for waves spreading out from a line-like source such as a radio linear antenna or a linear light source, cylindrical polar co-ordinates (Fig. 7.14*a*) would be appropriate. In cylindrical polar co-ordinates (ρ, ϕ, z), $\nabla^2\psi$ in Eq. (7.68) is transformed as

$$\nabla^2\psi = \frac{\partial^2\psi}{\partial\rho^2}+\frac{1}{\rho}\frac{\partial\psi}{\partial\rho}+\frac{1}{\rho^2}\frac{\partial^2\psi}{\partial\phi^2}+\frac{\partial^2\psi}{\partial z^2} \tag{7.71}$$

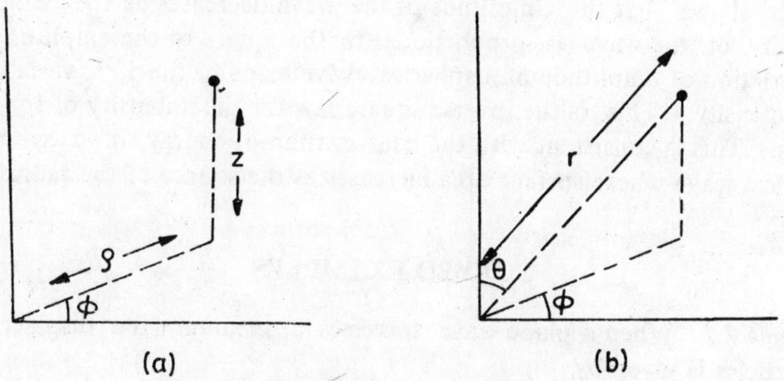

Fig. 7.14 (*a*) Cylindrical polar coordinates
(*b*) Spherical polar co-ordinates.

For waves spreading out from a point source, spherical polar co-ordinates (r, θ, ϕ) as shown in Fig. 7.14*b* would be appropriate. In terms of these co-ordinates $\nabla^2\psi$ is transformed as

$$\nabla^2\psi = \frac{\partial^2\psi}{\partial r^2}+\frac{2}{r}\frac{\partial\psi}{\partial r}+\frac{1}{r^2}\frac{\partial^2\psi}{\partial\theta^2}+\frac{\cot\theta}{r^2}\frac{\partial\psi}{\partial\theta}+\frac{1}{r^2\sin^2\theta}\frac{\partial^2\psi}{\partial\phi^2} \tag{7.72}$$

In some situations ψ may not vary in one or more co-ordinate directions and the full form of Eq. (7.72) might not be required. For example, in the special case of waves spreading out from a point source in a homogeneous isotropic medium, the waves have spherical symmetry and the wave function ψ depends only on r. In such a case Eq. (7.68) reduces to

$$\frac{\partial^2\psi}{\partial r^2}+\frac{2}{r}\frac{\partial\psi}{\partial r}=\frac{1}{v^2}\frac{\partial^2\psi}{\partial t^2}$$

This is the general form of the wave equation for spherically symmetrical waves. Since

$$\frac{\partial^2\psi}{\partial r^2}+\frac{2}{r}\frac{\partial\psi}{\partial r}=\frac{1}{r}\frac{\partial}{\partial r}(r\psi)$$

the wave equation becomes

$$\frac{\partial^2(r\psi)}{\partial r^2}=\frac{1}{v^2}\frac{\partial^2(r\psi)}{\partial t^2}$$

or

$$\frac{\partial^2(r\psi)}{\partial t^2}=v^2\frac{\partial^2(r\psi)}{\partial r^2}$$

which resembles the one-dimensional wave equation and has a general solution

$$r\psi = f(vt-r)+F(vt+r)$$

For harmonic waves we have

$$r\psi = A\exp i\,(\omega t\pm kr)$$

or

$$\psi = \frac{A}{r}\exp i\,(\omega t\pm kr)$$

This shows that the amplitude of the wave decreases as $1/r$. Since the intensity of the wave is proportional to the square of the amplitude, the $1/r$ variation of amplitude of a spherical wave leads to the $1/r^2$ variation of the intensity. This is the inverse square law for the intensity of spherical waves. This is consistent with the conservation of energy in a spreading spherical wave whose surface area increases as the square of the radius.

SOLVED EXAMPLES

Example 7.1 When a plane wave traverses a medium, the displacement of particles is given by

$$\psi(x, t) = 0.01\sin(4\pi t-0.02\pi x)$$

where ψ and x are expressed in metres and t in seconds. Calculate (*a*) the amplitude, wavelength, velocity and frequency of the wave; (*b*) the phase difference between two positions of the same particle at a time interval of 0.25 s; and (*c*) the phase difference, at a given instant of time, between two particles 50 m apart.

Solution

(*a*) The given equation can be rewritten in the form

$$\psi(x, t) = 0.01 \sin \left\{ \frac{2\pi}{100} (200t - x) \right\}$$

Comparing this with the wave equation,

$$\psi(x, t) = A \sin \left\{ \frac{2\pi}{\lambda} (vt - x) \right\}$$

We have,

amplitude $A = 0.01$ m

wavelength $\lambda = 100$ m

wavelocity $v = 200\ \text{ms}^{-1}$

and frequency $\nu = \dfrac{v}{\lambda} = 2.0$ Hz

(*b*) Phase change in a time interval of Δt is

$$\frac{2\pi}{T} . \Delta t = 2\pi\nu . \Delta t$$

or phase difference $= 2\pi \times 2 \times 0.25 = \pi = 180°$. In other words, the particle phase is reversed in a time 0.25 s which is obvious since its period

$$T = \frac{1}{\nu} = 0.5 \text{ s}$$

(*c*) Phase difference for path difference of Δx is

$$\frac{-2\pi}{\lambda} . \Delta x = -\frac{2\pi}{100} \times 50 = -\pi = -180°$$

In other words, the particle located 50 m (which is half the wavelength) ahead of another particle lags in phase by 180°.

Example 7.2 A transverse harmonic wave of amplitude 0.01 m is generated at one end ($x = 0$) of a long horizontal string by a tuning fork of frequency 500 Hz. At a given instant of time the displacement of the particle at $x = 0.1$ m is -0.005 m and that of the particle at $x = 0.2$ m is $+0.005$ m. Calculate the wavelength and the wave velocity. Obtain the equation of the wave assuming that the wave is travelling along the $+x$ direction and that the end $x = 0$ is at the equilibrium position at $t = 0$.

Solution

Since the wave is travelling along $+x$ direction and the displacement of the end $x = 0$ is zero at time $t = 0$, the general equation of this wave is

$$\psi(x, t) = A \sin \left\{ \frac{2\pi}{\lambda} (vt - x) \right\} \qquad (i)$$

where $A = 0.01$ m

(*a*) When $x = 0.1$ m, $\quad \psi = -0.005$ m,

$$-0.005 = 0.01 \sin\left\{\frac{2\pi}{\lambda}(vt - x_1)\right\}, \text{ where } x_1 = 0.1 \text{ m}$$

or
$$\sin\left\{\frac{2\pi}{\lambda}(vt - x_1)\right\} = -\tfrac{1}{2}$$

$\therefore$
$$\text{Phase } \phi_1 = \frac{2\pi}{\lambda}(vt - x_1) = \sin^{-1}(-\tfrac{1}{2})$$

$$= \frac{7\pi}{6} \qquad (ii)$$

(*b*) When $x = 0.2$ m, $\quad \psi = +0.005$ m,

$$0.005 = 0.01 \sin\left\{\frac{2\pi}{\lambda}(vt - x_2)\right\} \text{ where } x_2 = 0.2 \text{ m}$$

or
$$\text{Phase } \phi_2 = \frac{2\pi}{\lambda}(vt - x_2) = \sin^{-1}\left(\frac{1}{2}\right) = \frac{\pi}{6} \qquad (iii)$$

From Eqs. (*ii*) and (*iii*)

$$\Delta\phi = \phi_1 - \phi_2 = \pi$$

Now

$$\Delta\phi = -\frac{2\pi}{\lambda} \cdot \Delta x$$

i.e.
$$\pi = -\frac{2\pi}{\lambda}(0.1 - 0.2)$$

or
$$\lambda = 0.2 \text{ m}$$

Now frequency ν of the wave = frequency of the tuning fork = 500 Hz

Hence wave velocity $v = \nu\lambda = 100 \text{ ms}^{-1}$

Substituting for A, λ and v in Eq. (i) we get

$$\psi(x, t) = 0.01 \sin\{10\pi(100t - x)\}$$

This is the equation of the wave where ψ and x are in metres and t in seconds.

Example 7.3 Transverse wave are generated in two uniform steel wires A and B of diameters 10^{-3} m and 0.5×10^{-3} m respectively, by attaching their free end to a vibrating source of frequency 500 Hz. Find the ratio of the wavelengths if they are stretched with the same tension.

Solution

The density ρ of a wire of mass M, length L and diameter d is given by

$$\rho = \frac{4M}{\pi d^2 L} = \frac{4\mu}{\pi d^2} \text{ where } \mu = \frac{M}{L}$$

Now
$$v_A = \sqrt{\frac{T}{\mu_A}} \quad \text{and} \quad v_B = \sqrt{\frac{T}{\mu_B}}$$

$$\therefore \quad \frac{v_A}{v_B} = \sqrt{\frac{\mu_B}{\mu_A}} = \frac{d_B}{d_A}$$

but $v_A = \nu\lambda_A$ and $v_B = \nu\lambda_B$, ν being the frequency of the source.

Hence, $$\frac{\lambda_A}{\lambda_B} = \frac{v_A}{v_B} = \frac{d_B}{d_A} = \frac{0.5\times 10^{-3}}{10^{-3}} = 0.5 = \tfrac{1}{2}$$

Example 7.4 Calculate the velocity of sound in (*a*) water and (*b*) steel. Given density of steel = 7800 kgm^{-3}, Young's modulus of steel = 20×10^{10} Nm^{-2} and bulk modulus of water = 0.20×10^{10} Nm^{-2}.

Solution

(*a*) $$E_w = 0.20\times 10^{10}\ \text{Nm}^{-2}$$

$$\rho_w = 1000\ \text{kgm}^{-3}$$

$$\therefore \quad v_w = \sqrt{\frac{E_w}{\rho_w}} = 1414\ \text{ms}^{-1}$$

(*b*) $$Y_s = 20\times 10^{10}\ \text{Nm}^{-2}$$

$$\rho_s = 7800\ \text{kgm}^{-3}$$

$$\therefore \quad v_s = \sqrt{\frac{Y_s}{\rho_s}} = 5060\ \text{ms}^{-1}$$

Example 7.5 Compare the velocities of sound in hydrogen (H_2) and carbon dioxide (CO_2). The ratio (γ) of specific heats of H_2 and CO_2 are respectively 1.4 and 1.3.

Solution

$$v_{H_2} = \sqrt{\frac{\gamma_{H_2}P}{\rho_{H_2}}}\ ; \quad v_{CO_2} = \sqrt{\frac{\gamma_{CO_2}P}{\rho_{CO}}}$$

$$\therefore \quad \frac{v_{H_2}}{v_{CO_2}} = \sqrt{\frac{\gamma_{H_2}}{\gamma_{CO_2}}\cdot\frac{\rho_{CO_2}}{\rho_{H_2}}}$$

Since density of a gas is proportional to its molecular weight

$$\frac{\rho_{CO_2}}{\rho_{H_2}} = \frac{44.01}{2.016}.$$

$$\therefore \quad \frac{v_{H_2}}{v_{CO_2}} \sqrt{\frac{1.4}{1.3}\times\frac{44.01}{2.016}} \simeq 4.85$$

Velocity of sound in hydrogen is 4.85 times that in carbon dioxide.

Example 7.6 In a laboratory experiment (room temperature being 15°C) the wavelength of a note of sound of frequency 500 Hz is found to be 0.68 m. If the density of air at STP is 1.29 kgm^{-3}, calculate the ratio of the specific heats of air.

Solution

Velocity of sound (v_t) at 15°C $= \nu\lambda = 500\times0.68 = 340\ \text{ms}^{-1}$

Velocity (v_0) at 0°C $= 340\times\sqrt{\dfrac{273}{273+15}} = 331\ \text{ms}^{-1}$

At STP

$$\rho_0 = 1.29\ \text{kgm}^{-3}$$

$$P_0 = 0.76\times13.6\times10^3\times9.8\ \text{Nm}^{-2}$$

and

since $$v_0 = \sqrt{\frac{\gamma P_0}{\rho_0}};\quad \gamma = \frac{C_P}{C_V}$$

$$\gamma = v_0^2\frac{\rho_0}{P_0} = 1.39.$$

Example 7.7 The intensity of sound in a normal conversation at home is about $3\times10^{-6}\ \text{Wm}^{-2}$ and the frequency of normal human voice is about 1000 Hz. Find the amplitude of waves, assuming that the air is at standard conditions.

Solution

At standard conditions (STP)
Density (ρ) of air $= 1.29\ \text{kgm}^{3}$
Velocity of sound (v) $= 332.5\ \text{ms}^{-1}$
Using Eq. (7.45) we have

$$I = 2\pi^2\rho\nu^2A^2v$$

or $$A = \frac{1}{\pi\nu}\sqrt{\frac{1}{2\rho v}}$$

whence $$A = 1.88\times10^{-8}\ \text{m}$$

Note that the amplitude of sound waves in normal conversation is extremely small.

Example 7.8 A long uniform string of linear density 0.1 kg m^{-1} is stretched with a tension of 40 N. One end of the string ($x = 0$) is given transverse sinusoidal oscillations with an amplitude of 0.02 m and period of 0.1 s so that transverse waves are set up in the $+x$ direction. (*a*) Find the velocity and wavelength of the waves. (*b*) If at the driving end ($x = 0$) the displacement (y) at time $t = 0$ is 0.01m but the velocity at this end is negative at $t = 0$, what is the equation of the travelling waves. (*c*) What is the velocity of the string at a point $x = 10$ m at time $t = 1$ s?

Solution

Linear density (μ) $= 0.1\ \text{kg m}^{-1}$
Tension (T) $= 40\ N$
Time period $= 0.1$ s or frequency (ν) $= 10$ Hz
Amplitude (A) $= 0.02$ m

(*a*) Wave velocity $v = \sqrt{\frac{T}{\mu}} = 20 \text{ ms}^{-1}$

Wavelength $\lambda = \frac{v}{\nu} = 2 \text{ m}$

(*b*) It is given that, at the driving end ($x = 0$) the displacement at $t = 0$ is finite $= 0.01$ m but the velocity is negative. It is also given that waves are sinusoidal and they travel in $+x$ direction. These requirements are met if the equation of the waves is

$$y(x, t) = A \sin \left\{ \frac{2\pi}{\lambda}(x - vt) + \delta \right\}$$

where δ is the initial phase to be determined. Now at $x = 0$, $y = 0.01$ m at $t = 0$. Substituting these values in the above equation and remembering that $A = 0.02$ m we find that

$$\sin \delta = \frac{1}{2}$$

$$\therefore \quad \delta = \frac{\pi}{6}$$

Substituting the values of v, λ and δ in the wave equation we have

$$y(x, t) = 0.02 \sin \pi \left(x - 20t + \frac{1}{6} \right)$$

which is the required equation.

(*c*) The velocity of the string is obtained by differentiating $y(x, t)$ with respect to t. Thus

$$\frac{\partial y}{\partial t} = -0.4\pi \cos \pi \left(x - 20t + \frac{1}{6} \right)$$

Notice that at $x = 0$ and $t = 0$, the velocity is indeed negative. The velocity of the string at $x = 10$ m and $t = 1$ s is equal to

$$-0.4\, \pi \cos \left(-10\pi + \frac{\pi}{6} \right) = 1.088 \text{ ms}^{-1}$$

Example 7.9 A very long string of linear density 0.1 g cm^{-1} is stretched with a tension of 400 N. It is driven at one end in harmonic motion of amplitude 1 cm and frequency 100 Hz. Calculate the time-averaged energy flux in watts.

Solution

$$\mu = 0.1 \text{ g cm}^{-1} = 0.01 \text{ kg m}^{-1}$$

$$T = 400 \text{ N}$$

$$A = 1 \text{ cm} = 0.01 \text{ m}$$

$$\nu = 100 \text{ Hz}$$

The time-averaged energy flux is given by

$$\langle P(t)\rangle = 2\pi^2A^2\mu\nu^2 v$$

$$= 2\pi^2A^2\nu^2(\mu T)^{1/2} \qquad \left(\because \; v=\sqrt{\frac{T}{\mu}}\right)$$

$$= 39.48 \text{ W}$$

Example 7.10 The intensity of sunlight at normal incidence on the earth's surface is 1.35 kWm^{-2}. Calculate the amplitude of the electric and magnetic fields of the electromangetic wave at the earth's surface. Given $\mu_0 = 4\pi\times10^{-7}$ Hm^{-1} and $\varepsilon_0 = (36\pi\times10^9)^{-1}$ Fm^{-1}.

Solution

The electromagnetic waves emitted by the sun travel in free space of permeability μ_0 and permittivity ε_0. The average value of the Poynting vector measures the intensity of the wave. We know that the magnitude of the Poynting vector is given by

$$P = \tfrac{1}{2}\, v(\varepsilon_0E^2+\mu_0H^2) = v\varepsilon_0E^2 = v\mu_0H^2$$

where $$v = \frac{1}{\sqrt{\mu_0\varepsilon_0}}$$

The intensity I is given by (since E and H are harmonic function of time)

$$I = \langle P\rangle = \tfrac{1}{2}\, v\varepsilon_0E_0^2 = \tfrac{1}{2}\, v\mu_0H_0^2$$

where E_0 and H_0 are the amplitudes of E and H respectively. Thus

$$E_0 = \sqrt{\frac{2I}{v\varepsilon_0}}$$

and $$H_0 = \sqrt{\frac{2I}{v\mu_0}}$$

Now $$I = 1.35 \text{ kWm}^{-2} = 1350 \text{ Wm}^{-2}$$

$$\mu_0 = 4\pi\times10^{-7} \text{ Hm}^{-1}$$

$$\varepsilon_0 = \frac{1}{36\pi\times10^9} \text{ Fm}^{-1}$$

whence

$$E_0 = 1009 \text{ Vm}^{-1}$$

and $$H_0 = 2.7 \text{ Am}^{-1}$$

Example 7.11 The intensity of sunlight at the earth's surface 1.5×10^8 km away from the sun is 1.35 kWm^{-2}. Assuming that the total power lost by the sun is equal to the power received per unit area of earth's surface multiplied by the surface area of the sphere of radius equal to the distance of the earth from the sun, find out how much mass is converted per second to radiant energy in the sun.

Solution

Distance between the sun and the earth (R)

$$= 1.5\times 10^8 \text{ km} = 1.5\times 10^{11} \text{ m}$$

Intensity (i.e. power per unit area) (I) = 1.35 kWm^{-2} = 1350 Wm^{-2}

Energy radiated per second from the sun (E) = $I\times 4\pi R^2$

Using the mass-energy equivalence relation

$$E = mc^2$$

with $c = 3\times 10^8$ ms^{-1}, we can calculate the mass converted per second into radiant energy which is equal to

$$\frac{E}{c^2} = \frac{I\times 4\pi R^2}{c^2}$$

$$= \frac{1350\times 4\pi\times(1.5\times 10^{11})^2}{(3\times 10^8)^2}$$

$$= 4.24\times 10^9 \text{ kg}$$

Example 7.12 Lasers are light sources which give almost perfectly parallel beam of light. If a 2 kW laser beam is concentrated by a lens into a cross-sectional area 10^{-6} cm^2, what is the value of the amplitude of the electrical field and the magnitude of the Poynting vector?

Solution

The intensity of the concentrated laser beam is

$$I = \frac{2\times 10^3 \text{ W}}{10^{-6}\times 10^{-4}\, m^2} = 2\times 10^{13} \text{ Wm}^{-2}$$

But

$$I = \tfrac{1}{2}\, v\varepsilon_0 E_0^2$$

where $v = 3\times 10^8$ ms^{-1} and $\varepsilon_0 = (36\pi\times 10^9)^{-1}$ Fm^{-1}

Therefore

$$E_0 = \left(\frac{2I}{v\varepsilon_0}\right)^{1/2} = \left\{\frac{2\times 2\times 10^{13}\times(36\pi\times 10^9)}{3\times 10^8}\right\}^{\frac{1}{2}}$$

$$= 1.23\times 10^8 \text{ Vm}^{-1}$$

The Poynting vector P is given by

$$P = v\varepsilon_0 E_0^2 = 0.4\times 10^{14} \text{ Wm}^{-2}$$

Example 7.13 Sun radiates energy at a rate of 3.8×10^{26} W. The radius of the sun is 7×10^8 m. (*a*) Calculate the magnitude of the Poynting vector at the surface of the sun. (*b*) What is the intensity of the solar radiation incident on the earth which is 1.5×10^{11} m away from the sun?

Solution

(*a*) Radius of the sun (R) $= 7\times10^8$ m

Surface area of the sun $= 4\pi R^2 = 4\pi\times(7\times10^8)^2 = 6.16\times10^{18}$ m^2

Poynting vector (P) at the surface of the sun is given by

P = rate of energy radiated per unit surface area

$$= \frac{3.8\times10^{26}}{6.16\times10^{18}} = 6.17\times10^7 \text{ Wm}^{-2}$$

(*b*) Intensity of solar radiation on the earth $= \dfrac{3.8\times10^{26}}{4\pi\times(1.5\times10^{11})^2}$

$$= 1.34\times10^3 \text{ Wm}^{-2}$$

Example 7.14 Sunlight strikes the earth with an intensity of 2 cal cm^{-2} min^{-1}. How many watts of power must an electrical lamp radiate in order to produce, at 1 m, the brightness of sunlight?

Solution

Intensity (I) of sunlight on earth $= 2$ cal cm^{-2} min^{-1}

$$= \frac{2\times4.2}{10^{-4}\times60} = 1.4\times10^3 \text{ Wm}^{-2}$$

The power (P) of a lamp which produces an intensity of 1.4×10^3 Wm^{-2} at $r = 1$ m is given by

$$P = I\times4\pi r^2 = 1.4\times10^3\times4\pi\times(1)^2$$

$$= 17.6 \text{ kW}$$

Example 7.15 Assuming that all the energy from a 1000 W lamp is radiated uniformly, calculate the values of the electric and magnetic fields of the radiation at a distance of 2m from the lamp.

Solution

Energy flux per unit area per second $= \dfrac{1000}{4\pi\times(2)^2} = \dfrac{1000}{16\pi}$ Wm^{-2}

i e $$E_0H_0 = \frac{1000}{16\pi} \text{ Wm}^{-2}$$

Also impedance $$Z = \frac{E_0}{H_0} = \sqrt{\frac{\mu_0}{\varepsilon_0}} = 376.7\ \Omega$$

$\therefore$ $$E_0^2 = (E_0H_0)\times\frac{E_0}{H_0} = \frac{1000}{16\pi}\times376.7 = 7493$$

or $$E_0 = 86.6 \text{ Vm}^{-1}$$

And $$H_0^2 = (E_0H_0)\div\frac{E_0}{H_0} = \frac{1000}{16\pi\times376.7} = 0.0528$$

or $$H_0 = 0.23 \text{ Am}^{-1}$$

QUESTIONS

1. (a) What do you understand by wave motion?

   (b) Give two examples of a disturbance which can propagate as wave motion.

   (c) What are transverse and longitudinal waves? Give one example of each.

2. (a) Define the terms wave velocity, wavelength and frequency. How are they related?

   (b) What is a harmonic wave? Obtain an equation for the displacement when a plane harmonic wave travels in a medium in positive x direction.

3. (a) Deduce the expression for the velocity of transverse waves on a long stretched string.

   (b) Show that the characteristic impedance offered by a string to the travelling waves is given by

$$Z = \sqrt{\mu T}$$

   where μ is the linear density of the string and T is the tension with which it is stretched.

4. (a) Deduce the expression for the velocity of longitudinal waves in a column of a gas and hence obtain Newton's formula. What is Laplace's correction to Newton's formula?

   (b) Show that the characteristic impedance offered by a loss-free gas to the sound waves travelling in it is given by

$$Z = \sqrt{\rho E}$$

   where ρ is the density of the gas and E its bulk modulus.

5. (a) Deduce the expression for the velocity of current and voltage waves on an ideal transmission line.

   (b) Show that the characteristic impedance offered by an ideal transmission line to the current and voltage waves is given by

$$Z = \sqrt{\frac{L}{C}}$$

   where L is the inductance per unit length of the line and C is the capacitance per unit length.

6. Write down Maxwell's equations for a dielectric medium and explain the significance of each equation.

7. (a) Write down Maxwell's equations for free space and hence obtain the expression for the velocity of plane electromagnetic waves in free space.

   (b) What do you understand by plane polarized waves?

   (c) Show that the characteristic impedance of free space to electromagnetic waves is given by

$$Z = \sqrt{\frac{\mu_0}{\varepsilon_0}}$$

   where μ_0 is the magnetic permeability and ε_0 the electrical permittivity of free space.

8. (a) Obtain an expression for the total energy per wavelength in a string when transverse waves travel in it.

   (b) Show that the time-averaged input power of the source generating waves in a string is equal to the total energy per unit length of the string times the wave velocity.

9. (a) Obtain an expression for the energy density for a plane sound wave in a gas.

   (b) Show that the intensity of a plane sound wave in a gas is equal to the energy density times the wave velocity.

10. (a) What is Poynting vector? What does it represent?

    (b) Show that the magnitude of the Poynting vector of an electromagnetic wave is equal to the product of the energy density of the electromagnetic field and the velocity of the electromagnetic wave.

11. (a) How will you describe displacements in a two-dimensional wave?

    (b) Deduce the wave equation in the case of two-dimensional waves on a stretched membrane. Write down a solution of the wave equation.

12. (a) Obtain the wave equation in three dimensions for electromagnetic waves in free space.

    (b) Write down a solution of the wave equation in rectangular Cartesian co-ordinates.

    (c) Write down the wave equation in cylindrical polar co-ordinates and spherical polar coordinates.

    (d) Show that the intensity of a wave spreading from a point source in a homogeneous isotropic medium obeys the inverse square law of distance.

PROBLEMS

1. Determine the amplitude, period, frequency, speed and wavelength of the waves given by (all quantities are in SI units)

   (a) $\psi(x, t) = 0.01 \sin 2\pi(3t - 2x)$

   (b) $\psi(x, t) = 0.02 \sin(2t + 3x)$

   (c) $\psi(x, t) = 0.05 \cos \pi(6t - x)$

2. Write down the equation for a harmonic wave travelling in the $+x$ direction at an instant of time $t = 2T$ and compare it with the equation at $t = 0$; T being the period of the wave. Plot a graph of the wave form at $t = 0$ for x lying between 0 and 2λ, where λ is the wavelength.

3. Solve Problem 2 for $t = 3T/4$. Plot the graphs of the wave form at $t = 0$ and $t = 3T/4$ for x lying between 0 and 2λ. Compare the two wave forms.

4 Write down the equation of motion of a particle at $x = 2\lambda$ along a harmonic wave travelling in the $+x$ direction and compare it with the equation at $x = 0$; λ being the wavelength. Plot a graph showing the displacement of the particle at $x = 0$ for t lying between 0 and $2T$, where T is the period of the wave.

5. Solve Problem 4 for $x = 3\lambda/4$. Plot the graphs showing the displacement of the particle at $x = 0$ and $x = 3\lambda/4$ for t lying between 0 and $2T$. Compare the two graphs.
6. Solve Problems 2, 3, 4 and 5 for a wave travelling in the $-x$ direction.
7. The equation of a transverse wave on a string is

$$y(x, t) = 0.05 \sin (3t - 4x)$$

where y, x and t are in SI units. Answer the following questions:

(*a*) What is the wavelength, frequency and velocity of the wave?

(*b*) At $t = 0$, what is y at $x = 0$, 0.5 m, 1.0 m and $\pi/2$ m?

(*c*) At $x = 0.5$ m, what is y at $t = 0$, 0.5 s and $\pi/3$ s?

(*d*) What is the total energy per unit length of the string if mass per unit length of the string is 0.1 kg m^{-1}?

(*e*) Calculate the average rate at which energy is transferred along the string.

8. Two strings *A* and *B* are made of the same material. The cross-sectional area of *A* is half that of *B* while the tension on *A* is twice that on *B*. Compare the velocities of transverse waves on them.
9. The intensity of the sound of a barking dog is 10^{-3} Wm^{-2}. If the frequency is 1000 Hz, calculate the amplitude of the sound waves in air at standard conditions.
10. What will be the amplitude of the sound waves of Problem 9 if the air temperature were 35° C?
11. Calculate the characteristic impedance offered by air at standard conditions to sound waves travelling in it.
12. A wave of frequency 20 Hz has a velocity of 120 ms^{-1}.

(*a*) How far apart are two points whose displacements are 60° out of phase?

(*b*) At a given point, what is the phase difference between two displacements occurring at times separated by 0.01 s?

13. The velocity of sound is 330 ms^{-1} in air at standard conditions. Calculate the acoustic pressure for a painful sound of intensity 10 Wm^{-2}.
14. A sound of intensity 10^{-12} Wm^{-2} is barely audible in air. If the frequency of sound is 500 Hz, find the approximate value of the displacement amplitude of air molecules.
15. A radio station is radiating an average power of 10^6 W uniformly over a hemisphere concentric with the station Calculate (*a*) the magnitude of the Poynting vector and (*b*) the amplitudes of the electric and magnetic fields of the plane electromagnetic wave at a point distant 10 km from the station. Given $\mu_0 = 4\pi + 10^{-7}$ Hm^{-1} and $\varepsilon_0 = (36\pi + 10^9)^{-1}$ Fm^{-1}.

8

Reflection and Standing Waves

8.1 INTRODUCTION

In the preceding chapter we have discussed waves which travel uninterrupted in open or unbounded systems. We assumed that the medium was homogeneous and infinite in extent. This chapter is devoted to closed or bounded systems. We will describe some of the effects which take place when a wave travelling in a medium meets the boundary of another medium.

When a wave travelling in medium *1* meets the boundary of a medium *2* it is partly reflected and partly transmitted at the boundary. The reflected wave travels in medium *1* and the transmitted wave in medium *2*. In Chap. 7 we have seen that a medium offers a characteristic impedance to the waves travelling in it. The characteristic impedance depends on the properties of the medium. Since the two media offer different impedances, it will be interesting to find out how waves will respond to an abrupt change of impedance at the boundary separating the two media. We will analyse this problem for all the waves we have described in the preceding chapter; transverse waves on a string, longitudinal (sound) waves in a medium, voltage and current waves on a transmission line and electromagnetic waves in a dielectric medium.

8.2 REFLECTION AND TRANSMISSION OF TRANSVERSE WAVES AT A BOUNDARY BETWEEN TWO STRINGS

Let us suppose that two strings *1* and *2* of different linear densities are joined at a point to form a composite string. Let us assume that both strings are stretched with the same tension T. The characteristic impedances of the strings are $Z_1 = \mu_1 v_1$ and $Z_2 = \mu_2 v_2$, where μ_1 and μ_2 are the linear densities of the strings and v_1 $(= \sqrt{T/\mu_1})$ and v_2 $(= \sqrt{T/\mu_2})$ are the wave velocities in strings 1 and 2 respectively.

Let us suppose a wave (called the incident wave) is travelling in the $+x$ direction on string 1. The particled displacements of string *1* are given by,

$$y_i(x, t) = A_i \sin(\omega t - k_1 x)$$

where $k_1 = 2\pi/\lambda_1 = 2\pi\nu/v_1$ and A_i is the amplitude of the incident wave. When this wave reaches the boundary (which we shall take at $x = 0$) separating the two strings, it is partly reflected and partly transmitted at the boundary. The reflected wave travels on string *1* in the negative x direction and the transmitted wave travels on string *2* in the positive x direction (see Fig. 8.1). The particle displacements due to these waves are, therefore given by

$$y_r(x, t) = A_r \sin(\omega t + k_1 x)$$

$$y_t(x, t) = A_t \sin(\omega t - k_2 x)$$

where $k_2 = 2\pi/\lambda_2 = 2\pi\nu/v_2$ and A_r and A_t are the amplitudes of the reflected and transmitted waves respectively.

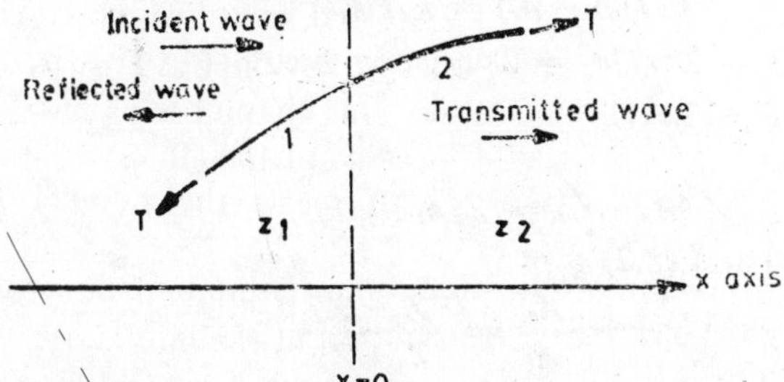

Fig. 8.1 Reflection and transmission of a wave at the boundary ($x = 0$) of characteristic impedances Z_1 and Z_2.

Boundary conditions. The point $x = 0$ at the boundary undergoes oscillations under the combined influence of the incident and reflected waves in string *1*. It then acts as a source of transmitted waves travelling in string *2*. The boundary conditions to be satisfied at $x = 0$ are

(1) The displacement is the same immediately to the left and to the right of the boundary at $x = 0$, i.e. $y(x, t)$ *is continuous across the boundary at* $x = 0$. Therefore, velocity $\frac{\partial y(x, t)}{\partial t}$ is also continuous.

(2) The restoring force or transverse component of tension $\left[-T \frac{\partial y(x, t)}{\partial x}\right]$ is continuous across the boundary at $x = 0$. To understand this, let us imagine that there is an infinitesimal element of mass at $x = 0$. If the gradient $\frac{\partial y}{\partial x}$ is not continuous, this would give rise to a net transverse force acting on this infinitesimally small mass of the string at $x = 0$. Consequently this element will have an infinite acceleration which is not permitted.

Recalling that $y(x, t)$ in string *1* is due to a superposition of the displacements due to the incident and reflected waves and $y(x, t)$ in string *2* is due only to the transmitted wave and using the boundary condition (1) we have

$$y_i(x, t)+y_r(x, t) = y_t(x, t)$$

or
$$A_i \sin (\omega t-k_1x)+A_r \sin (\omega t+k_1x) = A_t \sin (\omega t-k_2x)$$

Setting $x = 0$ we have

$$A_i \sin \omega t+A_r \sin \omega t = A_t \sin \omega t$$

for all t, which gives

$$A_i+A_r = A_t \tag{8.1}$$

Boundary condition (2) requires that at $x = 0$ we have

$$-T \frac{\partial y_i}{\partial x} -T \frac{\partial y_r}{\partial x} = -T \frac{\partial y_t}{\partial x}$$

or
$$A_ik_1T \cos (\omega t-k_1x)-A_rk_1T \cos (\omega t+k_1x) = A_tk_2T \cos (\omega t-k_2x)$$

Setting $x = 0$, we have (for all t)

$$k_1T(A_i-A_r) = k_2TA_t$$

Now
$$k_1T = 2\pi\nu T/v_1 = 2\pi\nu\mu_1v_1 = 2\pi\nu Z_1 \quad (\because T = \mu_1v_1^2 ; Z_1 = \mu_1v_1)$$

and
$$k_2T = 2\pi\nu Z_2$$

We then have

$$Z_1(A_i-A_r) = Z_2A_t \tag{8.2}$$

Equations (8.1) and (8.2) give

$$r_{12} = \frac{A_r}{A_i} = \frac{Z_1-Z_2}{Z_1+Z_2} \tag{8.3}$$

and
$$t_{12} = \frac{A_t}{A_i} = \frac{2Z_1}{Z_1+Z_2} \tag{8.4}$$

where r_{12} and t_{12} are the reflection and transmission amplitude coefficients when a wave travels from string *1* to string *2*.

The ratio A_r/A_i is the fraction of the incident amplitude reflected at the boundary and the ratio A_t/A_i is the fraction of the incident amplitude transmitted across the boundary. These fractions depend only on the impedances and are independent of the angular frequency ω of the incident wave.

If the string is rigidly fixed at $x = 0$, say, by attaching this end to a wall, then the medium *2* is infinitely massive which means that $Z_2 = \infty$. In this case $A_t/A_i = 0$ giving $A_t = 0$ indicating that there is no transmitted wave and $A_r/A_i = -1$ or $A_r = -A_i$ which means that the incident wave is completely reflected with a reversal in amplitude. The reversal of amplitude means a phase change of π on reflection. It is clear from Eq. (8.3) that if $Z_2>Z_1$, the ratio A_r/A_i will be negative indicating a phase change of π on reflection. Thus we conclude that if a wave travelling in a medium of lower impedance meets the boundary of a medium of a higher impedance (i.e. a medium in which the wave velocity is smaller), the wave reflected at the boundary undergoes a phase change of π.

On the other hand, if $Z_2 < Z_1$, the ratio A_r/A_i is positive. Thus, if a wave is reflected at the boundary of a lower impedance (i.e. a medium in which the wave velocity is higher), the reflected wave does not undergo any phase change.

Furthermore, Eq. (8.4) shows that the ratio A_t/A_i always remains positive independent of whether Z_2 is less than or more than Z_1. This shows that the transmitted wave does not undergo any phase change. This is illustrated in Fig. 8.2 which shows reflection and transmission of a pulse at a boundary between two strings. In the next chapter we shall discuss pulses in detail. For the moment, a pulse may be regarded as a particular shape or profile of a limited portion of a string.

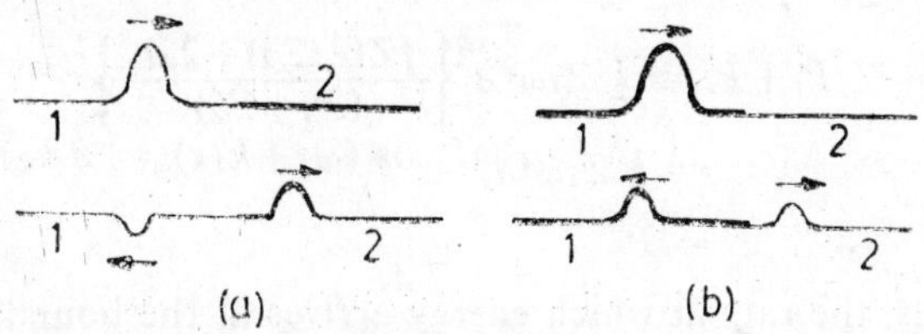

Fig. 8.2 (*a*) Reflection and transmission of a pulse when string 2 has a higher impedance. Note a phase of π in the reflected pulse,
(*b*) Reflection and transmission of a pulse when the second string has a lower impedance. Note that the reflected pulse does not undergo any phase change in this case, the transmitted pulse does not undergo any phase change.

If the wave travels from string 2 to string 1 the amplitude reflection and transmission coefficients are given by [see Eqs. (8.3) and (8.4)]

$$r_{21} = \frac{Z_2 - Z_1}{Z_2 + Z_1}$$

$$t_{21} = \frac{2Z_2}{Z_2 + Z_1}$$

Reflection and Transmission of Energy at the Boundary

Waves are a very useful mechanism for the transport of energy in a medium. It is interesting to consider what happens to the energy in a wave when it meets a boundary between two media of different impedances.

As the wave travels along the string, each part of the string is thrown into harmonic oscillations with the passage of time. In Chap. 7 we have seen that the rate at which energy is carried per unit length along the string is given by [see Eq. (7.43)]

$$P = \tfrac{1}{2}\,\mu v \omega^2 A^2 = \tfrac{1}{2}\,Z\omega^2 A^2$$

Let us now compute the rates at which energy is incident, reflected and transmitted at the boundary at $x = 0$. The rate of incident energy is the rate at which the energy is carried by the incident wave which is given by

$$P_i = \tfrac{1}{2} Z_1\omega^2A_i^2$$

Similarly the rates of reflected and transmitted energies respectively are

$$P_r = \tfrac{1}{2} Z_1\omega^2A_r^2$$

and
$$P_t = \tfrac{1}{2} Z_2\omega^2A_t^2$$

Using Eqs. (8.3) and (8.4) we find that

$$P_r = \tfrac{1}{2} Z_1\left(\frac{Z_1-Z_2}{Z_1+Z_2}\right) \omega^2A_i^2$$

$$P_t = \frac{Z_1Z_2}{(Z_1+Z_2)} \omega^2A_i^2$$

Notice that

$$P_r+P_t = \tfrac{1}{2} Z_1\omega^2A_i^2\left\{\frac{(Z_1-Z_2)+2Z_2}{(Z_1+Z_2)}\right\}$$

$$= \tfrac{1}{2} Z_1\omega^2A_i^2$$

$$= P_i$$

In other words, the rate at which energy arrives at the boundary with the incident wave is equal to the rate at which energy leaves the boundary with the reflected and transmitted waves. This is consistent with the conservation of energy at the junction of two media. Thus, energy is conserved. All the energy arriving at the boundary with the incident wave leaves the boundary with the reflected and transmitted waves. This is expected since we have assumed that there is no absorption of energy at the boundary.

The reflection and tramsmission energy coefficients are given by

$$\frac{\text{Reflected energy}}{\text{Incident energy}} = \frac{P_r}{P_i} = \frac{Z_1A_r^2}{Z_1A_i^2} = \frac{A_r^2}{A_i^2} = \left(\frac{Z_1-Z_2}{Z_1+Z_2}\right)^2$$

$$\frac{\text{Transmitted energy}}{\text{Incident energy}} = \frac{P_t}{P_i} = \frac{Z_2A_t^2}{Z_1A_i^2} = \frac{4Z_1Z_2}{(Z_1+Z_2)^2}$$

Notice that

$$\text{Reflection coefficient} + \text{Transmission coefficient} = 1$$

Also if $Z_1 = Z_2$, no energy is reflected and the impedances are said to be *matched.*

Standing Waves on a String of Fixed Length

Consider a string of length L stretched with a tension T along the x-axis with its ends $A(x = 0)$ and $B(x = L)$ rigidly fixed (Fig. 8.3). A transverse wave is produced by creating a harmonic disturbance near the end A. The wave travels along the string in the $+x$ direction and gets reflected at the fixed end B giving rise to a reflected wave which travels in the $-x$ direction. The interference between the two oppositely travelling waves gives rise to a standing wave on the string. Let us consider the simplest case of a monochromatic incident wave of a single frequency ω and amplitude A_i. The displacement at time t of a particle, say C, located at x due to the incident wave travelling in the $+x$ direction is given by

$$y_i(x, t) = A_i \sin(\omega t - kx) = A_i \sin\{k(vt - x)\} \tag{8.5}$$

where $k = 2\pi/\lambda$; λ being the wavelength of the wave and $v = \omega/k$ is the velocity of the wave.

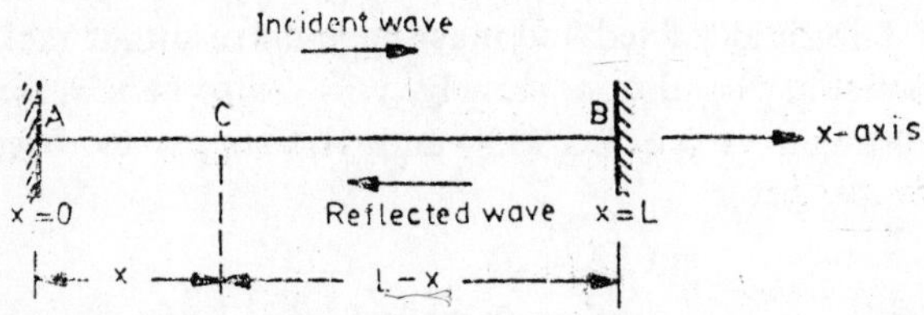

Fig. 8.3 Standing waves on a string fixed at both ends $x = 0$ and $x = L$.

Now we ask the question – what is the displacement at time t of the same particle C located at x due to the wave reflected from the fixed end B? Remember that the reflected wave is the wave created at A [which is the incident wave given by Eq. (8.5)] and reflected at B. Therefore, the reflected wave can be written as

$$y_r(x, t) = A_r \sin\{k(vt - x')\}$$

where A_r is the amplitude of the reflected wave and x' must be the distance travelled by the wave of Eq. (8.5) from A to B and then to C, i.e. $x = L + (L - x) = 2L - x$. Hence the displacement at time t of particle C at x due to the reflected wave must be given by

$$y_r(x, t) = A_r \sin\{k(vt + x - 2L)\} \tag{8.6}$$

The argument of the sine function involves $(vt + x)$ which represents a wave in the $-x$ direction, the direction of propagation of the wave reflected at B.

From the principle of superposition, the resultant displacement at time t at any point x is given by

$$\begin{aligned} y(x, t) &= y_i(x, t) + y_r(x, t) \\ &= A_i \sin\{k(vt - x)\} + A_r \sin\{k(vt + x - 2L)\} \end{aligned} \tag{8.7}$$

Since the ends $x = 0$ and $x = L$ are rigidly fixed, the boundary conditions are (*i*) $y = 0$ at $x = 0$ and (*ii*) $y = 0$ at $x = L$ at all times. Using condition (*ii*) in Eq. (8.7) gives

$$\begin{aligned} 0 &= A_i \sin\{k(vt - L)\} + A_r \sin\{k(vt - L)\} \\ &= (A_i + A_r) \sin\{k(vt - L)\} \end{aligned}$$

which is satisfied for all t values if

$$A_i + A_r = 0 \qquad \text{or} \qquad A_r = -A_i = -A \text{ (say)}$$

The reversal of amplitude of the reflected wave implies a phase change of 180° on reflection at the rigid end $x = L$. Setting $A_r = -A_i = -A$ in

Eq. (8.7) we have

$$y(x, t) = A\,[\sin \{k(vt-x)\} - \sin \{k(vt+x-2L)\}]$$

Recalling that $\sin\alpha - \sin\beta = 2\sin\frac{\alpha-\beta}{2}.\cos\frac{\alpha+\beta}{2}$, we get

$$y(x, t) = 2A.\sin\{k(L-x)\}\cos\{k(vt-L)\} \tag{8.8}$$

The correctness of Eq. (8.7) and hence of Eq. (8.8) can be immediately checked. Notice that at $x = L$, Eq. (8.8) gives $y = 0$ which is true since the end $x = L$ is rigidly fixed and must have no resultant motion.

The first boundary condition, namely, $y = 0$ also at $x = 0$ remains to be satisfied. Putting $x = 0$ in Eq. (8.8) and requiring y to become zero at all values of t, we get

$$\sin kL = 0$$

or

$$kL = 0, \pi, 2\pi, \ldots$$

or

$$k_n L = n\pi$$

or

$$k_n = \frac{n\pi}{L} \tag{8.9}$$

where n is an integer having values 0, 1, 2, 3,...

The allowed frequencies are ($\because\ \omega = kv$)

$$\omega_n = \frac{n\pi v}{L} \tag{8.10}$$

or

$$\nu_n = \frac{nv}{2L} = \frac{n}{2L}\sqrt{\frac{T}{\mu}} \tag{8.11}$$

Using Eq. (8.9) in Eq. (8.8) we have

$$y_n(x, t) = 2A\sin\left\{\frac{n\pi}{L}(L-x)\right\}\cos\left\{\frac{n\pi}{L}(vt-L)\right\}$$

$$= -2A\sin\left(\frac{n\pi x}{L}\right)\cos\left(\frac{n\pi vt}{L}\right) \tag{8.12}$$

$$= -2A\sin(k_n x)\cos(\omega_n t) \tag{8.13}$$

Equation (8.13) is the equation of a standing or stationary wave. It is obvious that this equation does not represent a travelling wave since it no longer has the characteristic form involving $(\omega t - kx)$ or $(\omega t + kx)$ in the argument of the trignometric function.

(c) *Modes of Vibration*. Notice that the frequencies given by Eq. (8.11) are just the frequencies of the normal modes of a string fixed at both ends which we have already come across in Chapter 5. The mode with $n = 0$ is unphysical since then $\nu = 0$ implying that there is no vibration at all.

The mode with $n = 1$ is called the fundamental mode. The frequency and particle displacements in this mode are given by [see Eqs (8.11) and (8.12)]

$$\nu_1 = \frac{1}{2L}\sqrt{\frac{T}{\mu}}$$

$$y_1(x, t) = -2A\sin\left(\frac{\pi x}{L}\right)\cos\left(\frac{\pi vt}{L}\right)$$

Notice that the points $x = 0$ and $x = L$ are permanently at rest. These points on the string are called *nodal points* or *nodes*. But the point $x = L/2$ has a maximum displacement. These points are called *antinodes*. Figure 8.4 (*a*) shows the particle displacement. in the fundamental mode.

The mode with $n = 2$ is called the second harmonic. The frequency and particle displacements in this mode are given by

$$\nu_2 = \frac{1}{L}\sqrt{\frac{T}{\mu}} = 2\nu_1$$

$$y_2(x, t) = -2A \sin\left(\frac{2\pi x}{L}\right) \cos\left(\frac{2\pi vt}{L}\right)$$

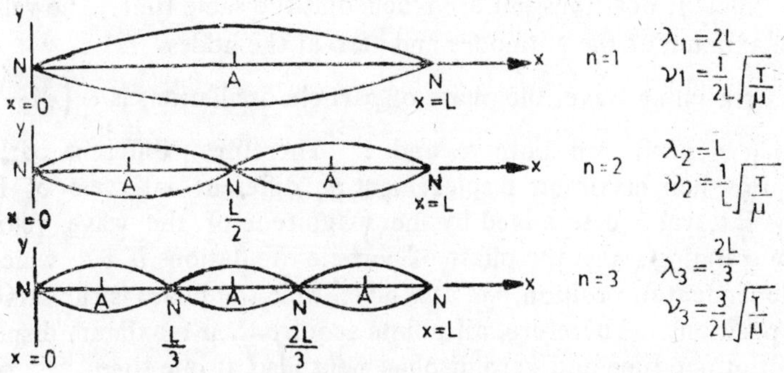

Fig. 8 4 Standing Waves on a string of length L fixed at both ends Modes of string: first three harmonics $n = 1$, 2 and 3.

Notice that, in addition to nodes at $x = 0$ and $x = L$, there is another node at $x = L/2$. The antinodes are at, $L/4$, $3L/4$. This mode is shown in Fig. 8.4 (*b*). Figure 8.4 (*c*) shows the third harmonic. We find that in the nth harmonic there are $(n-1)$ positions (between the fixed ends) equally spaced along the string where the displacement is always zero. These points marked N are the nodes in a standing wave pattern.

Characteristic Features of the Standing Wave. Standing waves are not waves in the true sense of the term; they can at most be called harmonic vibrations or modes with the following characteristic features.

(1) In a travelling wave the amplitude of particle oscillations is the same at all positions of the medium. On the other hand, in a stationary (standing) wave, the amplitudes of oscillations are different at different places. At nodes the amplitude is zero and at antinodes the amplitude is maximum, being equal to the sum of the amplitudes of the constituent waves. At other intermediate positions the amplitude varies between these two limits. The distance between two consecutive nodes or antinodes is half the wavelength of the constituent waves.

(2) In a travelling wave the particle velocity is given by

$$\frac{\partial y}{\partial t} = Akv \cos \{k(vt-x)\}$$

The velocity varies from place to place and from time to time. The maximum ($= \pm kAv = \pm A\omega$) and minimum ($= 0$) values of velocity are the same for all particles but are reached at different times. On the other hand, in a standing wave the particle velocity is given by

$$\frac{\partial y}{\partial t} = 2Akv \sin (kx) \sin (kvt)$$

The velocity varies from place to place and from time to time. The maximum and minimum values of particle velocity are different for different positions but are reached at the same time. The velocity is maximum at the antinodes and zero at the nodes.

(3) In a travelling wave, the phase of particle oscillations is $\omega\left(t-\frac{x}{v}\right)$ which depends on both x and t. Therefore, different particles acquire their maximum displacement at different intervals of time. This interval is determined by the magnitude of the wave velocity. In a standing wave, the phase of particle oscillations is ωt, which is independent of position, i.e. at a given time the phase is the same at all positions. Therefore, all points acquire their maximum displacement at one time and zero displacement also at one time.

(4) The most striking feature of a standing wave is that energy is not transported along the medium. The reason is the energy cannot conceivably flow past the nodes of the medium which are permanently at rest. There is nothing like a wave-front is standing wave, since the phase is the same at all points, changing only with time. Hence, there is no energy flux. A formal mathematical proof of this is obviously unnecessary. Further, in a standing wave, twice in each cycle, all particles of the medium have zero displacement so that the potential energy is zero and energy of the system is then entirely kinetic. A quarter of a period later, all particles of the medium simultaneously attain their maximum displacement and at this instant the kinetic energy is zero, the energy of the medium being entirely potential. At other instants the energy is partly potential and partly kinetic. Hence, in a standing wave, the energy periodically becomes entirely kinetic and entirely potential. In a travelling wave the average energy of the medium is half potential and half kinetic.

Standing Wave Ratio (SWR). Standing waves are formed due to a superposition of waves travelling in opposite directions. If the amplitudes of these travelling waves are equal and opposite (resulting from complete reflection at rigid boundaries) nodes (i.e. points of zero amplitude) will exist in the standing wave pattern. Often, however, the boundaries are not perfectly rigid resulting in incomplete or partial reflections.

If we have waves which are partially reflected (i.e. $|A_r| < |A_i|$) from a boundary, they combine with the incoming waves to form a standing wave pattern which is, however, superposed on travelling waves so that the amplitude at nodes is not zero. In such cases, we speak of *standing wave ratio* (SWR).

Consider two monochromatic waves of different amplitudes travelling in opposite directions in a medium. The displacement at a point x at time t due to the two waves are given by

$$y_i(x, t) = A_i \sin(\omega t - kx)$$

$$y_r(x, t) = A_r \sin(\omega t + kx)$$

The resultant displacement at a point x due to a superposition of the two waves is given by

$$\begin{aligned} y(x, t) &= y_i(x, t) + y_r(x, t) \\ &= A_i \sin(\omega t - kx) + A_r \sin(\omega t + kx) \\ &= (A_i + A_r) \cos kx \sin \omega t - (A_i - A_r) \sin kx \cos \omega t \end{aligned}$$

This equation may be recast as

$$y(x, t) = R \sin(\omega t - \phi)$$

where the resultant amplitude R and phase ϕ are given by

$$R^2 = (A_i^2 + A_r^2 + 2A_iA_r \cos 2kx)^{1/2}$$

and $$\tan \phi = \frac{(A_i - A_r)}{(A_i + A_r)} \tan kx$$

At antinodes (given by $\cos 2kx = +1$). The amplitude is maximum given by

$$R_{\max} = A_i + A_r$$

At nodes (given by $\cos 2kx = -1$) the amplitude is minimum given by

$$R_{\min} = A_i - A_r$$

The standing wave ratio (SWR) is defined as

$$\text{SWR} = \frac{R_{\max}}{R_{\min}} = \frac{A_i + A_r}{A_i - A_r} = \frac{1 + A_r/A_i}{1 - A_r/A_i} \tag{8.14}$$

In terms of amplitude reflection coefficient $r = A_r/A_i$, we have

$$\text{SWR} = \frac{1+r}{1-r}$$

or $$r = \frac{\text{SWR}-1}{\text{SWR}+1} \tag{8.15}$$

where we have assumed that the phase change on reflection is zero or π. The standing wave ratio is calculated by measuring the maximum and minimum values of amplitude R and the reflection coefficient is then determined from Eq. (8.15). This is the usual way of determining the reflection coefficient and hence of the impedance of a boundary. For a perfectly rigid boundary the reflection coefficient is unity and impedance infinite.

8 3 REFLECTION AND TRANSMISSION OF LONGITUDINAL (SOUND) WAVES AT A BOUNDARY BETWEEN TWO MEDIA

When a sound wave meets a boundary separating two media of different acoustic impedances, it is partly reflected and partly transmitted at the boundary. Consider a plane sound wave travelling in a medium *1* of density ρ_1 and incident normally on a plane boundary at $x = 0$ separating medium *1* from another medium *2* of density ρ_2. The acoustic impedances of the two media respectively are

$$Z_1 = \rho_1 v_1$$

$$Z_2 = \rho_2 v_2$$

where v_1 and v_2 are the sound speeds in medium *1* and medium *2* respectively (see Fig. 8.5). The incident, reflected and transmitted waves are respectively given by

$$\xi_i(x, t) = A_i \sin(\omega t - k_1 x)$$

$$\xi_r(x, t) = A_r \sin(\omega t + k_1 x)$$

$$\xi_t(x, t) = A_t \sin(\omega t - k_2 x)$$

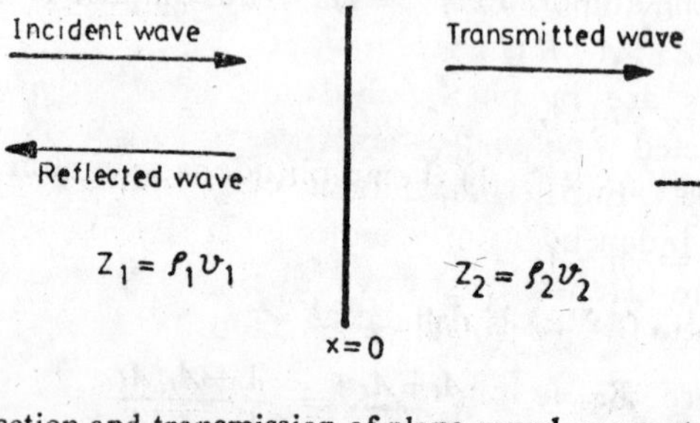

Fig 8.5 Reflection and transmission of plane sound waves at a plane boundary separating two media of acoustic impedances Z_1 and Z_2.

The boundary conditions are :—

(1) The particle displacement $\xi(x, t)$ is continuous across the boundary. Hence the particle velocity $\frac{\partial \xi}{\partial t}$ is also continuous.

(2) The excess pressure $\left(p = \Delta P = -E\dfrac{\partial \xi}{\partial x} = -\gamma P_0 \dfrac{\partial \xi}{\partial x}\right)$ is continuos across the boundary [(see Eq. 6.34 of Chap. 6)]. Here $\gamma = C_P/C_V$ and P_0 is the equilibrium pressure. Thus the two boundary conditions to be satisfied are (at $x = 0$)

$$\xi_i(x, t) + \xi_r(x, t) = \xi_t(x, t)$$

and

$$p_i + p_r = p_t$$

or

$$+\frac{\partial \xi_i}{\partial x} - \frac{\partial \xi_r}{\partial x} = -\frac{\partial \xi_t}{\partial x}$$

The two boundary conditions give

$$A_i + A_r = A_t \tag{8.16}$$

and

$$k_1(A_r - A_i) = k_2 A_t$$

But $k_1 = \dfrac{\omega}{v_1} = \dfrac{\omega}{\rho_1 v_1^2}\,\rho_1 v_1 = \dfrac{\omega Z_1}{\gamma P_0}$ $\left(\because Z_1 = \rho_1 v_1 ; v_1 = \sqrt{\dfrac{\gamma P_0}{\rho_1}}\right)$

and $k_2 = \dfrac{\omega}{\gamma P_0} Z_2$

Therefore, we have

$$Z_1(A_r - A_i) = Z_2 A_t \tag{8.17}$$

Equations (8.16) and (8.17) give

$$r_{12} = \frac{A_r}{A_i} = \frac{Z_1 - Z_2}{Z_1 + Z_2} \tag{8.18}$$

$$t_{12} = \frac{A_t}{A_i} = \frac{2Z_1}{Z_1 + Z_2} \tag{8.19}$$

These equations give the amplitude reflection and transmission coefficients. It is clear that if $Z_1 > Z_2$; A_r/A_i is positive indicating that the incident and reflected displacements are in phase. But if $Z_1 < Z_2$; A_r/A_i is negative showing that the reflected wave undergoes a phase change of π with respect to the incident wave. Furthermore, it is clear from Eq. (8.19) that A_t/A_i remains positive independent of whether Z_1 is less or more than Z_2 which means that the transmitted wave does not undergo any phase change.

At a rigid wall where Z_2 is infinity, $A_r = -A_i$ showing that the wave is completely reflected.

Reflection and Transmission of Sound Energy

The intensity of a sound wave of amplitude A and frequency ν travelling with a speed v in a medium of density ρ is given by [see Eq. (7.45) of Chap. 7].

$$I = 2\pi^2\nu^2 A^2 \rho v = 2\pi^2\nu^2 A^2 Z$$

where Z is the characteristic acoustic impedance offered by the medium. The intensity coefficients of reflection and transmission are, therefore, given by

$$\text{Reflection coefficient} = \frac{I_r}{I_i} = \frac{2\pi^2\nu^2 A_r^2 Z_1}{2\pi^2\nu^2 A_i^2 Z_1} = \frac{A_r^2}{A_i^2} = \left(\frac{Z_1 - Z_2}{Z_1 + Z_2}\right)^2 \quad (8.20)$$

$$\text{Transmission coefficient} = \frac{I_t}{I_i} = \frac{2\pi^2\nu^2 A_t^2 Z_2}{2\pi^2\nu^2 A_i^2 Z_1} = \frac{Z_2 A_t^2}{Z_1 A_i^2} = \frac{4Z_1 Z_2}{(Z_1+Z_2)^2} \quad (8.21)$$

Notice that $\frac{I_r}{I_i} + \frac{I_t}{I_i} = 1$ or $I_i = I_r + I_t$

which means that the energy is conserved.

Experiments show that there is almost total reflection of sound energy at the air-water interface, whereas, in the case of steel-water Interface, the reflection coefficient is about 0.85 or 85 per cent. Thus only about 15 per cent of sound energy is transmitted at a steel-water interface. This severely limits the transmission and detection devices used in submarines using ultrasonic waves.

Standing Waves in Pipes

Before we proceed to discuss standing waves in air columns in pipes; let us determine the reflection and transmission coefficients for particle pressures and velocities at a boundary. The particle velocities in incident, reflected and transmitted waves at the boundary at $x = 0$ are

$$\left(\frac{\partial \xi_i}{\partial t}\right)_{x=0} = V_i = A_i\,\omega \cos \omega t$$

$$\left(\frac{\partial \xi_r}{\partial t}\right)_{x=0} = V_r = A_r\,\omega \cos \omega t$$

$$\left(\frac{\partial \xi_t}{\partial t}\right)_{x=0} = V_t = A_t\,\omega \cos \omega t$$

$$\therefore \quad \text{Velocity reflection coefficient} = \frac{V_r}{V_i} = \frac{A_r}{A_i} = \left(\frac{Z_1 - Z_2}{Z_1 + Z_2}\right) \quad (8.22)$$

$$\text{Velocity transmission coefficient} = \frac{V_t}{V_i} = \frac{A_t}{A_i} = \frac{2Z_1}{Z_1+Z_2} \quad (8.23)$$

Similarly we have

$$\text{Pressure reflection coefficient} = \frac{p_r}{p_i} = -\frac{V_r}{V_i} = \left(\frac{Z_2 - Z_1}{Z_1 + Z_2}\right) \quad (8.24)$$

$$\text{Pressure transmission coefficient} = \frac{p_t}{p_i} = \frac{Z_2 V_t}{Z_1 V_i} = \frac{2Z_1}{Z_1+Z_2} \quad (8.25)$$

Thus we see that if $Z_1 > Z_2$ the incident and reflected particle velocities are in phase but the incident and reflected acoustic pressures are out of phase by 180°. On the other hand, if $Z_1<Z_2$, the acoustic pressures are in phase but the particle velocities are out of phase by 180°.

At a rigid wall where Z_2 is infinite we have

$$\frac{V_r}{V_i} = -1$$

$$\frac{p_r}{p_i} = +1$$

Standing Waves in a Closed Pipe

Consider a closed pipe of length L lying along the x-axis with its open end at $x=0$ and closed end at $x=L$. At the closed end, there is a rigid boundary (i e. Z_2 is infinite). Hence the particle velocity has a reflection coefficient -1 at the closed end which means that $A_i = -A_r = +A$ (say). Therefore, the incident wave (travelling along the tube in the $+x$ direction) and the reflected wave (travelling in the $-x$ direction) are given by

$$\xi_i(x, t) = A \sin\{k(vt-x)\}$$

$$\xi_r(x, t) = -A \sin\{k(vt+x-2L)\}$$

The superposition of these oppositely travelling waves gives standing waves in the pipe, which, as in the case of a string, are described by

$$\xi(x, t) = 2A \sin\{k(L-x)\} \cos\{k(vt-L)\} \tag{8.26}$$

$$\therefore \quad \frac{\partial \xi}{\partial x} = -2Ak \cos\{k(L-x)\} \cos\{k(vt-L)\}$$

The excess pressure at x at time t is given by

$$p = \Delta P = -\gamma P_0 \frac{\partial \xi}{\partial x} = 2\gamma P_0 Ak \cos\{k(L-x)\} \cos\{k(vt-L)\} \tag{8.27}$$

The open end at $x=0$ represents a condition of zero pressure variation during the oscillation This means that, at the open end, the excess pressure is zero, i.e. the pressure is equal to the equilibrium value P_0. Thus at $x=0$ we have $p=0$ for all time. Putting $x=0$ and $p=0$ in Eq. (8.27), we have

$$0 = 2\gamma P_0 Ak \cos kL \cos\{k(vt-L)\}$$

This is true for all values of t if

$$\cos kL = 0$$

or $$kL = \frac{\pi}{2}, \frac{3\pi}{2}, \frac{5\pi}{2}, \ldots$$

or $$k_m L = \frac{m\pi}{2}$$

or
$$k_m = \frac{m\pi}{2L} \tag{8.28}$$

or
$$\lambda_m = \frac{4L}{m}$$

where $m = 1, 3, 5, \ldots$

Hence the frequency of vibration of air column inside the pipe is given by

$$\nu_m = \frac{v}{\lambda_m} = \frac{m}{4L}\sqrt{\frac{E}{\rho}} \tag{8.29}$$

where $E = \gamma P_0$ is the modulus of elasticity of the gas. The particle displacements in the pipe are obtained by using Eq. (8.28) in Eq. (8.26) which gives

$$\xi(x, t) = 2A \sin\left\{\frac{m\pi}{2L}(L-x)\right\}\cos\left\{\frac{m\pi}{2L}(vt-L)\right\} \tag{8.30}$$

Notice that the frequencies given by Eq. (8.29) are just the frequencies of the normal modes of a pipe closed at one end (see Chap. 6). Figure 6.9(a) of chapter 6 shows the first three modes of the pipe. Remember that the particle displacements are longitudinal. We find that, a pipe closed at one end has only odd harmonics of frequencies 3, 5, 7, ... times the fundamental frequency (ν_1). All the even harmonics of frequencies $2\nu_1$, $4\nu_1$, etc. are absent. The frequency of the fundamental ($m = 1$) mode is given by

$$\nu_1 = \frac{1}{4L}\sqrt{\frac{E}{\rho}}$$

Standing Waves in an Open Pipe

Let us first discuss the mechanism of reflection at an open end of a pipe. Consider a pipe terminated by an opening into a large room. The equilibrium pressure P_0 of the air in the pipe is equal to the pressure of air in the room. At the open end of the pipe air can freely rush in or out. When a compressional wave reaches the open end, the particles emerge out of the pipe spreading freely in all directions Consequently expansion takes place just outside the pipe in all directions tending to bring the pressure of the emerging air equal to the equilibrium pressure P_0. This happens at a certain distance from the opening (called the end-correction). Right at the opening the pressure is not exactly equal to P_0. This difference in pressure just inside the pipe and outside gives rise to a force which causes the expansion of air which spreads out sideways. The spreading of air in the outside region produces rarefaction at the open end and hence to annul this rarefaction air from behind the compression must rush forward. This again results in a rarefaction in the back region where the air has rushed forward and so on. In this way a wave of rarefaction starts from the open end and travels backwards, i.e. in a direction opposite to the wave of compression. Thus a compression on arrival at the open end of a pipe is reflected back into the pipe as a rarefaction. Since the open end represents a place of zero pressure variation (this happens at a certain point just outside the opening), the excess pressure is zero at the opening. This means that the pressure reflection coefficient is -1 at the open end. It is clear from Eqs. (8.22) and (8.24) that the velocity or amplitude reflection coefficient is $+1$, showing that there is no reversal of amplitude as a result of reflection at the open end. Thus a compression is reflected from an open end as a rarefaction and vice versa without any reversal of sign of the amplitude.

We shall now discuss standing waves in a pipe of length L lying along the x-axis with its open ends at $x = 0$ and $x = L$. A sound wave of amplitude A_i and frequency ω ($= kv$) is sent in the $+x$ direction. The particle displacements due to this incident wave are given by

$$\xi_i(x,t) = A_i \sin \{k(vt-x)\}$$

The wave reflected at the end $x = L$ is given by

$$\xi_r(x, t) = A_r \sin \{k(vt+x-2L)\}$$

Since the amplitude reflection coefficient A_r/A_i at the open end is $+1$, we have $A_r = A_i = A$. Hence the resultant displacement is given by

$$\begin{aligned}\xi(x, t) &= A\,[\sin \{k(vt-x)\}+\sin \{k(vt+x-2L)\}\\ &= 2A \cos \{k(L-x)\} \sin \{k(vt-L)\}\end{aligned}$$

Therefore

$$\frac{\partial \xi}{\partial x} = 2kA \sin \{k(L-x)\} \sin \{k(vt-L)\}$$

The excess pressure is given by

$$p = -\gamma P_0 \frac{\partial \xi}{\partial x} = -2\gamma P_0 kA \sin \{k(L-x)\} \sin \{k(vt-L)\}$$

The excess pressure p at the other open end at $x = 0$ must also be zero.

Putting $p = 0$ at $x = 0$ gives

$$0 = -2\gamma P_0\, kA \sin kL \sin \{k(vt-L)\}$$

which is satisfied for all values of t if

$$\sin kL = 0$$

or

$$k_n L = n\pi \qquad (n = 1, 2, 3,\ldots)$$

The frequency of vibration of air column in the pipe is given by

$$\nu_n = \frac{n}{2L}\sqrt{\frac{E}{\rho}}$$

and the particle displacements are given by

$$\xi(x, t) = 2A \cos\left(\frac{n\pi x}{L}\right) \sin\left(\frac{n\pi v t}{L}\right)$$

Figure 6.9(b) of Chapter 6 shows the first three modes of a pipe open at both ends. Notice that all even and odd harmonics are present.

8.4 REFLECTION AND TRANSMISSION IN ELECTRICAL TRANSMISSION LINES

In Chap. 7 we have seen that current and voltage waves propagate along an ideal (lossless) electrical transmission line with a velocity v given by

$$v = \frac{1}{\sqrt{LC}}$$

where L is the inductance per unit length and C is the capacitance per unit length of the line. The characteristic impedance of the line is given by

$$Z = \sqrt{\frac{L}{C}}$$

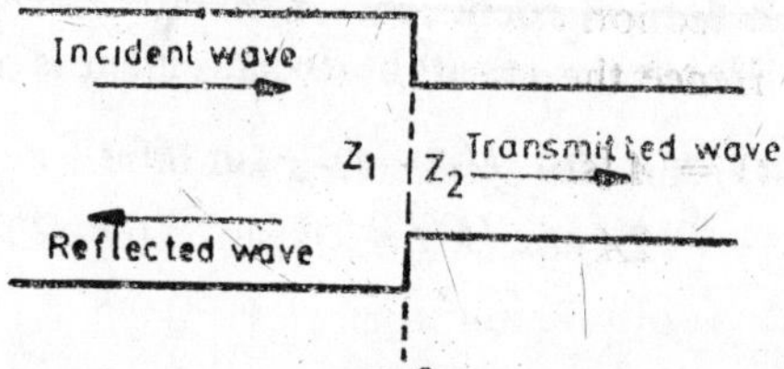

Fig. 8.6 Reflection and transmission at the junction between two loss-less transmission lines of impedances Z_1 and Z_2.

We will now consider reflection and transmission across a boundary between two connected lines. Consider two parallel transmission lines 1 and 2 of impedances Z_1 and Z_2 connected at a boundary at $x = 0$ (see Fig. 8.6). Suppose a current-voltage wave is travelling in line 1 in the $+x$ direction and is incident at the junction $x = 0$. The incident wave is partly reflected and partly transmitted at the junction. The incident reflected and transmitted waves are given by

$$V_i = V_{0i} \sin(\omega t - k_1 x)$$
$$I_i = I_{0i} \sin(\omega t - k_1 x)$$
$$V_r = V_{0r} \sin(\omega t + k_1 x)$$
$$I_r = I_{0r} \sin(\omega t + k_1 x)$$
$$V_t = V_{0t} \sin(\omega t - k_2 x)$$
$$I_t = I_{0t} \sin(\omega t - k_2 x)$$

The subscripts i, r and t respectively denote the parameters of the incident, reflected and transmitted waves.

The boundary conditions are :

(1) The voltage is continuous across the junction.

(2) The current is continuous across the junction.

In other words, the following conditions must be satisfied at the boundary at $x = 0$.

$$V_i + V_r = V_t$$

$$I_i + I_r = I_t$$

In addition, we have

$$\frac{V_i}{I_i} = Z_1, \quad \frac{V_r}{I_r} = -Z_1 \quad \text{and} \quad \frac{V_t}{I_t} = Z_2.$$

From these equations we obtain (it is not necessary to repeat the steps which are the same as in the case of strings or pipes).

$$\text{Voltage amplitude reflection coefficient} = \frac{V_{0r}}{V_{0i}} = \frac{Z_2 - Z_1}{Z_1 + Z_2} \tag{8.31}$$

$$\text{Voltage amplitude transmission coefficient} = \frac{V_{0t}}{V_{0i}} = \frac{2Z_2}{Z_1 + Z_2} \tag{8.32}$$

$$\text{Current amplitude reflection coefficient} = \frac{I_{0r}}{I_{0i}} = \frac{Z_1 - Z_2}{Z_1 + Z_2} \tag{8.33}$$

$$\text{Current amplitude transmission coefficient} = \frac{I_{0t}}{I_{0i}} = \frac{2Z_1}{Z_1 + Z_2} \tag{8.34}$$

These relations are in complete correspondence with those for strings and pipes.

Short-circuited Transmission Line (Zero Impedance)

If the right-hand end of line 2 is short-circuited by connecting across it a resistor of negligible resistance, then $Z_2 = 0$ and we have

$$V_{0t} = V_{0i} + V_{0r} = 0$$

so that $V_{0r} = -V_{0i}$. The incident voltage wave is completely reflected with a phase change of π. The voltage reflection coefficient is -1 at the short-circuited end. But the current reflection coefficient is $+1$. Thus a positive voltage is reflected as a negative voltage but a positive current is reflected as a positive current.

Standing Waves on a Short-circuited Transmission Line. The condition described above is exactly the condition for the existence of standing waves. At any position x on the transmission line the incident and reflected voltage waves are given by

$$V_i = ZI_i = V_{0i} \sin (\omega t - kx)$$

$$V_r = -ZI_r = V_{0r} \sin (\omega t + kx)$$

with $$V_{0i} = -V_{0r} = +V_0$$

The resultant voltage and current at x are

$$V = V_i + V_r = -2V_0 \sin kx \cos \omega t \tag{8.35}$$

and $$I = 2I_0 \cos kx \sin \omega t \tag{8.36}$$

From these equations we find that at any point x along the line the voltage V varies as sin kx and the current I varies as cos kx. Thus the voltage and current are out of phase by 90° at any point on the line. Equations (8.35) and (8.36) also tell us that the voltage changes with time as $-\cos \omega t$ and the current varies as $\sin \omega t$, indicating that the voltage lags the current by 90° in time. Thus the voltage and current are 90° out of phase in space (x) and in time (t). Hence the power factor $\cos \phi = \cos 90° = 0$; so that no power is consumed.

Equations (8.35) and (8.36) describe the standing waves of voltage and current in the transmission line which is short-circuited at one end. There is no net transfer of energy along the line. Nodes and antinodes of voltage and current are spaced equally along the line. The voltage node appears at the position of the current antinode and vice versa. Figure 8.7 shows standing current and voltage waves on a transmission line terminated at zero impedance i.e. short-circuited at one end.

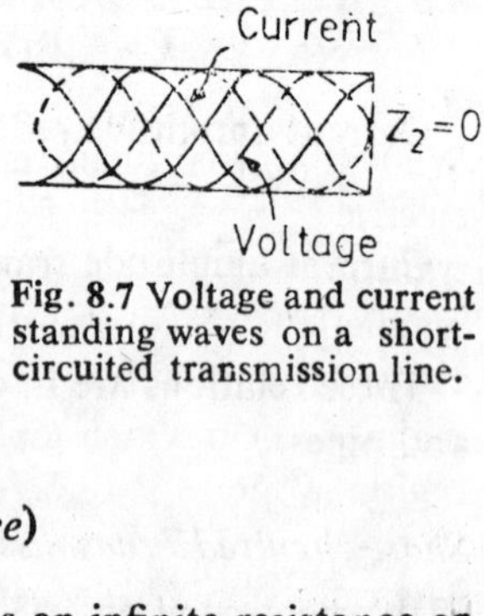

Fig. 8.7 Voltage and current standing waves on a short-circuited transmission line.

Open-circuited Transmission Line (Infinite Impedance)

If the right-hand end of the line is connected across an infinite resistance or left open with no resistance, the impedance Z_2 offered by the end is infinite. Consequently no net current can flow across from one conductor of the line to the other. Thus the current is always zero at the open-circuited end and the current reflection coefficient is —1. The voltage reflection coefficient is then +1. Thus a positive current is reflected as a negative current and a positive voltage is reflected as a positive voltage. In other words

$$I_{0t} = I_{0i} + I_{0r} = 0$$

so that $I_{0r} = -I_{0i}$. The incident current wave is completely reflected with a phase change of π but the incident voltage does not undergo any phase change on reflection at the open-circuited end. The standing waves on the line are described by [see Eqs. (8.35) and (8.36)]

$$I = -2I_0 \sin kx \cos \omega t$$

and
$$V = V_0 \cos kx \sin \omega t$$

Thus the voltage and current standing waves are 90° out of phase in space and time and no power is consumed. There is no net transfer of energy along the line.

It may be recalled that the above results hold only for a loss-less transmission line, i.e. a line consisting of only wattless components such as inductance and capacitance. Such an ideal line does not exist because there is always some finite resistance in the wires which is responsible for power losses. In the presence of a resistive component the amplitude of the travelling wave decays exponentially with distance. Our discussion of damped harmonic oscillations in Chap. 3 would predict this behaviour. Just as the presence of friction damps the motion of an oscillator, the presence of resistance damps the waves along the transmission line.

8.5 REFLECTION AND TRANSMISSION OF ELECTROMAGNETIC WAVES AT A BOUNDARY

We will now consider reflection and transmission of waves associated with electric and magnetic fields instead of those with voltage and current. In a transmission line, the electric field is proportional to voltage and the magnetic field is proportional to current. Associated with the harmonically varying current and voltage is the harmonically varying electromagnetic field which gives rise to electromagnetic waves travelling in the region around the line. Therefore, we would expect electromagnetic waves to reflect and transmit at a boundary just as voltage and current waves do. The reflection and transmission coefficients of the magnetic and electric fields of the wave should, therefore, be given by Eqs (8.31) to (8.34). This indeed is true as we shall see below.

Consider a plane monochromatic electromagnetic wave travelling in $+x$ direction in a medium of impedance Z_1 and incident *normally* on a plane boundary at $x = 0$ of a medium of impedance Z_2 (see Fig. 8.8). From electromagnetic theory the boundary conditions are that the components of field vectors **E** and **H** tangential or parallel to the boundary are continuous. Thus the conditions to be satisfied at the boundary are :

$$E_i + E_r = E_t$$

$$H_i + H_r = H_t$$

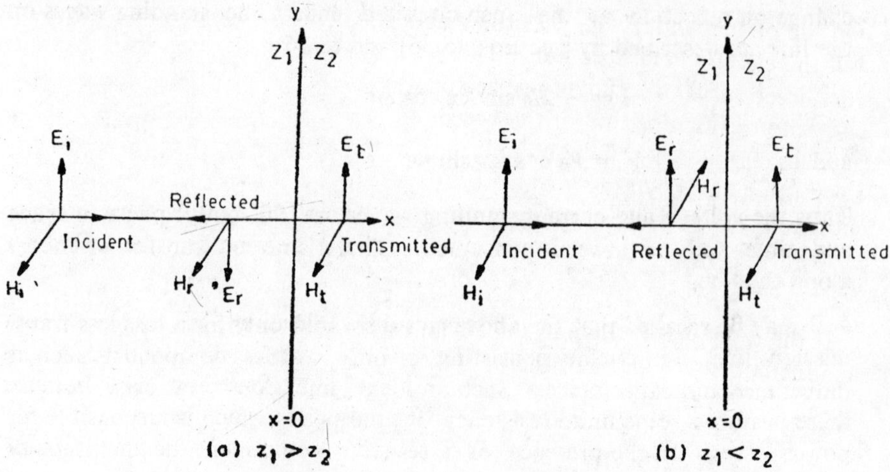

Fig. 8.8 Reflection and transmission of a plane electromagnetic wave incident normally on a plane boundary between two media of impedances Z_1 and Z_2. The Poynting vector of the reflected wave $(\mathbf{E}\times\mathbf{H})_r$ points in the $-x$ direction showing that (a) E_r reverses in sign if $Z_1 > Z_2$ and (b) H_r reverses in sign if $Z_1 < Z_2$.

Where E and H are respectively the y and z components (E_y and H_z) of vectors **E** and **H**. Furthermore (see Chap. 7)

$$\frac{E_i}{H_i} = Z_1, \quad \frac{E_r}{H_r} = -Z_1 \text{ and } \frac{E_t}{H_t} = Z_2$$

There is no need to repeat the steps we have used earlier, It is easy to see that the above relations give (refer to Sec. 8.4) for amplitude coefficients.

$$\text{Electric field reflection coefficient} = \frac{E_{0r}}{E_{0i}} = \frac{Z_2 - Z_1}{Z_1 + Z_2} \tag{8.37}$$

$$\text{Electric field transmission coefficient} = \frac{E_{0t}}{E_{0i}} = \frac{2Z_2}{Z_1 + Z_2} \tag{8.38}$$

$$\text{Magnetic field reflection coefficient} = \frac{H_{0r}}{H_{0i}} = \frac{Z_1 - Z_2}{Z_1 + Z_2} \tag{8.39}$$

$$\text{Magnetic field transmission coefficient} = \frac{H_{0t}}{H_{0i}} = \frac{2Z_1}{Z_1 + Z_2} \tag{8.40}$$

which agree with the coefficients we have found for other waves. These relations hold only for normal incidence.

If $Z_1 > Z_2$, it is clear from Eqs. (8.37) and (8 39) that E_{0r}/E_{0i} is negative but H_{0r}/H_{0i} is positive indicating that the electric field changes sign on eflection but the magnetic field does not. This is shown in Fig. 8.8(a).

On the other hand, if $Z_1 < Z_2$ the magnetic field changes sign on reflection but the electric field does not. This is shown in Fig. 8.8(b). Notice that the direction of the Poynting vector $(\mathbf{E}_r \times \mathbf{H}_r)$ of the reflected wave is opposite to the direction of the vector $(\mathbf{E}_i \times \mathbf{H}_i)$ of the incident wave. This satisfies the energy flow condition of the Poynting vector. Figure 8.8 shows that the Poynting vector of the reflected wave is reversed because either E or H is reversed in phase depending on whether

$$Z_1 > Z_2 \quad \text{or} \quad Z_1 < Z_2.$$

If the wave is travelling in the air and strikes normally on a plane perfect conductor (where $Z_2 = 0$) then

$$\frac{E_{0r}}{E_{0i}} = -1 \text{ and } \frac{E_{0t}}{E_{0i}} = 0$$

Thus the wave is completely reflected at a perfect conductor. Thus good conductors are also good reflectors of electromagnetic waves. These are just the conditions necessary for the existence of standing waves in a medium.

The reflection and transmission coefficients may also be written in terms of refractice indices of the media. The refractive index n of a dielectric of impedance Z is defined as

$$n = \frac{c}{v} = \sqrt{\frac{\mu\varepsilon}{\mu_0\varepsilon_0}} = \sqrt{\varepsilon_r} = \frac{Z_0}{Z} \qquad \left(\because \frac{\mu}{\mu_0} \simeq \mu_r \simeq 1\right)$$

where the quantities with subscript '0' refer to vacuum and c and v are the velocities of the electromagnetic wave in vacuum and dielectric respectively.

If an electromagnetic wave travelling in a dielectric medium of refractive index n_1 falls normally on a plane boundary of a medium of refractive index n_2, the reflection and transmission coefficients for the electric field are written as

$$\frac{E_r}{E_i} = \frac{Z_2 - Z_1}{Z_2 + Z_1} = \frac{Z_0/n_2 - Z_0/n_1}{Z_0/n_2 + Z_0/n_1} = \frac{n_1 - n_2}{n_1 + n_2}$$

$$\frac{E_t}{E_i} = \frac{2Z_2}{Z_1 + Z_2} = \frac{2Z_0/n_2}{Z_0/n_1 + Z_0/n_2} = \frac{2n_1}{n_1 + n_2}$$

If the first medium is air ($n_1 = 1$) and the second medium is a dielectric of refractive index n ($n_2 = n$) we have

$$\frac{E_r}{E_i} = \frac{1-n}{1+n}$$

$$\frac{E_t}{E_i} = \frac{2}{1+n}$$

If visible light travelling in air falls normally on a glass slab of refractive index 1.5 for visible light, we have

$$\frac{E_r}{E_i} = \frac{1-1.5}{1+1.5} = -\frac{1}{5}$$

The negative sign indicates that the electric field reverses sign on reflection. The reflection coefficient for intensity is given by

$$\frac{I_r}{I_i} = \frac{E_r^2}{E_i^2} = \frac{1}{25}$$

Thus, 4 per cent of incident light intensity is reflected at air glass interface at normal incidence.

8.6 SUPPRESSION OF REFLECTION : IMPEDANCE MATCHING

Reflections are often undesirable and need to be suppressed in many situations. Suppose we wish to transmit energy from one medium into another by means of travelling waves. When the wave reaches the boundary between the two media, it is partly reflected. Thus some energy is lost by reflection. If somehow the reflection could be prevented or suppressed, the energy would have transmitted in the second medium without any loss. Look at the following examples.

(1) Suppose we wish to study some characteristics of a travelling wave on a string. The string must end somewhere. At the free ends reflections will occur giving rise to a standing wave pattern superposed on travelling waves on the string, thus altering the travelling wave pattern.

(2) In a loudspeaker the sound energy has to be transferred from air in the loudspeaker to air in a room or hall without generating reflections. The reflections are undesiruble because they represent a loss and hence decrease the power output of an amplifier.

(3) Skindivers have a breathing apparatus and have their faces covered with masks made of glass. Suppose a skindiver wishes to talk to another skindiver under water, the sound of a diver is reflected at the glass of the mask and very little sound is transmitted in water. The reason is that the characteristic impedances of air, glass, and water are very different. If the impedances were equal, no reflections would occur and the divers could easily talk to each other.

(4) Long distance cables carry electrical energy from a power house to distant places. The cables have joints where reflections occur. The wastage of energy due to reflections must be avoided or minimised so that the power transfer from the generator is maximum.

(5) Suppose we wish to send travelling waves of visible light from air into a lens or a slab without generating a reflection. The reflections are undesirable because they decrease the intensity of the transmitted beam and also cause glare in some other part of the apparatus.

Transferring energy by travelling waves from one medium to another without reflection is called *impedance matching*. We have seen that if a wave of any kind (transverse wave on a string, sound wave in a medium, voltage and current wave in a transmission line and electromagnetic wave in a medium) travelling in medium *1* of characteristic impedance Z_1 meets the boundary of medium 2 of characteristic impedance Z_2, the amplitude reflection and transmission coefficients are given by

$$r_{12} = \frac{A_r}{A_i} = \frac{Z_1 - Z_2}{Z_1 + Z_2}$$

$$t_{12} = \frac{A_t}{A_i} = \frac{2Z_1}{Z_1 + Z_2}$$

On the other hand, if a wave travelling in medium 2 meets the boundary of medium *1*, these coefficients become

$$r_{21} = \frac{A_r}{A_i} = \frac{Z_2 - Z_1}{Z_1 + Z_2}$$

$$t_{21} = \frac{A_t}{A_i} = \frac{2Z_2}{Z_1 + Z_2}$$

It is clear from these relations that if $Z_1 = Z_2$, no reflection takes place and the wave is completely transmitted. The impedances of the two media are said to be *matched*.

Impedance matching is a very important practical problem in many branches of optics and engineering where transfer of energy from one medium into another is desired. Electrical energy is carried over long distance cables which have joints. The impedances across the joints must be accurately matched to avoid wastage of energy by reflection. A generator delivers power across a load. For maximum power transfer the impedance of the load must match that of the generator. For the same reason the impedance of a loudspeaker has to be matched with the impedance of the power output of an amplifier.

How is impedance matching achieved if Z_1 is not equal to Z_2? We shall discuss two methods : one involves the insertion of a coupling element called the anti or non-reflecting layer between two mismatched impedances and other involves what is called 'tapering'.

Antireflecting Layer

This method has wide ranging applications in engineering and optics. In this method we insert between two media of mismatched impedances Z_1

and Z_3 a section or a film of a transparent medium of appropriate intermediate impedance Z_2 and of thickness L equal to $\lambda_2/4$ where λ_2 is the wavelength of the wave in medium 2.

Suppose medium 1 extends from $x = -\infty$ to $x = 0$. An impedance matching element (medium 2) extends from $x = 0$ to $x = L$ Medium 3 extends from $x = L$ to $x = \infty$ (see Fig. 8.9). We wish to match impedances between media 1 and 3 for waves of angular frequency ω. In other words, we do not want any reflected wave when a wave of frequency ω travels in medium *1* in the $+x$ direction and is incident on the boundaries of the intermediate layer of medium *2*. Thus we want all the incident energy to transmit in medium 3.

There is no way by which reflections at discontinuities at $x = 0$ and $x = L$ could be avoided. There are multiple reflections within the film and if a superposition of these waves gives a resultant wave in medium 1 with zero amplitude, then loss of energy by reflection is prevented and the impedances Z_1 and Z_3 of media 1 and 3 are matched Thus our problem is to find the values of L and Z_2 so that impedances Z_1 and Z_3 are matched.

Let us assume that $Z_1 < Z_2 < Z_3$ (we could also assume that $Z_1 > Z_2 > Z_3$). Let a wave of amplitude A travelling in medium 1 in $+x$ direction fall normally on the boundary of medium 2 at $x = 0$. It is partly reflected and partly transmitted at the boundary. The amplitude of the reflected wave is Ar_{12} where r_{12} is the amplitude reflection coefficient when the wave travelling in medium 1 is reflected at the boundary of medium 2. We have seen that

$$r_{12} = \frac{Z_1 - Z_2}{Z_1 + Z_2} \tag{8.41}$$

This reflected wave is labelled 1 in Fig. 8.9. The amplitude of the wave transmitted in medium 2 is At_{12} where

$$t_{12} = \frac{2Z_1}{Z_1 + Z_2} \tag{8.42}$$

This wave of amplitude At_{12} is reflected at $x = L$ giving rise to a reflected wave of amplitude $At_{12}r_{23}$ travelling in medium 2 in the $-x$ direction. Here r_{23} is the amplitude reflection coefficient when a wave travelling in medium *2* is reflected at the boundary of medium *3* and is given by

$$r_{23} = \frac{Z_2 - Z_3}{Z_2 + Z_3} \tag{8.43}$$

When this wave of amplitude $At_{12}r_{23}$ meets the boundary at $x = 0$, it is transmitted into medium 1 with amplitude $At_{12}r_{23}t_{21}$ (marked number 2 in Fig. 8.9) and reflected back into medium 2 with an amplitude $At_{12}r_{23}r_{21}$

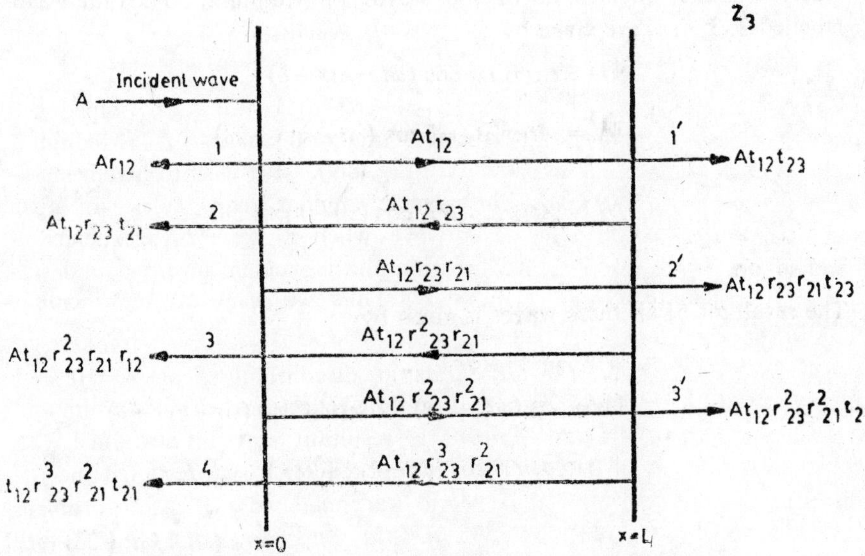

Fig. 8.9 Multiple reflections in an intermediate matching layer

where r_{21} and t_{21} are respectively the amplitude reflection and transmission coefficients when a wave travelling in medium 2 is reflected and transmitted at the boundary of medium 1. We know that they are given by

$$r_{21} = \frac{Z_2 - Z_1}{Z_1 + Z_2} \tag{8.44}$$

$$t_{21} = \frac{2Z_2}{Z_1 + Z_2} \tag{8.45}$$

The rest of the diagram is self explanatory. The waves reflected back into medium 1 are labelled 1, 2, 3, 4...; they all travel in the $-x$ direction. If the resultant of all these waves is a wave of zero amplitude ; no energy is reflected and the impedances Z_1 and Z_3 are matched.

Let the wave marked 1 be given by (a sine function is equally good)

$$\psi_1 = Ar_{12} \cos (\omega t - k_1 x)$$

where $k_1 = 2\pi/\lambda_1$; λ_1 being the wavelength in medium 1. Now the path difference between waves 1 and 2 is $2L$ traversed in medium 2. Hence the phase difference between waves 1 and 2 is

$$\delta = \frac{2\pi}{\lambda_2} \times 2L = \frac{4\pi L}{\lambda_2} \tag{8.46}$$

where λ_2 is the wavelength of the waves in medium 2. Thus the waves labelled 2, 3, 4,... are given by

$$\psi_2 = At_{12}r_{23}t_{21}\cos(\omega t - k_1x + \delta)$$

$$\psi_3 = At_{12}r_{23}^2t_{21}r_{21}\cos(\omega t - k_1x + 2\delta)$$

$$\psi_4 = At_{12}r_{23}^3t_{21}r_{21}^2\cos(\omega t - k_1x + 3\delta)$$

and so on.

The resultant of all these waves is given by

$$\begin{aligned}\psi &= \psi_1 + \psi_2 + \psi_3 + \psi_4 + \ldots \\ &= A\,[r_{12}\cos(\omega t - k_1x) + t_{12}r_{23}t_{21}\cos(\omega t - k_1x + \delta) \\ &\quad + t_{12}r_{23}^2t_{21}r_{21}\cos(\omega t - k_1x + 2\delta) + t_{12}r_{23}^3t_{21}r_{21}^2 \\ &\quad \cos(\omega t - k_1x + 3\delta) + \ldots] \\ &= A\,\mathrm{Re}\left[e^{i(\omega t - k_1x)}\left\{r_{12} + t_{12}r_{23}t_{21}e^{i\delta} + t_{12}r_{23}^2t_{21}r_{21}e^{2i\delta}\right.\right. \\ &\quad \left.\left. + t_{12}r_{23}^3t_{21}r_{21}^2e^{3i\delta} + \ldots\right\}\right] \\ &= A\,\mathrm{Re}\,[e^{i(\omega t - k_1x)}\{r_{12} + S\}] \end{aligned} \tag{8.47}$$

where

$$\begin{aligned} S &= t_{12}t_{21}r_{23}e^{i\delta}\left(1 + r_{21}r_{23}e^{i\delta} + r_{21}^2r_{23}^2e^{2i\delta} + \ldots\right) \\ &= \frac{t_{12}t_{21}r_{23}e^{i\delta}}{1 - r_{21}r_{23}e^{i\delta}} \end{aligned}$$

This can be done because

$$\frac{1}{1-\alpha} = 1 + \alpha + \alpha^2 + \ldots$$

Now from Eqs. (8.41), (8.42), (8.44) and (8.45) we have

$$r_{12} = -r_{21}$$

and $$r_{12}^2 + t_{12}t_{21} = 1 \quad \text{or} \quad t_{12}t_{21} = 1 - r_{12}^2$$

Thus we have

$$S = \frac{\left(1 - r_{12}^2\right)r_{23}e^{i\delta}}{1 + r_{12}r_{23}e^{i\delta}}$$

Hence

$$r_{12}+S=\frac{r_{12}+r_{23}\,e^{i\delta}}{1+r_{12}r_{23}\,e^{i\delta}} \tag{8.48}$$

Substitution of Eq. (8.48) in Eq. (8.47) gives

$$\psi = AR_e\left[e^{i(\omega t-k_1 x)}\frac{(r_{12}+r_{23}e^{i\delta})}{(1+r_{12}r_{23}e^{i\delta})}\right]$$

which may be written as

$$\psi = BR_e\left\{e^{i(\omega t-k_1 x)}\,e^{i\theta}\right\} = B\cos(\omega t-k_1x+\theta) \tag{8.49}$$

where

$$Be^{i\theta}=\frac{r_{12}+r_{23}e^{i\delta}}{1+r_{12}r_{23}e^{i\delta}}$$

The complex conjugate is

$$Be^{-i\theta}=\frac{r_{12}+r_{23}e^{-i\delta}}{1+r_{12}r_{23}e^{-i\delta}}$$

The product of these two equations gives the amplitude B of the resultant wave.

$$B^2 = Be^{i\theta}\times Be^{-i\theta}$$

$$=\frac{r_{12}^2+r_{23}^2+2r_{12}r_{23}\cos\delta}{1+r_{12}^2r_{23}^2+2r_{12}r_{23}\cos\delta} \tag{8.50}$$

where we have used the identity

$$\cos\delta=\tfrac{1}{2}\,(e^{i\delta}+e^{-i\delta})$$

Thus the resultant wave is given by Eq. (8.49) with an amplitude B given by Eq. (8.50). If $B = 0$, there would be no wave travelling in $-x$ direction in medium 1, i.e. all the incident energy is transmitted. It is clear from Eq. (8.50) that $B = 0$ if [see Eq. (8.46)]

$$\delta=\frac{\pi}{2},\ \text{i.e. } L=\frac{\lambda_2}{4}$$

and $\quad r_{12} = r_{23}$

which from Eqs (8.41) and (8.43) gives,

$$Z_2=\sqrt{Z_1Z_3}$$

Thus we see tnat if the impedance of the coupling medium is the harmonic mean of the two impedances to be matched and the thickness of the medium is $\lambda_2/4$, then all the energy in the incident wave at angular frequency

ω will be transmitted with zero reflection. Here λ_2 is the wavelength in the medium of impedance Z_2 of the wave of frequency ω. The same result is obtained if $Z_1 > Z_2 > Z_3$. Figure 8.9 also shows the amplitudes of waves marked 1, 2, 3,...etc. transmitted in medium 3. The energy of the resultant of all these waves will be equal to the energy of the incident wave when Z_1 and Z_3 are matched, i.e. when $Z_2 = \sqrt{Z_1 Z_3}$ and $L = \frac{\lambda_2}{4}$.

An important application of non-reflecting layers is in transmission lines which are matched to loads by inserting quarter-wavelength stubs of lines with appropriate impedance. Non-reflecting coatings are used on lenses and other optical refracting devices to eliminate reflections from surfaces. A quarter-wave layer of a material of refractive index $n_2 = \sqrt{n_1 n_3}$ provides an impedance match between media of refractive indices n_1 and n_3. Such non-reflecting (or more appropriately anti-reflecting) films are applied to glass lenses by heating and then evaporating a dielectric substance on their surfaces. Lenses are often coated with a layer of magnesium fluoride which has a refractive index of $n_2 = 1.38$. When a glass lens of refractive index $n_3 = 1.9$ placed in air of refractive index $n_1 = 1$ is coated with a quarter-wave layer of mangesium fluoride the matching condition

$$n_2 = \sqrt{n_1 n_3}$$

is satisfied because $(1.38)^2 = 1.9$. But when applied to lenses of refractive index between 1.5 and 1.7, a useful but incomplete reduction in reflection is achieved.

Tapering

It is clear that the quarter-wave non-reflecting films are fully effective for waves of a particular frequency. For example, when such a film is applied to glass to give the best performance for light in the middle of the visible spectrum it reflects appreciably more in the blue and the red ends of the spectrum. The film, therefore, shows a purple bloom by reflected light. The application of non-reflecting films is often called 'blooming'.

It is possible to obtain a fairly good impedance matching between two media over a wide frequency range by using an intermediate medium in which the wave velocity and impedance Z_2 vary smoothly and slowly between Z_1 and Z_3, the impedances of the two media. Such carefully graduated systems called *tapered sections* are frequently used in acoustics.

SOLVED EXAMPLES

Example 8.1 Two strings of linear densities μ_1 and μ_2 are joined together and stretched with tension T. A transverse wave is incident on the boundary. Find (a) the fraction of the incident amplitude reflected and transmitted at the boundary (b) the fraction of the incident energy reflected and transmitted at the boundary if $\mu_2/\mu_1 = 4$ and $\frac{1}{4}$.

Solution

The characteristic impedances of the strings are

$$Z_1 = \mu_1 v_1 = \mu_1 \sqrt{\frac{T}{\mu_1}} = \sqrt{\mu_1 T}$$

$$Z_2 = \mu_2 v_2 = \mu_2 \sqrt{\frac{T}{\mu_2}} = \sqrt{\mu_2 T}$$

$$\therefore \quad \frac{Z_1}{Z_2} = \sqrt{\frac{\mu_1}{\mu_2}}$$

Case 1 : $\frac{\mu_2}{\mu_1} = 4$. In this case

$$\frac{Z_1}{Z_2} = \frac{1}{2}$$

(a) The fraction of the incident amplitude reflected at the boundary is given by

$$\frac{A_r}{A_i} = \frac{Z_1 - Z_2}{Z_1 + Z_2} = \frac{Z_1/Z_2 - 1}{Z_1/Z_2 + 1} = -\frac{1}{3}$$

The fraction of incident amplitude transmitted across the boundary is

$$\frac{A_t}{A_i} = \frac{2Z_1}{Z_1 + Z_2} = \frac{2Z_1/Z_2}{Z_1/Z_2 + 1} = \frac{2}{3}$$

(b) The reflection and transmitted coefficients of energy are

$$\frac{P_r}{P_i} = \left(\frac{Z_1 - Z_2}{Z_1 + Z_2}\right)^2 = \frac{1}{9}$$

$$\frac{P_t}{P_i} = \frac{4Z_1Z_2}{(Z_1 + Z_2)^2} = \frac{8}{9}$$

Case 2 : $\frac{\mu_2}{\mu_1} = \frac{1}{4}$. In this case

$$\frac{Z_1}{Z_2} = 2$$

(a) The reflected and transmitted amplitude coefficients are

$$\frac{A_r}{A_i} = \frac{1}{3} \text{ and } \frac{A_t}{A_i} = \frac{4}{3}$$

(a) The reflected and transmitted energy coefficients are

$$\frac{P_r}{P_i} = \frac{1}{9} \text{ and } \frac{P_t}{P_i} = \frac{8}{9}$$

Example 8 2 A point mass m is concentrated at a point on a string of linear density μ. A transverse wave of angular frequency ω travels in the $+x$ direction with a velocity v and is partly reflected and transmitted at the mass The boundary conditions are :

(i) The string displacements just to the left and right of the mass are equal, and

(ii) The difference between the transverse force to the left and right of the mass is equal to the product of the mass and its acceleration.

(a) If A_i, A_r, and A_t are respectively the incident, reflected and transmitted wave amplitudes, show that

$$\frac{A_r}{A_i} = \frac{-i\alpha}{1+i\alpha} \text{ and } \frac{A_t}{A_i} = \frac{1}{1+i\alpha}$$

where $\alpha = \dfrac{\omega m}{2\mu v}$ and $i = \sqrt{-1}$

(b) Calculate the reflected and transmitted energy coefficients and show that the energy is conserved.

Solution

(a) Let T be the tension on the string and let the mass m be concentrated at point $x = 0$. Reflection and transmission takes place at the discontinuity at $x = 0$. Using the exponential notation, the incident, reflected and transmitted waves can be written as

$$y_i(x,t) = A_i\, e^{i(\omega t - kx)}$$

$$y_r(x,t) = A_r\, e^{i(\omega t + kx)}$$

$$y_t(x,t) = A_t\, e^{i(\omega t - kx)}$$

where $k = \omega/v$ and $i = \sqrt{-1}$. Notice that k is the same on both sides of the mass. These equations give

$$\frac{\partial y_i}{\partial x} = -kA_i\, e^{i(\omega t - kx)}$$

$$\frac{\partial y_r}{\partial x} = +kA_r\, e^{i(\omega t + kx)}$$

$$\frac{\partial y_t}{\partial x} = -kA_t\, e^{i(\omega t - kx)}$$

$$\frac{\partial^2 y_i}{\partial t^2} = -\omega^2 A_i\, e^{i(\omega t-kx)}$$

$$\frac{\partial^2 y_r}{\partial t^2} = -\omega^2 A_r\, e^{i(\omega t+kx)}$$

$$\frac{\partial^2 y_t}{\partial t^2} = -\omega^2 A_t\, e^{i(\omega t-kx)}$$

To the left of the mass there are two waves on the string ; the incident wave and the reflected wave. To the right of the mass we have only the transmitted wave on the string. The resultant displacement of the string to the left of the mass is given by

$$y(x,t) = y_i(x,t) + y_r(x,t)$$

The first boundary condition is that at $x = 0$, we have

$$y(0,t) = y_i(0,t) + y_r(0,t) = y_t(0,t)$$

or $$A_i + A_r = A_t \qquad \text{(i)}$$

Now the transverse forces on the string to the left of the mass are $-T\dfrac{\partial y_i}{\partial x}$ and $-T\dfrac{\partial y_r}{\partial x}$ and the transverse force on the right of the mass is $-T\dfrac{\partial y_t}{\partial x}$.

Thus the second boundary condition is that at $x = 0$, we have

$$-T\left(\frac{\partial y_i}{\partial x} + \frac{\partial y_r}{\partial x}\right)_{x=0} + T\left(\frac{\partial y_t}{\partial x}\right)_{x=0} = m\frac{\partial^2}{\partial t^2}\left(y_i + y_r\right)_{x=0}$$

or $$-ikT(-A_i + A_r) - ikTA_t = -m\omega^2(A_i + A_r)$$

or $$A_i - A_r = \left(1 - \frac{m\omega^2}{ikT}\right)A_t = \left(1 + \frac{im\omega^2}{kT}\right)A_t \qquad \text{(ii)}$$

Using (i) and (ii) we get

$$\frac{A_r}{A_i} = \frac{-i\alpha}{1+i\alpha} \text{ and } \frac{A_t}{A_i} = \frac{1}{1+i\alpha}$$

where $$\alpha = \frac{m\omega^2}{2kT} = \frac{\omega m}{2\mu v}\left(\because k = \frac{\omega}{v} \text{ and } v = \sqrt{\frac{T}{\mu}}\right)$$

(b) Since the medium on both sides of the discountinuity is the same, the characteristic impedance is the same to the left or right of the discontinuity. Hence energy reflection and transmission coefficients E_r/E_i and E_t/E_i are equal to the squares of the amplitude reflection and transmission coefficients, which are just the products of the amplitude coefficients (which are complex) with their complex conjugates. Thus we have

$$\frac{E_r}{E_i} = \left(\frac{A_r}{A_i}\right)\left(\frac{A_r}{A_i}\right)^* = \frac{-i\alpha}{(1+i\alpha)} \cdot \frac{+i\alpha}{(1-i\alpha)} = \frac{\alpha^2}{1+\alpha^2}$$

$$\frac{E_t}{E_i} = \left(\frac{A_t}{A_i}\right)\left(\frac{A_t}{A_i}\right)^* = \frac{1}{(1+i\alpha)}\,\frac{1}{(1-i\alpha)} = \frac{1}{1+\alpha^2}$$

Notice that $\dfrac{E_r}{E_i} + \dfrac{E_t}{E_i} = \dfrac{\alpha^2}{1+\alpha^2} + \dfrac{1}{1+\alpha^2} = 1$

or $E_i = E_r + E_t$. Thus the energy is conserved.

Example 8.3 A plane sound wave in air of density 1.29 kg m^{-3} falls on a water surface at normal incidence. The speed of sound in air is 334 ms^{-1} and in water the speed of sound is 1480 ms^{-1}. (a) What is the ratio of the amplitude of sound wave that enters water to that of the incident wave? (b) What fraction of the incident energy flux enters the water?

Solution

The characteristic impedances of air and water respectively are

$$Z_1 = \rho_1 v_1$$

$$Z_2 = \rho_2 v_2$$

$$\therefore \quad \frac{Z_1}{Z_2} = \frac{\rho_1 v_1}{\rho_2 v_2} = \frac{1.29 \times 334}{1000 \times 1480} = 0.000291$$

(a) $$\frac{A_t}{A_i} = \frac{2Z_1}{Z_1+Z_2} = \frac{2Z_1/Z_2}{Z_1/Z_2+1}$$

Substituting the value of Z_1/Z_2 we get

$$\frac{A_t}{A_i} = 5.82 \times 10^{-4}$$

(b) The fraction of the incident energy flux entering water is

$$\frac{I_t}{I_i} = \frac{4Z_1Z_2}{(Z_1+Z_2)^2} = \frac{4Z_1/Z_2}{(Z_1/Z_2+1)^2} = 1.16 \times 10^{-3}$$

Example 8.4 Sound waves are incident normally on water-steel interface. If the characteristic impedances of water and steel are 1.43×10^6 kg m^{-2} s^{-1} and 3.90×10^7 kg m^{-2} s^{-1} respectively, what percentage of the incident energy is reflected at the interface?

Solution

$$Z_1 = 1.43 \times 10^6 \text{ kg m}^{-2}\text{s}^{-1}$$

$$Z_2 = 3.90 \times 10^7 \text{ kg m}^{-2}\text{s}^{-1}$$

$$\text{Energy reflection coefficient} = \left(\frac{Z_1-Z_2}{Z_1+Z_2}\right)^2 \simeq 0.86$$

Thus 86% of the incident energy is reflected at the interface.

Example 8.5 Standing waves are produced in a short-circuited loss-free transmission line of length L and characteristic impedance Z_0 by sending current and voltage waves of wavelength λ. Show that the net (or input) impedance of the line is given by

$$Z = iZ_0 \tan\left(\frac{2\pi L}{\lambda}\right).$$

What is the impedance of a short-circuited quarter-wavelength loss-free line ? What is the use of such a line element ?

Solution

The incident voltage and current waves on the line are given by (using exponential notation)

$$V_i = V_{0i}e^{i(\omega t-kx)}$$

$$I_i = -I_{0i}e^{i(\omega t-kx)}$$

where $k = 2\pi/\lambda$. The reflected voltage and current waves travel in the $-x$ direction and are given by

$$V_r = V_{0r}e^{i(\omega t+kx)}$$

$$I_r = -I_{0r}e^{i(\omega t+kx)}$$

Now, for a short-circuited line $V_{0i} = -V_{0r} = V_0$ and $I_{0i} = I_{0r} = I_0$. Thus the net voltage and current at a point x on the line at time t is given by

$$V = V_0e^{i(\omega t-kx)} - V_0e^{i(\omega t-kx)}$$

$$= V_0e^{i\omega t}(e^{ikx} - e^{-ikx}) = -2iV_0e^{i\omega t}\sin kx$$

Similarly

$$I = -2I_0e^{i\omega t}\cos kx$$

Hence the impedance of the line is given by

$$Z = \frac{V}{I} = i\frac{V_0}{I_0}\tan kx$$

or $$Z = iZ_0 \tan\left(\frac{2\pi x}{\lambda}\right)$$

For a line of length L we have (putting $x = L$)

$$Z = iZ_0 \tan\left(\frac{2\pi L}{\lambda}\right)$$

For a quarter-wavelength line, $L = \lambda/4$. Hence the impedance of such a line element is $Z = iZ_0 \tan \pi/2 = \infty$. But if $L = \lambda/2$ the impedance is zero since $\tan\left(\frac{2\pi L}{\lambda}\right) = \tan \pi = 0$. We know that the fundamental mode has $\lambda = 4L$ and the second harmonic has $\lambda = 2L$. Thus when a quarter-wavelength short-circuited line is bridged across another transmission line, it will have no effect on the fundamental frequency but will short-circuit any undesirable second harmonic.

Example 8.6 Light at normal incidence passes through a glass plate of refractive index 1.5 suffering only one air to glass and one glass to air reflection. What percentage of the incident light intensity is lost due to reflections ?

Solution

We know that when light travelling in air falls normally on a plane surface of refractive index n, the ratio of the transmitted intensity to the incident intensity is

$$\frac{I_t}{I_i} = \frac{4n}{(1+n)^2}$$

At air to glass reflection, the transmitted intensity is given by

$$I_t = \frac{4nI_i}{(1+n)^2}$$

When this light suffers glass to air reflection, the intensity of the emergent light is given by

$$I_t' = \frac{4n\, I_t}{(1+n)^2} = \frac{16n^2\, I_i}{(1+n)^4}$$

or $$\frac{I_t'}{I_i} = \frac{16n^2}{(1+n)^4} = \frac{16\times(1.5)^2}{(1+1.5)^4} \simeq 0.92$$

Thus 92% of the incident light is transmitted and 8% is lost due to reflections.

Example 8.7 Three strings *1*, *2* and *3* are joined together and the composite string is stretched with a tension of 10N. The linear densities of strings 1 and 3 are 1×10^{-3} kgm^{-1} and 4×10^{-3} kgm^{-1} respectively. A transverse wave of frequency 100 Hz is produced in string 1. Calculate the linear density and length of the intermediate string 2 so that the wave is completely transmitted through the composite string without any loss due to reflections at the boundaries.

Solution

$$T = 10 \text{ N}$$

$$\mu_1 = 1\times10^{-3} \text{ kgm}^{-1}$$

$$\mu_3 = 4\times10^{-3} \text{ kgm}^{-1}$$

The wave velocities in strings 1 and 3 are

$$v_1 = \sqrt{\frac{T}{\mu_1}} = 100 \text{ ms}^{-1}$$

$$v_3 = \sqrt{\frac{T}{\mu_3}} = 50 \text{ ms}^{-1}$$

The characteristic impedances of the strings are

$$Z_1 = \mu_1 v_1$$

$$Z_2 = \mu_2 v_2$$

$$Z_3 = \mu_3 v_3$$

The condition for impedance matching is

$$Z_2^2 = Z_1 Z_3$$

or $$\mu_2^2 v_2^2 = \mu_1 \mu_3 v_1 v_3$$

or $$\mu_2^2 \frac{T}{\mu_2} = \mu_1 \mu_3 \frac{T}{\sqrt{\mu_1 \mu_3}}$$

or $$\mu_2 = \sqrt{\mu_1 \mu_3}$$

$$= (1\times10^{-3}\times4\times10^{-3})^{1/2} = 2\times10^{-3} \text{ kgm}^{-1}$$

If L is the length of string 2, the second condition of impedance matching is

$$L = \frac{\lambda_2}{4}$$

where λ_2 is the wavelength in string 2. Now, since frequency cannot change on reflection and transmission we have

$$v_1 = \nu\lambda_1$$

$$v_2 = \nu\lambda_2$$

$$\therefore \qquad \lambda_2 = \lambda_1 \frac{v_2}{v_1}$$

But $$\lambda_1 = \frac{v_1}{\nu} = \frac{100 \text{ ms}^{-1}}{100 \text{ Hz}} = 1 \text{ m}$$

and $$\frac{v_2}{v_1} = \sqrt{\frac{\mu_1}{\mu_2}} = \frac{1}{\sqrt{2}}$$

Hence,

$$L = \frac{\lambda_2}{4} = \frac{1}{4}\lambda_1 \frac{v_2}{v_1} = 0.177\text{m} = 17.7 \text{ cm}$$

Thus, for transmission without reflection the intermediate string must have a length of 17.7 cm and a linear density of 2×10^{-3} kg m^{-1}.

Example 8.8 Light travelling in free space enters a glass lens which has a refractive index of 1.69 for a free space wavelength of 5200 Å. The lens is coated with a non-reflecting film to avoid reflections at this wavelength. Calculate the refractive index and thickness of the non-reflecting film.

Solution

Let n_1, n_2 and n_3 respectively be the refractive indices of free space the film (coating) and lens for light of wavelength λ_1 = 5200 Å in free space.

Here $$n_1 = 1$$
$$n_3 = 1.69$$

The refractive index n_2 of the coating so that reflections are avoided is given by

$$n_2 = \sqrt{n_1 n_3}$$
$$= \sqrt{1 \times 1.69} = 1.3$$

The thickness L of the film must be

$$L = \frac{\lambda_2}{4}$$

where λ_2 is the wavelength of light in the film of refractive index 1.3.

$$\lambda_2 = \frac{\lambda_1}{n_2} = \frac{5200 \text{ Å}}{1.3} = 4000 \text{ Å} = 4 \times 10^{-5}\text{m}$$

Hence $$L = 1.0 \times 10^{-5} \text{ cm}$$

The lens is placed in a vacuum chamber in which the coating material is heated in a crucible until it evaporates. Molecules of the material fly in all directions and evenly coat the lens on the side facing the crucible. Heating is stopped when the coating of the required thickness is obtained.

QUESTIONS

1. A transverse wave travels in the $+x$ direction in a string of characteristic impedance Z_1 and meets the junction of another string of characteristic impedance Z_2. The composite string is stretched with a tension T. Deduce the expressions for the reflection and transmission coefficients of amplitude and energy.

2. Deduce the expressions for the reflection and transmission amplitude and energy coefficients when a transverse wave travelling in the $-x$ direction in the string of impedance Z_2 meets the junction of a string of impedance Z_1.

3. (a) What is a standing wave?

   (b) Derive the equation that describes a standing wave on a string of length L fixed rigidly at both ends.

   (c) State the characteristic features which distinguish a standing wave from a travelling wave.

4. A transverse wave incident on a boundary is reflected with a reduced amplitude. Obtain an expression for the standing wave ratio in terms of the amplitude reflection coefficient. What is the significance of this ratio?

5. Obtain the amplitude and energy reflection and transmission coefficients when a plane sound wave travelling in a medium of impedance Z_1 is incident normally on a plane boundary of a medium of impedance Z_2.

6. Discuss analytically standing waves in (a) an open pipe and (b) a closed pipe.

7. Two parallel lossless transmission lines of impedances Z_1 and Z_2 are joined. A voltage-current wave travelling in one line meets the junction of the other line.

   (a) Deduce the expressions for the voltage and current reflection and transmission coefficients.

   (b) Discuss the cases of (i) a short-circuited line and (ii) an open-circuited line.

8. A plane monochromatic electromagnetic wave travelling in the $+x$ direction in a dielectric medium of impedance Z_1 is incident normally on a plane boundary of another medium of impedance Z_2. Obtain the reflection and transmission coefficients of the electric and magnetic fields if

   (a) $Z_1 > Z_2$

   (b) $Z_1 < Z_2$.

9. Write down the coefficients of Q. 8 in terms of the refractive indices of the two media.

10. (a) What is impedance matching? Why is it necessary?

    (b) Two media of impedances Z_1 and Z_3 are separated by a medium of intermediate impedance Z_2. A wave travelling in the first medium is incident normally on the plane boundary of the second medium. Deduce the condition which must be satisfied so that all the incident energy is transmitted without reflection.

    (c) Describe two important applications of non-reflecting layers.

11. (a) How is impedance matching achieved by a tapering section?

    (b) What do you understand by blooming?

12. Show that when light travelling in free space is incident normally on the surface of a dielectric of reflective index n, the reflected and transmitted

intensity coefficients are given by

$$R = \left(\frac{1-n}{1+n}\right)^2$$

and

$$T = \frac{4n}{(1+n)^2}$$

PROBLEMS

1. Two strings of linear densities μ_1 and μ_2 are joined together and the composite string is stretched with a certain tension. A transverse wave travelling in the first string is incident on the junction separating the second string. Calculate the fraction of the incident amplitude reflected and transmitted at the junction if μ_1/μ_2 is (a) 0 (b) 0.25 (c) 4 and (d) ∞.
2. Calculate the reflection and transmission energy coefficients in each case in Problem 1.
3. Stationary waves are produced by a superposition of two waves which are given by

$$\psi_1 = 0.02 \sin \pi(t-2x)$$

$$\psi_2 = 0.02 \sin \pi (t+2x)$$

where all quantities are measured in MKS units.

(a) Determine the amplitude, period, wavelength and velocity of each wave.

(b) Prove that the resultant displacement of a particle at x at time t is given by $\psi = 0.04 \cos 2\pi x \sin \pi t$.

(c) At what value of x is ψ zero for all values of t ?

(d) What is the difference between the two nearest values of x at which $\psi = 0$? Is this difference related to the wavelength of either wave ? If so, how ?

4. In Example 8.2, if $\alpha = \tan\theta$, show that the reflected and transmitted energy coefficients are represented by $\sin^2\theta$ and $\cos^2\theta$ respectively.
5. Sound waves produced under water are incident normally on water-air interface. If the speed of sound in water is 1500 ms^{-1} and in air 350 ms^{-1}, calculate the percentage of incident sound energy transmitted out into air. Density of air is 1.3 kg m^{-3}.
6. Yellow light of wavelength 5890 Å is incident normally on two plane glass slabs placed one after the other. if the refractive index of glass for yellow light is 1.5, what percentage of light is transmitted through the slabs.

   (Note : There are four transmissions ; two for each slab).
7. Compare the amplitude and intensity reflection coefficients r and R when light falls normally on a plane glass surface of refractive index 1.5 if light is incident from (a) air to glass and (b) glass to air.
8. Light travelling in vacuum enters a glass lens which has a refractive index of 1.5 for a vacuum wavelength of 5500 Å. The lens is coated with a non-

reflecting layer of a transparent material to avoid loss of light due to reflections. What should be the refractive index and thickness of the non-reflecting layer ?

9. Impedance matching is achieved between two strings 1 and 3 by inserting between them a string 2 of intermediate impedance. A transverse wave of frequency 100 H_z is generated in s tring 1. If the speed of the wave in string 1 is 100 ms^{-1} and in string 3 the speed is 25 m s^{-1}, what is the speed of the wave in the intermediate string ? What is the length of this string ?
10. A longitudinal wave travelling in a rod encounters a discontinuity where the Young's modulus suddenly doubles, the density remaining the same. Calculate the reflection and transmission amplitude coefficients .

9

Modulations, Wave Groups and Pulses

9.1. INTRODUCTION

So far we have confined ourselves only to monochromatic waves, i.e. waves of a single frequency and wavelength. One very useful application of waves is that they can be used to send a signal or message. This cannot be done with a harmonic travelling wave involving only a single frequency. The reason is that such a wave travels unchanged ; each cycle just like the preceding cycle. If we want to send a message with a wave, we must bring about a change (or *modulation*) in the wave at the transmitting station in such a way that it can be decoded at the receiving station. Three different modulations are possible. We may modulate either the amplitude or the frequency or the phase constant of the wave. These ways are respectively called amplitude modulation (AM), frequency modulation (FM) and phase modulation.

We know that a harmonic driving force generates harmonic waves in a medium. Thus a modulated wave cannot be produced by a harmonic driving force. In Chapt. 2 we have discussed the superposition of two harmonic oscillations of slightly different frequencies which results in the phenomenon of beats. The equation describing the beat phenomenon was not harmonic ; it was essentially anharmonic. A general name for beats is 'modulations'. We have seen that the anharmonic motion has a modulated amplitude which is repeated with time at a frequency called the beat or modulation frequency. In this chapter we shall study beats in time as well as in space. We will show that these modulations propagate as travelling waves. These travelling modulations (consisting of a mixture or group of harmonic oscillations) are called *wave groups* or *wave packets* which carry energy as they travel at a velocity called the *group velocity*. We shall, later in the chapter, use the method of Fourier analysis and show that a wave group can always be expressed as a linear combination of the constituent harmonic components.

9.2 GROUP AND PHASE VELOCITIES

We shall begin by constructing a special form of a modulated wave from just two components and determine its behaviour. We shall consider what happens if we have two harmonic waves of slightly different frequencies ω_1 and ω_2 travelling in the same direction in a dispersive medium, i e. a medium in which waves of different frequencies travel with different speeds. For simplicity, we will assume that the two waves have the same amplitude, say A. These waves are described by the following equations:

$$\psi_1 = A\cos(\omega_1 t - k_1 x)$$

$$\psi_2 = A\cos(\omega_2 t - k_2 x)$$

where $k_1 = 2\pi/\lambda_1$ and $k_2 = 2\pi/\lambda_2$; λ_1 and λ_2 being the wavelengths of waves of angular frequencies ω_1 and ω_2 respectively. The speeds of the waves are

$$v_1 = \frac{\omega_1}{k_1} = \nu_1\lambda_1$$

$$v_2 = \frac{\omega_2}{k_2} = \nu_2\lambda_2$$

where ν_1 and ν_2 are the frequencies of the waves. We could have taken the amplitudes to be different. It turns out that this only adds to the mathematical complexity without giving any new information. We could also consider waves described by a sine function rather than a cosine function.

The superposition of the two waves gives

$$\begin{aligned}\psi &= \psi_1 + \psi_2 \\ &= A\{\cos(\omega_1 t - k_1 x) + \cos(\omega_2 t - k_2 x)\} \\ &= 2A\cos\left\{\frac{(\omega_1 - \omega_2)t}{2} - \frac{(k_1 - k_2)x}{2}\right\}\cos\left\{\frac{(\omega_1 + \omega_2)t}{2} - \frac{(k_1 + k_2)x}{2}\right\}\end{aligned}$$

or $\qquad \psi = A_m \cos(\omega_a t - k_a x) \qquad (9.1)$

where $\qquad A_m = 2A\cos(\omega_m t - k_m x) \qquad (9.2)$

with $\qquad \omega_m = \frac{1}{2}(\omega_1 - \omega_2)$

$$k_m = \frac{1}{2}(k_1 - k_2) \qquad (9.3)$$

$$\omega_a = \frac{1}{2}(\omega_1+\omega_2)$$
$$k_a = \frac{1}{2}(k_1+k_2)$$

Equation (9.1) describes a wave system with an average frequency ω_a which is close to the frequency of either component (since ω_1 is not very different from ω_2) and a modulated amplitude A_m changing in space and time. Notice that the wave described by Eq. (9.1) is not harmonic as its amplitude A_m varies in space and time. This wave system is shown in Fig. 9.1. The wave envelope of modulated frequency ω_m advances in space

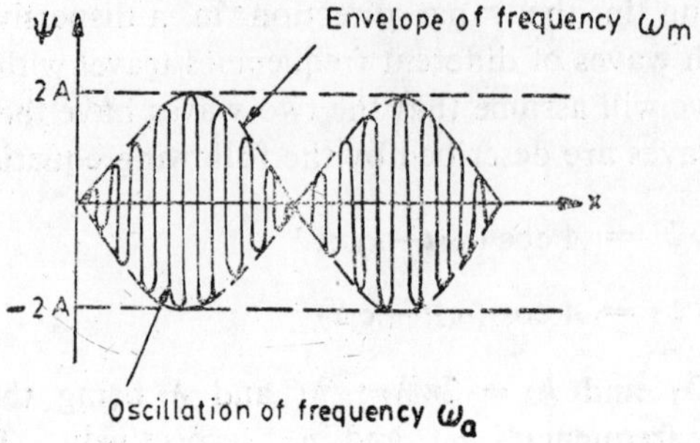

Fig 9.1 A group formed by a superposition of two waves of slightly different frequencies ω_1 and ω_2. The faster oscillation takes place at the average frequency $\omega_a = \frac{1}{2}(\omega_1+\omega_2)$. The slowly changing group envelope has a frequency $\omega_m = \frac{1}{2}(\omega_1-\omega_2)$.

and time and travels at a certain speed which is called the modulation or group velocity.

Group Velocity and Phase Velocity

With what velocity do the modulations propagate ? To find the answer to this question let us say that

$$\omega_2 = \omega \qquad \text{and} \qquad \omega_1 = \omega+\delta\omega$$

where $\delta\omega$ is a small difference between ω_1 and ω_2, so that

$$k_2 = k \qquad \text{and} \qquad k_1 = k+\delta k$$

Equation (9.1) may then be written as

$$\psi = 2A\cos\{\tfrac{1}{2}(\delta\omega t-\delta kx)\}\cos\{(\omega t-kx)\} \tag{9.4}$$

This expression can be interpreted as a high frequency (ω) and short wavelength $\lambda(=2\pi/k)$ wave modulated in amplitude by an envelope (see Fig. 9.1) of frequency $\delta\omega$ varying slowly in time ($\because$ $\delta\omega$ is small). Both these wavelike disturbances $\cos(\omega t-kx)$ and $\cos\frac{1}{2}(\delta\omega t-\delta kx)$

move. We can therefore identify two velocities in expression (9.4) by inspection. One of these velocities is associated with the wave $\cos \frac{1}{2}(\delta\omega t - \delta kx)$ and the other with the wave $\cos(\omega t - kx)$.

Group Velocity

The velocity associated with the wave $\cos \frac{1}{2}(\delta\omega t - \delta kx)$ is the velocity with which the modulating envelope moves (see Fig. 9.1). Because this envelope encloses a group of short waves (wavelength $\approx \delta\lambda$) the velocity of the envelope is called *group velocity* denoted by v_g.

To find the velocity of the envelope we have to find the velocity with which a constant value of the modulated amplitude A_m moves. Let us focus our attention on the crest of the modulation wave (i.e. place where $A_m = +1$). We need to maintain a constant value of the argument $(\delta\omega t - \delta kx)$. Thus when t increases by δt, x must increase by δx so that the increment $(\delta\omega\delta t - \delta k\delta x)$ of $(\delta\omega t - \delta kx)$ remains zero. Hence,

$$\delta\omega\delta t - \delta k\delta x = 0$$

or

$$\frac{\delta x}{\delta t} = \frac{\delta\omega}{\delta k}$$

or

$$v_g = \frac{\delta\omega}{\delta k}$$

If ω_1 and ω_2 are nearly equal, $\delta\omega$ and δk are infinitesimally small so that

$$v_g = \frac{d\omega}{dk} \tag{9.5}$$

Phase Velocity

The velocity associated with the wave $\cos(\omega t - kx)$ is called the *phase velocity* and is denoted by v. It is the velocity with which a crest belonging to the average wave number k moves which is obtained by requiring

$$\omega\delta t - k\delta x = 0$$

or

$$v = \frac{\delta x}{\delta t} = \frac{\omega}{k} \tag{9.6}$$

Relation Between Group Velocity and Phase Velocity in a Dispersive Medium

We have seen (refer to Fig. 9.1) that a signal or modulation propagates not at the phase velocity $v = \omega/k$ but at the group velocity $v_g = \dfrac{d\omega}{dk}$. Thus energy transport in a wave takes place at the group velocity. The group velocity, therefore, is of great physical importance.

We shall now obtain the relation between the two velocities in a *dispersive* medium, i.e. a medium in which the phase velocity (ω/k) is frequency (or wavelength) dependent. For example, lights of different wavelengths (and therefore different frequencies) travel with different speeds in glass. This phenomenon is called dispersion. A ***dispersion relation*** expresses a relation between the angular frequency ω and wave number k or between frequency ν and wavelength λ.

Since $\omega = kv$, where v is the phase velocity, we have

$$v_g = \frac{d\omega}{dk} = \frac{d}{dk}(kv) = v + k\frac{dv}{dk} \tag{9.7}$$

which relates the group velocity to the phase velocity. In terms of wavelength λ, this relation becomes ($\because k = 2\pi/\lambda$)

$$v_g = v + \frac{2\pi}{\lambda}\frac{dv}{d\lambda}\frac{d\lambda}{dk}$$

But $$\frac{d\lambda}{dk} = \frac{d}{dk}\left(\frac{2\pi}{k}\right) = -\frac{2\pi}{k^2} = -\frac{\lambda^2}{2\pi}$$

Hence $$v_g = v - \lambda\frac{dv}{d\lambda} \tag{9.8}$$

The relation between v_g and v may be written is yet another way.

$$\frac{1}{v_g} = \frac{dk}{d\omega} = \frac{d}{d\omega}\left(\frac{\omega}{v}\right) = \frac{1}{v} - \frac{\omega}{v^2}\frac{dv}{d\omega} \tag{9.9}$$

or, since $\omega = 2\pi\nu$

$$\frac{1}{v_g} = \frac{1}{v} - \frac{\nu}{v^2}\frac{dv}{d\nu} \tag{9.10}$$

If we refer especially to light waves and write $v = c/n$ where n is the refractive index of the medium in which light travels with a speed v ; c being the speed of light in vacuum, we have

$$\frac{1}{v_g} = \frac{1}{v} + \frac{\nu}{c}\frac{dn}{d\nu} \tag{9.11}$$

or (since $c = \nu\lambda$)

$$\frac{1}{v_g} = \frac{1}{v} - \frac{\lambda}{c}\frac{dn}{d\lambda} \tag{9.12}$$

where λ is the vacuum wavelength of light.

To summarize, the relationship between the group and phase velocities may be expressed in any one of the following forms (whichever is the most convenient for a particular situation) ;

$$v_g = v + k\frac{dv}{dk}$$

$$= v - \lambda\frac{dv}{d\lambda} \quad (9.13)$$

$$\frac{1}{v_g} = \frac{1}{v} - \frac{\omega}{v^2}\frac{dv}{d\omega}$$

$$= \frac{1}{v} - \frac{\nu}{v^2}\frac{dv}{d\nu}$$

For light waves

$$\frac{1}{v_g} = \frac{1}{v} + \frac{\omega}{c}\frac{dn}{d\omega}$$

$$= \frac{1}{v} + \frac{\nu}{c}\frac{dn}{d\nu} \quad (9.14)$$

$$= \frac{1}{v} - \frac{\lambda}{c}\frac{dn}{d\lambda}$$

Dispersion and Group Velocity

In the case of a non-dispersive medium the phase velocity v $\left(=\frac{\omega}{k}=\nu\lambda\right)$ is independent of ω (or ν) and k (or λ), i.e.

$$\frac{dv}{dk} = \frac{dv}{d\lambda} = \frac{dv}{d\omega} = \frac{dv}{d\nu} = 0$$

which means that $v_g = v$. For example, light waves in free space and sound waves in a medium are non-dispersive. In such cases, the group velocity equals the phase velocity.

In the case of a dispersive medium the phase velocity depends upon frequency or wavelength. If $\frac{dv}{d\lambda}$ is positive $\left(\text{or } \frac{dv}{dk} \text{ is negative}\right)$, $v_g < v$. This is called *normal dispersion*. On the other hand, if $\frac{dv}{d\lambda}$ is negative, then $v_g > v$. This is called *anomalous dispersion*. Figure 9.2 shows the behaviour of the group velocity in which the angular frequency ω is plotted against the wave number k for a medium which shows no dispersion, normal dispersion and anomalous dispersion. The wave or phase velocity at a particular point on the curve is ω/k and the group velocity at that point is the slope of the curve. In the case of normal dispersion, the group velocity is less than the phase velocity for any point on the curve. In the case of anomalous dispersion, the group velocity is greater than the phase velocity for any point on the curve.

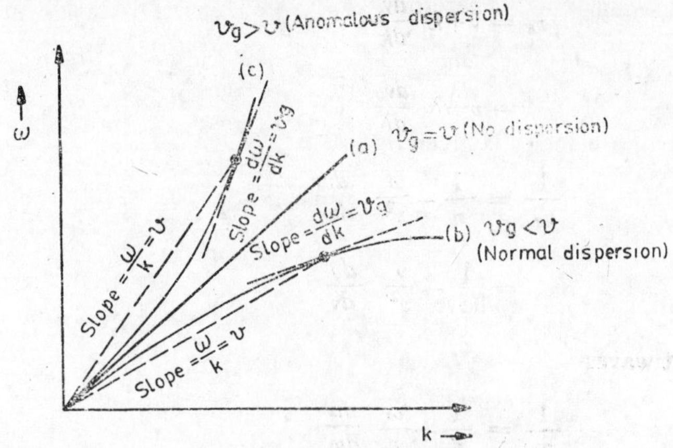

Fig 9.1 Angular frequency ω and wave number k relationship for (a) a non-dispersive medium $v_g = v$; (b) a medium showing normal dispersion $v_g < v$ and ; and (c) a medium showing anomalous dispersion $v_g > v$.

Our discussion on group velocity has been restricted to a group formed by a superposition of only two harmonic waves of slightly different frequencies. It turns out that our definition of the group velocity $v_g = \dfrac{d\omega}{dk}$ is valid not for just two waves but for a whole spectrum of any number of superposing waves. We shall treat this problem in Sec 9.3 where we will consider how a pulse is formed by a superposition of a large number of harmonic oscillations with frequencies lying in a certain range. The pulse, so formed, travels in space and time in form of a group at a velocity given by $v_g = \dfrac{d\omega}{dk}$.

Demonstration of Group Velocity

The most familiar example of dispersion (i.e. the difference between phase and group velocities) is provided by waves in a liquid. The simplest kind are the so-called 'gravity waves' which are short wavelength waves in water of depth much larger than the wavelength. The restoring force on a portion of water is provided by gravity and it is found that the wave velocity (which we are now calling the phase velocity) is given by

$$v = \sqrt{\frac{g\lambda}{2\pi}}$$

where g is the acceleration due to gravity and λ is the wavelength. Thus

$$v = C\sqrt{\lambda}$$

where C is a constant. Therefore,

$$\frac{dv}{d\lambda} = \frac{1}{2} C\lambda^{-1/2}$$

Now the group velocity is given by

$$v_g = v - \lambda \frac{dv}{d\lambda}$$

Substituting for $\frac{dv}{d\lambda}$ we have

$$v_g = v - \frac{1}{2} C\lambda^{1/2} = \frac{v}{2}$$

Thus the group velocity is half the phase velocity. The gravity waves are, therefore, strongly dispersive.

The difference between the group and phase velocities is beautifully demonstrated when a small stone is dropped in the still water in a big tub. The group of waves is seen to spread out from the origin, i.e. the place where the stone is dropped. If you observe carefully and concentrate your attention on a particular wave, it will be seen to travel almos ttwice as fast as the group or packet of tiny wavelets. The faster wave is seen to go through the group from back to front. When it reaches the leading edge of the group, it dwindles away, and new wave appears at the rear end of the group. The waves travel at the phase velocity. The wave group or packet as a whole travels at the group velocity.

Examples of Nondispersive and Dispersive Waves

Electromagnetic Waves in Vacuum

Light waves and other electromagnetic radiations in vacuum are non-dispersive. The dispersive relation is given by

$$\omega = kc$$

where c is the velocity of the waves in vacuum. The phase and group velocities are given by

$$v = \frac{\omega}{k} = c$$

$$v_g = \frac{d\omega}{dk} = c$$

Thus $v = v_g = c$. Modulations travel in vacuum at the speed of light.

Transverse Waves on Strings

The relation between ω and k for transverse waves on a uniform string is

$$\omega = k\sqrt{\frac{T}{\mu}}$$

which gives

$$v = \frac{\omega}{k} = \sqrt{\frac{T}{\mu}}$$

$$v_g = \sqrt{\frac{T}{\mu}} = v$$

Thus transverse waves on strings are nondispersive.

Sound Waves in a Gas

For sound waves, we have

$$\omega = k\sqrt{\frac{\gamma P}{\rho}}$$

which gives $v = v_g = \sqrt{\frac{\gamma P}{\rho}}$. Thus sound waves in a gas (or other elastic media) are nondispersive. In other words, the speed of sound does not depend upon the frequency of the source of sound. High and low frequency sounds travel at the same speed in a given medium. Imagine the state of affairs if, while listening to an orchestra, the sounds of different frequencies emitted by different musical instruments were to reach our ears at different times and at different speeds.

Light Waves in Glass

The dispersion relation for visible light in glass is given by

$$\frac{c^2}{v^2} = A + \frac{B}{\lambda^2}$$

or

$$n^2 = A + \frac{B}{\lambda^2}$$

where c is the velocity of light in vacuum and λ is the wavelength. A and B are constants. This relation is known as Cauchy's dispersion formula. It tells us that the refractive index n of glass increases with the decrease in wavelength. Since the wavelength of blue light is less than that of the red component of visible white light, the refractive index of glass is more for blue light than for red light. This agrees with the experimental result that a glass prism bends blue component more than the red component.

This is normal dispersion for which the group velocity is less than the phase velocity.

From Cauchy's formula we have

$$\frac{dn}{d\lambda} = -\frac{B}{n\lambda^3}$$

Using Eq. (9.14) the group velocity is given by

$$\frac{1}{v_g} = \frac{1}{v} - \frac{\lambda}{c}\frac{dn}{d\lambda}$$

$$= \frac{1}{v}\left(1 + \frac{v}{c}\frac{B}{n\lambda^2}\right)$$

$$= \frac{1}{v}\left(1 + \frac{v^2}{c^2}\frac{B}{\lambda^2}\right) \qquad \left(\because n = \frac{c}{v}\right)$$

Now $v = \nu\lambda$ and $c = \nu\lambda_0$ where λ is the wavelength in glass and λ_0 is the vacuum wavelength. Hence

$$\frac{1}{v_g} = \frac{1}{v}\left(1 + \frac{B}{\lambda_0^2}\right)$$

or $v_g < v$ indicating that the dispersion of visible light in glass is normal.

Electromagnetic Waves in a Dielectric

The velocity of electromagnetic waves in a dielectric is given by

$$v = \frac{1}{\sqrt{\mu\varepsilon}}$$

where μ is the permeability and ε is the permittivity of the dielectric. In vacuum, the velocity of the electromagnetic waves is the given by

$$c = \frac{1}{\sqrt{\mu_0\varepsilon_0}}$$

The refractive index of the dielectric is

$$n = \frac{c}{v} = \sqrt{\frac{\mu\varepsilon}{\mu_0\varepsilon_0}} = \sqrt{\mu_r\varepsilon_r}$$

where $\mu_r = \frac{\mu}{\mu_0}$ and $\varepsilon_r = \frac{\varepsilon}{\varepsilon_0}$. For most dielectrics the relative permeability $\mu_r \simeq 1$ but the relative permittivity (or dielectric constant) depends upon frequency. Thus the velocity v (and refractive index n) are frequency dependent. Now

$$n = \sqrt{\varepsilon_r}$$

$$\therefore \quad \frac{dn}{d\omega} = \frac{1}{2}(\varepsilon_r)^{-1/2}\frac{d\varepsilon_r}{d\omega}$$

The group velocity is given by [see Eq. (9.14)]

$$\frac{1}{v_g} = \frac{1}{v} + \frac{\omega}{c}\frac{dn}{d\omega}$$

Substituting for $\frac{dn}{d\omega}$ we have

$$\frac{1}{v_g} = \frac{1}{v}\left(1 + \frac{\omega}{2\epsilon_r}\frac{d\varepsilon_r}{d\omega}\right)$$

In terms of wavelength λ this becomes $\left(\because \quad \omega = 2\pi\nu = \frac{2\pi c}{\lambda}\right)$

$$\frac{1}{v_g} = \frac{1}{v}\left(1 - \frac{\lambda}{2\varepsilon_r}\frac{d\varepsilon_r}{d\lambda}\right)$$

Now every dielectric has one (or more) resonant frequency (ω_0) which depends upon how strongly the electrons in the dielectric are bound to their atoms. For glass ω_0 lies in the extreme ultraviolet region (the corresponding wavelength $\lambda_0 = \frac{c}{\nu_0} = \frac{2\pi c}{\omega_0}$ is about 1000 Å). If the frequency ω of the electromagnetic waves is much smaller or much larger than the resonant frequency ω_0 of glass, i.e. if $\omega \ll \omega_0$ (or $\lambda \gg \lambda_0$) and $\omega \gg \omega_0$ (or $\lambda \ll \lambda_0$), $\frac{d\varepsilon_r}{d\omega}$ is positive $\left(\text{or } \frac{d\varepsilon_r}{d\lambda} \text{ is negative}\right)$. In these cases $v_g < v$ indicating that the dispersion is normal.

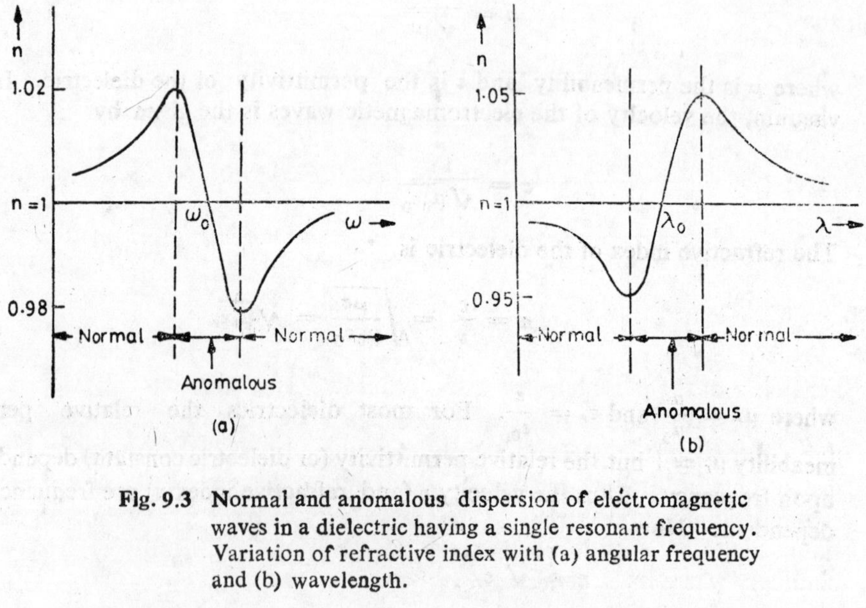

Fig. 9.3 Normal and anomalous dispersion of electromagnetic waves in a dielectric having a single resonant frequency. Variation of refractive index with (a) angular frequency and (b) wavelength.

If $\omega \simeq \omega_0$ (or $\lambda \simeq \lambda_0$), i.e. near resonance, the wave is strongly absorbed and $\frac{d\epsilon_r}{d\omega}$ is negative $\left(\text{or } \frac{d\epsilon_r}{d\lambda} \text{ is positive}\right)$. Thus near resonance, $v_g > v$ and the dispersion is anomalous. Figure 9.3 shows the variation of refractive index with frequency (and wavelength) in the region of normal and anomalous dispersion.

9.3 WAVE PULSES

We have so far been considering harmonic waves, i.e. waves described by sine or cosine function of space and time. Such waves continue on and on for ever involving a whole succession of crests and troughs. Harmonic disturbances are not the only ones that travel in a medium. In fact, we come across a variety of situations in which a single disturbance of a limited duration travels from one place to another in a medium. Such a non-harmonic disturbance of a limited duration is called a *pulse*. For example, a single clap of hands or a single word of command produces a pulse of sound which travels from one place to another. Pulses of this kind can be set up in strings or springs by producing a local deformation. Figure 9.4 shows a pulse in a string produced by jerking the hand up and bringing it down to the earlier position and holding it still. The subsequent

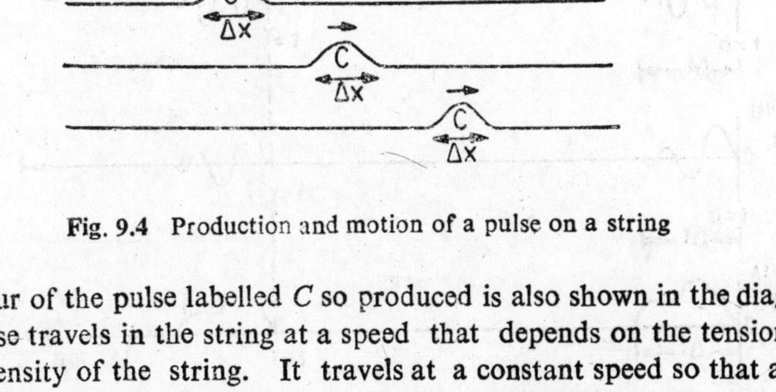

Fig. 9.4 Production and motion of a pulse on a string

behaviour of the pulse labelled *C* so produced is also shown in the diagram. The pulse travels in the string at a speed that depends on the tension and linear density of the string. It travels at a constant speed so that at any instant only a limited region of the string is disturbed ; the rest of the string remaining undisturbed. The pulse will keep on travelling, with its shape unaltered, until it reaches the far end of the string and meets a boundary where it is reflected exactly as a harmonic wave does. If it meets a boundary of a higher impedance (fixed end), the pulse undergoes a phase change, i.e. a positive pulse is reflected as a negative pulse.

Figure 9.5 shows a compressional pulse in a spring produced by jerking the piston to the right and bringing it back to the earlier position and holding it still. The compression marked C travels along the spring at a constant speed and preserves its shape. It may be remarked that in a dispersive medium a pulse spreads as it moves undergoing a change in its shape.

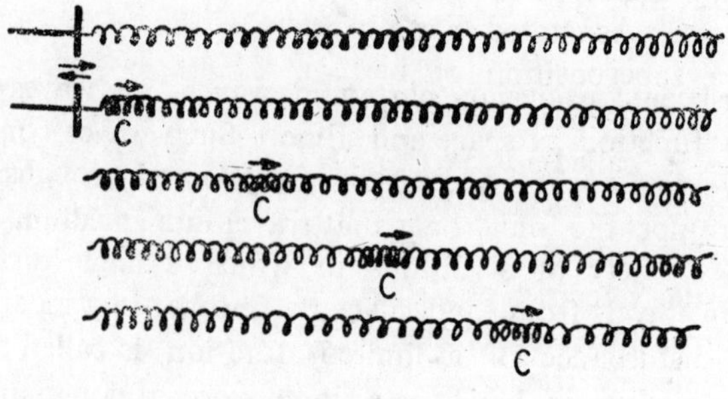

Fig. 9.5 Production and motion of a compressional pulse in a spring

Thus a pulse is a disturbance $\psi(t)$ which is finite only during a limited time interval called the *duration time* of the pluse. We shall show below that a pulse is formed by a superposition of harmonic oscillations of different frequencies. Although these harmonic oscillations individually continue for ever, the absence of the resultant disturbance outside the duration time is achieved by an appropriate combination of harmonics

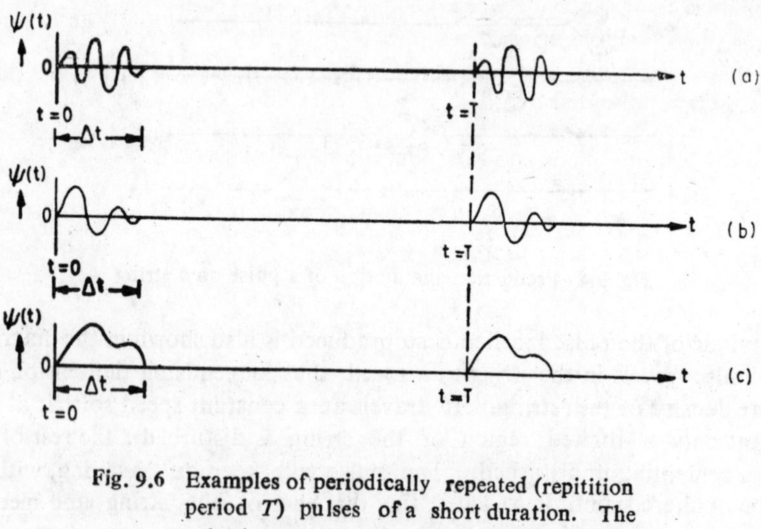

Fig. 9.6 Examples of periodically repeated (repitition period T) pulses of a short duration. The pulse is zero over most of the repitition period.

resulting in complete cancellation at times earlier and later than the duration time but building up to give the desired nonzero disrurbance during a particular time interval (called the duration time). Such pulses may be formed in time by imparting a number of simultaneous harmonic oscillations to a certain point (say $x = 0$) of a medium. If this procedure is repeated at a later time, another pulse is formed. These pulses may be periodic or non-periodic. Periodic pulses are repeated after a fixed time interval called the *repetition time* (T) of the pulse that is long compared with the duration time (Δt) of the pulse. Figure 9.6 shows three examples of periodic pulses. As stated earlier, each of these and any other pulse is formed by a superposition of harmonic disturbances of appropriate frequencies and amplitudes. These pulses travel in space as a group. A travelling pulse is called a wave group or a wave packet. In Sec. 9.2 we have studied a pulse formed by a superposition of just two harmonic disturbances of frequencies ω_1 and ω_2. We have seen that this pulse has two characteristic velocities-phase and group velocities. We shall now study a pulse formed by a superposition of a large number (N) of harmonic oscillations of different frequencies and deduce a very important theorem (which is true for pulses of all kinds) called the *bandwidth theorem.* We shall later learn how such a pulse can be Fourier analysed into its constituent harmonic disturbances.

Construction of a Pulse by a Superposition of *N* Harmonic Oscillations : The Bandwidth Theorem

We shall now obtain the shape or profile of a pulse $\psi(t)$ formed by a linear superposition of N different harmonic components of angular frequencies distributed uniformly between ω_1 and ω_2 with a frequency spacing of $\delta\omega$ between two neighbouring components. For simplicity, we will assume that they have equal amplitudes A and zero phase constant. The superposition is represented by

$$\psi(t) = A\cos\omega_1 t + A\cos(\omega_1+\delta\omega)\,t + A\cos(\omega_1+2\delta\omega)\,t + \ldots + A\cos\omega_2 t \tag{9.15}$$

where $\omega_2 = \omega_1 + (N-1)\,\delta\omega$, so that

$$\delta\omega = \frac{\omega_2-\omega_1}{N-1} = \frac{\Delta\omega}{N-1} \tag{9.16}$$

where $\Delta\omega = \omega_2 - \omega_1$ is called the *angular frequency bandwidth.* The *frequency bandwidth* is

$$\Delta\nu = \nu_2 - \nu_1$$

where ν_1 and ν_2 are the frequencies of the two extreme components.

In Chap. 2 we have already obtained the result of the superposition given, in Eq. (9.15). There is no need to carry out the mathematical steps again. The result is [refer to Eq. (2.37] of Chap. 2].

$$\psi(t) = A\frac{\sin(\frac{1}{2}N\delta\omega t)}{\sin(\frac{1}{2}\delta\omega t)}\cos\omega_a t$$

or
$$\psi(t) = A_m(t)\cos\omega_a t \tag{9.17}$$

where
$$A_m(t) = A\frac{\sin(\frac{1}{2}N\delta\omega t)}{\sin(\frac{1}{2}\delta\omega t)} \tag{9.18}$$

and $\omega_a = \frac{1}{2}(\omega_2+\omega_1) = \omega_1+\frac{1}{2}(N-1)\,\delta\omega$ is the average of the two extreme frequencies.

We shall now consider a limiting case when N becomes large and the frequency spacing $\delta\omega$ between neighbouring harmonic components becomes small so that the frequencies (instead of increasing discretely in steps of $\delta\omega$) are distributed uniformly between the two extreme values ω_1 and ω_2.

For a large N we can write $N \simeq N-1$ so that

$$N\delta\omega \simeq (N-1)\,\delta\omega = \omega_2-\omega_1 = \Delta\omega$$

Also, if $\delta\omega$ is very small and t does not go to infinity, we may write

$$\sin(\tfrac{1}{2}\delta\omega t) \simeq \tfrac{1}{2}\,\delta\omega t$$

In this limit, the modulated amplitude given by Eq. (9.18) can be written as

$$A_m(t) = A\frac{\sin(\frac{1}{2}\Delta\omega t)}{\frac{1}{2}\delta\omega t}$$

$$= AN\frac{\sin(\frac{1}{2}\Delta\omega t)}{\frac{1}{2}\Delta\omega t} \qquad (\because \Delta\omega = N\delta\omega)$$

$$= AN\frac{\sin\alpha}{\alpha}$$

where $\alpha = \frac{1}{2}\Delta\omega t.$ (9.19)

Notice that at time $t = 0$, $\alpha = 0$, so that $\sin\alpha/\alpha = 1$. Hence

$$A_m(0) = AN$$

Therefore,
$$A_m(t) = A_m(0)\frac{\sin\alpha}{\alpha}$$

where $A_m(0)$ is the value of $A_m(t)$ at $t = 0$. Thus we have

$$\psi(t) = A_m(0)\frac{\sin(\frac{1}{2}\Delta\omega t)}{\frac{1}{2}\Delta\omega t}\cos\omega_a t$$

$$= A_m\ (0)\ \frac{\sin \alpha}{\alpha} \cos \omega_a t \tag{9.20}$$

where $\alpha = \frac{1}{2}\ \Delta\omega t$ is half the phase difference between the extreme components at time t.

The time behaviour of the pulse $\psi\ (t)$ given by Eq. (9.20) can be obtained as shown in Fig. 9.7. We first plot the variation of the fast oscillation

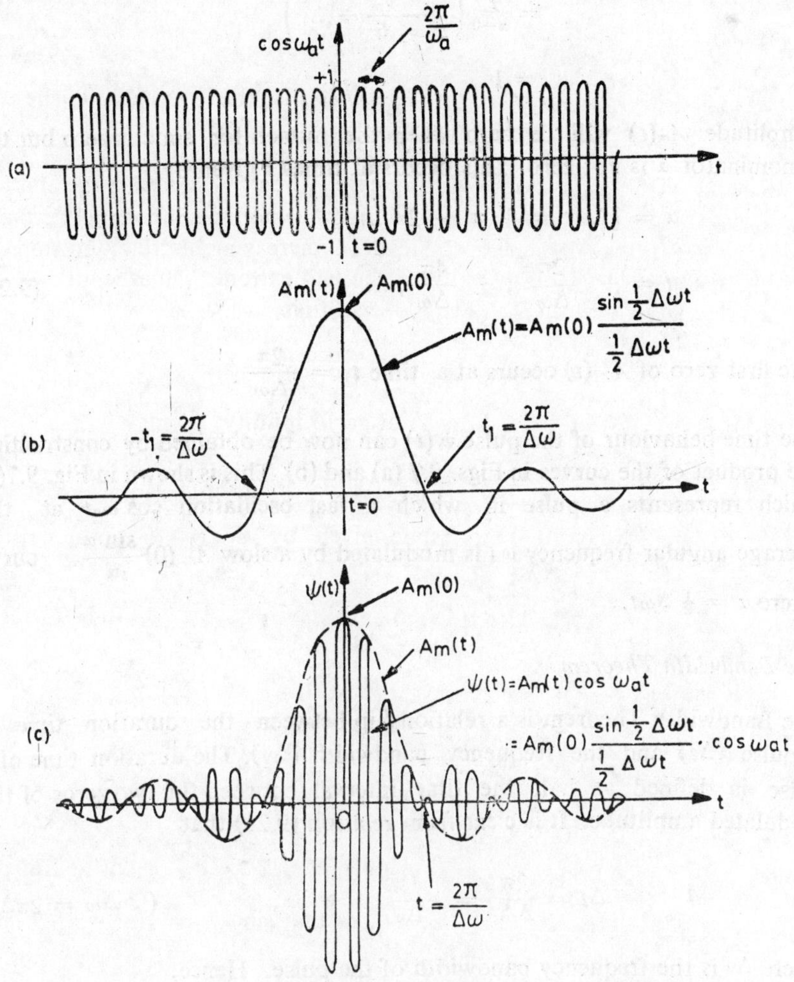

Fig. 9.7 (a) Variation of fast oscillation $\cos \omega_a t$ with time.
(b) Variation of modulated slowly changing amplitude $A_m(t)$ with time.
(c) Representation [obtained by the product of curves (a) and (b)] of the pulse. Cosine curve of average frequency ω_a modulated by $A_m(t)$ $= A_m\ (0)\ \dfrac{\sin \frac{1}{2}\Delta\omega t}{\frac{1}{2}\Delta\omega t}$ curve.

cos $\omega_a t$ with time. This is shown in Fig. 9.7 (a). Then the slowly changing modulated amplitude A_m (t) is plotted against time. This is shown in Fig. 9.7 (b). Notice that at time $t = 0$, $A_m(t) = A_m(0) = AN$, the maximum value of the modulated amplitude. This is because

$$\underset{t\to 0}{Lt}\ \frac{\sin \alpha}{\alpha} = \underset{\alpha\to 0}{Lt}\ \frac{1}{\alpha}\left(\alpha - \frac{\alpha^3}{3!} + \ldots\right)$$

$$= \underset{\alpha\to 0}{Lt}\left(1 - \frac{\alpha^2}{6} + \ldots\right)$$

$$= 1$$

Amplitude $A_m(t)$ will be zero when the numerator $\sin \alpha$ is zero but the denominator α is not zero. This occurs at times t given by

$$\alpha = \tfrac{1}{2}\Delta\omega t = \pm\,\pi, \pm\,2\pi, \ldots$$

or
$$t = \pm\,\frac{2\pi}{\Delta\omega}, \pm\,\frac{4\pi}{\Delta\omega}, \ldots \tag{9.21}$$

The first zero of A_m (t) occurs at a time $t_1 = \dfrac{2\pi}{\Delta\omega}$.

The time behaviour of the pulse ψ (t) can now be obtained by constructing the product of the curves in Figs. 9.7 (a) and (b). This is shown in Fig. 9.7(c) which represents a pulse in which a fast oscillation cos $\omega_a t$ at the average angular frequency ω_a is modulated by a slow $A_m(0)\dfrac{\sin \alpha}{\alpha}$ curve where $\alpha = \frac{1}{2}\Delta\omega t$.

The Bandwidth Theorem

The Bandwidth Theorem is a relationship between the duration time of a pulse (Δt) and the frequency bandwidth $(\Delta\nu)$. The duration time of a pulse is defined as half the time interval between the two zeros of the modulated amplitude. It is clear from relation (9.21) that

$$\Delta t = \frac{2\pi}{\Delta\omega} = \frac{1}{\Delta\nu} \qquad (\because \Delta\omega = 2\pi\Delta\nu)$$

where $\Delta\nu$ is the frequency bandwidth of the pulse. Hence,

$$\Delta\omega\ .\ \Delta t = 2\pi$$

or
$$\Delta\nu\ .\ \Delta t = 1 \tag{9.22}$$

This relation is known as the Bandwidth Theorem. It states that the components of the pulse of frequency bandwidth $\Delta\nu$ will superpose to give a disturbance ψ (t) significantly different from zero only for a time Δt after which the pulse decays due to a destructive interference between the superposed components. The greater is the frequency bandwidth $\Delta\nu$, the shorter is the duration Δt of the pulse Alternatively, the shorter the duration of the pulse, the wider must be the range $\Delta\nu$ of the frequencies required to represent it. If $\Delta\nu = 0$, we have a monochromatic wave of a single frequency. For such a wave Δt is infinity. Hence a monochromatic wave of a single frequency must (at least theoretically) have an infinitely long time span.

As stated earlier, when a pulse travels in a medium it is called a wave group or packet and $\psi(x, t)$ depends on space x as well as time t. We may therefore work in terms of parameters Δk and Δx rather than $\Delta\nu$ and Δt. Notice that the products $(\Delta\omega . \Delta t)$ and $(\Delta k . \Delta x)$ are both dimensionless. Replacing ω by k and t by x, the Bandwidth Theorem becomes

$$\Delta x . \Delta k = 2\pi$$

or

$$\Delta x . \Delta\left(\frac{1}{\lambda}\right) = 1 \qquad (\because k = 2\pi/\lambda)$$

For a monochromatic wave $\Delta k = 0$ giving $\Delta x = \infty$, i.e. a monochromatic wave is an infinitely long wavetrain. Hence, theoretically, a monochromatic wave of a single frequency is an infinitely long wavetrain with an infinitely long time span. It continues on and on unaltered in space as well as in time (see also Sec. 9.5)

We shall now compute the decrease in amplitude and energy of the pulse from its maximum value at time $t = 0$ to a time at the end of the interval Δt, i.e. at time $t = t_1/2$ where $t_1 = 2\pi/\Delta\omega$. The amplitude at $t = t_1/2$ is

$$A_m (t = t_1/2) = A_m (0) \frac{\sin \pi/2}{\pi/2} \qquad (\because \tfrac{1}{2}\Delta\omega t_1 = \pi/2)$$

$$= \frac{2}{\pi} A_m(0) \simeq 0.64 \, A_m (0)$$

Thus at the beginning and the end of the interval Δt the amplitude $A_m(t)$ falls by a factor of $2/\pi \simeq 0.64$ from its maximum value. The energy stored in the pulse is proportional to $A_m^2 (t)$. Thus the energy is maximum at $t = 0$ (the centre of the pulse) and falls by a factor of $(2/\pi)^2 \simeq 0.406$ at the beginning and end of the interval Δt We may, therefore, define the duration time Δt of the pulse also as the time interval during which the pulse has at least 40 per cent of its maximum stored energy.

Fourier Integral Representation of the Pulse

The superposition given in Eq. (9.15) is a Fourier series consisting of discrete frequencies. In the limit $\delta\omega \to 0$ this series would represent a continuous harmonic superposition and would then transform into a Fourier integral as shown below.

We know that (for large N)

$$A = \frac{A_m(0)}{N} = \frac{A_m(0)\,\delta\omega}{N\,\delta\omega} = \frac{A_m(0)}{\Delta\omega}\cdot\delta\omega$$

The superposition Eq. (9.15) can now be expressed as

$$\psi(t) = \frac{A_m(0)}{\Delta\omega}\left\{\cos\omega_1 t.\ \delta\omega + \cos(\omega_1+\delta\omega)t.\ \delta\omega + \ldots + \cos\omega_2 t.\ \delta\omega\right\}$$

In the limit $\delta\omega \to 0$, the expression in brackets is just the integral $\int_{\omega_1}^{\omega_2} \cos\omega t\, d\omega$ integrated from $\omega = \omega_1$ to $\omega = \omega_2$. Thus we have

$$\psi(t) = \frac{A_m(0)}{\Delta\omega}\int_{\omega_1}^{\omega_2} \cos\omega t\, d\omega$$

or

$$\psi(t) = B\int_{\omega_1}^{\omega_2} \cos\omega t\, d\omega \qquad (9.23)$$

where $B = A_m(0)/\Delta\omega$ and has the dimensions of amplitude of $\psi(t)$ divided by angular frequency. For the pulse under consideration B is a constant called the Fourier coefficient of the Fourier integral given in Eq. (9.23). We then have a superposition of harmonic components whose frequencies increase continuously (not discretely) from ω_1 to ω_2. Such a continuous harmonic superposition can always be represented by a Fourier integral.

The pulse represented by integral (9.23) has been simplified by assuming that all frequency components have the same amplitude B. A more general case is a pulse in which the amplitude is frequency dependent with the frequencies uniformly distributed anywhere between zero and infinity. Such a pulse is written as

$$\psi(t) = \int_0^{\infty} B(\omega)\cos\omega t\, d\omega$$

The most general case must also include $\sin\omega t$ terms in addition to the $\cos\omega t$ term. It turns out, as we saw in Chap. 6, that any time-dependent pattern of displacement $\psi(t)$ can be expressed as a continuous Fourier superposition of the general form

$$\psi(t) = \int_0^{\infty} B(\omega)\cos\omega t\, d\omega + \int_0^{\infty} C(\omega)\sin\omega t\, d\omega \qquad (9.24)$$

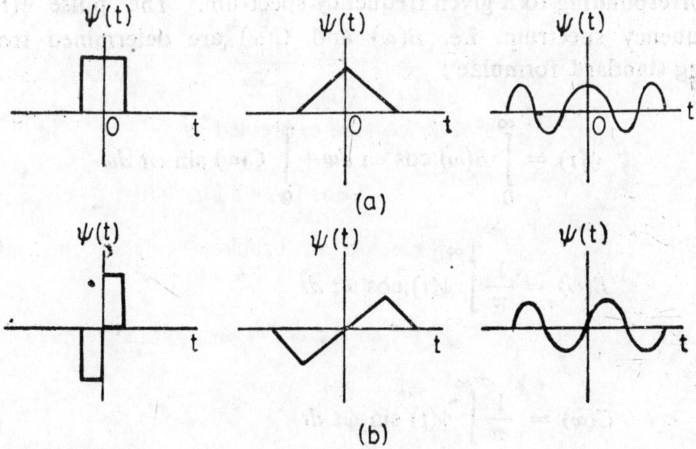

Fig. 9.8 (a) Pulses with even symmetry in time
(b) Pulses with odd symmetry in time

The continuous functions $B(\omega)$ and $C(\omega)$ are called *Fourier coefficients* of $\psi(t)$. Pulses with even time symmetry, i.e. $\psi(t) = +\psi(-t)$ are described by cosine functions [see Fig. 9.8(a)] while those with odd time symmetry, i.e. $\psi(t) = -\psi(-t)$ are described by sine functions [see Fig. 9.8(b)]. Any arbitrary pulse must include both sine and cosine functions.

Frequency Spectrum

A graph of the Fourier coefficients $B(\omega)$ and $C(\omega)$ versus ω is called the *frequency spectrum* of the pulse. Our pulse described by Eq. (9.22) and depicted in Fig. 9.7 has even symmetry in time. Thus for our pulse

$$\begin{aligned} C(\omega) &= 0 \quad \text{for all } \omega \\ B(\omega) &= 0 \quad \text{for } \omega < \omega_1 \quad \text{and } \omega > \omega_2 \\ &= \frac{A_m(0)}{\Delta\omega} \quad \text{for } \omega \text{ lying between } \omega_1 \text{ and } \omega_2 \end{aligned}$$

The frequency spectrum of our pulse is shown in Fig. 9.9(b). It is flat [i.e. $B(\omega)$ is constant] over a limited frequency band of width $\Delta\omega$ and is zero outside the band. Such a spectrum is called a *square spectrum.* Fig. 9.9(a) shows the corresponding discrete frequency spectrum for which $\delta\omega$ is small but finite not tending to zero.

9.4 FOURIER ANALYSIS OF PULSES

We shall now Fourier analyse some simple pulses. We will show how we can find the frequency spectrum of a given pulse and how we can find the pulse corresponding to a given frequency spectrum. The pulse $\psi(t)$ and the frequency spectrum, i.e. $B(\omega)$ and $C(\omega)$ are determined from the following standard formulae :

$$\psi(t) = \int_0^{\infty} B(\omega) \cos \omega t \, d\omega + \int_0^{\infty} C(\omega) \sin \omega t \, d\omega \tag{9.25}$$

$$B(\omega) = \frac{1}{\pi} \int_{-\infty}^{+\infty} \psi(t) \cos \omega t \, dt \tag{9.26}$$

$$C(\omega) = \frac{1}{\pi} \int_{-\infty}^{+\infty} \psi(t) \sin \omega t \, dt \tag{9.27}$$

The Fourier coefficients $B(\omega)$ and $C(\omega)$ are obtained by multiplying Eq. (9.25) by cos ωt (or sin ωt) and carrying out the integrations. We will now apply these formulae to analyse a few simple pulses.

Pulse having a Square Frequency Spectrum

Let us find the pulse $\psi(t)$ whose frequency spectrum is flat and square as

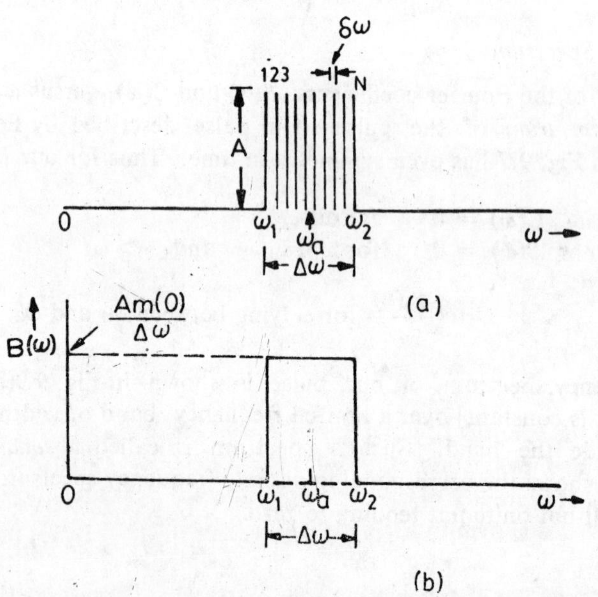

Fig. 9.9 (a) Discrete frequency spectrum of pulse $\psi(t)$ given by Eq. (9.15) ; $\delta\omega \neq 0$
(b) Continuous frequency spectrum of pulse $\psi(t)$ given by Eq. (9.23) ; $\delta\omega \to 0$,

shown in Fig. 9.9(b). We have already solved this problem analytically in the preceding section. We will show how this can be done very simply by the Fourier method. Suppose

$$C(\omega) = 0 \qquad \text{for all } \omega$$

and
$$B(\omega) = 0 \qquad \text{for } \omega < \omega_1 \quad \text{and} \quad \omega > \omega_2$$

$$= \frac{K}{\Delta\omega} \quad \text{for } \omega \text{ lying between } \omega_1 \text{ and } \omega_2$$

where $\Delta\omega = \omega_2 - \omega_1$ and K is a constant whose dimensions are those of $\psi(t)$. Thus B is zero everywhere else except in the bandwidth $\Delta\omega$ where it has a constant value. Substituting for $B(\omega)$ and $C(\omega)$ in Eq. (9.25) we get

$$\psi(t) = \int_0^\infty C(\omega) \sin \omega t \, d\omega + \int_0^\infty B(\omega) \cos \omega t \, d\omega$$

$$= 0 + \int_{\omega_1}^{\omega_2} \frac{K}{\Delta\omega} \cos \omega t \, d\omega = \frac{K}{\Delta\omega} \left. \frac{\sin \omega t}{t} \right|_{\omega_1}^{\omega_2}$$

$$= \frac{K}{\Delta\omega t} (\sin \omega_2 t - \sin \omega_1 t)$$

$$= \frac{2K}{\Delta\omega t} \sin \{\tfrac{1}{2}(\omega_2 - \omega_1)t\} \cos \{\tfrac{1}{2}(\omega_1 + \omega_2)t\}$$

or
$$\psi(t) = K \frac{\sin (\frac{1}{2}\Delta\omega t)}{(\frac{1}{2}\Delta\omega t)} \cos \omega_a t$$

Thus $\psi(t)$ is a fast oscillation at average frequency ω_a with a slowly varying modulated amplitude $A_m(t)$:

$$\psi(t) = A_m(t) \cos \omega_a t$$

where
$$A_m(t) = K \frac{\sin (\frac{1}{2}\Delta\omega t)}{(\frac{1}{2}\Delta\omega t)}$$

This result is identical with Eq. (9.20) obtained in Sec. 9.3. We now recognize that the constant K is the value of $A_m(t)$ at time $t = 0$, i.e. $K = A_m(0)$. Figure 9.9(b) shows the frequency spectrum and the corresponding pulse is shown in Fig. 9.7(c).

Frequency Spectrum of a Square Pulse

Suppose we have a pulse $\psi(t)$ which is constant during a time interval $\Delta_t = t_2 - t_1$ and is zero at times earlier than t_1 and later than t_2. The pulse is given by

$$\psi(t) = \frac{A}{\Delta t} \quad \text{for } t_1 \leqslant t \leqslant t_2\,;$$

$$= 0 \quad \text{for } t \leqslant t_1 \text{ and } t \geqslant t_2$$

where A is a constant equal to the area under the pulse as shown in Fig. 9.10(a) ; the centre of the pulse is at time $t = t_0$. To find the Fourier coefficients $B(\omega)$ and $C(\omega)$ we will use a trick which saves a lot of labour. We know that if $\psi(t)$ is an *even* function of t, so that $\psi(-t) = +\psi(t)$ for any t, then the Fourier analysis requires *cosine* functions only. On the other hand, if $\psi(t)$ is an *odd* function of t, so that $\psi(-t) = -\psi(t)$, then only *sine* functions are sufficient. The trick is that we shift the time origin and achieve even symmetry in pulse. This procedure is depicted in Fig. 9.10(b). The centre of the pulse is at $t = 0$ and $t_1 = -\tau$ and $t_2 = +\tau$ so that $\Delta t = t_2 - t_1 = 2\tau$. Now the shifted pulse is an even function of time and only cosine term in Eq. (9.25) is required. Thus

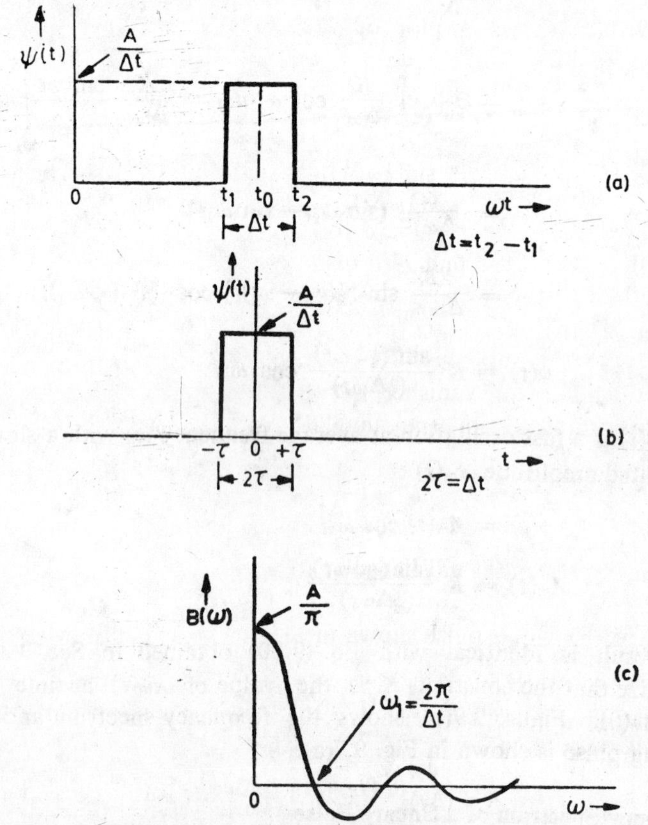

Fig. 9.10 (a) Square pulse $\psi(t)$ not centred at $t = 0$.
(b) Shifting the time origin to achieve symmetry in pulse. Now $\psi(t)$ is an even function of time.
(c) Frequency spectrum of a square pulse.

$$C(\omega) = 0$$

because $\sin \omega t$ is an odd function.

and
$$B(\omega) = \frac{1}{\pi}\int_{-\infty}^{+\infty} \psi(t) \cos \omega t \, dt$$

$$= \frac{1}{\pi}\int_{-\tau}^{+\tau} \frac{A}{\Delta t} \cos \omega t \, dt = \frac{A}{\pi \Delta t} \left. \frac{\sin \omega t}{\omega} \right|_{-\tau}^{+\tau}$$

$$= \frac{2A}{\pi \omega \Delta t} \sin \omega \tau$$

or
$$B(\omega) = \frac{A}{\pi} \frac{\sin (\frac{1}{2}\omega \Delta t)}{(\frac{1}{2}\omega \Delta t)}$$

Figure 9.10(c) shows a plot of $B(\omega)$ versus ω. This is the frequency spectrum of the pulse.

Notice that when $\omega = 0$,

$$\underset{\omega \to 0}{\text{Lt}} \frac{\sin(\frac{1}{2}\omega \Delta t)}{(\frac{1}{2}\omega \Delta t)} = \underset{\alpha \to 0}{\text{Lt}} \frac{\sin \alpha}{\alpha} = 1$$

and $B(0) = A/\pi$. The first zero of $B(\omega)$ occurs when

$$\frac{\omega \Delta t}{2} = \pi$$

so that the numerator vanishes, but the denominator does not. Thus the first zero of $B(\omega)$ occurs at an angular frequency ω' given by

$$\omega' = \frac{2\pi}{\Delta t}$$

Frequency Spectrum of a Triangular Pulse

Consider a triangular pulse shown in Fig. 9.11(a). The pulse is given by

$$\psi(t) = \frac{4A}{(\Delta t)^2} (t - t_1) \; ; \; t_1 \leqslant t \leqslant t_0$$

$$= \frac{4A}{(\Delta t)^2} (t_2 - t) \; ; \; t_0 \leqslant t \leqslant t_2$$

where $t_0 = \frac{1}{2}(t_1 + t_2)$ is the centre of the pulse and A is a constant which represents the area under the pulse. Notice that at $t = t_0$; $t_0 - t_1 = \frac{1}{2}\Delta t$ and $\psi(t = t_0) = 2A/\Delta t$. Also $t_2 - t_0 = \frac{1}{2}\Delta t$.

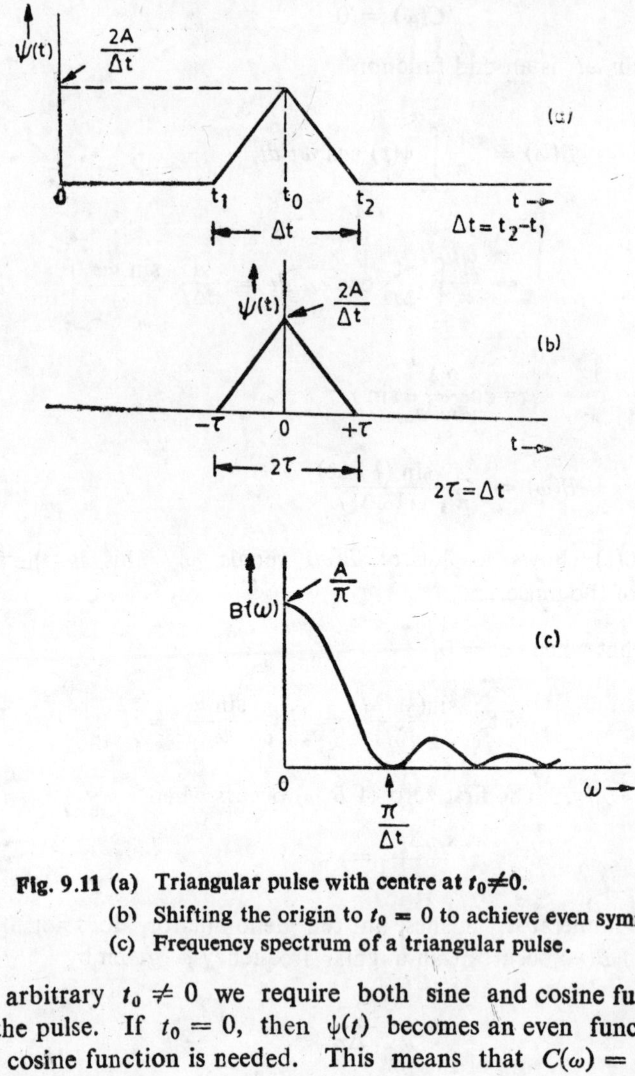

Fig. 9.11 (a) Triangular pulse with centre at $t_0 \neq 0$.
(b) Shifting the origin to $t_0 = 0$ to achieve even symmetry.
(c) Frequency spectrum of a triangular pulse.

For an arbitrary $t_0 \neq 0$ we require both sine and cosine functions to describe the pulse. If $t_0 = 0$, then $\psi(t)$ becomes an even function of t, then only cosine function is needed. This means that $C(\omega) = 0$ and we need to evaluate only $B(\omega)$. Figure 9.11(b) shows how the pulse is symmetrized by shifting the origin. With the shifted origin the pulse is described by ($\because$ $2\tau = \Delta t$)

$$\psi(t) = \frac{A}{\tau^2}(t+\tau) \; ; -\tau \leqslant t \leqslant 0$$

$$= \frac{A}{\tau^2}(\tau - t) \; ; \; 0 \leqslant t \leqslant \tau$$

The Fourier coefficient $B(\omega)$ is given by

$$B(\omega) = \frac{1}{\pi}\int_{-\infty}^{+\infty} \psi(t) \cos \omega t \, dt$$

$$= \frac{A}{\pi\tau^2}\left\{\int_{-\tau}^{0}(t+\tau)\cos\omega t\, dt + \int_{0}^{\tau}(\tau-t)\cos\omega t\, dt\right\}$$

Now since

$$\int_{-\tau}^{0}\cos\omega t\, dt = \int_{0}^{\tau}\cos\omega t\, dt$$

and

$$\int_{-\tau}^{0} t\cos\omega t\, dt = -\int_{0}^{\tau} t\cos\omega t\, dt$$

we have

$$B(\omega) = \frac{2A}{\pi\tau^2}\left(\tau\int_{0}^{\tau}\cos\omega t\, dt - \int_{0}^{\tau} t\cos\omega t\, dt\right)$$

Integrating by parts we get

$$B(\omega) = \frac{2A}{\pi\tau^2\omega^2}(1-\cos\omega\tau)$$

$$= \frac{A}{\pi}\frac{\sin^2(\frac{1}{2}\omega\tau)}{(\frac{1}{2}\omega\tau)^2}$$

$$= \frac{A}{\pi}\frac{\sin^2(\omega\Delta t)}{(\omega\Delta t)^2}$$

The frequency spectrum of the pulse, i.e. a plot of $B(\omega)$ versus ω is shown in Fig. 9.11(c).

Fourier Analysis of a Periodic Pulse

We have so far Fourier analysed non-periodic pulses having a continuous frequency spectrum. We have seen that such pulses are formed by a superposition of harmonic components whose frequencies increase continuously within a bandwidth. We shall now discuss periodic pulses, i.e. pulses which repeat periodically in a certain time interval called the *repetition period* T of the pulse of a certain *duration time* τ. Such pulses are formed by a superposition $\psi(t)$ of harmonic components whose frequencies are multiples of a fundamental frequency $\omega_1 = 2\pi/T_1$. Equation

(6.41) is an example of such a superposition. Since $k_n = nk_1$ and $\omega_n = n\omega_1$, this superposition Eq. (6.41) may be written as

$$y(x, t) = \sum_{n=1}^{\infty} \sin(nk_1x)(B_n \cos n\omega_1 t + C_n \sin n\omega_1 t)$$

where ω_1 is the angular frequency of the fundamental. As explained in Chap. 6, this function is doubly periodic; it has a spatial periodicity $\lambda_1 = 2\pi/k_1$ and a temporal periodicity $T_1 = 2\pi/\omega_1$. We have learnt how we can Fourier analyse such a function in space (x). We shall now Fourier analyse it in time (t).

If we focus our attention on a particular value of x, say $x = x_0$, this function becomes a function only of time and we can write it as

$$y(t) = \sum_{n=1}^{\infty} (B_n \cos n\omega_1 t + C_n \sin n\omega_1 t)$$

where the constant $\sin(nk_1x_0)$ has been absorbed in constants B_n and C_n. Writing $\psi(t)$ for $y(t)$ the function

$$\psi(t) = \sum_{n=1}^{\infty} (B_n \cos n\omega_1 t + C_n \sin n\omega_1 t) \tag{9.29}$$

represents a periodic pulse with a repetition period $T_1 = 2\pi/\omega_1$. We could obtain it directly as follows:

Consider a superposition of a large number of harmonic components of amplitudes $A_1, A_2, A_3, \ldots, A_n$, phase constants $\phi_1, \phi_2, \phi_3, \ldots, \phi_n$ and angular frequencies $\omega_1 = \omega$, $\omega_2 = 2\omega$, $\omega_3 = 3\omega, \ldots, \omega_n = n\omega$ respectively. The superposition is given by

$$\psi(t) = A_1 \cos(\omega_1 t + \phi_1) + A_2 \cos(\omega_2 t + \phi_2) + \ldots + A_n \cos(\omega_n t + \phi_n)$$

$$= \sum_{n=1}^{\infty} A_n \cos(\omega_n t + \phi_n)$$

or

$$\psi(t) = \sum_{n=1}^{\infty} A_n \cos(n\omega t + \phi_n)$$

The above equation may be recast in the form

$$\psi(t) = \sum_{n=1}^{\infty} (B_n \cos n\omega t + C_n \sin n\omega t) \tag{9.30}$$

where the constants B_n and C_n are related to constants A_n and ϕ_n as

$$B_n = A_n \cos \phi_n$$

$$C_n = -A_n \sin \phi_n$$

The function $\psi(t)$ given by Eq. (9.30) is periodic. This is because all the components are harmonics of the fundamental frequency $\omega_1 = \omega$. So, by the time the fundamental has completed one cycle, the second harmonic $(\omega_2 = 2\omega)$ has completed two cycles, the third harmonic $(\omega_3 = 3\omega)$ has completed three cycles, and so on. Thus, during the second cycle of the fundamental, the motion is an exact repetition of the first cycle. In other words, the function $\psi(t)$ is periodic in time having a temporal periodicity of $T = 2\pi/\omega$. It may be mentioned that, although $\psi(t)$ is periodic, it is not a harmonic function of t. The reason is that the resultant of a superposition of harmonic components of different frequencies is not a harmonic function. Such a function, described by Eq. (9.30) forms a periodic pulse with a repetition period $T = 2\pi/\omega$. This means that

$$\psi(t+T) = \psi(t)$$

This can be easily checked. At time $t+T$, Eq. (9.30) gives $(\because T = 2\pi/\omega)$

$$\psi(t+T) = \sum_{n=1}^{\infty}\{B_n \cos n\omega(t+T)+C_n \sin n\omega\,(t+T)\}$$

$$= \sum_{n=1}^{\infty}\{B_n \cos (n\omega t+2n\pi)+C_n \sin (n\omega t+2n\pi)\}$$

$$= \sum_{n=1}^{\infty}(B_n \cos n\omega t+C_n \sin n\omega t)$$

$$= \psi\,(t)$$

Thus T is the repetition time of the pulse $\psi\,(t)$ given by Eq. (9.30). The series (9.30) is called a Fourier series and B_n and C_n are the Fourier coefficients. These coefficients are determined as discussed in chapter 6 except that integrations have to be performed over time T. Multiplying Eq. (9.30) by $\cos (n\omega t)$ and integrating over a repetition time T gives B_n. Multiplication of Eq. (9.30) by $\sin (n\omega t)$ and integrating over T gives C_n. The results of these integrations are :

$$B_n = \frac{2}{T}\int_0^T \psi\,(t) \cos (n\omega t)\, dt \tag{9.31}$$

$$C_n = \frac{2}{T}\int_0^T \psi\,(t) \sin (n\omega t)\, dt \tag{9.32}$$

We shall apply Eqs. (9.30) to (9.32) to a simple periodic pulse. An infinite train of square pulses of unit height has a pulse duration τ and a repetition period T. The pulse is given by

$$\psi(t) = \begin{cases} 0 \text{ for } -\frac{1}{2}T < t < \frac{1}{2}\tau \\ 1 \text{ for } -\frac{\tau}{2} < t < \frac{\tau}{2} \\ 0 \text{ for } \frac{\tau}{2} < t < \frac{1}{2}T \end{cases}$$

and $\qquad \psi(t+T) = \psi(t)$

This pulse train is shown in Fig. 9.12 (a). It is clear that the pulse $\psi(t)$ is an even function of time because the pulse shape remains unchanged. when t is replaced by $-t$. Hence $C_n = 0$ (since $\sin(n\omega t)$ is an odd function). B_n is obtained from Eq. (9.31). Note that the limits of integrations are $-\frac{\tau}{2}$ to $+\frac{\tau}{2}$ because outside these limits $\psi(t) = 0$; hence the integrand is zero. Therefore, from Eq. (9.31) we have

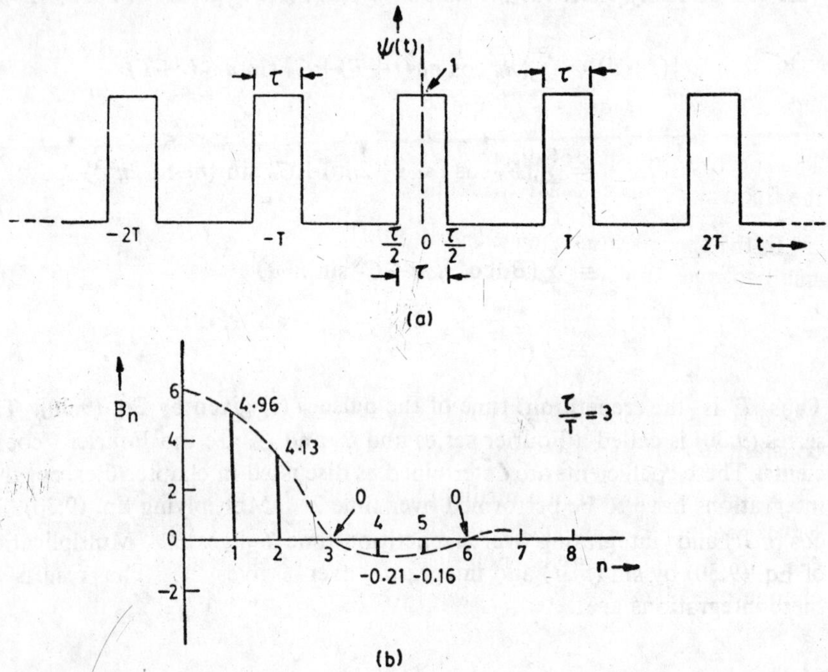

Fig. 9.12 (a) An infinite train of square pulses of unit height, repetition period T and pulse duration τ.

(b) Discrete frequency spectrum consists of vertical lines for $\tau/T = 1/3$.

$$B_n = \frac{2}{T}\int_{-\tau/2}^{+\tau/2} 1 \,.\, \cos(n\omega t)\, dt$$

$$= \frac{2}{Tn\omega} \sin(n\omega t) \Big|_{-\tau/2}^{\tau/2}$$

giving $$B_n = \frac{2}{n\pi} \sin\left(\frac{n\pi\tau}{T}\right) = \frac{2T}{\tau} \frac{\sin\left(\frac{n\pi\tau}{T}\right)}{\left(\frac{n\pi\tau}{T}\right)} = \frac{2T}{\tau} \frac{\sin \alpha_n}{\alpha_n}$$

where $\alpha_n = n\pi\tau/T$.

We shall plot the frequency spectrum for a typical value of $\frac{\tau}{T} = \frac{1}{3}$. In this case

$$B_n = 6 \frac{\sin\left(\frac{n\pi}{3}\right)}{\left(\frac{n\pi}{3}\right)} \quad ; n = 1, 2, 3, \ldots$$

$$= 0 \text{ for } n = 3, 6, 9, \ldots$$

The spectrum consists of discrete set of lines of heights shown in Fig. 9.12(b); the frequencies 3ω, 6ω, 9ω,... being absent from the spectrum.

In the special case when $\tau \ll T$ and n is not too large, α_n becomes very small so that $\sin \alpha_n \simeq \alpha_n$ and

$$B_n = \frac{2T}{\tau}$$

which is independent of n. Thus the spectrum consists of a set of lines of constant height $2T/\tau$ [see also Fig. 9.9 (a)].

A pulse will travel in a medium and such a travelling pulse is called a wave group. If the speed of the pulse is independent of the frequency (i.e. if the medium is non-dispersive) the pulse will travel without any change of shape.

9.5 TRAVELLING PULSES OF CONSTANT SHAPE

Suppose a vibrator (transmitter) at $x = 0$ produces a number of harmonic oscillations of different frequencies in a band of width $\Delta\omega$. The superposition of these harmonic components forms at pulse $\psi(t, x = 0)$. If the transmitter is coupled to a medium (which it always is) and it vibrates for a limited interval of time, a pulse of a limited duration time will travel in the medium. Such a travelling pulse $\psi(x, t)$ is called a wave group or wave packet. The wave group travels with the group velocity ; hence the pulse of a limited duration will also be of a limited extent in the medium. The existence of an angular frequency bandwidth $\Delta\omega$, therefore, implies an existence of a corresponding band Δk of wave numbers (and corresponding band $\Delta\lambda$ of wavelengths). The reason is that angular frequency ω and wave number k are related to each other through the dispersion relation $k = f(\omega)$ where f is an undefined function of ω. Then Δk is given by

$$v_g = \frac{\Delta\omega}{\Delta k}$$

or
$$\Delta k = \frac{\Delta\omega}{v_g} \qquad (9.33)$$

Now a packet of length Δx, travelling at the group velocity v_g, passes a given point x in time interval Δt given by

$$\Delta t = \frac{\Delta x}{v_g}$$

or
$$\Delta x = v_g \, \Delta t \qquad (9.34)$$

From Eqs. (9.33) and (9.34) we get

$$\Delta k . \Delta x = \Delta\omega . \Delta t$$

According to the frequency bandwidth theorem,

$$\Delta\omega . \Delta t = 2\pi$$

Hence
$$\Delta k . \Delta x = 2\pi \qquad (9.35)$$

This is the wave number bandwidth theorem obtained earlier in Sec. 9.3.

It should be noted that the length Δx of the wave group does not remain constant if the group travels in a dispersive medium. This is because the group velocity v_g depends upon wave number k and v_g will change within the band Δk. From Eq. (9 34) this will lead to a spreading of the wave packet as it progresses in the medium. Due to this spreading of the wave packet, the relation (9.35) cannot hold for all values of t in the case of a dispersive medium. Thus, a pulse will travel with a constant shape only in a non-dispersive medium in which the phase velocity v equals the group velocity v_g.

Equation of Motion of a Pulse Travelling in a Non-dispersive Medium

We will now obtain the equation of motion of an arbitrary pulse $\psi(x,t)$ travelling in a non-dispersive homogeneous medium, say, in the $+x$ direction. Suppose at a point A (which we shall call $x = 0$), the pulse is described by an equation

$$\psi(x = 0, t) = f(t)$$

If the function $f(t)$ is known, the exact shape of the pulse is known. If the pulse as a whole is travelling at a velocity v (which equals v_g, since the medium is non-dispersive), then at a later time $(t+\frac{x}{v})$, the displacement that originally existed at A ($x = 0$) will now exist at a point B (at x). Thus (see Fig. 9.13),

Displacement at x at time t = displacement that existed at $x = 0$ at an earlier time $t' = t-\frac{x}{v}$

Hence $$\psi(x,t) = f(t') = f\left(t-\frac{x}{v}\right)$$

which may also be written as ($\because$ v is constant)

$$\psi(x, t) = f(vt-x) \tag{9.36}$$

Similarly a pulse travelling in the $-x$ direction is described by

$$\psi(x, t) = g\,(vt+x) \tag{9.37}$$

It is easy to see that our general equations (9.36) and (9.37) both satisfy the same differential wave equation. Writing

$$z \equiv vt-x$$

in Eq. (9.36) we have

$$\psi(x, t) = f(z)$$

Differentiating with respect to x we have

$$\frac{\partial\psi}{\partial x} = \frac{\partial f}{\partial z}\cdot\frac{\partial z}{\partial x} = -\frac{\partial f}{\partial z} \qquad \left[\because \frac{\partial z}{\partial x} = \frac{\partial}{\partial x}(vt-x) = -1\right]$$

Differentiating again we have

$$\frac{\partial^2\psi}{\partial x^2} = \frac{\partial^2 f}{\partial z^2} \tag{9.38}$$

Similarly differentiating with respect to t we have

$$\frac{\partial\psi}{\partial t} = \frac{\partial f}{\partial z}\cdot\frac{\partial z}{\partial t} = v\frac{\partial f}{\partial z}$$

$$\therefore \frac{\partial^2\psi}{\partial t^2} = v^2\frac{\partial^2 f}{\partial z^2} \tag{9.39}$$

Comparing Eqs (9.38) and (9.39), we get

$$\frac{\partial^2\psi(x, t)}{\partial t^2} = v^2\frac{\partial^2\psi(x, t)}{\partial x^2} \tag{9.40}$$

which is called the classical wave equation. It is easy to see that the function (9.37) also satisfies the same equation (9.40). We have come across this equation earlier in chapter 7 where it represents a harmonic travelling wave of a single frequency. Here it represents a travelling pulse where $\psi(x, t)$ represents any one of the harmonic travelling waves in the superposition which gives the pulse. Since each component of the pulse satisfies (Eq. 9.40) ; the entire superposition i.e. the total wave function $\psi(x, t)$ satisfies Eq. (9.40). Thus Eq. (9.40) does not represent only harmonically varying $\psi(x, t)$ but any $\psi(x, t)$ provided that it is expressible as function of $(vt \pm x)$. This function need not be a sine or a cosine function characteristic of travelling harmonic waves. For example we could define a certain shape of pulse, moving in the $+x$ direction, by the equation

$$\psi(x, t) = \frac{h^3}{h^2+(vt-x)^2}$$

Figure 9.13 shows the shape of this pulse at time $t = 0$. The peak of the pulse is of height h and this peak passes through the point $x = 0$ at time $t = 0$. If we differentiate with respect to x we have

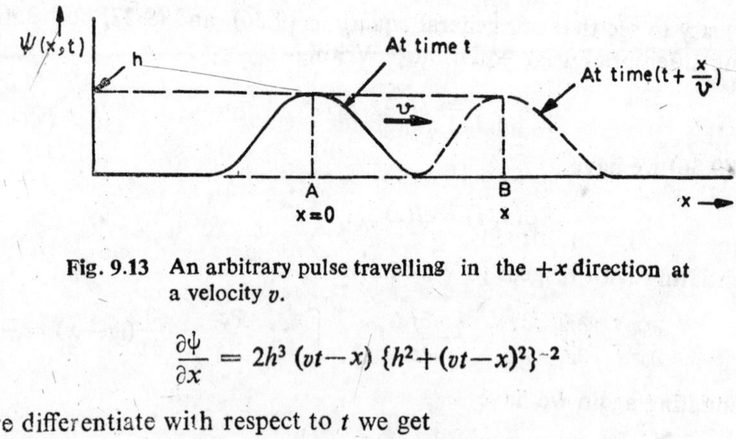

Fig. 9.13 An arbitrary pulse travelling in the $+x$ direction at a velocity v.

$$\frac{\partial \psi}{\partial x} = 2h^3 (vt-x) \{h^2+(vt-x)^2\}^{-2}$$

If we differentiate with respect to t we get

$$\frac{\partial \psi}{\partial t} = -2vh^3 (vt-x) \{h^2+(vt-x)^2\}^{-2}$$

These equations give

$$\frac{\partial \psi}{\partial t} = -v \frac{\partial \psi}{\partial x}$$

Hence

$$\frac{\partial^2 \psi}{\partial t^2} = -v \frac{\partial}{\partial t}\left(\frac{\partial \psi}{\partial x}\right) = -v \frac{\partial}{\partial x}\left(\frac{\partial \psi}{\partial t}\right)$$

$$= -v \frac{\partial}{\partial x}\left(-v \frac{\partial \psi}{\partial x}\right)$$

or $$\frac{\partial^2\psi}{\partial t^2} = v^2 \frac{\partial^2\psi}{\partial x^2}$$

which is Eq. (9.40). In fact we could write down any number of other possible shapes of pulses by using powers, exponentials and trignometric functions (other than sine and cosine) of $(vt \pm x)$. All such pulses will be governed by Eq. (9.40) and will travel at a speed v with their shape unchanged provided only that the medium is homogeneous and non-dispersive. Equation (9.40) does not apply in the case of dispersive media.

SOLVED EXAMPLES

Example 9.1

A wave group is formed by a superposition of two waves of slightly different wavelengths λ and $\lambda+\delta\lambda$ where $\delta\lambda/\lambda$ is small. Show that the number of wavelengths contained in a distance between two successive zeros of the modulating envelope is $\simeq \lambda/\delta\lambda$.

Solution

From Eq. (9.4) the modulated amplitude is given by

$$A_m = 2A \cos \{\tfrac{1}{2} (\delta\omega t - \delta kx)\}$$

At time, say $t = 0$, it reads

$$A_m = 2A \cos (\tfrac{1}{2}\, \delta kx)$$

where $$\delta k = k_1 - k_2 = 2\pi \left(\frac{1}{\lambda_1} - \frac{1}{\lambda_2} \right) = 2\pi \frac{(\lambda_2 - \lambda_1)}{\lambda_1 \lambda_2} \simeq 2\pi \frac{\delta\lambda}{\lambda^2}$$

The modulated amplitude will be zero at values of x given by

$$\cos (\tfrac{1}{2}\, \delta kx) = 0$$

i.e. $$\tfrac{1}{2}\, \delta kx = \frac{\pi}{2}, \frac{3\pi}{2}, \frac{5\pi}{2}, \ldots$$

or $$\delta kx = \pi, 3\pi, 5\pi, \ldots$$

Therefore, the distance d between two successive zeros of A_m is

$$d = \frac{2\pi}{\delta k} \simeq \frac{\lambda^2}{\delta\lambda}$$

Now, by definition, in a distance λ, there is exactly one wave. Hence the number of waves of wavelength λ contained in a distance d is $= \dfrac{d}{\lambda} = \dfrac{\lambda}{\delta\lambda}$.

Example 9.2

The phase velocity of deep-water waves is given by

$$v^2 = \frac{g\lambda}{2\pi} + \frac{2\pi\sigma}{\rho\lambda}$$

where $g = 9.8\ \text{ms}^{-2}$, $\rho = 1000\ \text{kg m}^{-3}$ and $\sigma = 7\,2 \times 10^{-2}\ \text{Nm}^{-1}$ (surface tension of water). (a) Determine the value (λ_0) of the wavelength of the waves which do not disperse in water. What are the phase and group velocities at this wavelength ? (b) Show that for wavelengths $\lambda \ll \lambda_0$; $v_g = 1.5\ v$ and for $\lambda \gg \lambda_0$; $v_g = v/2$.

Solution

(a) For no dispersion, $v_g = v$, i.e. $\dfrac{dv}{d\lambda} = 0$. Differentiating the expression

$$v^2 = \frac{g\lambda}{2\pi} + \frac{2\pi\sigma}{\rho\lambda}$$

with respect to λ and setting $\dfrac{dv}{d\lambda} = 0$ and $\lambda = \lambda_0$ gives

$$\lambda_0 = 2\pi\sqrt{\frac{\sigma}{\rho g}}$$

Substituting the values of σ, ρ and g yields

$$\lambda_0 = 1.7 \times 10^{-2}\ \text{m} = 1.7\ \text{cm}.$$

Thus waves of average wavelength of 1.7 cm do not disperse in water. The group and phase velocities at λ_0 are equal to each other.

$$v = v_g = \left(\frac{g\lambda_0}{2\pi} + \frac{2\pi\sigma}{\rho\lambda_0}\right)^{1/2}$$

Substituting the values of g, σ, ρ and λ_0 yields

$$v = v_g = 23\ \text{cms}^{-1}$$

(b) For $\lambda \ll \lambda_0$, the surface tension term dominates, so that

$$v^2 = \frac{2\pi\sigma}{\rho\lambda}$$

or $$v = \sqrt{\frac{2\pi\sigma}{\rho\lambda}}$$

Now $$v_g = v - \lambda\frac{dv}{d\lambda}$$

$$= v - \lambda\left(\frac{2\pi\sigma}{\rho}\right)^{1/2}(-\tfrac{1}{2})\,\lambda^{-3/2}$$

$$= v + \tfrac{1}{2}\left(\frac{2\pi\sigma}{\rho\lambda}\right)^{1/2}$$

$$= v + \tfrac{1}{2}\,v = 1.5\,v$$

This is the speed with which ripples (short-wavelength waves) propagate in water. For such waves $v_g > v$, indicating that they show anomalous dispersion

On the other hand, if $\lambda \gg \lambda_0$, the gravity term dominates. The phase velocity of these long wavelength waves is, therefore, given by

$$v = \left(\frac{g\lambda}{2\pi}\right)^{1/2}$$

which yields

$$v_g = \frac{v}{2}$$

as shown in the text. These waves show normal dispersion, i.e. their phase velocity decreases with decrease in wavelength.

Example 9.3

The velocity of microwaves of angular frequency ω in the ionosphere is given by

$$v = \frac{c}{\left(1-\dfrac{\omega_p^2}{\omega^2}\right)^{1/2}}$$

where c = velocity of light in free space and $\omega_p^2 = \dfrac{4\pi Ne^2}{m}$ is a constant depending on N, the number of electrons per unit volume in ionosphere, and m and e are electronic mass and charge. Show that, although the phase velocity v exceeds c, the group velocity is always less than c.

Solution

The phase velocity is given by

$$v = c\left(1-\frac{\omega_p^2}{\omega^2}\right)^{-1/2}$$

showing that $v > c$, i.e. the phase velocity exceeds the velocity of light in free space. This result is not in conflict with the basic assumption of relativity theory. In fact, phase velocity can have any value including infinity ; it can even be a complex quantity (as in the case of electromagnetic waves in a conductor). In the case of an ionized gas, the physical origin of phase velocity exceeding c lies in the crucial phase relation which exists between the electron osciliation $\psi(x)$ and the driving harmonic force $eE(t) = eE_0 e^{i\omega t}$ of the electromagnetic wave. Electrons have a natural frequency (ω_0) called the resonant frequency. If the frequency ω of the electric field of the wave exceeds the resonant frequency ω_0 of electron oscillations, the displacement cannot keep up with the driving force. This results in an increased restoring force and hence an increased phase velocity exceeding c. On the other hand if $\omega < \omega_0$, the electron displacement is in phase with the field and builds up a field which tends to cancel the original field. The decreased field gives a decreased restoring force and hence a reduced phase velocity less than c.

But the group velocity can never exceed c ; because energy or a signal is transmitted at the group velocity and not at the phase velocity. From Eq. (9.13) the group velocity is given by

$$\frac{1}{v_g} = \frac{1}{v} - \frac{\omega}{v^2}\frac{dv}{d\omega}$$

$$= \frac{1}{v} - \frac{\omega}{v^2}\left(-\frac{1}{2}\right)\left(1-\frac{\omega_p^2}{\omega^2}\right)^{-3/2}\frac{2\omega_p^2}{\omega^3}$$

$$= \frac{1}{c}\left(1-\frac{\omega_p^2}{\omega^2}\right) + \frac{\omega}{c^2}\left(1-\frac{\omega_p^2}{\omega^2}\right) c\left(1-\frac{\omega_p^2}{\omega^2}\right)^{-3/2}\frac{\omega_p^2}{\omega^3}$$

$$= \frac{1}{c}\left(1-\frac{\omega_p^2}{\omega^2}\right)^{-1/2}$$

i.e.
$$v_g = c\left(1-\frac{\omega_p^2}{\omega^2}\right)^{1/2}$$

Thus $v_g < v$; as it should be.

The fact that the phase velocity $v > c$ indicates that the refractive index (n) of the ionosphere is less than unity (since $n = c/v$). Thus a microwave signal sent up from the earth, travelling in air of refractive index equal to unity meets an ionospheric layer of a lower refractive index. This satisfies the condition of total internal reflection. The signal is, therefore, reflected back to earth where it can be received. In fact, experiments have been done to explore the ionosphere by sending a radio-wave signal and observing the reflected signal at a receiving station. The time interval between the sending and receiving of the signal is controlled by the group velocity and is much longer than that expected from the phase velocity. This has been confirmed by timing experiments, and the height of the reflecting layer (called the Heaviside layer) of the ionosphere has been computed.

Example 9.4 The spectral lines have a finite width. The finite spectral line-width is due to the fact that the radiations from radiating atoms have a finite frequency bandwidth $\Delta\omega$. Therefore the optical wave-trains emitted by radiating atoms constitute pulses of average frequency ω_0 and a finite duration time τ. The pulse is given by

$$\psi(t) = A\cos\omega_0 t \qquad \text{for } -\frac{\tau}{2} \leqslant t \leqslant \frac{\tau}{2}$$

$$= 0 \qquad \text{for } t \text{ outside the interval } \pm \tau/2,$$

where A is a constant amplitude. Determine the frequency spectrum of the radiation. Hence show that $\Delta\nu = \dfrac{1}{\tau}$ where $\Delta\nu$ is the frequency difference between the maximum and the first zero of frequency spectrum.

Solution

$\psi(t)$ is an even function of t; hence $C(\omega) = 0$. The frequency spectrun is determined only by Fourier coefficient $B(\omega)$ which is given by

$$B(\omega) = \frac{1}{\pi}\int_{-\infty}^{+\infty} \psi(t)\cos\omega t\, dt$$

$$= \frac{A}{\pi}\int_{-\tau/2}^{+\tau/2} \cos\omega_0 t \cos\omega t\, dt$$

$$= \frac{2A}{\pi}\int_{-\tau/2}^{+\tau/2} \left\{ \cos(\omega+\omega_0)t\,dt + \cos(\omega-\omega_0)t\,dt \right\}$$

$$= \frac{A}{\pi}\left\{ \frac{\sin(\omega+\omega_0)\tau/2}{(\omega+\omega_0)} + \frac{\sin(\omega-\omega_0)\tau/2}{(\omega\ \ \omega_0)} \right\}$$

The values of ω lie between the two extreme angular frequencies ω_1 and ω_2 with angular frequency bandwidth $\Delta\omega = \omega_2 - \omega_1$ and average angular frequency $\omega_0 = \frac{1}{2}(\omega_1 + \omega_2)$. For small bandwidth $\omega = \omega_2 \simeq \omega_1 \simeq \omega_0$ so that the denominator of the first term in the brackets is much larger than that of the second term. Since the maximum value of the numerator of either term can only be unity; it is clear that the first term is very small compared to the second term [since $(\omega + \omega_0) \gg (\omega - \omega_0)$] and may, therefore, be neglected. Thus

$$B(\omega) = \frac{A}{\pi}\,\frac{\sin(\omega-\omega_0)\tau/2}{(\omega-\omega_0)}$$

$$= \frac{A\tau}{2\pi}\,\frac{\sin(\omega-\omega_0)\tau/2}{(\omega-\omega_0)\tau/2}$$

or
$$B(\omega) = \frac{A\tau}{2\pi}\,\frac{\sin\alpha}{\alpha}$$

where
$$\alpha = (\omega-\omega_0)\tau/2 = \omega^*\tau/2$$

with
$$\omega^* = \omega-\omega_0.$$

$B(\omega)$ is maximum when $\alpha = 0$ or $\omega^* = 0$. The first zero of $B(\omega)$ occurs at $\alpha = \pi$ or $\omega^*\tau/2 = \pi$ or $\omega^* = 2\pi/\tau$. The angular frequency difference between the maximum of $B(\omega)$ and its first zero is $\Delta\omega^* = 2\pi/\tau$, which gives

$$\Delta\omega = \frac{2\pi}{\tau} \qquad (\because\ \omega_0 = \text{constant})$$

or
$$2\pi\Delta\nu = \frac{2\pi}{\tau}$$

or
$$\Delta\nu = \frac{1}{\tau}$$

which is the Bandwidth Theorem.

Example 9.5 The coherence length l of the light pulse in Example 9.4 is defined as $l = c\tau$, where c is speed of light.

(a) If the half-width of the central maximum of the energy spectrum of Example 9.4 is $\Delta\nu$, show that

$$l = \frac{\lambda_0^2}{\Delta\lambda}$$

(b) For a ruby laser, it is found that $\Delta\nu = 10^4$ Hz and $\lambda_0 = 6.936 \times 10^{-7}$ m. Show that $\Delta\lambda = 1.6\times10^{-17}$ m and the coherence length l of the laser beam is 3×10^4 m.

Solution

(a) The energy spectrum $E(\omega)$ is obtained by squaring the frequency amplitude $B(\omega)$, i.e.

$$E(\omega) = K\times B^2(\omega)$$

where K is a constant. Therefore

$$E(\omega) = K\left(\frac{A\tau}{2\pi}\right)^2 \frac{\sin^2\alpha}{\alpha^2}$$

$$= E(0)\frac{\sin^2 \alpha}{\alpha^2}$$

where $E(0)$ is the maximum value of $E(\omega)$ when $\alpha = 0$ or $\omega^* = 0$. The first zero of $E(\omega)$ occurs at $\alpha = \pm\pi$ or $\omega^*\tau/2 = \pm\pi$ or $\omega^* = \pm 2\pi/\tau$. Hence the total width $2\Delta\omega$ of the central maximum of $E(\omega)$ is given by

$$2\Delta\omega = \frac{4\pi}{\tau}$$

or

$$\Delta\nu = \frac{1}{\tau}$$

or

$$\tau = \frac{1}{\Delta\nu}$$

Now

$$l = c\tau = \frac{c}{\Delta\nu}$$

But

$$\Delta\nu = \nu - \nu_0$$

$$= \frac{c}{\lambda} - \frac{c}{\lambda_0} = c\frac{(\lambda_0 - \lambda)}{\lambda_0\lambda} \simeq \frac{c\Delta\lambda}{\lambda_0^2} \qquad (\because \lambda \simeq \lambda_0)$$

Hence

$$l = \frac{c}{\Delta\nu} = \frac{\lambda_0^2}{\Delta\lambda}$$

(b) For a ruby laser

$$\Delta\nu = 10^4 \text{ Hz}$$

$$\lambda_0 = 6.936 \times 10^{-7} \text{ m}$$

Now

$$\Delta\lambda = \Delta\nu\,\frac{\lambda_0^2}{c}, \text{ where } c = 3 \times 10^8 \text{ ms}^{-1}$$

Substituting the numerical values, we get

$$\Delta\lambda = \frac{10^4 \times (6.936 \times 10^{-7})^2}{3 \times 10^8} = 1.6 \times 10^{-17} \text{ m}$$

The coherence length of the laser beam, is

$$l = \frac{\lambda_0^2}{\Delta\lambda}$$

$$= \frac{(6.936 \times 10^{-7})^2}{1.6 \times 10^{-17}}$$

$$= 3 \times 10^4 \text{ m}$$

QUESTIONS

1. (a) What do you understand by the terms wave group and group velocity ?

   (b) A wave group is formed by a superposition of two harmonic waves of equal amplitudes but slightly different frequencies travelling in the same direction in a dispersive medium. Obtain the expressions for the phase and group velocities.

   (c) Deduce a relation between phase velocity and group velocity.

   (d) Show that, for light waves of frequency ν travelling in a dispersive medium of refractive index n, the group velocity v_g is given by

$$\frac{1}{v_g} = \frac{1}{v} - \frac{\lambda}{c}\frac{dn}{d\lambda}$$

   where v is the phase velocity, c is the velocity of light in vacuum and λ is the vacuum wavelength.

2. (a) Give two examples each of dispersive and non-dispersive waves.

   (b) Distinguish between normal and anomalous dispersion.

3. (a) What is a pulse ?

   (b) What are periodic and non-periodic pulses ?

   (c) Distinguish between the duration time and repetition time of a periodic pulse.

4. (a) Deduce the expression for the pulse $\psi(t)$ formed by a linear superposition of a large number of harmonic components of angular frequencies uniformly distributed within a frequency bandwidth. Assume that the harmonic components have equal amplitudes and zero phase constants. Draw a graph showing the time behaviour of the pulse.

   (b) Deduce the frequency bandwidth theorem and explain its implications.

5. Use the method of Fourier analysis and obtain the frequency spectrum of (a) a square pulse of a unit height and duration time τ and (b) a triangular pulse of maximum height A and duration time τ.

6. (a) How will you Fourier analyse a pulse of duration time τ and repetition time T. The pulse is formed by a linear superposition of n harmonic components of different amplitudes and phase constants and frequencies which are multiples of the fundamental frequency of the first component.

   (b) An infinite train of pulses of unit height and duration time 2τ and repetition time $2T$ is expressed as

$$\begin{aligned}\psi(t) &= 0 \text{ for } -T<t<-\tau\\ &= 1 \text{ for } -\tau<t<\tau\\ &= 0 \text{ for } \tau<t<T\end{aligned}$$

   show that the Fourier coefficients of the pulse are given by

$$B_n = \frac{2}{n\pi}\sin\left(\frac{\pi n\tau}{T}\right)$$

7. Each half-cycle of a periodic pulse is identical except in sign with the previous half-cycle. Show that the Fourier frequency spectrum of the pulse contains only odd order frequency components. What is the physical significance of this result ?

8. Draw wave forms of three pulses with (a) even symmetry and (b) odd symmetry. Give two examples of achieving symmetry in a pulse by shifting the time origin.
9. Deduce the classical wave equation for any arbitrary pulse travelling in the $-x$ direction in a homogeneous non-dispersive medium.

PROBLEMS

1. A wave group is formed by a superposition of two waves of slightly different wavelengths 0.98 cm and 1.02 cm. What is the distance between two successive crests of the group ?
2. The motion of ripples of wavelength $\lambda = 1$ cm on water is controlled by surface tension σ. The phase velocity of such ripples is given by

$$v = \left(\frac{2\pi\sigma}{\rho\lambda}\right)^{1/2}$$

where ρ is the density of water. If $\sigma = 7.0\times10^{-2}$ Nm^{-1}, calculate the group velocity of the ripples.

3. The phase velocity v of transverse waves of wavelength λ in a crystal is given by

$$v = v_0 \frac{\sin(\pi d/\lambda)}{(\pi d/\lambda)}$$

where d is the average interatomic separation and v_0 is a constant.

(a) What is the ratio of phase velocity v and group velocity v_g?

(b) Calculate this ratio for long wavelengths $\lambda \gg d$.

4. The refractive index n of a gas for waves of wave number k is given by

$$n^2 = \frac{c^2}{v^2} = \alpha + \beta k^2 - \frac{\gamma}{k^2}$$

where α, β and γ are constants, v is the phase velocity and c is the velocity of light in vacuum. Show that the group velocity is given by

$$v_g = \frac{v}{n^2}\left(\alpha - \frac{2\gamma}{k^2}\right).$$

What is ratio v_g/v in the limiting case of very short wavelengths ?

5. The Fourier frequency spectrum of a rectangular pulse of height h and duration time τ is given by

$$B(\nu) = h\tau \frac{\sin\alpha}{\alpha}; \text{ where } \alpha = \pi\nu\tau.$$

Obtain the expression for the Fourier frequency spectrum $B(\nu)$ of an isosceles triangular pulse of height h and duration time 2τ. Express $B(\nu)$ in terms of h, τ and α.

ANSWERS TO PROBLEMS

Chapter 1

1. $0.0272\ ms^{-1}$, $-0.0493\ ms^{-2}$
2. $-0.136\ ms^{-1}$
3. 2.83 cm, $-8.89\ cms^{-1}$, $-28\ cms^{-2}$
4. 2.5 s, $4.53\ cms^{-1}$
5. (a) $\phi = 0, \delta = \frac{\pi}{2}$; (b) $\phi = \pi, \delta = \frac{3\pi}{2}$; (c) $\phi = \frac{\pi}{3}, \delta = \frac{\pi}{6}$;

   (d) $\phi = \frac{\pi}{2}, \delta = 0$
6. (a) $\phi = 0, \delta = \frac{3\pi}{2}$; (b) $\phi = \pi, \delta = \frac{\pi}{2}$; (c) $\phi = \frac{\pi}{3}, \delta = \frac{7\pi}{6}$;

   (d) $\phi = \frac{\pi}{4}, \delta = \frac{5\pi}{4}$
7. 0.5 s, 1.95×10^{-3} J
8. $\frac{3}{4}$, $\frac{10}{\sqrt{2}}$ cm.
9. Amplitude = 0.01 m, period = 2 s, phase difference = π
10. 0.1 m, 0.7 s
11. 0.635 s
12. 5 Hz (approximately)
13. No ; the frequency of oscillations in the second case (i.e. after the spring is cut into two) is $\sqrt{2}$ times that in the first case.
14. $3k$, $\frac{3}{2}k$
15. (a) $2k$, (b) $\frac{k}{2}$; (c) $2k$
16. The clock slows down by 43.2 s per day
17. 0.2%
18. 83.7 kg wt, 36.3 kg wt.
19. (a) 2.5 cm (b) 1.25 cm
20. $\pi : 2\sqrt{2}$

Chapter 2

1. 7.74 mm, 11.4°
2. $x \simeq 0.16 \sin(2\pi t + 0.287\pi)$
3. $x = A\cos(9\pi t + \delta)$; with $A = \sqrt{2}\times10^{-2}\{17 + 15\cos(2\pi t)\}^{1/2}$ m
   $\delta = \tan^{-1}(0.25 \tan \pi t)$; beat period = 1 s
4. 2 Hz
5. 499 Hz, 501 Hz

6. (*a*) 1 Hz, (*b*) 5.25 Hz, (*c*) $\frac{1}{4\pi}(3+\pi)$ Hz, (*d*) ν Hz

10. 1.52 cm, 19.23°

Chapter 3

1. (*a*) 43.3 s, (*b*) 680, (*c*) 197.4 Nm^{-1}
2. (*a*) 0.4 N s m^{-1}, (*b*) 4.97, (*c*) 1.8×10^{-3}
3. (*a*) 0.014 N s m^{-1}, (*b*) 68 approx., (*c*) 4×10^{-5} Hz
4. 4643
5. 1.59 s
6. 53 s
7. (a) 4.6×10^{-3} N s m^{-1}, (b) 4.35 s, (c) 13.7, (d) 0.115, (e) 1.0007 Hz, (f) 2.12 cm, (g) 7.9×10^{-5} J
8. Yes, 356 Hz
9. 632 ℧
10. 100.5Ω, 9936 Hz if R = 0

Chapter 4

1. (a) 1.02×10^{-3} m, (b) 10 m, (c) 4.16×10^{-3} m
2. (a) 200, (b) 4 s, (c) 0.25 rad s^{-1}
3. (a) 0.5 m, (b) 0.25 W
4. $\pi\rho\omega A^2$
5. (a) 50, (b) $50F_0^2/\omega_0^2$, (c) $50P_m/\omega_0$, (d) $50P_m/\omega_0$, (e) $0.02\omega_0$ (f) $\exp(-0.04\pi)$, (g) $50/\omega_0$, (h) $200/\pi$
6. (a) $\sqrt{2}\omega_0$, (b) $100/\sqrt{2}$, (c) 2γ, (d) P_m, (e) $E_0/2$
7. 50 m H, 2 A
8. (a) 50Ω, (b) 2 A, (c) 20 W

Chapter 5

1. **1.42 s (approx.), 1.38 s (approx.)**
2. **50 s (approx.)**
3. $\omega_1 = \{(k_1+k_2+k_3)/m\}^{1/2}$, $\omega_2 = (k_3/m)^{1/2}$
4. 0.35 Hz (approx.), 0.61 Hz (approx.)
5. 3.85 s (approx.)
6. $\nu = \frac{1}{2\pi}\sqrt{\frac{K}{m}}\,(1\pm\mu)^{-1/2}$
7. 408 Hz, 707 Hz
8. 5.8 Hz
9. 3 : 1
10. 9.18×10^{13} Hz
11. (a) $\left(\frac{3+\sqrt{5}}{3-\sqrt{5}}\right)^{1/2}$ or $\left(\frac{\sqrt{5}+1}{\sqrt{5}-1}\right)$

    (b) $\frac{1}{2}(\sqrt{5}-1)$, $\frac{1}{2}(\sqrt{5}+1)$

Chapter 6

1. 1.566 Hz
2. 2×10^{11} N m^{-2}
3. 4 beats per second
4. 1.97×10^{-3} J
5. 5.26
6. 2420 Hz, 1.8×10^{11} N m^{-2}

8. $$y(x, t) = \frac{9h}{\pi^3} \sum_{n=1}^{\infty} \frac{1}{n^2} \sin\left(\frac{n\pi}{3}\right) \sin\left(\frac{n\pi x}{L}\right) \cos\left(\frac{n\pi vt}{L}\right); \quad v = \sqrt{\frac{T}{\mu}}$$

9. $$y(x, t) = y_0 \sin\left(\frac{\pi x}{L}\right) \cos\left(\frac{\pi vt}{L}\right), \quad v = \sqrt{\frac{T}{\mu}}$$

Chapter 7

1. (a) 0.01 m, 1/3 s, 3 Hz, 1.5 m s^{-1}, 0.5 m
   (b) 0.02 m, 3.14 s, 0.32 Hz, 0.67 m s^{-1}, 2.09 m
   (c) 0.05 m, 1/3 s, 3 Hz, 6 m s^{-1}, 2 m
7. (a) $\frac{\pi}{2}$ m, $\frac{3\pi}{2}$ Hz, 0.75 m s^{-1}
   (b) 0, —0.0455 m, 0.0378 m and zero
   (c) —0.0455 m, —0.024 m, 0.042 m and 0.0455 m
   (d) 0.1096 J
   (e) 0.082 W
8. 2 : 1
9. 3.44×10^{-7} m
10. 3.33×10^{-7} m
11. 427 N m^{-3} s
12. (a) 1 m, (b) 72°
13. 6.5×10^{-4} of an atmosphere or 65 N m^{-2}
14. 10^{-10} m
15. (a) 1.59×10^{-3} W m^{-2}, (b) 1.09 V m^{-1}, 2.9×10^{-4} A m^{-1}

Chapter 8

1. (a) —1, 0 ; (b) $-\frac{1}{3}$, $\frac{2}{3}$; (c) $\frac{1}{3}$, $\frac{4}{3}$; (d) 1, 2
2. (a) 1, 0 ; (b) $\frac{1}{9}$, $\frac{8}{9}$; (c) $\frac{1}{9}$, $\frac{8}{9}$; (d) 1, 0
3. (a) 0.02 m, 2 s, 1 m, 0.5 m s^{-1} ; (c) $x = \frac{1}{4}$ m, $\frac{3}{4}$ m, $\frac{5}{4}$ m, etc.
   (d) 0.5 m, this difference = $\lambda/2$.
5. 0.1% approx.
6. 84.9%
7. (a) $r = -\frac{1}{5}$, $R = \frac{1}{25}$ (b) $r = +\frac{1}{5}$, $R = \frac{1}{25}$

8. 1.22, 1.12×10^{-5} cm
9. 50 m s^{-1}, 12.5 cm
10. —0.17, 0.83

Chapter 9

1. 25 cm
2. 31.5 cm s^{-1}
3. (a) $\dfrac{v}{v_g} = \dfrac{\tan(\pi d/\lambda)}{(\pi d/\lambda)}$, (b) $\dfrac{v}{v_g} = 1$
4. $\dfrac{v}{v_g} = \dfrac{\alpha}{\alpha + \beta k^2}$
5. $B(\nu) = 4h^2\tau^2 \dfrac{\sin^2 \alpha}{\alpha^2}$

REFERENCES

1. Crawford, (Jr.), F.S., *Waves*, Berkeley Physics Course—Vol. 3. McGraw-Hill Book Company, 1968.
2. French, A.P., *Vibrations and Waves*, M.I.T. Introductory Physics Series, Arnold—Heinemann India, 1973.
3. Pain, H.J., *The Physics of Vibrations and Waves*, John Wiley and Sons Ltd., 1979.
4. Anderson, R.A., *Fundamentals of Vibrations*, Amerind Publishing Co. Pvt. Ltd., 1967.
5. Braddick, H.J.J., *Vibrations, Waves and Diffraction*, McGraw-Hill Publishing Co. Ltd , 1965.
6. Seto, W.W., *Theory and Problems of Mechanical Vibrations*, Schaum Outline Series McGraw-Hill International Book Co., 1983.
7. Barker, J.R., *Mechanical and Electrical Vibrations*, Methuen Monograph, Methuen, London, 1964.
8. Bishop, R.E.D., *Vibration*, Cambridge University Press, New York, 1965.
9. Booker, H.G., *A Vector Approach to Oscillations*, Academic, New York, 1965.
10. Coulson, C.A., *Waves*, Oliver and Boyd, Edinburgh, 1941.
11. McLachlan, N.W., *Theory of Vibrations*, Dover, New York, 1951.
12. Morse, P.M., *Vibrations and Sound*, McGraw-Hill, New York, 1948.
13. Pearson, J.M., *A Theory of Waves*, Allyn and Bacon, Boston, 1966.
14. Stephens, R.W.B. and A.E. Bate, *Wave Motion and Sound*, Edward Arnold and Co , London, 1950.
15. Waldron, R.A., *Waves and Oscillations*, Momentum Books, Van Nostrand Reinhold, New York, 1964.
16. Wood, A., *Acoustics*, Blackie, London, 1940.
17. Kock, W., *Sound Waves and Light Waves*, Anchor Books, Doubleday and Co. Inc. New York, 1965.
18. Den Hartog, J.P., *Mechanical Vibrations*, McGraw-Hill, New York, 1956.
19. Thomson, W.T., *Vibration Theory and Applications*, Englewood Cliffs, N.J. Prentice—Hall Inc., 1965.

Index